HUMAN GEOGRAPHY

About the Companion Website

The *Human Geography: An Essential Introduction* Second Edition companion website includes a number of resources created by the author that you will find helpful when reading this book.

www.wiley.com/go/boyle

For students:

- Student tutorial exercises

For instructors:

- Specimen module outline for Human Geography 101 courses
- Additional essay-style exam questions
- A multiple-choice exam paper
- PowerPoint slide lectures for each chapter
- Comprehensive Mp4 and Mp3 recorded lectures by the author for each chapter.

HUMAN GEOGRAPHY

MARK BOYLE

An Essential Introduction

Second Edition

WILEY Blackwell

Registered Offices
John Wiley & Sons, Inc., 111 River Street, Hoboken, NJ 07030, USA
John Wiley & Sons Ltd, The Atrium, Southern Gate, Chichester, West Sussex, PO19 8SQ, UK

Editorial Office
9600 Garsington Road, Oxford, OX4 2DQ, UK
For details of our global editorial offices, customer services, and more information about Wiley products visit us at www.wiley.com.

Wiley also publishes its books in a variety of electronic formats and by print-on-demand. Some content that appears in standard print versions of this book may not be available in other formats.

Library of Congress Cataloging-in-Publication Data

Names: Boyle, Mark, author.
Title: Human geography : an essential introduction / Mark Boyle.
Description: Second edition. | Hoboken, NJ : Wiley-Blackwell, 2021. |
　Includes bibliographical references and index.
Identifiers: LCCN 2020042464 (print) | LCCN 2020042465 (ebook) | ISBN
　9781119374718 (paperback) | ISBN 9781119374725 (adobe pdf) | ISBN
　9781119374695 (epub)
Subjects: LCSH: Human geography.
Classification: LCC GF41 .B69 2021 (print) | LCC GF41 (ebook) | DDC
　304.2–dc23
LC record available at https://lccn.loc.gov/2020042464
LC ebook record available at https://lccn.loc.gov/2020042465

Cover Design: Wiley
Cover Image: © Tetra Images/Getty Images

C075991_160321

Take up the White Man's burden
Send forth the best ye breed
Go send your sons to exile
To serve your captives' need
To wait in heavy harness
On fluttered folk and wild
Your new-caught, sullen peoples,
Half devil and half child.

Rudyard Kipling, "The White Man's Burden:
The United States and the Philippine Islands" (1899)

O wad some power the giftie gie us
To see oursels as ithers see us!
It wad frae mony a blunder free us,
An' foolish notion:
What airs in dress an' gait wad lea'e us,
An' ev'n devotion!

Robert Burns, "To a Louse: On Seeing
One on a Lady's Bonnet, at Church" (1786)

Plate x.1   To infinity and beyond! "Broad, wholesome, charitable views of men and things cannot be acquired by vegetating in one little corner of the earth all of one's lifetime." - Mark Twain. Source: Matthew Henry on Unsplash.

Contents

Contents

List of Figures

List of Plates

List of Maps

List of Tables

Acknowledgments

Six years have elapsed since *Human Geography: An Essential Introduction* (first edition title: Human Geography: A Concise Introduction) was first published in 2015. I would like to thank the outstanding team at Wiley for their support and dedication to the project of developing this second edition. In particular, my gratitude goes to publisher Justin Vaughan for commissioning this new edition. Since 2015, the world and its geographies have changed in astonishing ways – the Covid-19 global pandemic has capstoned a period marked by breathless turbulence. There has been so much to update. This book has taken longer to write than the first edition (in truth, four years from start to finish; the first edition took me two years!) and has been extensively rewritten from the first word to the last. I have left no stone unturned. This has been a painstaking labor of love; for some reason a particularly epic undertaking. Justin's patience and support throughout have provided space and time to reflect and write. I would also like to thank Merryl Le Roux and Clelia Petracca (Senior Editorial Assistants) and Charlie Hamlyn (Managing Editor) at Wiley for their stellar work in helping the manuscript into production. Special commendation must go to Anandan Bommen (Content Refinement Specialist) for going the extra mile in support of editorial refinement.

I have taught human geography modules for over a quarter of a century now, first at the University of Strathclyde in Glasgow, Scotland; Maynooth University in County Kildare, Ireland; and most recently at the University of Liverpool. This book could not have been written without these prior teaching experiences. I would like to acknowledge the contribution of students who have taken these modules; in digesting, engaging in, challenging, and reframing lecture and tutorial content, they have helped me to sharpen my understanding of topics and clarify how best to communicate ideas. I would also like to thank inspirational colleagues at Strathclyde University, Maynooth University, and the University of Liverpool from whom I have learned much about this most unusual of trades. My ongoing collaborations and friendships with Audrey Kobayashi, Elaine Ho, James Sidaway, Guy Baeten, Chris Philo, Michael Parkinson, Georgina Endfield, Brendan Gleeson and Rob Kitchin continue to nourish. I would also like to acknowledge the guidance offered by Allan Findlay, Elspeth Graham, Ronan Paddison, Mary Gilmartin, John Hasse, Gerry Kearns, Neil Brenner, Kevin Cox, Peter Shirlow, Peter North, and Andy Davies on various subject matters pertaining to specific chapters in the book.

I completed this book as the horrors of the global Covid-19 pandemic unfolded and whilst in lockdown in a small rural village in County Kildare, Ireland. Like others, I spent the year battling to stay positive - and in awe of front line workers in the foundational economy and in particular health care workers whose generosity, expertise and courage cannot be overstated. Writing about far-off worlds enabled me to escape temporarily the brutal realities of heightened precarity, illness and death. It also helped me to escape entrapment. Covid-19 has limited horizons. I have been able to write about many of the places in this book as I have visited them. Of course safety must come first always and until zero-carbon transportation becomes mainstreamed we must always be conscious of our travel-related ecological footprint. But hopefully in time student geographers will be able to explore the world once more, for this is the only way one truly gets to know it. In the interim, I hope this book will serve as a worthy substitute (for first hand exploration) and therapeutic companion for all restless and nomadic souls who have struggled to cope with sedentary isolation.

My mother and family in Scotland and Australia undoubtedly helped to sustain the endeavor. I thank my wife Deborah and sons Patrick and Joseph for their forbearance. As ever they exist as the prime casualties of my career choices, publication schedule, and distaste for military timing. I was only able to complete this book because of the love and support they furnished me with as we bunkered down.

Mark Boyle
Liverpool and Kildare, January 2021

Plate x.2 Rotterdam Centraal Station, Rotterdam, the Netherlands. Photo by Jurriaan Snikkers on Unsplash.

List of Abbreviations

AAG	Association of American Geographers
AR6	IPCC sixth assessment report
ASEAN	Association of Southeast Asian Nations
BCE	Before the Common Era
BP	Before the Present Day
Brici	Brazil, Russia, India, China, and Indonesia
CBD	Convention on Biological Diversity
CBR	crude birth rate
CDC	Centers for Disease Control and Prevention
CDR	crude death rate
CE	Common Era
CMP	Capabilities Measurement Project
COP	Conference of Parties
DALY	Disability-Adjusted Life Year
DDT	dichlorodiphenyltrichloroethane
DNA	deoxyribonucleic acid
EC	European Commission
ECB	European Central Bank
EM-DAT	Emergency Events Database
EU	European Union
FAO	UN Food and Agricultural Organization
FDI	foreign direct investment
FEMA	Federal Emergency Management Agency
GCC	Global Commodity Chain
GDP	Gross Domestic Product
GGS	Guns, Germs and Steel
GIS	Geographical Information Systems
GIScience	Geographical Information Science
GNI	Gross National Income (formerly Gross National Product)
GPN	Global Production Network
GVA	Gross Value Added
GVC	Global Value Chain

HDI	Human Development Index
HFA	Hyogo Framework for Action
HFCs	hydrofluorocarbons
HIHD	Historical Index of Human Development
ICD	International Classification of Disease
ICT	information and communications technology
IDA	International Development Aid
IDP	internally displaced person
IGU	International Geographical Union
IHDI	Inequality-adjusted Human Development Index
IHME	Institute for Health Metrics and Evaluation
IMF	International Monetary Fund
In3	instrumentation, interconnection, and intelligence
IO	international organization
IOM	International Organization for Migration
IPCC	Intergovernmental Panel on Climate Change
IPL	International Poverty Line
IR4	Fourth Industrial Revolution
IUCN	International Union for Conservation of Nature
MDGs	Millennium Development Goals
Mercosur	Mercado Común del Cono Sur (Southern Cone Common Market)
MNC	multinational corporation
MPI	Multidimensional Poverty Index
MYBP	Million Years Before the Present Day
NAFTA	North American Free Trade Agreement
NATO	North Atlantic Treaty Organization
NGO	nongovernmental organization
NIDL	New International Division of Labor
OECD	Organisation for Economic Co-operation and Development
OIDL	Old International Division of Labor
OPEC	Organization of Arab Petroleum Exporting Countries
PD	post-development
PPP	purchasing power parity
RBA	rights-based approach
RGS(IBG)	Royal Geographical Society (with the Institute of British Geographers)
RIMA-II	Resilience Index Measurement and Analysis
SAP	Structural Adjustment Programme
SDGs	Sustainable Development Goals
SHDI	Subnational Human Development Index
SOE	state-owned enterprise
TFR	total fertility rate
TNC	transnational corporation
UBI	Universal Basic Income
UBS	Universal Basic Services
UDHR	Universal Declaration of Human Rights
UK	United Kingdom
UN	United Nations
UNDP	United Nations Development Programme
UNFCCC	United Nations Framework Convention on Climate Change

UNHCR	United Nations High Commissioner for Refugees
UNISDR	United Nations International Strategy for Disaster Risk Reduction
UNPD	United Nations Population Division
UNU	United Nations University
US	United States
USSR	Union of Soviet Socialist Republics
WB	World Bank
WEF	World Economic Forum
WHO	World Health Organization
WID	World Inequality Database
WTO	World Trade Organization
WW1	World War I
WW2	World War II
YLDs	years of life lost to disability
YLLs	years of life lost

A Guide to Reading the Second Edition of Human Geography: An Essential Introduction

For Whom This Book Is Written and Why

The aim of this book is to provide undergraduate students who are embarking upon geography programs in universities throughout the world with a concise and essential introduction to human geography.

For nearly thirty years now, I have taught human geography modules to undergraduate and postgraduate students at the University of Strathclyde in Glasgow, Scotland; Maynooth University in County Kildare, Ireland; and the University of Liverpool in northwest England. In so doing I have come to realize that there already exists a historically unprecedented range of high-quality introductory textbooks in the field of human geography. But still I have felt compelled to write, rewrite, and update this book. Why? Part of the answer resides in my belief that there is scope for introductory books to do better in three main areas:

1) **Level**: Whilst written for Human Geography 101 modules, this book will not patronize students. Alongside case study illustrations, it will address theories, concepts, ideas, and debates that are often avoided in introductory courses because they are difficult to summarize in a basic and digestible form. It will introduce students to seminal thinkers and influential texts. Although designed for the beginner student, it will provide you with insights gleaned from the most recent cutting-edge human geographical research being published. This will be an introductory book with sophisticated ideas communicated clearly and concisely.

2) **Narrative**: Students learn best when a book has a strong organizing framework. Penned by a single author, this book has been written with a clear organizing framework in mind: human geography is best introduced in and through the story of the rise, reign, and faltering of Western civilization from the fifteenth century. In an important way, to study the principal demographic, social, political, economic, cultural, and environmental processes that are unfolding in any region of the world today is to study how that region has figured and does figure in the story of the emergence, reign, and dethroning of the West. History makes geography and geography history. Each of the chapters will put flesh on the skeleton of this organizing framework.

3) **Interdisciplinarity**: Human geographers draw from and contribute to cognate subjects in the social sciences, humanities, and even the natural sciences. Indeed, some of the best human geographical writing can be found in related subjects such as sociology, anthropology, political science, international relations, economics, regional studies, archaeology, cultural studies, environmental science, and so on. Any introduction to human geography will fail in its mission if it refuses to cast its net wider than literature that is narrowly defined as "human geography."

Although not the first book to register and respond to these concerns, this book is perhaps unique in terms of its commitment to fostering student engagement with current "live" research, canonical texts, and seminal thinkers and their ideas; its privileging of an overarching narrative and in particular a narrative that takes a historical perspective and that recognizes the ways in which the ascent, dominance, and stumbling of Western civilization has forged the world in which we live today (and of course the ways in which space place, landscape and nature in turn have been centrally implicated in forging the story of the ascent, dominance, and stumbling of Western civilization); and its insistence that human geography is practiced by many disciplines and not just by human geographers. It is also distinctive in that it provides a relatively concise introduction to human geography. It has been written in an economical way: tight, concise, crisp, and with a brisk pace. It will attempt to present students with a distilled introduction, cutting through the vast terrain of human geography, separating the wheat from the chaff, and furnishing students with an appreciation of the most essential ideas, debates, and case study examples.

About the Second Edition

In publishing this second edition (2021), I have taken the opportunity to substantially rewrite and reorder material in light of the truly astonishing changes that have taken place in the world in the past five years and the world historical Covid-19 global pandemic. In many ways, this is an entirely new book rather than a second edition. Specifically, I have:

- Extended (breadth and depth) coverage of key theories, concepts, ideas, and debates.
- Incorporated relevant research literature and data sets that have been published since 2014.
- Engaged key authors and texts I neglected to give sufficient attention to in the first edition, and added influential new voices that have only recently emerged.
- Updated all case studies and incorporated a wide range of new case studies.
- At critical junctures, added important points of detail to develop, clarify, and strengthen arguments. Chapters are longer, richer, and better layered.
- Added a brand-new chapter, Chapter 7, "(Under)Developed: Challenging Inequalities Globally," and added a new "Coda on Covid-19."
- Added new deep dive boxes, figures, tables, plates, and maps to enliven the text.
- Added comprehensive new learning resources both in the text and to the companion site.
- Updated and significantly expanded reference lists and guidance on further reading.

Locating Seminal Thinkers

As you will discover in this book, human geography is in no small way a child of European civilization and as such has developed thus far largely as a quintessential Western academic subject. In many ways, human geography itself continues to bear the stamp of its emergence in and through the rise of the West from the fifteenth century, its embroilment in the West's dominance of global affairs for over 500 years, and its entanglement in the West's faltering in the twentieth and twenty-first centuries. Reflecting its origins and center of gravity and the skill set of the author, primarily this book will provide you with an introduction to Western and in particular Anglo-American thinkers, ideas, and debates. But it also considers one of its duties to be to help you achieve a greater consciousness of the origins and limitations of the approaches to which you are being introduced. A variety of strategies are used throughout the book to assist you, and both Chapter 2 and Chapter 13 will explicitly address the question of the virtues and vices of the Western-centricity of much existing human geographical writing. Here, though, you might like simply to note that this book will explicitly identify the nationality and disciplinary identity of the various authors whose work is under discussion. When authors have significant and formative ties to various locations (place of birth, place of past and present residence, etc.) and subjects (geography, sociology, anthropology, political science, international relations, economics, regional studies, archaeology, cultural studies, environmental science, etc.), these multiple locations and specialties will be noted.

Learning Supports

To assist instructors and students, this book has incorporated a number of learning supports. Use of these learning supports will help you to make the most of the book.

Chapter Learning Objectives and Checklist of Key Ideas: To clarify for students the core points to be alert to, each chapter will begin with a box summarizing "Chapter Learning Objectives" and will end with a "Checklist of Key Ideas." Of course, these boxes will cover only the bare bones, and it is hoped that you will learn much more besides. But they will provide you with a useful initial orientation as to what is expected and what ought to be given preferential attention.

Deep Dive Boxes: Throughout each chapter, you will encounter a number of clearly focused "Deep Dive Boxes." These boxes are designed to aid your understanding in three principal ways. First, some boxes will incorporate content that merits more detailed elaboration and commentary than is possible within the main body of the text. Second, other boxes will provide you with a guided tour of some of the most important, seminal books and pioneering thinkers in the discipline. Finally, a group of boxes will include case study examples to illustrate important processes and patterns.

Chapter Essay Questions: Each chapter will conclude with a list of four essay questions students should be able to attempt after reading the chapter. The completion of these essays will enable you to gauge how far you have understood the contents of the chapter and are able to engage critically with the material contained therein.

Want to Read More? Students will find "References and Guidance on Further Reading" sections at the end of each chapter.

Glossary: A number of key technical terms will be used throughout the book. The meaning of these terms will most often be explained in full in the main text. Nevertheless, to assists students to better understand and further clarify particularly terms, a glossary (organized alphabetically) has been included at the end of the book.

Companion Site Resources: This book is supported by a Companion Site http://www.wiley.com/go/boyle. There instructors will find additional resources including a specimen Human Geography 101 module outline, essay-style exam questions, a multiple-choice exam paper, PowerPoint slides for each lecture, and comprehensive Mp4 and Mp3 lectures delivered by the author for each chapter.

Nomenclature on Dates

This book frequently references key dates in human history. When surveying a long sweep in human history, the titles MYBP and YBP are used to refer to "millions of years before the present day" and "years before the present day," respectively. When surveying recent developments in human history, the title Common Era (CE) will be used. Although based upon the Gregorian Calendar, CE is preferred because it is less invested with Western and Christian connotations and is therefore open to more universal adoption. Readers should note that CE is equivalent to Anno Domino (AD), while BCE, which stands for Before the Common Era, is equivalent to Before Christ (BC). 500 BCE is equivalent to 500 BC, while AD 500 is equivalent to 500 CE.

Where on Earth Is the West Today? The OECD World as a Proxy for the West

The label "West" will be used throughout this book. Of course, this is a slippery idea, for some too slippery to be taken literally. It is certainly best understood as a vernacular region. But surely it is more than just an abstract imagined space? Surely it manifests itself materially? With caution, we will use it in this book to refer in the first instance to the rise of a number of European nation-states and empires (embryonic developments can be traced to, say, 1418 when Prince Henry the Navigator established a school of navigation at Cape Saint Vincent in Portugal, or 1492 when Christopher Columbus "discovered" the Americas), incorporating countries we recognize today as Portugal, Austria, Belgium, France, Germany, Italy, the Netherlands, Spain, Russia, and the United Kingdom. For 500 years, these countries have dictated world affairs. Through mass emigration from Europe to the New World, they were subsequently to be joined, and in some instances trumped, by the settler states they birthed (including the United States, Canada, Australia, and New Zealand). Japan, an outlier, then began its Datsu-A Ron ("leaving of Asia" or "goodbye Asia") and became Asia's Western jewel. But as the West has expanded its influence across the world and many countries have embraced Western ways, it is pertinent to ask: where is the West today? A working guide for you to follow in this book is to refer to the membership base of the Organization for Economic Co-operation and Development (OECD). Established in 1948 to oversee the US-backed Marshall Plan for the reconstruction of war-torn Europe, today the OECD exists to help countries better cooperate so as to enhance their economic growth. From its base in Paris, the OECD has grown to incorporate 37 member countries (see Map x.1). These countries are, for the most part, rich countries in the Global North, but countries like Mexico, Chile, Slovenia, Turkey, and Estonia are also full members of the OECD. In addition, the OECD is currently examining applications for membership from Argentina, Brazil, Bulgaria, Croatia, Peru, and Romania.

Biographies of Leading Human Geographical Thinkers

You will encounter a large number of well known geographers throughout this book. It is often helpful to learn a little about the background of these key thinkers so that their ideas might be set into context. Fortunately, there exist many good online and in-print resources charting the biographies of key geographers.

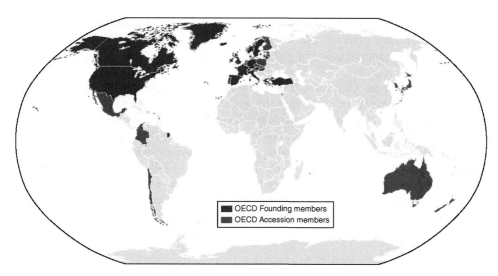

Map x.1   Where on earth is the West today? The OECD world as a proxy for the West. Source: OECD.

A good introduction to key thinkers in human geography is provided in:
Hubbard, P., Kitchin, R., and Valentine, G. (eds.) (2022 – see also second edition 2011) *Key Thinkers in Space and Place*, 3e. London: Sage.

Biographies of key thinkers can also be found in:
Gregory, D., Johnson, R., Pratt, G., et al. (eds.) (2009) *The Dictionary of Human Geography*, 5e. Chichester: Wiley.
Rogers, A., Castree, C., and Kitchin, R. (2013) *A Dictionary of Human Geography*. Oxford: Oxford University Press.
H. Lorimer and C. Withers also provide insights into the biographies of key geographers in their series *Biobibliographical Studies (Geographers)* (London: Academic, Bloomsbury). So far, they have produced nearly 40 edited volumes.

The Geographers on Film (GOF) series was produced by Maynard Weston Dow (Plymouth State University) and Nancy Freeman Dow. Interviews with respected geographers were conducted between 1974 and 2002. This series was supported in part by the Association of American Geographers (AAG). The story of GOF is recounted in:
Martin, G.J. (2013) Maynard Weston Dow (1929–2011) and "Geographers on Film", Annals of the Association of American Geographers, 103:1, 1–4. The US Library of Congress now hold the GOF archive which contains 306 interviews of 274 geographers – https://www.loc.gov/collections/geographers-on-film/.

One World, Many Environments: The Köppen Geiger Climate Classification System

For historical reasons, human geographers are wary of attributing variations from place to place in human activity and behavior to environmental influences. In the past, so-called "environmental determinism" was used perniciously to make the spurious claim that white Europeans were a superior people by dint of the temperate – green and pleasant – environments in which they lived. Nevertheless, it remains the case that the physical environment does set the stage upon which human geographies are carved (Plate x.3). Some understanding of global environments, then, might be useful before departure.

The Köppen Geiger Climate Classification System remains the most useful and widely employed system for classifying the world's climate regimes. It was first introduced by Russian and German climatologist Wladimir Köppen in 1884. Köppen went on to further develop his classification system in 1918 and 1936. Subsequently German climatologist Rudolf Geiger worked with Köppen to further elaborate on the system. Additional classification schemas were developed by Leslie Holdridge (in 1947, Holdridge Life Zones System) and by Glenn Trewartha in the 1950s and 1960s (Trewartha climate classification). But the basic Köppen Geiger climate classification system remains to this day the most widely consulted.

The Köppen Geiger Climate Classification System is built around three layers of classification.

Layer 1: Large-Scale Regions
Five broad climate regions are recognized based upon prevailing temperature and precipitation patterns, denoted by a capital letter.

A   Tropical Humid climates
B   Dry climates
C   Mild Midlatitude climates
D   Severe Continental Midlatitude climates
E   Polar climates

Layer 2: Climate Subtypes Within Large-Scale Regions (by Precipitation)
Six major subclimatic types are recognized based on seasonal trends in precipitation, denoted by a second lowercase or uppercase letter.

f   Moist with adequate precipitation in all months and no dry season (in regions A, C, and D)
m   Rainforest climate in spite of short, dry season in monsoon cycle (in region A)
s   Includes a dry season in the summer of the respective hemisphere (high-sun season) (in regions C and D)
w   Includes a dry season in the winter of the respective hemisphere (low-sun season) (in regions A, C, and D)
W   Desert (in region B)
S   Steppe (in region B)

Layer 3: Climate Subtypes Within Large-Scale Regions (by Temperature)
Finally, six major subclimatic types are recognized based on seasonal trends in temperature, denoted by a third lowercase letter.

a   Hot summers where the warmest month is 22 °C or higher (in regions C and D)
b   Warm summer with the warmest month below 22 °C (in regions C and D)
c   Cool, short summers with less than four months over 10 °C (in regions C and D)
d   Very cold winters with the coldest month below −38 °C (−36 °F) (in region D)

f   Polar Frost (in region A)
h   Dry-hot with a mean annual temperature over 18 °C (in region B)
k   Dry-cold with a mean annual temperature under 18 °C (in region B)
t   Polar Tundra (in region A)

Based on this system, the Köppen Geiger Climate Classification system organizes the world into the following climate regions (See Map x.2).

A) **Tropical Humid climates are climates in which all months have average temperatures of 18 °C and above.** They normally extend from the equator to about 10 to 30° of latitude, both north and south. Four subclimates exist. In Tropical Wet Rainforest (AF) climates, heavy precipitation occurs all year long. In Tropical Monsoon (AM) climates, annual rainfall is just as high but rain falls mainly in the seven to nine hottest months of the year. In Tropical Savannah climates, annual rainfall is usually lower and falls either in the winter (As) or summer (AW) seasons.

B) **Dry climates are climates with extremely low levels of precipitation.** They are usually found between 20 and 35° north and south of the equator, and in continental regions and mountainous areas. Two subclimates exist. Dry Arid (BWh, BSK) climates are desert climates. Dry Semiarid climates (BSh, BSk; sometimes called Steppe climates) enjoy more precipitation.

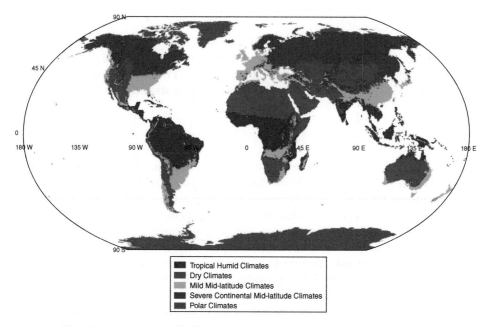

Map x.2   Global environments: The Koppen Geiger Climate Classification System world map. Source: Based on Beck, H.E., Zimmermann, N. E., McVicar, T. R., Vergopolan, N., Berg, A., & Wood, E. F., CC BY 4.0.

C) **Mild Midlatitude climates usually extend between 30° and 50° north and south of the equator and on the western or eastern extremes of continental landmasses.** Warm and humid summers are followed by mild but wet winters. Three subclimates exist. Mediterranean (Csa, Csb) climates are characterized by warm arid summers and mild winters with moderate levels of rainfall. Midlatitude Dry Winter (CWa, Cwb, Cwc) climates are characterized by mild conditions and little winter precipitation. Finally, Humid Subtropical and Maritime (Cfa, Cfb, Cfc) climates are found in coastal regions and entail warm dry summers and mild but very wet winters. In some cases, hot muggy summers and mild winters are interrupted by cyclones.

D) **Severe Continental Midlatitude or Snow climates are found both north and south of 50° latitude.** They are characterized by warm (sometimes very hot) to cool summers and very cold winters characterized by icy winds and extensive snowfall. Three subtypes exist: those with dry winters (Dw: Dwa, Dwb, Dwc, Dwd), those with dry summers (Ds: Dsa, Dsb, Dsc, Dsd), and those with year long precipitation (Df: Dfa. Dfb, Dfc, Dfd). High-latitude Taiga forests fall within this category.

E) **Polar climates suffer from extremely cold winters and summers.** Temperatures rarely exceed 10° C, and precipitation is low. Two subcategories exist. Polar Tundra (ET) climates support a small growing season for hardy species of vegetation during the summer months. Polar Ice Cap (EF) climates, in contrast, are permanently capped by snow and glaciers. In addition, Highland Climates (H) are often grouped into this final category. These comprise climates that do not follow the above patterns A to D because high-elevation mountains distort normal temperature and precipitation regimes.

But, of course, our climate does not stay the same over time. Global warming and climate change mean that regions that once experienced one type of climate regime might now be transitioning to a different climate regime. This means that it is necessary to constantly update the Köppen Geiger Climate Classification System world map.

How to Read This Book

Three points should be borne in mind while reading this book. First, this manuscript has been written with a particular chronological sequence in mind (see Figure x.1). But in no sense should you assume that the sequencing of chapters settled on implies a simple causal chain, with processes examined in any particular chapter being "explained" by processes introduced in prior chapters. Indeed, the sequence presented need not be strictly followed by instructors or students. Chapter 1 naturally should be read first, as it sets the scene for the book. An introduction to the past, present, and future of human geography as an academic discipline provides a suitable foundation for the rest of the book (but see below). (Chapter 2). Moreover, it is my view that "Big History" and the rise and fall of prior civilizations (Chapter 3) helps to set the emergence of the West into relief. There then follows four chapters looking at the rise of the European capitalist economy (Chapter 4); the rise of European nation states and empire (Chapter 5); the rapacious march of Western culture to the four corners of the world (Chapter 6); and uneven geographical development and global, regional, and urban inequalities in human flourishing (Chapter 7). In turn, this bundle of chapters helps to inform understandings of the modern rise in world population (Chapter 8), the unprecedented pressures that

Plate x.3 The Great Plains of the United States: Breadbasket to the world.

humanity is now placing on the earth's resources and ecosystems (Chapter 9), the urbanization of the surface of the earth (Chapter 10), the growing scale of international migration and the routes traversed by such migrants (Chapter 11), and the uneven vulnerability of people in different parts of the world to natural hazards (Chapter 12). And the above together make the conclusion reached in the final chapter (Chapter 13) possible. The Coda on Covid-19 constitutes a fitting capstone. But no single process is necessarily antecedent or more of a progenitor of history than any other, and no chapter is intrinsically better as a point of departure or logically prior to any other. Moreover, if you find it difficult to master any single chapter, fear not; each chapter is in a sense autonomous from the rest and can be read and digested on its own merits.

Second, Chapters 3–12 have been written so as to introduce you to core themes within systematic branches of human geography. And so, for instance, in addition to providing you with a snapshot overview of key watersheds in human history, Chapter 3 will also familiarize you with core ideas within environmental history. In turn, Chapter 4 will furnish you with an induction on economic geography; Chapter 5, political geography; Chapter 6, cultural geography; Chapter 7, development geography; Chapter 8, population geography; Chapter 9, environmental/resource geography; Chapter 10, urban geography; Chapter 11, the geography of migration; and, finally, Chapter 12, the geography of hazards. Because these systematic branches of human geography are being introduced in and through the story of the rise, reign, and faltering of Western civilization, in no sense will a comprehensive coverage of each be presented. Moreover, while incorporated in different ways into different chapters, no single chapter has been dedicated to such systematic branches of human geography as social geography, rural geography, the geography of health, or the geography of tourism. Nevertheless, the intention is that by the end of the book, you will know more about the key ideas that undergird some of human geography's most important subfields.

Finally, insofar as it places the story of the past, present, and future of human geography under scrutiny, Chapter 2 provides a fitting introduction for the book. I believe it is important that an introductory textbook should furnish students with an opportunity to take stock of human geography's heritage, biography, and future aspirations. But Chapter 2 treats human geography itself as an object of study in a way that is consistent with the overall framework of the book. In my view, a solid understanding of the rise, reign, and faltering of Western civilization is necessary before it is possible to appreciate the history and philosophy of the discipline of human geography. Human geography is both a product of and has contributed variously to the ascendance, dominance, and stumbling of the West from the fifteenth century. Today, its central challenge is to become less ethnocentric and Western-centric and more capable of rendering the world intelligible from other vantage points. Locating this chapter at the outset, then, is a little problematic. You may need to understand Chapter 1 and Chapters 3–13 if you are to properly grasp the story of the birth, development, and future aspirations of human geography. For this reason, Chapter 2 is perhaps best read as a floating chapter, to be engaged with repeatedly as you digest each subsequent chapter. Certainly, it ought to be revisited at the end of the book when its full meaning might be culled.

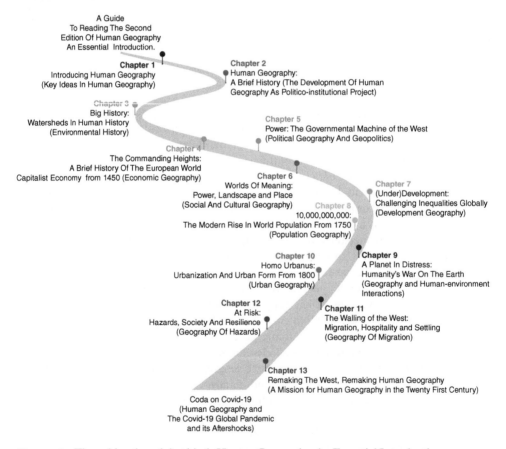

Figure x.1   The golden thread that binds Human Geography: An Essential Introduction.

Chapter 1

Introducing Human Geography

Chapter Table of Contents

Chapter Learning Objectives

By the end of this chapter you should be able to:
- Attain a heightened awareness of your geographical imagination and recognise the value of improving it through formal education.
- Provide a working definition of human geography, and identify the three central concerns of human geography.
- Appreciate the importance of the ascent of the West from the fifteenth century, its global dominance across a subsequent 500-year period, and its recent crisis in confidence, in the making of the contemporary world.
- Identify and comment upon the merits of four overarching theories human geographers use to explain the rise, reign, and faltering of the West from the fifteenth century.

Human Geography: An Essential Introduction, Second Edition. Mark Boyle.
© 2021 John Wiley & Sons Ltd. Published 2021 by John Wiley & Sons Ltd.
Companion website: www.wiley.com/go/boyle

Introduction

In the twenty-first century, humanity will be forced to confront a number of questions of epic consequence.

How many people will exist on planet earth in the year 2100? Where will these people live? What challenges will be posed by an aging population? Will uneven geographical development persist? Will there always be a developing world and lagging and poorer city-regions? Or will inequalities and the blight of world poverty finally be ended? Can the earth support continued population and economic growth? How serious is climate change, and what might its consequences be? How can societies make better usage of renewable energy resources? Can we build a low-carbon future? Will the sun finally set on the West's half-millennium dominance of global affairs? Will the United States or China or perhaps Russia or India or even Brazil or Indonesia or Nigeria rule the late twenty-first century? Or will the world in the future be governed by supranational entities like the European Union instead of nation states? Why are cities across the world expanding so rapidly, and what are the implications of this rapid urbanization of the earth's surface? Are infectious diseases re-emerging, and what can we do to remediate Covid-19 and its aftershocks and build back better so we are more prepared for Covid-20 and Covid-21? How will we govern the sprawling megacities of the future? Why are people migrating in ever-larger numbers, where are they moving from and to, and what might the consequences be for both sending and receiving countries? Why do cultures clash with one another, and how might intercultural understanding and dialogue be promoted? Why are natural hazards seemingly occurring with increasing frequency, and why do they tend to affect vulnerable people dwelling in poorer countries most? How can we build greater resilience to nature's extremes?

If these questions exercise your mind too, then congratulations! You have made a good start already by choosing to study human geography. The aim of this book is to provide you with an essential introduction to human geography. The purpose of this chapter is to provide you with a concise introduction to this essential introduction.

The chapter begins by furnishing you with a working definition of human geography and an understanding of the discipline's three key concerns: *society and space, place and the cultural landscape,* and *human–environment interactions.* A distinctive feature of this book is the world historical perspective it employs as its overarching organizing framework and analytical approach – crudely summarized, our focus will be upon the role of the rise, reign, and faltering of Western societies in the making of the contemporary world. The key argument advanced is that variations across the earth's surface in cultures, polities, societies, and economies can be traced in no small way to the rise of the West from the fifteenth century; the West's subsequent dominance over a 500-year period, expressed most visibly in its great empires; and the collapse of these empires from the mid-twentieth century and progressive erosion of Western hegemony from the 1970s. This chapter will introduce you to this story and will explain why it is an essential point of departure for any introduction to human geography. It will also alert you to the existence of a number of ways of narrating it. How we tell the story of the West's ascent, imperialism, and stumbling matters - to our understanding of how the world got to be as it is today, the challenges it currently faces and the direction in which it is heading.

What Is Human Geography?

So What Is Human Geography?

Human geography is a branch of knowledge that seeks to *venture descriptions of and explanations for the uneven distribution of human activity across the surface of the earth*. Or, to phrase it slightly differently, human geography works to *describe and explain variations from place to place in the ways in which human beings have inhabited the face of the earth*. Or, more simply, human geography concerns itself with *what is where, why, and with what consequence*. Human geography is built around three key concerns: *society and space* (the spatial distribution of human activity, or why and how societies deposit certain spatial arrangements and patterns, and why and how in turn these arrangements and patterns both enable and constrain what societies are able to do and to become), *place and the cultural landscape* (the anointment by human beings of some locations as symbolically significant, emotionally loaded, and meaningful, and the inscription of human meaning onto the built environment through the imprint of the cultural landscape), and *human–environment interactions* (the two-way relationships between people and the natural environment) (see Deep Dive Box 1.1).

Deep Dive Box 1.1 Awaken Your Geographical Imagination! Around the World in Eighty Days (Or Even Eighty Hours?)

Let us begin as we mean to continue: with some homework! Go to your nearest library, then find and read a copy of Jules Verne's famous book *Around the World in Eighty Days*. Although initially serialized in the media as if a real travel diary, this fictional work was eventually published as a novel in 1873. It was written against the backdrop of nineteenth-century British and French colonial and imperial expansion, aided and abetted by revolutionary changes in transportation technology.

Having accepted a wager of £20 000 made by a fellow member of the Reform Club (a "gentleman's club") in Pall Mall in London that he could not travel around the world in eighty days, the book follows the travails of Phileas Fogg and his French valet Jean-Passepartout as they attempt to circumnavigate the earth's circumference in that time or less. Departing London by train on October 2nd, Fogg's mission was to return to the Reform Club by December 21st, transiting through Suez, Bombay, Calcutta, Hong Kong, Yokohama, San Francisco, New York, and Liverpool; crossing therein the Mediterranean, Red, South China, and East China Seas and the Pacific and Atlantic Oceans by steamboat, and journeying across continents by rail and, it transpires, elephants.

Take stock of Verne's vivid depictions of Fogg's adventures and encounters in strange, faraway, foreign, and exotic lands.

Now indulge your imagination for a second, and try to visualize the sorts of people, places, and landscapes you might encounter were you to follow in the footsteps of Phileas Fogg today. Only chart your own itinerary. Allow yourself to wander off the beaten track, to take the roads less traveled. Picture yourself moving

(Continued)

Box 1.1   *(Continued)*

across and between the world's oceans and continents. What mode of transportation are you using? Phileas Fogg arrived back in London on time and collected his wager. How quickly could you complete the journey today? Given modern transport technology, could you circumnavigate the world in only eighty hours? Perhaps consciously slow down a little, and pay more attention to the view *en route*. Journey through and immerse yourself in the mosaic of cultures and environments you might encounter. What do you see in your mind's eye? A tsunami of social change has occurred around the world since the late nineteenth century; how far do you think the places Fogg visited have changed? Are they recognizable at all today? Nurture a sense of wonderment, puzzlement, and curiosity about the arrangement, texture, shape, and color of the varied human landscapes that mark the face of the earth today. If you have a vivid imagination, perhaps you will be able to convince yourself that you can actually smell, touch, taste, hear, as well as see certain places!

Although geographical literacy varies from one individual to the next, everyone has at the very least a rudimentary geographical imagination – that is, a mental map of the physical environments and different ways in which human beings occupy the surface of the earth in different parts of the world. Human geography is a branch of knowledge that invites you to bring your geographical imagination to the forefront of your consciousness, no matter how imperfect and impartial it is, and whatever its stage of development. It asks that you place this imagination under critical scrutiny, challenge and interrogate its accuracy and objectivity, and cultivate and develop it through formal education. We find amusing and perhaps a little excruciating Phileas Fogg's clumsy and unschooled encounters with the peoples, cultures, and technologies he meets on his voyage; his mental map was evidently shaped in the gentleman's clubs of Victorian London! Fortunately, human geography stands ready to support you to do much better. In its theories, concepts, methods, and case studies, you will find the resources you need to move toward a more self-aware, informed, improved, and rigorous understanding of the world.

Society and space: How are human activities distributed across the face of the earth, and why does it matter to the functioning of societies? Human geography concerns itself with the location of human beings and their activities, specifically the causes and implications of the uneven distribution of both over the earth's surface.

A cursory glance from the window of a plane flying at 30 000 ft reveals a key truth; humans occupy the surface of the earth in uneven ways. Populations are concentrated in some areas and dispersed widely in others. Some human activities (and associated land uses – services, industries, agriculture, forests, houses, transport hubs and links, recreational spaces, and so on) are found here, others there. Moreover, the distribution of human beings and their activities changes over time. There can be a drift to the north, south, east, and/or west. Centripetal forces can drive human beings together and create clusters, concentrations, and agglomerations, while centrifugal forces can disperse human beings and distribute human activities over a wider area. Innovations, technological breakthroughs, diseases, fashions, and fads originate in one place and steadily fan out or diffuse in uneven ways to other places.

Ever Wonder Why?

Human geographers believe that spatial patterns are rarely random and instead are ordered by underlying processes. They work to uncover these processes so that existing spatial conditions and geometries can be explained. In turn, they believe that the spatial organization of any society both enables and constrains what that society is capable of doing. By dint of the way they are organized – their core cultural, social, political, and economic institutions – societies carve and etch unique built environments onto the earth's surface, much like a pattern on a carpet. These patterns – or, as some refer to them, spatial fixes – help them accomplish key tasks, but they also limit what can be done too. When these limitations become too much of a hindrance, or in some cases directly at odds with the ability of societies to progress, human beings work to destroy old arrangements and new spatial conditions arise. Patterns can dissolve and recrystallize. Human geographers speak about the role of "creative destruction" in making spatial conditions anew. Just like a kaleidoscope, a jolt can destroy an old pattern but pave the way for another shape or morphology to appear (see Deep Dive Box 1.2).

Deep Dive Box 1.2 Society and Space: The Causes and Implications of the Spatial Structure of the United States

Plate 1.1 depicts a night picture taken by NASA of the United States from space. Describe the patterns of human activity – a kind of urban galaxy – you see from the distribution of night lights. Why are people spread so unevenly

Plate 1.1   Black marble: The United States by night. Source: NASA Worldview.

(Continued)

Box 1.2   (*Continued*)

across the country? Why do they cluster in particular locations, and why are some clusters bigger than others? How have these concentrations of population changed through time? In what ways is the fact that the United States is a liberal democracy and advanced capitalist economy with a federal political system important? Do such societies have unique human geographies, and if so, why? In turn, how do the geographical patterns you observe enable the United States to work as it does? Would the United States work in the same way if people were scattered evenly across the land surface? Can the United States continue to prosper given its current spatial structure? In what ways could a different human geography pave the way for a better and more prosperous future? What would a night picture of the United States from space look like if you could reengineer the arrangement of humans and their activities across the territory? What would you hope to achieve by turning the dial on the kaleidoscope?

Place and the cultural landscape: How do human beings convert the surface of the earth into a home? Human geography is interested in the ways in which human beings turn otherwise empty spaces and soulless landscapes into places by investing meaning in particular locations, and the way they project meaning onto space by designing, planning, and organizing the built environment in particular ways – they speak about these architectural expressions in terms of the idea of the cultural landscape.

We encounter places with all the senses our bodies have to give. Human geographers use the idea of sense of place to capture the ways in which particular locations are felt, experienced, and sensed. Specifically, human beings attach emotions, significance, and values to places, and in so doing turn barren and anonymous locations and environments into intensely meaningful places. They can develop a deep existential love for a place – called topophilia. In this way, they convert patches of the earth's surface into a home. The concept of territoriality is used to explain why and how humans often seek to claim ownership of land and assert their exclusive right to govern over a particular turf. Human geographers use the idea of the cultural landscape to capture the ways in which human cultures etch their imprints onto the face of the earth. They try to "read" the cultural landscape, to discern what particular societies are trying to communicate in and through their buildings, streetscapes, public spaces, field patterns, place names, environmental engineering projects, and so on. The cultural landscape projects the power of societies' elites onto the built environment (think of the Gardens of Versailles or the streetscapes of Manhattan), but it also symbolizes the struggles and resistances of marginalized minorities (think of Chinatowns or Native American reservations) (see Deep Dive Box 1.3). Familiar architecture, it seems, creates a sense of belonging, while unfamiliar landscapes can create a sense of estrangement and alienation. Feeling at home or out of place are powerful human emotions.

Human–environment interactions: How do humans interact with the natural environment? Human geographers use concepts such as environmental limits and hazards to capture the impact of nature on human possibilities, resources and pollution to apprehend the ecological footprint left by humans on nature, and sustainability and resilience to better understand how societies protect themselves from nature's extremes and at the same time manage more carefully their relationships with nature.

Deep Dive Box 1.3   Place and the Cultural Landscape: The Meaning of the Red Light District in Amsterdam

Plate 1.2 displays a picture of the Red Light District in Amsterdam. What kind of cultural landscape is evident? How is the iconography of the landscape to be read? Who is to provide the reading? Notice the streetscapes, building frontages, light displays, and admixture of land uses. Reflect upon the extent to which the landscape can be read differently at night compared to day. Does the Red Light District symbolize a progressive, liberal, and tolerant neighborhood in a sexually enlightened and relaxed world city? Or does it embody a criminal and immoral niche, a deviant ghetto marked by hedonism and defiling the ambience of a European city dripping otherwise in imperial history and grandeur? Should this cultural landscape be defended in the name of freedom? Should it be erased in the name of decency? Should human sexuality be visible in the cultural landscape, when, and under what circumstances? Why is sexuality hidden from view in mainstream city landscapes? How would you feel walking through this area, say, at noon or at midnight? Does this amount to a majoritarian view that sexuality is best located in the private realm of the home? Should minority views on the importance of displaying sexuality in the public realm be tolerated? If so, why? If not, why not?

Plate 1.2   DeWallen, the Red Light District in Amsterdam, the Netherlands. Source: Photo by Zoltan Tasi on Unsplash.

Societies continue at be at risk from nature's extremes. Natural hazards (such as earthquakes, tsunamis, volcanoes, hurricanes, floods, forest fires, and droughts) can become social, political, economic, and cultural disasters. Moreover, there would appear to be a growing relationship between global warming and freak weather: we will witness more dangerous hazard events as climate change escalates. Human geographers concern themselves with how the social organization of society mediates the impact of nature on human populations. Why are poorer people, women, ethnic minorities, children, and those suffering from a disability always more vulnerable to environmental hazards? Societies in turn exist by metabolizing resources (water, food, energy resources, minerals, etc.) from their surrounding natural environments. In so doing, they deplete the earth's finite supply of nonrenewable resources and change the ways in which ecosystems work. In transforming nature for their own ends, human beings also pollute the environment with their waste (through air, water, and land pollution). Fossil fuel–driven, human-induced climate change is a pressing example, but other problems include poisoned rivers, acid rain, salinated soils, reduced biodiversity, desertification, and contaminated water tables (see Deep Dive Box 1.4). Building resilience to natural hazards, including climate change–induced extreme weather events, and embracing sustainable development, circular (recyclable-based) economies, and a low-carbon future are critical priorities for humanity.

Deep Dive Box 1.4   Human–Environment Interactions: The Relationship Between Shanghai and China's Eastern Coastline

Plate 1.3 displays a picture of Shanghai Bund under smog. How does Shanghai interact with its surrounding environment? Is the city exposed to natural hazards? Has the city's phenomenal growth contaminated coastal habitats in east China? To what extent is the city pursuing sustainable development and building environmental resilience? Shanghai remains at risk from flooding – and the development of overcrowded neighborhoods with poor-quality housing, to house a surging migrant population, has deepened the vulnerability of this population. With a surging economy and large population, Shanghai has an enormous ecological footprint. It is a thirsty city and draws water primarily from the surrounding delta, including the Yangtze River. Antiquated infrastructure, however, ensures that supply systems are beset by wasteful leakage. An engine in the booming Chinese economy, Shanghai produces a significant quantity of greenhouse gases. It continues to rely on nonrenewable energy sources and fossil fuels. Air pollution and smog exist as severe problems. The city is contributing negatively to global climate change. Shanghai also secretes enormous amounts of waste; landfill and incineration continue to be overused solutions and present environmental risks. But things are changing. As Shanghai becomes a prosperous global city, it will lead Chinese efforts in the future to curb fossil fuel emissions and transition to a low-carbon economy. It will harness the latest technology to develop effective flood-planning and mitigation strategies. And it will embrace a circular economy, recycling and reusing its existing assets and materials whenever possible.

Plate 1.3 Father and child walk through smog near the Shanghai Bund. Source: Getty Images News / Getty Images.

We might say, then, that the mission of human geography is to describe and explain how and why human beings: locate themselves and their activities unevenly over the earth's surface, why these patterns change over time, and how they make a difference to the functioning of societies; turn arid spaces into humane places, create distinctive and meaningful places and landscapes at various scales; and accommodate best to the varied physical environments in which they live, while addressing the degree to which they are scarring those environments through overuse of resources and pollution.

One Planet, Many Cultures, Unconscionable Inequality

The Origins of Our Unequal World: The Rise, Reign, and Faltering of the West

The mission of human geography is to describe and explain variations from place to place in the ways in which human beings have inhabited the face of the earth. But how might one go about making sense of these variations? What explanatory frameworks might guide us?

Denis E. Cosgrove was a British historical and cultural geographer. According to Cosgrove, the rise of the West was a world historical event that ought to be central to geographical enquiry. Consider his opening remarks in his 1984 book, *Social Formation and Symbolic Landscape*:

> Between 1400 and 1900 much of Europe and the society it founded in North America were progressing towards a characteristic form of social and economic organisa-tion which we now term capitalist.... In developing a capitalist mode of production Europeans established and achieved a dominance over a global economy and a global division of labour which remains a critical determinant of our present social and

economic geography. The European transition from societies dominated by feudal social relations and their associated cultural assumptions to capitalist centrality in a worldwide system of production and exchange is a phenomenon of central historical importance in making sense of our own world. We understand a great deal about many of the fundamental features of the change: its associated demographic trends, alterations in agricultural and commodity production, the political reorganisation of peoples and territories, and the changing relations between individuals, groups, and classes. Whatever the specific focus of historical attention, it is the internal reorganisation and outward expansion of European societies, gathering pace throughout the period, which insistently compels historical enquiry and demands historical understanding.   (Cosgrove, 1984, pp. 2–3)

Like Cosgrove, this book too insists that historical enquiry is central to human geography. Like Cosgrove, it too argues that to understand the human geography of the world today, it is necessary to study the ways in which the rise of European society from the fifteenth century has profoundly shaped (but clearly not determined) the fate of all world regions. Only we will insist that the sun is now setting on this world historical event and that we are moving toward a new world dynamic.

To some, this starting point might appear a little strange. This is, after all, a book about human geography and not human history. That is, until it is realized that it is impossible to make sense of how human beings have inhabited the surface of the earth (their location and distribution over the earth's surface, the places they have crafted, and their vulnerability to natural hazards and ecological footprints) without paying attention to the West's ascent to the pinnacle of world history from the fifteenth century, its subsequent supremacy for over 500 years, and its precarious dominance over global affairs today (see Deep Dive Box 1.5). In important ways, it remains the case that to study the historical development of any world region, including its principal demographic, social, cultural, economic, technological, political, and environmental features, is to study how that region figured in the story of the rise, reign, and stumbling of the West as a global superpower.

Deep Dive Box 1.5   The Course of Empire

One way to visualize and engage critically and thoughtfully with the narrative around which this book is written is to refer to United Kingdom/United States artist Thomas Cole's famous five paintings (1833–1836) charting the course of empire (Plates 1.4–1.8). At the heart of these paintings is the idea that empires pass through a number of cycles, beginning with great promise but almost preordained to end in decay and collapse. The five stages in the cycle Cole identifies are: *The Savage State, The Arcadian or Pastoral State, The Consummation of Empire, Destruction,* and *Desolation.* Cole was a leading member of the Hudson River School, a mid-19th century group of landscape artists who, influenced by Romanticism, fixated on the iconography of human–environment interactions in what was then a new, youthful and expanding nation.

Look at these paintings. How do they relate to the story of the rise, reign, and faltering of Western civilization? What kinds of geographical imaginations do you think existed at each stage? What kinds of spatial patterns and landscapes might human beings have imprinted on the earth's surface at each point? How would you describe the shifting nature of human–environment interactions as the cycle unfolds? Where is the West in the cycle? Is it really inevitable that all great civilizations collapse? Why was Cole so pessimistic? How useful are these paintings in helping us to understand the world as it presents itself today?

Plate 1.4   Thomas Cole's *The Course of Empire: The Savage State* (1834). Source: Wikimedia Commons, https://commons.wikimedia.org/wiki/File:Cole_Thomas_The_Course_of_Empire_ The_Savage_State_1836.jpg. Public Domain.

Plate 1.5   Thomas Cole's *The Course of Empire: The Arcadian or Pastoral State* (1834). Source: Wikimedia Commons, https://commons.wikimedia.org/wiki/File:Cole_Thomas_ The_Course_of_Empire_The_Arcadian_or_Pastoral_State_1836.jpg. Public Domain.

(Continued)

Box 1.5   *(Continued)*

Plate 1.6   Thomas Cole's *The Course of Empire: The Consummation of Empire* (1836). Source: Wikimedia Commons, https://commons.wikimedia.org/wiki/File:Cole_Thomas_The_Consummation_The_Course_of_the_Empire_1836.jpg. Public Domain.

Plate 1.7   Thomas Cole's *The Course of Empire: Destruction* (1836). Source: Wikimedia Commons, https://commons.wikimedia.org/wiki/File:Cole_Thomas_The_Course_of_Empire_Destruction_1836.jpg. Public Domain.

Plate 1.8 Thomas Cole's *The Course of Empire: Desolation* (1836). Source: Wikimedia Commons, https://commons.wikimedia.org/wiki/File:Cole_Thomas_The_Course_of_ Empire_Desolation_1836.jpg. Public Domain.

We will certainly have cause to question the integrity and coherence of the very idea of the West throughout this book. Indeed, arguably there is no more slippery word in the English language. There are, it seems, as many Wests as there are commentators (see Deep Dive Box 1.6). But this should not stop us from recognizing that from the fifteenth century, a number of dramatic intellectual, social, cultural, political, and economic changes coalesced in Europe and propelled European, and then later other countries to a position of global dominance. These changes were given first life by the European Renaissance, the Protestant Reformation, and the Age of Reason or European Enlightenment. They bore witness to the rise of European modernity and a new tradition of Western utopian thought. At Westphalia (in Germany), they stimulated the formation of sovereign nation states and, through the Age of Revolutions, witnessed the rise of liberal democracy (where political rulers are voted in and out in free and open elections and on the basis of mass enfranchisement). At their heart were the development of free-market capitalist economic systems (based upon private ownership of property and systems of production, distribution, and consumption), the Industrial Revolution, and the rise of a global capitalist economy. Through the European Age of Exploration and the Age of European Empires, they presided over the rapacious march of European countries to the four corners of the world.

As these developments unfolded, a number of powers (incorporating countries we recognize today as Portugal, Austria, Belgium, France, Germany, Italy, the Netherlands, Spain, Russia, and the United Kingdom) steadily broke from the pack and collectively climbed to the summit of world history. For 500 years, these countries came to dominate world affairs. Through exploration, colonization, and exploitation of the Americas, Asia, Africa, Australasia, and the polar regions, they expanded their prowess and exerted their will over the entire world. The fate of other civilizations became inextricably intertwined

Deep Dive Box 1.6   The Strengths and Limitations
of Focusing upon the Story of the West and Its Impact
on the World

This book places front and center the rise, reign, and faltering of the West, from the fifteenth century to the present, as the single most important historical dynamic that has shaped the world in which we live today. It is difficult to understand the nature and trajectory of any region of the world without reflecting upon how that region has been differentially embroiled in the story of the ascent, dominance, and stumbling of the West.

But, by thinking about the world in this way, are we falling prey to self-centered and blind arrogance? Is it parochial and an unwelcome hangover from the colonial era to see the rise of the West as the principal world historical force at play across the past 500 years? Is our analytical approach Eurocentric – does it say more about the author, a white European male, than it does about reality? Is this book not unwittingly underpinned by flawed European assumptions about the significance of the West in world historical affairs? Surely we must refuse the idea that the history of non-Western peoples was somehow jump-started only when the West established contact? Surely each region is energized by its own unique historical dynamic? Surely we must tell the story of different places without the calibrating backdrop of a canonical West protruding on our geographical imaginations?

Here we can return to British historical and cultural geographer Denis E. Cosgrove for guidance. Cosgrove in fact was concerned not only with the role of the ascendance of the West, and in particular the European capitalist economy, in the making of the actual world but also with the significance of this historical development in the forging of the most pervasive cultural frames through which the world is now commonly understood. Passed off as universal and truthful, Cosgrove constantly thought to *provincialize* (or reveal the parochial status of) these understandings as uniquely Western and emergent from Western economic and political developments. His musings focused initially upon cultures in early modern Europe (his early work, for instance, sought to understand views of landscape in sixteenth-century Renaissance Venice), but toward the end of his life his focus broadened to include cultural imaginings in twentieth-century Rome and twentieth-century scientific cultures, for example astronautical images of the earth from space. Cosgrove deemed it necessary to walk a tightrope: approaching the rise of the West as at once a world historical event that has profoundly shaped all world regions *and* a source of stories about how the world has developed in the way in which it has, which we must approach with caution and a degree of suspicion.

We must too walk the tightrope.

As with most of contemporary human geography, our geographical imagination is indeed a very Western one. It is dominated by a geographical lexicon that lingers as a powerful legacy of the past. We jump to certain images of the world all too readily. These images continue to serve us well, but the world is changing rapidly, and they are becoming ever less relevant and eventually they may well even be eclipsed and rendered obsolete. Stubborn assumptions about the character of the world's regions will need to be challenged and kept under constant review. Nevertheless, this book holds that the world that the West has built will not easily be undone, and so the geographical imaginations on which this world has long pivoted will continue to serve a purpose. For the foreseeable future, we will continue to benefit from concepts such as the "first" "second," and "third" worlds, "more developed" and "less developed" countries, and the "Global North" and "Global South." But as these concepts come under increased stress and yield ever-diminishing insights, we may need to invent an entirely new language and to think about geographical differences anew. This point is near but perhaps not imminent as yet.

with their fate. Through mass emigration from Europe to the New World, they were subsequently to be joined, and in some instances trumped, by their offshoots, the newer Western countries of the United States, Canada, Australia, and New Zealand. Meanwhile, from the eighteenth century, and under the influence of the Netherlands, Japan began its Datsu-A Ron ("leaving of Asia" or "goodbye Asia") and became Asia's leading pioneer of Occidentalism. Thereafter, through all sorts of complex and tortuous routes, and to varying degrees, other countries too embraced the ways of the West.

From the mid-twentieth century onward, the ascent of the West began to peter out and, indeed, according to some, the West began to degenerate. Perhaps the most visible sign of the West's reversal in fortunes was the collapse of the European imperial project. Decolonization, which had begun in Latin America from the late eighteenth century, swept across Asia and Africa, and Europe was forced into retreat. From the mid-1970s onward, on virtually every measure of development, the gap between the West and most of Latin America, Asia, and even parts of Africa began to narrow. Today the West's flagship nation, the United States, would appear to be losing ground to the Brici (Brazil, Russia, India, China, and Indonesia) countries. Perhaps the second half of the twentieth century will belong to China. Meanwhile, the West is being forced in a variety of ways to reap the harvest of its past misdeeds in Latin America, Asia, and Africa. All the while, anticolonial movements, perhaps wearing the clothes of radical Islam, continue to remind the West that many around the world perceive it to be a rapacious and abusive power.

It is undoubtedly premature to announce the death of the West, and in no sense is the decline of the West irreversible or inevitable. But in an important sense, our world is a world currently coming to terms with the fate of a civilization that for 500 years directed global affairs but is now struggling to maintain its dominance (see Deep Dive Box 1.7).

Four Theories Explaining the Rise, Reign, and Faltering of the West

There are many ways to tell the story of the rise, reign, and faltering of the West from the fifteenth century. In 2004, British-born and US resident Marxist geographer David Harvey delivered the eighth Alfred Hettner Lecture at the University of Heidelberg in Germany. Hettner (1859–1941) was a seminal figure in the rise of the discipline of geography in Germany. He was interested in mapping and explaining variations between regions over the surface of the earth and had himself undertaken studies of regions in Europe, Colombia, Chile, and Russia. Meanwhile, Harvey's intellectual career has been dedicated toward studying the root causes of uneven geographical development and what might be done to tackle inequality, injustice, and poverty and create a fairer world.

Deep Dive Box 1.7   An Image of the World Today?
The Monument to the Discoveries in Lisbon, Portugal

Standing proudly on the northern bank of the River Tagus in the city of Lisbon is the famous Portuguese Monument to the Discoveries (Plate 1.9).

Initially designed as a showpiece for the World Fair held in Lisbon in 1940, this monument was eventually built in its present form in the late 1950s and unveiled in 1960, during the Estado Novo, Portugal's so-called Second Republic. At the head of the monument sits Prince Henry the Navigator (1394–1460), inventor of modern seafaring navigation and founding father of the European Age of Discoveries. Among the many other Portuguese and other luminaries featured on the monument are: Vasco da Gama (1460–1524, pioneer of shipping routes between Europe and India), Pedro Álvares Cabral (1467–1520, the first European to discover Brazil), Bartolomeu Dias (1451–1500, the first European to sail through the Cape of Good Hope), and Diogo Cão (1452–1485, whose explorations in Africa revealed to Europeans the existence of the Congo River). Portuguese exploration raised awareness of the potential riches that might be procured from the colonization and domination of lands that lay beyond the horizon. Perhaps not unexpectedly, Portugal emerged thereafter as a global power, presiding over a vast empire that reached into the four corners of the world. Portugal's most famous colonies included Brazil, Angola, Mozambique, Guinea-Bissau, Cape Verde, São Tomé and Príncipe, Goa, East Timor, and Macau.

What does this monument symbolize? How are we to read it?

We might say that the monument simply commemorates Portugal's pioneering role in the European Age of Discovery: that period commencing in the fifteenth century when European explorers began to venture forth and to discover the existence of lands in the Americas, Asia, Oceania, and the Polar Regions. It was designed as a beacon to project to the world an image of Portuguese innovation and greatness. It spoke to the capacity of Portugal to lead a global empire.

The monument was commissioned under the reign of António de Oliveira Salazar, prime minister of Portugal from 1932 to 1968. Salazar presided over a Roman Catholic, conservative, authoritarian, and nationalistic regime that feared the loss of Portuguese influence in the world. Salazar's mission was to ensure that Portugal's pioneering role in the rise of Western civilization was front and center and that the country continued to be seen as one of the world's great powers.

Salazar believed that the Portuguese presided over a superior empire than that constructed by other European nations. He subscribed to "Lusotropicalism," a doctrine coined by Brazilian sociologist Gilberto Freyre to denote the distinctive character of Portuguese imperialism overseas. This doctrine held that the Portuguese were better colonizers because they were innately disposed to treating indigenous populations humanely and fairly. Because it was founded through genetic mixing and miscegenation, Portugal itself was a multicultural, multiracial, and pluri-continental nation. And so it alone was uniquely able to govern over a culturally and ethnically diverse empire in a way that respected and embraced local differences and rights.

But let us dig a little deeper.

By 1960, many colonies had gained independence, and other countries in Europe were actively preparing to downsize their empires. Salazar was determined to stand alone and to defend what remained. Few Portuguese colonies shared his views on Portuguese exceptionalism or accepted his belief in Lusotropicalism. Salazar displayed a dogged refusal to cede any of the remaining Portuguese colonies, particularly those in Africa. He believed it was his historical destiny to continue to spread Portuguese "civilization" to the world. His quest was to be in vain. By 1975, Portugal had lost its last two jewels in the imperial crown, Angola and Mozambique. The winds of change could not be arrested or reversed.

The Monument to the Discoveries, then, might be reinterpreted, not as an icon of confidence in Portuguese greatness or a celebration of seafaring navigation and exploration or a projection of the advanced standing of the Portuguese nation and its fitness to lead a global family, but as an anxious statement of defiance in the wake of a crumbling empire. It sought to revel in the genius of Portuguese explorers at precisely that point in time when the sun was setting on the Portuguese empire and the nation was faltering in authority. What was designed to be a powerful declaration of reality has become a repository of memory for a bygone era and a gathering point for nostalgia.

Interrogating the meaning of this monument is instructive. At once a celebration of Portugal's ascendance and global dominance and a desperate act of vanity by a country in the throes of decline, in an important sense, the story of the Monument to the Discoveries captures exactly the state of the West as it presents itself today.

Plate 1.9   Monument to the Discoveries, Lisbon, Portugal. Source: ullstein bild / Getty Images

Summing a life's work, Harvey entitled this lecture "Towards a General Theory of Uneven Geographical Development" (Harvey 2006). Reflecting Hettner's interest in regional differences across the globe, Harvey fastened on the crunch question: Why do different parts of the world develop at different rates, and why at any point in time does uneven geographical development exist? Specifically, he asked, why do global regions like the West break from the pack and develop more rapidly than other regions?

Harvey argued that there exist at least four possible ways of making sense of uneven geographical development. We might title these explanations: *Only in the West because of favorable environmental endowments*; *First in the West, then elsewhere*; *Because in the West, not elsewhere*; and *The West versus the Rest*. It matters which of these stories you find most compelling. Each leads to a particular way of thinking about the ascent, hegemony, and dethroning of the West. Each results in a different image of how the contemporary world is organized and in which direction it is moving. And each circumscribes the kinds of political interventions we might want to make as we work to produce a better and fairer world.

1. The mantra of the first explanation is *Only in the West because of favorable environmental endowments*. Uneven geographical development arises because there exist different physical environments in different parts of the globe, and these different physical environments endow the societies that dwell therein with different levels of opportunity. Deeply flawed and racist versions of this argument suggest that Europe's more temperate climate bred a superior species of humanity: with a better intellect, Europeans were able to build more civilized societies. Europe succeeded because its climate bequeathed a more advanced civilization. It is little surprise that this civilization went on to dominate the entire world. Humans who, by dint of their formation in harsher environments, were debilitated by inferior intellects built only primitive societies and were no match for the Europeans. This logic simply has zero basis in fact. More respectable versions of this tradition of scholarship prioritize the resources nature provided for Europeans, resources that, if presented to any race, would allow them to stride ahead.

    If you endorse the environmental explanation, you might imagine the West to be reaching the end of a process that has been thousands of years in the making. The capacity of the West to continue to harvest the prior advantages afforded to it by nature is now encountering the law of diminishing returns and running its course. Indeed, through overextraction of natural resources and pollution, you might speculate that the West has sabotaged any environmental advantages it might once have enjoyed. You might also consider the extent to which the physical environment is playing an active role in shaping winners and losers today. And you might speculate on the claim that the civilizations that will succeed in the future will be ones that are currently most able to treasure, conserve, and sustain their natural habitats.

2. The mantra of the second explanation is *First in the West, then elsewhere*. History unfolds down a single pathway, during which one civilization streaks ahead of the others, and following which lagging civilizations steadily submit to the same inevitable process. The history of humanity from the fifteenth century is the story of human innovation in the West followed by a staggered catching-up by other world regions. The West is best. It alone figured out how to bring humanity to a new level. With the help of the West, the rest can copy the winning formula and make the same transition. There is only one pathway to modernization; the West has trodden this pathway and shown the rest how it is to be done. All regions are predestined to follow the

same course of development. "Westernization" and "development" (and other ideas such as "modernization," "progress," and "civilization") are one and the same thing.

If you accept this explanation, you might conclude that our world is reaching the end of a process of modernization that began in Europe in the fifteenth century and steadily entrained other regions thereafter. Some of these regions have accelerated through the process and are beginning to eclipse the West. Some are even becoming better at being the West than the West itself. You might ponder why the original West has allowed the rest to catch up. Believing that West is best, you might conclude that the original West has faltered because of ignorance of, and a loss of faith in, the central pillars that brought it to greatness. All is not lost, however. As long as it rediscovers what made it best and returns to its core values, the original West will rebound.

3. The mantra of the third explanation is *Because in the West, not elsewhere*. The West only became a global superpower because it was able to oppress, exploit, and plunder the resources of other world regions. The rest only became the rest because they were subordinated, impoverished, and pillaged by the West. The haves get richer only because the have-nots are forced to have even less. Underdevelopment in no way reflects cultural backwardness or inferiority. Instead, it is a product of a historical process in which the rapaciousness of one civilization has been serviced only at great cost to other civilizations. This active development of uneven development began with rise of the European-centered world capitalist economy from the fifteenth century. It was consolidated during the Age of European Empires, when Europe annexed vast tracts of Latin America, Asia, and Africa.

    If you believe in this approach, you might conclude that the West will remain best only if it continues to be able to accumulate wealth from the labor of other regions. No longer benefiting from global empires to oversee this extraction of resources from elsewhere, today the West might be recognized as engaged in a series of neocolonial strategies and struggles, desperate efforts to continue to extract value from the rest for its own gain. Our world, then, is a volatile and dangerous one precisely because the West, recognizing that it sits on the brink of losing its dominance, is thrashing to maintain its supremacy. Meanwhile, you might reflect upon the postcolonial trajectories of former European colonies. In what ways and to what extent have these colonies escaped the long arm of the West? How are they making their way in the world and, in so doing, contributing to a fairer but more uncertain world? Could China be accused of acting as a "neocolonial" power in Africa? Could China grow today if it could not capture African resources?

4. The mantra of the final explanation is *The West versus the Rest*. Uneven geographical development arises when competition between different territorially organized powers (cities, regions, states, and continents) creates winners and losers. Ongoing competition, however, means winners need to stay constantly alert; their success is not preordained to continue forever. For the past 500 years, it has been the West that has proven the most formidable competitor. But the West's dominance has always been precarious, and it has had to actively strive to sustain its supremacy. Rival competitors have never left the field of play, and through time they too have learned to be more dynamic and innovative. By securing a new competitive edge, it is entirely possible that losers will eventually become winners, that the West will be usurped, and that the world will be turned on its head.

    If you accept this approach, you might view the world as one reeling from 500 years of Western dominance and in the throes of a volatile process of rebalancing. While

European colonization and annexation of the Americas, Asia, Africa, Australasia, and the Polar Regions stacked the world decisively in Europe's favor, other civilizations refused to succumb entirely to Western dominance and influence. Our world is a product of the hurdles the West has encountered as it has marched to the four corners of the world, and the struggles and conflicts that this march has generated. These struggles are now entering a new phase. Countries are discovering alternative and historically new development formulae. Exactly what will emerge remains up for grabs. For this reason, it is only possible to say that the post-Western world into which we are heading will, for a while at least, be an unstable, unpredictable, and potentially violent world.

All four explanations will be mobilized in many guises throughout this book. We will reject outright deeply offensive and flawed ideologies that trace the success of the West to the innate or biological superiority of the people of Europe. Patently this is nonsense. Otherwise, we will be cosmopolitan in our coverage, accommodating authors with a range of different views. My purpose is not to encourage you to settle upon any one or any particular combination of these four contending images of the contemporary world, but simply to bear witness to the ways in which each has contributed to human geographical studies of our complex and restless world. Harvey (2006) himself sees virtues and vices in each. He refuses to fully endorse or fully dismiss any particular one, even though he finds certain variants within each abhorrent. He sets about putting all four to work in what he terms a "unified field theory" of uneven geographical development. Once you have read this book, you yourself might begin to favor one of these explanations over the others. Or perhaps, like Harvey, you might see merit in each and agree that they all ought to be combined into a unified field theory. Either way, this book will provide you with sufficient fodder to reach a considered choice.

Conclusion

Everyone has a geographical imagination, that is, a mental map of the different ways in which human beings have occupied the surface of the earth in different parts of the world. Becoming conscious of one's own geographical imagination is a prelude to strengthening and cultivating that imagination through formal education. Human geography seeks to help you register, question, nourish, expand, and fortify your geographical imagination. The mission of human geography is to describe and explain the irregular and changing distribution of human activity over the face of the earth (society and space), the variety of places that humans create and the ways in which human beings turn earth into a home (place and the cultural landscape), and the ways in which humans defile the environment through pollution, remain vulnerable to hazards, and seek to build sustainable and resilient communities (human–environment interactions). In seeking to describe and explain the ways in which human beings currently inhabit planet Earth, this book takes seriously the role of the rise, reign, and faltering of the West from the fifteenth century in the fashioning of the modern world. Historical enquiry is central to human geography. To understand why our world can be described as unequal and unfair but amid a process of rebalance and change, it is necessary to situate the present with respect to events that have unfolded in the past or at least whose roots lie in the past.

Checklist of Key Ideas

- Everyone has a geographical imagination, that is, a mental map of the different ways in which human beings have occupied the surface of the earth in different parts of the world. Becoming conscious of one's own geographical imagination is a prelude to strengthening and cultivating that imagination through formal education.
- Human geography is at root a discipline that seeks to describe and explain the differentiation of human activity across the face of the earth. Human geography has three central concerns: the relationship between society and space, place and the cultural landscape, and human–environment interactions.
- This book places front and center the role of the rise, reign, and faltering of the West from the fifteenth century in the making of the modern world. Much of the human geographical knowledge that circulates today emerged in and through this world historical event. This knowledge continues to give us a powerful handle on how the world works, but this handle is rooted in a particular politico-intellectual world that is in the throes of change. Human geography must be careful not to allow the West to mark its own homework. Continuing to harness the power of human geographical theories, concepts, ideas, and debates to make sense of the world while striving to decolonize human geographical knowledge will be a tension that human geographers will need to live with for now. Fortunately, this tension is proving productive and is yielding fruitful geographical imaginations.
- There are at least four different ways of narrating the story of the rise, reign, and faltering of the West. We can title these explanations: *Only in the West because of favorable environmental endowments*; *First in the West, then elsewhere*; *Because in the West, not elsewhere*; and *The West versus the Rest*. Which story you chose to prefer matters – to how you understand the structure of the world as it presents itself today and the future prospects of different world regions.

Chapter Essay Questions

1) Write an essay entitled: The limits of my geographical imagination and why they matter.
2) Provide a definition of human geography, and outline and comment upon human geography's three key concerns. Using these three key concerns, describe the human geography of the area in which you live.
3) The story of the rise of the West from the fifteenth century and its role in the making of the modern world is of central interest to human geographers. Discuss.
4) Write an essay entitled: The iconography of the Monument to the Discoveries in Lisbon and the stories this monument neglects to tell.

References and Guidance for Further Reading

An excellent book tackling the fundamental question "What is human geography?" is:
Bonnett, A. (2008). *What Is Geography?* London: Sage.

Although Bonnett's book provides you with a short introduction to human geography, longer introductions to the subject can be found in:
Cloke, P., Crang, M., and Goodwin, M. (eds.) (2014). *Introducing Human Geographies*, 3ee. London: Routledge.
Daniels, P., Bradshaw, M., Shaw, D. et al. (2016). *An Introduction to Human Geography: Issues for the 21st Century*, 5e. Upper Saddle River, NJ: Pearson.
Fouberg, E.H., Murphy, A.B., and de Blij, H.J. (2020). *Human Geography: People, Place, and Culture*, 12e. Chichester: Wiley.
Knox, P. and Marston, S. (2016). *Places and Regions in Global Context: Human Geography*, 7ee. Upper Saddle River, NJ: Pearson.
Murphy, A.B. (2018). *Geography: Why It Matters*. Cambridge: Polity Press.
Norton, W. and Mercier, M. (2016). *Human Geography*, 9e. Toronto: Oxford University Press.
Rubenstein, J.M. (2018). *Contemporary Human Geography*, 4e. Upper Saddle River, NJ: Pearson.
Rubenstein, J.M. (2019). *The Cultural Landscape: Introduction to Human Geography*, 12e. Upper Saddle River, NJ: Pearson.

Other valuable short introductions to human geography that adopt a very different approach from the one adopted in this book can be found in:
Jones, A. (2012). *Human Geography: The Basics*. London: Routledge.
Short, J.R. (2017). *Human Geography: A Short Introduction*. Oxford: Oxford University Press.

An excellent introduction to human–environment interactions from a geographical perspective is provided in:
Moseley, W.G., Perramond, E., Hapke, H.M., and Laris, P. (2014). *An Introduction to Human–Environment Geography*. Chichester: Wiley.

Excellent resources for students wishing to explore the full breadth of human geography include:
Dorling, D. and Lee, C. (2016). *Geography: Ideas in Profile*. London: Profile Books.
Hubbard, P., Kitchin, R., and Valentine, G. (eds.) (2011). *Key Thinkers in Space and Place*, 2ee. London: Sage.
Kobayashi, A. (2020). *International Encyclopedia of Human Geography*, 2e. Boston: Elsevier.
Kuby, M., Harner, J., and Gober, P. (2013). *Human Geography in Action*, 6e. Chichester: Wiley.
Lee, R., Castree, N., Kitchin, R. et al. (eds.) (2014). *The SAGE Handbook of Human Geography*. London: Sage.
Richardson, D., Castree, N., and Goodchild, M. (2017). *International Encyclopedia of Geography: People, the Earth, Environment and Technology*. London: Wiley Blackwell.

An interesting and popular book exploring human geographical questions is:
Marshall, T. (2019). *Prisoners of Geography: Ten Maps That Tell You Everything You Need to Know*. London: Elliott & Thompson Limited.

For definitions and discussions of key ideas in human geography, students might find it useful to consult:
Gregory, D., Johnson, R., Pratt, G. et al. (eds.) (2009). *The Dictionary of Human Geography*, 5ee. Chichester: Wiley.
Mayhew, S. (2016). *A Dictionary of Geography*. Oxford: Oxford University Press.
Rogers, A., Castree, C., and Kitchin, R. (2013). *A Dictionary of Human Geography*. Oxford: Oxford University Press.

This chapter engages two key readings:

Cosgrove, D.E. (1984). *Social Formation and Symbolic Landscape*. London: Croom Helm.

Harvey, D. (2006). *Spaces of Global Capitalism: Towards a Theory of Uneven Geographical Development*. London: Verso.

Website Support Material

A range of useful resources to support your reading of this chapter are available from the Wiley *Human Geography: An Essential Introduction* Companion Site http://www.wiley.com/go/boyle.

Chapter 2

Human Geography: A Brief History

Chapter Table of Contents

Chapter Learning Objectives

By the end of this chapter you should be able to:
- Distinguish between "internalist" and "contextualist" approaches to the writing of the history of human geography; with reference to the latter, propose a framework through which the story of the history of human geography might be recounted.

Human Geography: An Essential Introduction, Second Edition. Mark Boyle.
© 2021 John Wiley & Sons Ltd. Published 2021 by John Wiley & Sons Ltd.
Companion website: www.wiley.com/go/boyle

- Describe and comment upon the ways in which human geography developed in the premodern era.
- Describe and comment upon the ways in which human geography developed in the modern era; distinguish between the mission and concerns of human geography in each of the early modern, modern, and late modern periods.
- Describe and comment upon the ways in which human geography is developing in the postmodern era.
- Identify nascent developments in human geography in the twenty-first century. Discuss what is meant by postfoundational human geographies, and explain why some human geographical traditions can be thought of as both postfoundational but also anti-relativist. Discuss the implications of computerized data analytics and the data revolution for human geography. Explain what is meant by Geocomputation and Spatial Data Science.
- Contemplate the meaning and implications of the claim that geography is a quintessential Western academic subject, and as such is bestowed with powerful but also parochial ways of making sense of the world; comment critically upon the project of decolonizing geography; and comment critically upon the project of decolonizing historiographies of geography.

Introduction

The purpose of this chapter is to provide you with a brief history of key thinkers in human geography and their ideas, and the rise of human geography as a distinctive branch of knowledge. Institutional Human Geography emerged within Europe and across the past century has been dominated by Anglo-American scholarship. As such, my focus will be upon the history of a hegemonic geographical tradition. This is not of course to say that other voices (those of non Western indigenes, women and children for example) are unimportant nor indeed that they have failed to impact and influence the development of European-Anglo-American human geography. Indeed I will conclude that the work of retrieving and revalorising non Western geographical imaginations constitutes a key challenge for twenty first century historiographies of human geography. But it is to note that just as the West rose to overpower the rest of the world so too western human geography rose to overpower (but not extinguish) non Western human geographical traditions. Undoubtedly, our narrative will be insufficiently inclusive. 1 acknowledge from the outset that like others, my historiography of human geography will be inescapably situated and partial. It will nevertheless be attentive to this fact and relentlessly conscious of its implications.

Why awaken geography's dead?

The story of the history of human geography is of more than mere historical curiosity and antiquity. In their efforts to describe and explain variations from place to place in the ways in which human beings have inhabited the surface of the earth, human geographers have conspired to create a variety of approaches to description and explanation. It is essential that you understand something about these various human geographical traditions and schools of thought. Some have long since been recognized as flawed and have been rendered obsolete and junked. Still, these discarded perspectives remain worthy of scrutiny lest mistakes made in the past be repeated. But others remain insightful and continue to serve a useful purpose. They intrude on the present even if they are no longer as fashionable as they once were. Human Geography, then, exists today as a plural and contested subject, mobilizing older geographical traditions that remain (at least to some) persuasive and innovating and experimenting with fresh schools of thought. To discriminate between the rival merits of different approaches in human geography – and perhaps even to develop a preference for any one – you must first gain an appreciation of the origins and development of these approaches.

But there exists a further reason why it is essential that we study human geography's past. We need to understand the history of geography so that we can appreciate more fully the fact that the geography of human geography matters.

Geography, it turns out, has more parochial roots than we might care to admit. Reflecting the overarching framework that guides this book, the provocation underpinning this chapter is that the birth of geography and the many trials and tribulations it has faced over the course of its history have been inextricably wound up with the rise, reign, and faltering of the West from the fifteenth century. This chapter will attempt to show you the ways in which both developments necessarily unfolded together. Human geography is a child of Western civilization and continues to this day to exist as a quintessential Western academic subject. Equally, the West's climb to the summit of world history would be unthinkable without geographical knowledge, much of which was produced by the formal academic discipline of geography.

If it is to prosper in the twenty-first century, human geography will need to face up to the virtues and vices of its European and Anglo-American parentage and become more self-aware of the strengths and limitations of the powerful but also largely Western collection of theories, concepts, analytical tools, and methodologies it uses to make sense of the world.

Telling the Story of the History of Human Geography

It is common to speak today of physical geography and human geography as two distinctive specialisms. But it has to be remembered that both physical geography and human geography have their origins in the unified discipline of geography. The date at which it was deemed necessary to break geography into two streams, physical and human, remains the subject of debate. For some, the divorce was never that clear cut, and it is still meaningful to speak of the existence of an overarching subject called geography that bridges both. For others, physical and human geography are now on such different paths that never the twain shall again meet. Either way, it is impossible to tell the story of the history of human geography without spilling over into the wider history of the discipline of geography.

How might we tell the story of the history of geography? Here we might usefully distinguish between "internalist" and "contextualist" histories.

One way to differentiate these approaches is to recognize a distinction between "analytical reason" and "dialectical reason." The former, a remnant of the "long seventeenth century" European Enlightenment, construes reason as positivist and independent of any particular rational system; the latter, rooted in historical materialism, holds reason to be historically embedded, relative to socially constructed systems of logic, and therefore always provincial. Glibly, one might say internalist accounts treat approaches within human geography as instances of analytical reason, while contextualist accounts treat them as instances of dialectical reason. But this is too simplistic. In truth most histories of geography, whether they be internalist or contextualist in orientation, recognize ultimately that all geographical reason is produced *within history*: lacking in self-understanding, analytical reason is simply a form of reason that is unconscious of its historicity. What really distinguishes internalist from contextualist accounts is the extent to which the former takes more seriously disputes over the internal logic and illogic of analytical geographical reason (on occasion losing sight of historical context), while the latter gives priority to historical and societal conditions (and so, for some, fails to afford geographers agency as active makers of their discipline).

Let us illustrate this claim with reference to two books that have proved seminal in shaping the telling of the story of the history of geography.

An Internalist Classic

The spring of 2016 saw the publication of the seventh edition (first edition: 1979) of the landmark "internalist" history of the discipline, *Geography and Geographers: Anglo-American Human Geography Since 1945* by British-based geographer Ron J. Johnston and British-born Singapore-based geographer James D. Sidaway. This book is structured around a short analytical, interpretive, and framing "outer part" and a broadly chronological and more detailed "inner part" marching through the substantive twists and turns taken by human geography in the Anglo-American world, principally from 1945. At the heart of the outer part is the "internalist" paradigm theory of disciplinary change proposed by US physicist, historian, and philosopher of science Thomas Samuel Kuhn in his 1962 book *The Structure of Scientific Revolutions*. Kuhn contended that academic disciplines develop over time largely at the behest of processes internal to their own parochial communities and oscillate between periods of calm when particular paradigms (worldviews or ways of making sense of the world) dominate and periods of revolution when these worldviews no longer appear capable of solving the problems that disciplines wish to address (Kuhn 1962). The outer book stands somewhat apart from the inner book, leaving the latter to mostly describe and recount (rather than to scrutinize the wider determinants of) the rise and fall of particular geographical ideas and debates. The inner part bears witness to the unfolding in sequence of various schools of thought – or paradigms. Geographers, it seems, suffer from commitment issues! Over time and habitually, they create, critique, and then move on from particular philosophies, approaches, and scholarly traditions (see Deep Dive Box 2.1). Johnston and Sidaway apply paradigm theory to the history of Anglo-American geography and find it to be only partially accurate: the trajectory of the subject perhaps follows a paradigm model until the late 1960s, after which point geography spawns a plurality of rival traditions of thought. Moreover, the development of geography is best understood as being the outcome of three interacting processes: alongside the internal drivers of the prevailing occupational structure (the academic career structure) and the organizational framework that guides research (how universities incentivize and reward research), the external environment (societal dramas of the day) has proven formative too.

A Contextualist Classic

Meanwhile, British geographer David Livingstone's (1992) *The Geographical Tradition: Episodes in the History of a Contested Enterprise* proved key in advancing the case for historicizing more fully geographical philosophies and practices – indeed, something called geography period – and giving scholarly and eloquent expression to the kinds of archival methods, analytic instruments, and modalities of narration such a contextualist project might demand. Livingstone detected two critical flaws in existing historiographical accounts like those just identified: "presentism," or interpreting and even judging past geographical ideas by the (scientific, moral, and aesthetic) standards of today; and

Deep Dive Box 2.1   Common Approaches or "Paradigms" Identified in "Internalist" Histories of Geography

The most popular and accessible "internalist" history of Anglo-American geography remains Johnston and Sidaway's (2016) book *Geography and Geographers: Anglo-American Human Geography Since 1945* (7th edition). But *Geography and Geographers* now finds itself in the company of many excellent historiographical accounts that are similarly "internalist" in inspiration. Included in this tradition are works largely by British and North American geographers, such as Cloke, Philo, and Sadler's (1991) edited book, *Approaching Human Geography*; Unwin's (1992) *The Place of Geography*; Peet's (1998) *Modern Geographical Thought*; Nayak and Jeffrey's (2011) *Geographical Thought: An Introduction to Ideas in Human Geography*; Cresswell's (2013) *Geographic Thought: A Critical Introduction*; Couper's (2014) *A Student's Introduction to Geographical Thought*; Cox's (2014) *Making Human Geography*; Martin's (2015) *American Geography and Geographers: Toward Geographical Science*; and Holt Jensen's (2018) *Geography: History and Concepts*.

The weight given to particular key thinkers and their approaches varies from one author to the next and over time. Nevertheless, all of the following approaches commonly feature in these books (note: dates are approximate and for guidance only).

- Environmental determinism (peaking 1850s to 1920s, holding that the environment determines human culture)
- Environmental possibilism (peaking 1900s to 1930s, asserting that the environment conditions but does not determine human culture)
- Cultural geography (peaking 1910s to 1940s, assuming a two-way interaction between the natural environment and human culture)
- Regional geography (peaking 1930s to 1950s, providing formal methods through which distinctive regions might be identified rigorously)
- Spatial science and the first quantitative revolution (peaking 1950s to 1970s, drawing upon neoclassical economics and founding a new science of space and location)
- Behavioral geography (peaking 1960s to 1970s, aligning geography with psychology and explaining spatial behavior in terms of human cognition, mental maps, attitudes, and perceptions)
- Humanistic geography (peaking 1970s and 1980s, aligning geography and the humanities and fine arts, and bringing to the fore human emotions, meanings, senses, and existential attachments to place and landscape)
- Structural approaches, critical realism, and Marxism (peaking 1970s to 1990s, aligning geography and the social sciences and exploring the nexus between society and space or social structures and spatial patterns)
- Poststructuralism (from the 1980s onward, inspired by a "cultural turn" in the discipline and promoting a new focus upon power, language, semiotics,

and the cultural politics of geographical representations of "other" peoples and places)

- Various associated postfoundational geographies (such as, from the 1980s onward, postmodern, postcolonial, feminist, queer, black, children's, disability, affective, anarchist, relational, and nonrepresentational geographies, which attempt to understand the world from alternative positions and use novel modes of knowing and sensing the world)
- Applied geography (which has grown in popularity, especially from the 1970s, promoting the relevance of geography and its capacity to address key social, economic, and environmental problems)
- A second quantitative revolution embracing Geographical Information Systems (GIS), geocomputation, big data, and data-driven science (which uses big data and new data analytic techniques to map the world in real time)

A variety of visualizations of these histories has been presented by geographers; perhaps the two most famous are Peet's (1998) and Philo's (2008) depictions (Figure 2.1a and 2.1b).

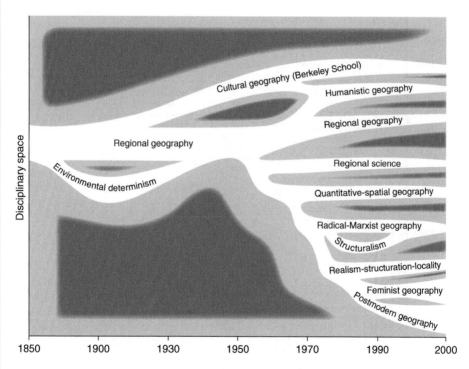

Figure 2.1a Richard Peet's "octopus" depiction of changing approaches in human geography. Source: Peet R (1998) Modern Geographic Thought (Blackwell, Oxford) © 1998, John Wiley & Sons.

(Continued)

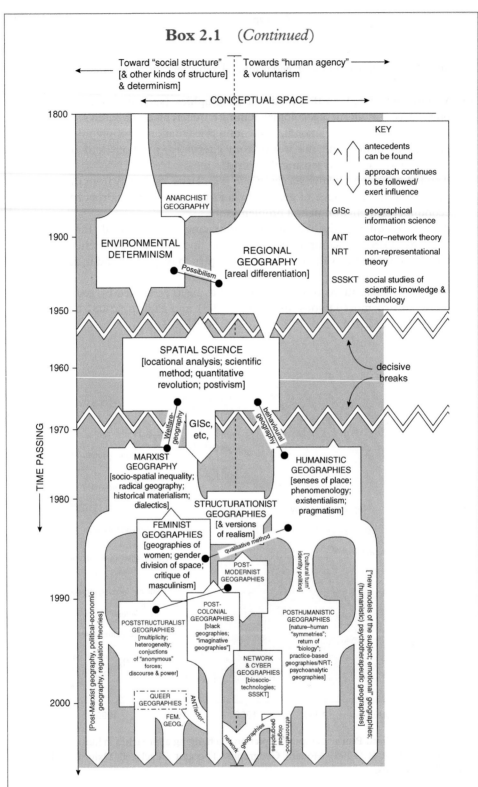

Figure 2.1b   Disciplinary spaces: a map of changing approaches within human geography by Chris Philo. Source: Philo P (ed) (2008) Theory and Methods: Critical Essays in Human Geography (Routledge, London).

"internalism," or construing the evolution of the discipline in terms of interior drivers of change (such as scholarly fraternities, their champions, spats, alignments, and plays). Refusing to label and police the boundaries of his alternative approach too strictly, he invoked the simple yet powerful idea of "situated messiness": all intellectual endeavor, including the practice of creating, disseminating, and ingesting geographical knowledge, is best thought of as a situated social practice inextricably embroiled in the wider social, economic, political, and intellectual dramas of the day. In particular, the history of geography could not be told apart from the history of European exploration, racism, empire, and science (see Deep Dive Box 2.2). There followed a polymathic reading of over 500 years of "episodes in the history of a contested enterprise," incorporating *inter alia* geography's entanglements in: the age of reconnaissance; the scientific revolution and its alter ego, alchemy; the European Enlightenment and Age of Reason; the pre-Darwinian expedition tradition; Darwinism and the institutional politics of the nineteenth-century European academy; European imperial expansion, colonization, and scientific racism; shifting understandings of the nature–culture nexus; and technological change and the quantitative revolution.

So which approach will we adopt here?

This chapter takes as its cue contextualist accounts of the history of the discipline, that is, it assumes that it is wider societal factors more so than drivers internal to human

---

Deep Dive Box 2.2   European Empires and Geography as a Colonial Politico-Institutional Project

Of all the contexts that have played a role in shaping the development of the discipline, arguably it has been the entanglement of the subject in the march of European empires to the four corners of the world that has most occupied the attention of historians of geography. In the 1990s and 2000s, a number of influential texts emerged insisting that human geography was a colonial-born academic subject and existed only because the knowledge it produced once served powerful European and Western imperial interests.

In addition to Livingston's (1993) *The Geographical Tradition*, books such as Godlewska and Smith's (1994) *Geography and Empire*, Edney's (1997) *Mapping an Empire: The Geographical Construction of British India, 1765–1843*, Driver's (1999) *Geography Militant: Cultures of Exploration in the Age of Empire*, Wither's (2001) *Geography, Science and National Identity: Scotland Since 1520*, Gregory's (2004) *The Colonial Present: Afghanistan, Palestine, Iraq*, Bell's (2005) *Geography and Imperialism 1820 to 1940*, Rothenburg's (2007) *Presenting America's World: Strategies of Innocence in National Geographic Magazine 1888–1945*, and Benton's (2009) *A Search for Sovereignty: Law and Geography in European Empires, 1400–1900*, among others, have all served to ignite interest in human geography's controversial past.

Not only did navigators and cartographers pave the way and guide would-be colonizers but also, through the doctrine of environmental determinism, geographers validated variants of "scientific racism" that promoted the view that when compared with their European masters, "natives" and "indigenes" were subhuman by dint of the inhospitable environments in which they dwelled. This enabled imperial powers to present invasion and annexation as a philanthropic civilizing mission (Plate 2.1).

(Continued)

Box 2.2   (*Continued*)

Plate 2.1   Cartographies of power: Mapping Britain's Empire, 1886. Source: Walter Crane, Wikimedia Commons, https://commons.wikimedia.org/wiki/File:Imperial_Federation,_Map_of_the_World_Showing_the_Extent_of_the_British_Empire_in_1886_(levelled).jpg. Public Domain

geography that have motivated the subject to develop as it has over time. We will propose a contextualist framework through which the history of geography might be figured by using the ideas of "modernity" and the "West." As used in this book, the term "modern" refers to that period from the fifteenth century onward when, from its cradle in Europe, "the West" as a distinctive civilization first appeared and began its climb to the summit of world history. This watershed in human history was also a watershed in the history of geography. Without the rise of the West, it is possiblec <lowercase>that the discipline of geography would never have been born at all; certainly, it is inconceivable that the subject would have taken the form it now does. And in an important sense, the story of the faltering of the West in the twentieth and twenty-first centuries has determined the trajectory taken by human geography too.

It is therefore useful to divide the emergence of geography into three phases, which we will term here the premodern era (before the West, when civilizations tackled geographical questions without the aid of a formal discipline called geography), the modern era (when the West emerged as a dominant civilization, and geography became established as a professional academic subject), and the postmodern era (when confidence that "West is best" ebbed, and faith was lost in the capacity of human geography to render the world intelligible). Furthermore, the modern era itself can be further subdivided into the early modern period, the modern period, and the late modern period (see Figure 2.2), reflecting the rise, reign, and faltering of human geography as a professional academic discipline). Wwithin the meaning, precise dates, and relative

Principal eras and periods		Indicative dates	Key concerns of (human) geography during these eras/periods
Premodern era		Early civilizations, Greco-Roman civilization (circa 600 BCE to 600 CE), Middle Ages (fifth to fifteenth centuries CE).	The search for basic knowledge concerning the size, shape, and physics of planet earth. The "discovery" of and accumulation of knowledge concerning "foreign" regions of the world. The development of rudimentary cartography.
Modern era	Early modern period	From the Age of Discovery (fifteenth century CE) to Ritter's *Erkunde* (early 1800s CE) and Von Humboldt's *Kosmos* (mid-1800s CE).	The European "discovery" of the remaining unknown regions of the world and the accumulation of scientific knowledge concerning these regions. The development of modern cartography. The first attempts to define geography as a distinctive science predicated upon the integration and synthesis of both environmental and social data pouring into Europe from explorations.
	Modern period	Establishment of geography as a formal university discipline in 1874 and the reign of environmental determinism from 1874 to the 1920s.	Geography succeeded in establishing itself as an autonomous discipline within the European university system and spread quickly to universities throughout the world. Geography cohered strongly around the big idea that environmental influences shape the development of human culture in powerful ways. Geography was favored by governments on account of the role it might play in the European colonial adventure.
	Late modern period	From the rise of regional geography in the 1920s to the rise of skepticism in the mid-to late 1980s.	With the demise of environmental determinism as the big idea geography used to make sense of the uneven inhabitation of the face of the earth, human geographers pondered on the existence of an alternative big idea. Environmental possibilism, culture as a superorganism, regional geography, spatial science, radical/marxist/feminist geography, behavioral and humanistic geography, structuration theory and humanistic Marxism are offered as candidates.
Postmodern era		From the late 1980s to the present: the postmodern challenge and the rise of post-foundational but also anti-relativist human geographies	Today, there is widespread skepticism that human geographers will ever unearth one big idea that will be capable of making sense of the world. The key question now is, how can human geography continue to serve as a purposeful discipline when there is no shared agreement on the way the world works? Although human geographers have answered this question in many ways, postmodern geography has emerged as a new and uncomfortable challenge for human geography. In the twenty first century human geography looks set to develop along two pathways. First, there are arising human geographies which are both postfoundational and anti-relativist (poststructuralism, pragmatism, critical geographies, postcolonial geographies, feminist geographies, queer studies, children's geographies, anarchist geographies, geographies of disability, black geographies, relational geographies, non-representational geographies, more than representational geographies and affective geographies). Second a new quantitative revolution based upon geocomputation and spatial science is giving birth to 'real time' human geographies.

Figure 2.2   A framework for making sense of the history of human geography. Source: (By the author).

significance of these eras and periods are matters of ongoing debate, this framework does permit a basic grasp of key moments in the subject's development.

Human Geography in the Premodern Era

Arguably, geographical understanding in one form or another pervaded all prior civilizations. But it was Greek scholars and thereafter scholars from the Greco-Roman world who first conceived the idea of "geography" and who achieved the most impressive breakthroughs in geographical understanding. Greek scholarship recognized few disciplinary boundaries and did not consider geography *per se* to be a separate branch of knowledge. Problems of a geographical character emerged only in the context of the wider puzzles that occupied Greek society. These problems tended to center upon the shape, size, and physics of planet earth, and the map-making techniques needed to depict its landmasses (see Deep Dive Box 2.3).

Following the collapse of the Roman Empire, premodern geography progressed rather more slowly. Concern with the wider world lessened, and former Greco-Roman societies became more inward-looking. It is for this reason that the Middle Ages (c. the

---

Deep Dive Box 2.3　Key Geographers in the Greco-Roman Period

Although in no way representative of the entire tradition, a sense of Greco-Roman geography can be gleaned from the work of a number of its more famous alumni:

Eratosthenes (c. 275–195 BCE): Eratosthenes was a native of Cyrene, modern-day Libya, then part of the Greek Empire. He is commonly credited with introducing the word "geography" into the Greek language. Eratosthenes was the first person to create a system of latitude and longitude, and using this system he ventured a speculative map of the world. He was the first person to calculate the circumference of the earth. He also pioneered understanding of the tilt of the earth's axis and was first to estimate the distance between earth and the sun.

Strabo (c. 60 BCE–24 CE): Strabo was born in Pontus, modern-day Turkey, then under the control of the Roman Empire. Strabo's major work *Geographica* represented a summation of Greco-Roman knowledge of the world at that time. Strabo drew the first detailed map of Europe. He described the character of different regions. He attempted to explain differences between people he regarded as civilized and those he considered more barbaric with reference to the effects of the physical environment on human culture.

Ptolemy (c. 90–168 CE): Ptolemy was a citizen of the Roman Empire who lived in Egypt. He pioneered new techniques in the field of cartography. In his famous manuscript *Geographia*, Ptolemy developed a system through which the earth, as a three-dimensional sphere, could be reproduced on a map that was two-dimensional. This task is now referred to as map projection. Ptolemy devised a fresh system of latitude and longitude. He used this new grid to draw a speculative two-dimensional map of the world. He then populated this map with content derived from gazetteers and written by early explorers and travel writers.

fifth century to the fifteenth century CE) is often taken to be one of the darkest periods in the intellectual history of Europe. Nevertheless, religious rivalries (between, in particular, the Christian West and Islamic East) did generate some important geographical breakthroughs. Christian missions and crusades created new understandings of lands beyond Europe, and in particular territories in what is now called the Middle East. Meanwhile, Islamic missions and Islamic cartography created better understanding of the West from the perspective of populations living in the East. Hellenistic mapping techniques were advanced significantly in particular during the Islamic Golden Age (8th century and 16th century) and not least by the 12th century Spanish-Arab cartographer Muhammad al-Idrisi.

The Middle Ages came to an end in the fifteenth century, the moment when Western civilization entered the historical arena. From an unpromising and unlikely start, a number of European countries began what was to be a steep ascent to the summit of world history. For over 500 years now, these countries have dominated global affairs. Leading the pack were territories that incorporate countries we now identify as Portugal, Austria, Belgium, France, Germany, Italy, the Netherlands, Spain, Russia, and the United Kingdom. As European migrants fled to the New World, they spread Europe's influence, and within a short period of time the United States, Canada, Australia, and New Zealand emerged as Western powers too. Meanwhile, from the eighteenth century, Japan began its Datsu-A Ron ("leaving of Asia" or "goodbye Asia") and became Asia's leading evangelist of Western ways. These developments were to radically transform society's thirst for particular kinds of geographical knowledge. A new era for geography beckoned.

Human Geography in the Modern Era

The European Enlightenment effectively gave life to the formal discipline of geography. In the modern era, geography became, in its ideals and its mission, a quintessential European branch of knowledge. The early modern period begins in the fifteenth century with the Age of European exploration and Age of Reason, and ends in the mid-1800s with the influential works of German scholars Alexander von Humboldt and Carl Ritter. The modern period begins in Prussia in 1874 with the establishment of geography as a university subject, incorporates the Age of European Empires, and ends in the 1920s with geography parading itself as a beacon of European Enlightenment ideals, and an indispensable partner in the building of European empires. The late modern period begins with early twentieth-century questioning of geography's identity and purpose, and ends in the late 1980s when confidence in the discipline's mission and trajectory reaches a new low. It was arguably only during this final period that physical and human geography separated into distinctive camps.

Early Modern Period

With the rise of the West came the Age of European exploration when pioneers, seafarers, and adventures began to develop new navigation skills, traverse the world, and encounter the "elsewheres" which lay beyond the horizon. In 1418, the Portuguese explorer Prince Henry the Navigator established the world's first geographical research center, the Escola de Sagres, at Lagos in Portugal. Successive waves of Portuguese explorers, and later explorers from other European countries, were inducted in the science of exploration. Safe and reliable passageways to other parts of the world became established. Major explorations were conducted. The lands of the Americas, Asia, Africa, the Arctic, the Antarctic, and Oceania were "discovered" by Europeans and mapped.

European exploration marked a dramatic broadening of the European mind. Of course, the newly discovered lands and their peoples were experienced as exotic by European explorers, and these pioneers captured and recounted their encounters with indigenous peoples and natives in ways that fascinated and intrigued the European public. The knowledge collected during expeditions was fed back to eager and excited European audiences in part through the establishment of new national geographical societies. But as with all travel writing, the tales regaled by explorers contained only a kernel of truth. With hindsight, depictions of the world often said more about the Europeans who were doing the looking than it did about the "foreign" cultures that were being looked at.

The early modern period reached its zenith with the works of the founding fathers of modern geography, the German scholars Alexander von Humboldt (1769–1859) and Carl Ritter (1779–1859). Both men recognized that 400 years of European exploration had produced a gigantic corpus of knowledge concerning the character of different world regions. This knowledge was scattered in the form of published lecture manuscripts, scientific books, geographical gazetteers, travel writer guidebooks, scientific log books, artist sketches, field diaries, and oral histories. They too traveled and amassed voluminous insights about other places, their climates, and their cultures (Map 2.1). Nineteenth-century Europe was literally drowning in data, and this deluge of information was overwhelming. This data needed to be collated, archived, interpreted, digested, and put to effective use.

Von Humboldt and Ritter looked at the disciplines that already existed in the natural sciences, social sciences, and humanities. They argued that although each of these subjects explained something about the physical and human geography of the earth, none had the capacity to bring all the data together and make sense of the result. What was missing was a method to understand how natural processes (soils, climates, landscapes, and vegetation) and social processes (society, politics, culture, and the economy) functioned together as an integrated system. In von Humboldt's major work *Kosmos*, published in five volumes between 1845 and 1862, and likewise Ritter's definitive work *Die Erdkunde*, published in 19 volumes between 1816 and 1859, geography was to serve as the integrating discipline: synthesizing the strengths of all subjects, specialisms, fields, disciplines, and schools within the natural sciences, the social sciences, and the humanities to describe and explain variations in physical environments and human activity over the face of the planet. Geography was conceived as the science of the earth in relation to *both* nature and the history of humankind.

Von Humboldt and Ritter called for geography to be formally enshrined as a distinctive new university subject, and they were to get their wish sooner than they expected.

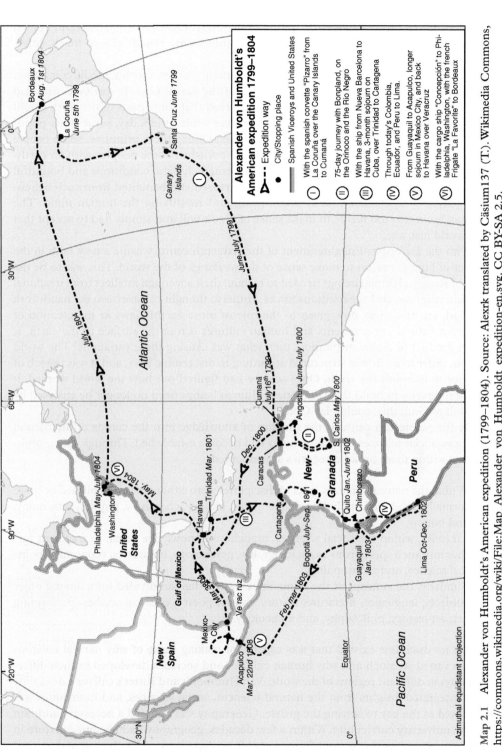

Map 2.1 Alexander von Humboldt's American expedition (1799–1804). Source: Alexrk translated by Cäsium137 (T). Wikimedia Commons, https://commons.wikimedia.org/wiki/File:Map_Alexander_von_Humboldt_expedition-en.svg. CC BY-SA 2.5.

Modern Period

European exploration exposed the existence of different natural ecosystems and human cultures, and laid down a challenge. New questions were raised: why so much difference? Why is what where?

Of course, for most of human history, human beings attributed differentiation across the face of planet earth to the maker of the earth: God. It was God who had created the earth, and it was God who had decreed that some parts of the earth would be warm while others were cold, some parts wet while others dry, some parts populated while others empty, some cultures civilized while others primitive, some people black while others white. It was God who had punished some peoples with an earthquake or a famine and rewarded others with benign climatic conditions and bountiful harvests. Organized religion and cults both nurtured and benefited from such beliefs, and played a significant role in infantilizing and mystifying the human mind. The human brain was too feeble to make sense of God's will and simply had to accept that the world just was.

With the European Enlightenment of the sixteenth century came a new faith in the power of human reason to make sense of the workings of the world. This was to be the Age of Reason. Human beings needed to reclaim their sovereign intellect from irrational cultish, religious, and superstitious forces. Hitherto thought to have been the handiwork of God, attention was now given to the role of more earthly laws in the creation of different natural environments and human cultures across the surface of the earth. If not a product of divine inspiration, then what was causing these variations? The world had an order to it – it was structured according to discernible laws, and it was the job of science to establish the truth. Once science had figured out how the world worked, it could intervene to build the world anew. Human beings need no longer be enslaved by the will of God; they could play God.

By the nineteenth century, the division of knowledge into the camps of the natural sciences, social sciences, and humanities had become established. Through time, subdivisions within these broad categories also emerged:

- Within the natural sciences, a set of disciplines were either newly established or more formally developed, including mathematics, biology, chemistry, physics, astronomy, and botany.
- Likewise, within the social sciences, branches of knowledge began to fragment further into such specialisms as sociology, law, psychology, linguistics, economics, political science, and anthropology.
- Similarly, the concept of the humanities was further subdivided into, among other subjects, languages, literature, history, music, poetry, classical studies, performing art, art history, philosophy, and archeology.

But no discipline existed that was capable of making sense of why natural environments varied so much and why human cultures and societies developed in such different ways in different regions of the world. Von Humboldt and Ritter's call for a discipline that integrated insights from the natural sciences, social sciences, and humanities was accepted as the key to solving the puzzle. Geography was deemed a necessary addition to the university curriculum. Within a few decades, geography had become a fixture in leading universities throughout Europe, and from there it spread to the rest of the world (see Deep Dive Box 2.4).

Deep Dive Box 2.4   On the Origins and Dispersal of Geography

The book *All Possible Worlds: A History of Geographical Ideas* was first published in 1972 by US geographer Preston E. James and updated in a fourth edition by Association of American Geographers archivist Geoffrey J. Martin in 2005 (Martin and James 2005). This book presents a full account of the emergence of geography from antiquity to the present, and traces the origins and diffusion of the discipline as a full and independent subject within the university system, beginning with its establishment as an autonomous university subject in Prussia in 1874.

Led by Prussian geographers such as Friedrich Ratzel (1844–1904), Ferdinand von Richthofen (1833–1905), Albrecht Penck (1858–1945), and Alfred Hettner (1859–1941), geography became established first in German universities in Munich, Bonn, Tübingen, Göttingen, Cologne, Leipzig, and Berlin. Through the pioneering efforts of Paul Vidal de la Blache (1845–1918) in Paris, Halford Mackinder (1861–1947) in Oxford, Pyotr Seminov Tyan-Shanski (1827–1914) in St. Petersburg and Moscow, and William Morris Davis (1850–1934) at Harvard in Boston, geography then developed throughout France, Great Britain, Russia, and the United States.

In a cascading effect:

- Prussian/German geography was then to influence the emergence of professional geography in Sweden, Norway, Finland, Denmark, the Netherlands, Switzerland, and Austria.
- French geography, in turn, informed the institutionalization of the discipline in Belgium, Italy, Spain, Portugal, Latin American countries (especially Brazil), and French-speaking Canada. (Plates 2.2a–2.2c show Portuguese events and publications from the era.)
- British influence stretched geography into Australia and New Zealand, English-speaking Canada, India, Pakistan, Egypt, Africa, and the West Indies.
- Finally, Russian/Soviet geography prompted and conditioned the emergence of the discipline in Poland, Hungary, Czechoslovakia, Romania, Bulgaria, Yugoslavia, and East Germany.

Martin and James argue that, as an official university subject, geography is firmly of European parentage and that geography only spread to other countries because of Europe's various historical dealings and associations with those countries – and especially its colonial and postcolonial ties.

Pioneered by German geographer Friedrich Ratzel in his two-volume *Anthropogeographie*, published in 1882 and 1891, and propagated in the United States by his student Ellen Churchill Semple in her book *Influences of Geographic Environment: On the Basis of Ratzel's System of Anthropo-Geography*, published in 1911, the role of the physical environment in molding human culture in different regions came to the fore as the defining feature of the geographical approach. People in deserts, tropical

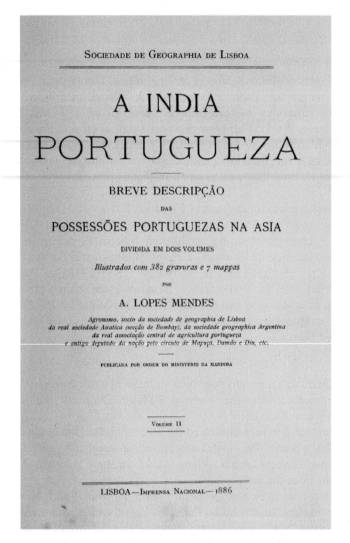

Plate 2.2a   Front cover of an 1886 publication by the Sociedade de Geografia de Lisboa (SGL) on Portuguese colonies in India. Source: A India portugueza; breve descripção das possessões portuguezas na Asia, dividida em dois volumes. Publicada por ordem do Ministerio da Marinha By Lopes Mendes, António, 1835-1894; Sociedade de Geografia de Lisboa.

rainforests, temperate lands, and the polar regions lived as they lived because the environment imposed limitations on what was possible. The natural environment increasingly came to be seen as a determinant of the uneven development of human culture and society over space. Climate, soils, landforms, vegetation, and geology all combined not only to determine skin color, but also to produce a mosaic of human types, cultures, and civilizations. In the same way it determines vegetation type, pedigree, and quality, so too the quality of the natural environment determines the type, pedigree, and quality of people. This new explanatory framework was labeled environmental determinism.

PAGODE DE CHANDRENATE

Plate 2.2b   Plate taken from the same 1886 SGL publication as in 2.2a showing Portuguese civilization of tropical Goa. Source: A India portugueza; breve descripção das possessões portuguezas na Asia, dividida em dois volumes. Publicada por ordem do Ministerio da Marinha By Lopes Mendes, António, 1835-1894; Sociedade de Geografia de Lisboa.

That geography offered a fresh way to address one of the most perplexing questions facing human kind in the nineteenth century, however, was only part of the reason European (and then other) governments looked upon it favorably. Soon after the European Age of Discovery began, European countries began to claim political sovereignty over, settled, and governed over vast tracks of land in the Americas, Africa, Asia, Oceania, and the Polar Regions. By the early twentieth century, most of the earth's surface had been or was currently governed by Europe.

It is no accident that geography prospered during the Age of European Empire. Indeed, it could be argued that human geography was made possible only because of Europe's colonization and settlement of vast territories in the Americas, Asia, and Africa

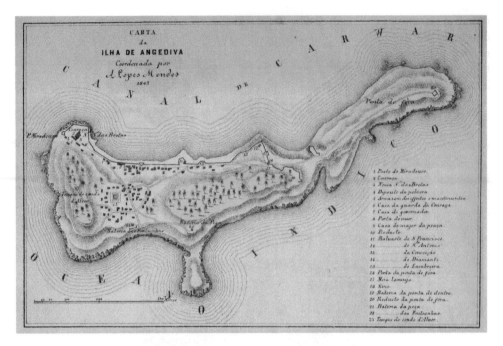

Plate 2.2c   Map of the Island of Angediva, Goa, in the same 1886 SGL publication as in figure 2.2a. Source: A India portugueza; breve descripção das possessões portuguezas na Asia, dividida em dois volumes. Publicada por ordem do Ministerio da Marinha By Lopes Mendes, António, 1835-1894; Sociedade de Geografia de Lisboa.

in particular. Human geographers were active stakeholders in the European colonial endeavor and provided a vital resource for would-be colonizers. Human geographers provided governments who were busy building empires overseas with maps of seafaring routes and of other lands, including their soils, mineral resources, forests, climates, crops, topography, cultural patterns, transport routes, disease regimes, and so on. Human geographers assisted European colonizers to find, settle, subdue, and exploit colonies with rich bounties. Cartography became a science dedicated toward improving colonial planning, conquest, and administration. British geographer Halford J. Mackinder's famous 1904 address on "The Geographical Pivot of History" at the Royal Geographical Society in London offered the new subdiscipline of geopolitics to European colonizers (Kearns 2009). Described as "Woodrow Wilson's geographer" and later "Franklin D. Roosevelt's geographer," US geographer Isaiah Bowman (1878– 1950) was to continue this tradition, placing human geography in the service of the building of an American empire predicated upon a liberal foreign policy, economic globalization, procurement of industrial minerals from European colonies, and capture of the spoils arising from the rebuilding of Germany after the Second World War (Smith 2004).

More importantly, geography's overarching explanatory framework – environmental determinism – became used and abused to justify European annexation of lands across the world.

This doctrine provided the intellectual basis for what came to be labeled "scientific racism" – the use and abuse of "science" to assert that some cultures were innately more advanced, civilized, intelligent, organized, creative, and noble than others. Populations that lived in more hospitable environments, the temperate lands of Europe for instance, developed better intellects and more civilized cultures. In contrast, populations that resided in harsher environments, like tropical or desert environments, lagged behind in their mental faculties and led more barbaric and primitive existences. Not only was it legitimate for Europe to colonize the less-developed parts of the world; it was their divine duty to do so. Blessed with higher levels of culture, Europe had a moral obligation to help progress members of the human species of lesser stature.

Late Modern Period

In the late modern period, confidence in the power of human reason to figure out the laws that governed nature and society began to ebb. Human reason, it turned out, was not as dependable as first considered. Great historical experiments conducted in the name of human reason often resulted in disastrous outcomes. Europe considered itself to be the cradle of the Enlightenment, and yet it was Europe, and more broadly the West, that had presided over exploitative empires, World War I and World War II, oppressive totalitarian political regimes, the Great Depression of the 1930s, the atom bomb, abuses of psychiatry, and environmental destruction. Western civilization, which was hitherto seen as impregnable and invincible, now began to look less certain and dominant. Perhaps the West did not have all the answers after all.

Mirroring declining faith in the European Age of Reason, European scholars began to lose confidence in the explanatory frameworks they used to make sense of the world and its workings. Questioning was in, and geography's explanatory framework was not to be exempted.

By the 1920s and 1930s, environmental determinism was starting to fall out of favor. More than any other species, the human species had creative and intellectual faculties. Human beings were not particularly bound by environmental constraints. Indeed, the development of technology meant that the environment was becoming less relevant as a cause of geographical variations in human activities between places. Cities could be built in deserts, in tropical rainforests, on mountain tops, and in polar regions, with all the amenities one might need to live a charmed existence. If human beings occupied planet earth differently in different places, something other than the natural environment had to be responsible for these differences.

Without now relying upon the physical environment to explain areal differentiation, human geographers attempted to find other explanatory frameworks through which variations in societies and cultures from one place to another might be understood. As the twentieth century unfolded, a number of rival contenders to replace environmental determinism emerged. It was at this time that human geography and physical geography began to separate and go their different ways.

In 1918, French geographer Paul Vidal de la Blache published his book *Principles of Human Geography* in which the idea of environmental possibilism was popularized. The physical environment did not determine all that occurred in an area, albeit it did constrain what human beings were able to do in that area. But ultimately human culture trumped nature. The environment provided the clay, but human ingenuity was free to mold this clay how it chose. For de la Blache, variations from place to place reflected these interactions between culture and nature. Over time, regions came to take on a unique personality or to support particular *genres de vie* (ways of life). Through fieldwork in his native France, de la Blache claimed to identify a mosaic of *pays* (pockets of culture that were unique to each local environment).

In his famous 1925 essay "The Morphology of Landscape," US geographer and leader of the Berkeley School of Geography Carl Sauer set out to define the field of geography. According to Sauer, geography was best thought of as the study of the cultural landscape (Sauer 1925). Sauer rejected the idea that human culture was shaped principally by the natural environment. He puzzled over the different cultures that existed on either side of the US–Mexican border. The physical environment immediately north and south of the border was identical – in both cases, semi-desert – and therefore nature could not explain why distinctive American and Mexican cultures had emerged in such close proximity. Culture had to be behaving independently of the environment. Sauer believed that culture could be conceived of as a superorganic entity. Cultures were real existing entities in and of themselves, and were guided by their own internal laws and workings. Sauer called upon geographers to place primary attention upon the material deposits that particular cultural groups impress on the landscape. Although not determined by nature, cultures fashioned the natural environments they were located in and etched onto them human imprints. Sauer's famous maxim was that "culture is the agent, nature is the medium, the cultural landscape is the result."

At the same time as some human geographers were taking an interest in interactions between culture and nature, others were becoming more interested in the development of systematic branches of the discipline. Instead of becoming an expert in a particular area or region, human geographers, it was argued, would be better advised to develop expertise in particular topics, such as past societies, population trends, how economies work, how cities function, how social groups live, how culture is practiced, how poor areas might develop, the extent to which the environment might support the human species, how politics operates, human health, the plight of rural areas, and so on. Geography ought to consist of systematic branches, such as historical geography, population geography, economic geography, urban geography, social geography, cultural geography, development geography, environmental geography, political geography, health geography, rural geography, and so on. These systematic branches might then be applied to particular regions to establish how their general principles manifest themselves in particular places.

Initially, both sets of geographers worked cooperatively. A branch of geography entitled regional geography was the product. Perhaps the clearest statement outlining the purposes of regional geography was that provided by US geographer Richard Hartshorne in his 1939 *The Nature of Geography* (Hartshorne 1939). Hartshorne reiterated that the objective of geography was to describe "areal differentiation across the surface of the earth." This differentiation should first be mapped and then explained. Here systematic branches of the discipline could be useful. Geographers with expertise in history, populations, industry, cities, societies, cultures, development processes, environmental systems, politics, health, and the countryside could work with specialists in particular regions, and together they could explain why specific regions were unique and different from the rest. Regional geographers, then, sought to combine regional expertise with systematic branches of the discipline to explain the idiosyncratic features of different areas.

But through time, some systematic branches of the discipline came to believe that the processes and mechanisms they were studying were general and not unique to each place. Entering into a public dispute with Hartshorne, US geographer Fred Kurt Schaefer called for a "quantitative revolution" in the discipline. Regional geographers were wrong to assume that each region was distinctive. Regional geographers were presiding over a geography that was unscientific. Geography too ought to be a science: a science of space. In the 1950s and 1960s, and pioneered by geographers from the University of Washington in Seattle and University of Bristol in the United Kingdom, geography began to contemplate its future as a spatial science. Spatial scientists believed that there was a science to the arrangement of human activities across the earth's surface. A set of universal laws determined, for instance, land-use patterns within the city, the hierarchy of settlement sizes in any country, the distribution of crop types over fields, the location of industries, and the volume of migration between settlements. These laws were as fixed as the laws of gravity and held in all regions. Differences in the imprint of different societies in the landscape were only apparent and not real; they embodied only "local noise" or variations around a common mean. A quantitative turn in the discipline – more use of hard statistics – would allow spatial laws to be tested and affirmed as true.

Of course, appearances can be deceptive: the cloak of science should not conceal the ideology behind spatial patterns. Spatial science served geography well; it underscored its scientific credentials in the age of US McCarthyism demonstrated its utility to the then surging practice of spatial planning I will note later the inextricable link between science and ideology, politics and the production of space, and spatial geometries and statecraft. For now we might simply note that the paradigmatic case was German geographer. Walter Christaller's 1933 Central Place Theory (CPT): while presenting as a theory optimizing the efficiency of resource allocation by structuring service centers into a hierarchy (see Figure 2.3), in reality CPT emerged as an instrument of Nazi spatial planning, and Christaller has come to be known as "Hitler's geographer."

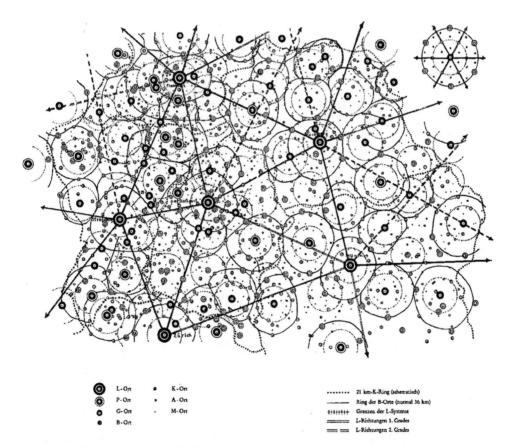

◉ L-Ort	• K-Ort	········	21 km-K-Ring (schematisch)
◎ P-Ort	. A-Ort	———	Ring der B-Orte (normal 36 km)
○ G-Ort	. M-Ort	++++++++	Grenzen der L-Systeme
• B-Ort		======	L-Richtungen 1. Grades
		====	L-Richtungen 2. Grades

Figure 2.3 Spatial Science or Nazi Spatial Planning? Christaller's Central Place Theory (1933), applied to Süddeutschland, Germany. Source: Wiki Commons.

Spatial science generated two counterreactions. The first reaction rallied against the notion that human activity was organized according to a set of laws and that human beings could be studied in the same way as natural scientists studied the laws of nature. Initially, a tradition of behavioral geography was proposed: insights from psychology were to be brought to bear in human geography, and spatial behavior was to be understood as the product of attitudes, cognitions, perceptions, and mental maps. Subsequently, a branch of geography entitled humanistic geography emerged, which argued that human geography ought to engage with the humanities more so than the natural sciences. Humanistic geographers such as Irish geographer Anne Buttimer, Chinese-born and US-based geographer Yi Fu Tuan, and Canadian geographer Edward Relph sought to redefine the purpose of geography. Instead of studying variations in human activity from place to place, humanistic geographers became interested in human attachments to places and environments. Humans are complex beings; they have complicated psychologies, they have feelings and emotions, they attach meanings to things, and they can be conscious in their dealings with the world. According to humanistic geographers, human geography is best thought of as the study of the ways in which the existential, emotional, and psychological makeup of human beings leads them to make and experience different environments and places in different ways.

The second reaction accepted that there are processes that organize human activities over space in ways that are coherent and predictable, but placed emphases upon society's relationship with space rather than with laws that determine spatial patterns. A branch of structural geography emerged, which argued that human geography ought to engage with the social sciences more so than the natural sciences. Critical realism was the philosophy of choice for structural geography; the key to understanding the spatial organization of society was to study the ways in which social, political, and economic structures created spatial patterns and processes. These structures were real and exerted material influence on the organization of space. Structural geography was pioneered most successfully by British-born but US-based Marxist geographer David Harvey and British feminist and Marxist geographer Doreen Massey. Marxist geography believes that the mode of production that exists in each society determines the geographical layout of these societies. Societies with different political, social, and economic institutions produce different human geographies. Marxist geographers are particularly interested in the geographies that the capitalist mode of production creates, including urbanization, transportation systems, regional variations in development, poverty, migration, and so on. Beyond class, structures of patriarchy, racism, ableism, ageism, heterosexism, sectarianism, and so on were considered to play a central role in shaping space.

As these twin trajectories developed, efforts were made to hold the discipline together by fusing the fruits of each. Human geographers turned to British Sociologist Anthony Gidden's Structuration Theory which sought to transcend the dualism between social structure and human agency by speaking about the duality of social life: structures structured but they also presented the very conditions through which human praxis was made possible. There emerged a tradition of locality studies which, informed by Structuration Theory, sought to understand the ways in which de-industrialising British and American cities were both vulnerable to structural change in the global capitalist economy but also capable actors with agency to fashion their own postindustrial and regeneration strategies. In a similar vein, scholars such as British geographer Denis Cosgrove and Canadian geographer Audrey Kobayashi pointed to the tradition of Western Humanistic or Existential Marxism, as a fruitful analytic tradition through which structure and agency might be held in productive tension.

In its search for a new explanatory framework and clearer sense of mission, identity, and purpose, ironically human geography lost much of its coherence and began to fragment. Human geographers disagreed as to whether regional geography, spatial science, humanistic geography, or structural geography provided the best route forward. In their 2016 book *Geography and Geographers: Anglo American Human Geography Since 1945*, British born geographers Ron J. Johnston and James D. Sidaway (2016) describe the late modern period as a time when human geography witnessed splintering, schisms, skirmishes, and factionalism like never before. There existed no shared agreement as to what human geography was, what human geographers do, and why human geographers do what they do. Virtue was made of pluralism in the discipline; according to some, it was healthy to foster different ways of looking at the world. But many human geographers feared that the discipline had lost its sense of unity and purpose and was in danger of coming apart at the seams. And by the late 1980s, those fears were to reach their zenith.

Human Geography in the Postmodern Era

Postmodern Human Geography: On Relativisers and Responsibility

For over a century now, a growing number of voices have sought to claim that Western civilization has passed its peak and is now in terminal decline. The audibility of these voices has risen to a crescendo in the last 40 years. The fall of the great European empires throughout the twentieth century is taken to be the clearest evidence that the West is no longer best. Europe no longer governs the Americas, Asia, and Africa from afar. The global oil crisis of 1973 marked an abrupt end to the 30 glory years of capitalism (1940s to 1970s) that followed World War II. Lurching from one economic crisis to another, the West seems unable to figure out a way to create sustainable economic growth, address environmental challenges, ameliorate poverty, and support ongoing human development. Moreover, from the 1970s onward, in terms of virtually every measurement of prosperity and development, the West stopped pulling ahead of the rest, and the gap between the West and rest began to narrow. The twenty-first century, it is said, will see the final demise of the West and the emergence of new global superpowers and a new world order.

Evidenced in the rise of populism and in the age of Trump and Brexit, the West is thrashing to "make itself great again." But it is doing so, it seems, from a position of insecurity, fear, and anger; having lost its world historical advantage, it is no longer sure that its formula for success is valid after all. Globalization, free and liberal trade, and neoliberal market conditions are not delivering for a growing number of "left behind" people and places. Authoritarianism and nationalism are back; hints of fascism threaten. Borders are to be closed. Protectionism is to be championed. Transnational corporations are to "come home." In many ways, the political populism that is sweeping the West today constitutes a *cri de couer* of sorts from those who yearn for the days of empire, who mourn the fading of past glories and victories, and who are persuaded that a "real" and revived West can once again be rekindled. But it is no longer clear what the real West really is; populism and nationalism on the one hand, and globalization and traditional liberal prescriptions for success on the other hand, are diametrically opposed. The West cannot be best when it itself is suffering from internal contradiction and an acute existential identity crisis.

Accordingly, confidence in the West and its ways of making sense of the world has ebbed to an all-time low. With the faltering of the West, doubt in the power of human reason to understand and dominate over the natural and social worlds has reached its apogee. Many claim that the modern era (with all its certainties, truths, bold visions, and big stories) is now giving way to a postmodern era (marked by uncertainty, doubt, rejection of grand stories about how the world works, and hesitancy). A child of the West, it is not surprising that human geography too has lost its sense of mission. From the late 1980s, as a reflection of growing self-doubt within wider culture, a less certain and more questioning human geography can be said to have entered a new postmodern era.

Postmodern human geography approaches the explanatory frameworks that modern geography birthed with a degree of incredulity (Harvey 1989). The age of the master narrative or meta-theory is over. It would simplify matters greatly were it possible to find a single overarching explanation for the varied ways in which human beings have inhabited the face of the earth. But experience instructs us to approach all explanatory frameworks with caution and distrust. The explanatory traditions that geography has invented to make sense of the world have invariably proved to be wrong, if not downright dangerous. With each generation has come a revolution in geographical thought; ways of understanding the world, which were once believed to have figured it all out, are dispensed with, and trust is placed in an entirely fresh approach. These revolutions eventually become wearisome. Faith is lost in the very possibility of finding an approach that is more than a passing fad. It becomes more sensible to conclude that no single explanatory framework will suffice.

But postmodern geography itself has been met with fierce opposition. A number of important questions are now being asked. Should we give up on the ideals of modern geography so readily? What is so wrong with looking at the world through a Western lens? Is the world really that parochial? Is postmodern geography not tantamount to throwing in the towel? Does not a sense of paralysis automatically follow? Surely not all explanatory frameworks deserve an equal hearing? Is everyone's version of what the world looks like equally valid? These questions continue to be debated by human geographers, and how to do geography in the postmodern era remains an unresolved puzzle.

So, what next? Where does human geography go from here? What intellectual tools will human geography need as it strives to make sense of twenty-first-century human geographical processes and patterns? What approaches to human geography might prove most illuminating?

There have emerged two possibilities that merit particular reflection. The first accepts postmodernism's claim that all knowledge be treated as postfoundational but refuses its claim that radical relativism is the only logical pathway to follow. The second argues that with developments in artificial intelligence (AI), digital technologies, big data, geocomputation, computerized data analytics and geo-visualization, it is time for a second quantitative revolution in human geography and a return to spatial science, but now in the form of a spatial data science.

Postfoundational and Anti-Relativist Human Geographies

In the early 1970s, against the backdrop of a global oil crisis and economic recession, civil rights marches on the streets of Western cities, and the destruction of the Amazon rainforest, many human geographers began to question the relevance of their discipline. What did geography have to say about these great issues facing humanity? Very little, some believed! A "relevance debate" followed – should human geography be applied to (that is, put more actively to the service of) solving the pressing social, economic, and environmental problems of the day (Pacione 1999)? How can human geography prove itself useful and have impact?

Consciously and unconsciously, many human geographers have come to adopt the principle of efficacy as truth, first established in the US philosophical tradition of classical pragmatism. This philosophy suspends judgments about the truthfulness of any theory, preferring instead to ask: does it work? For as long as it helps to address a particular problem, a specific theory can be judged to remain vital and worthy of application. Once it stops addressing the problem, its validity can be called into question. The arbiter is the capacity of a theory to solve a given problem. Doing is better than thinking. Any fix will always be temporary, and so any theory can only ever be said to be temporarily useful. All humans are fallible, and therefore all theory is fallible. Something unexpected will always crop up to render a pet theory impotent.

Efficacy is a laudable basis upon which to judge the worth of a particular theory, and there is much value in placing pragmatism at the heart of a new human geography. But pragmatism does not resolve fully the central challenge posed by postmodern relativism. Works for whom? Works on what terms? Works at what cost?

As titles go, few are as arresting – or as recited – as David Harvey's "What Kind of Geography for What Kind of Public Policy?" The occasion in which it came to press was a themed intervention, sparked by Terry Coppock's Presidential Address at the Annual IBG Conference held at the University of East Anglia (in the UK) in January 1974, and published in *Transactions of the Institute of British Geographers* (also in 1974) as "Geography and Public Policy: Challenges, Opportunities and Implications." Harvey staked out what for critical geographers has become a time-honored lament; lest their intellectual labor be appropriated, academic freedom impaired, and capacity for criticality compromised, there must exist a clear distance if not dissonance between scholars and the corporate state, construed as a "proto-fascist" technocratic instrument to preserve and strengthen the status quo. The supreme priority for the geographer was to reclaim sovereign control over intellectual labor, in part by critiquing uncritical liberal humanist-inspired applied or public service geography.

Poststructuralist human geographies can be said to be "postfoundational" in that they refute the claim that there is any solid or reliable datum point from which to know the world objectively and accurately. Instead, they focus upon the sheer strangeness of the ways in which human geographers have conventionally sought to understand areal differentiation; human geographical knowledge to date has been more skewed, biased, and provincial than purists might like to believe. They point to the fact that far from being neutral, these ways of knowing are peculiar and deeply ethnocentric – indeed, fundamentally Western and masculinist. Poststructuralists believe that language creates reality rather than merely places a mirror up to the world. They emphasize the role of language, signs, symbols, and images in the production of representations, depictions, stereotypes, and portrayals of other people and other places. They hold that Western ways of being in the world and by implication Western ways of knowing the world privilege the visual (sight – the all-seeing or "God's eye") capacities of human beings more so than any of the other four senses; sound, touch, smell, and taste. We know the world through looking; the primary focus is upon what things look like.

Poststructuralists thus concern themselves with the relationship between power, visualization, language, and knowledge. They refuse to ask if this or that discourse about, or representation of, other people and other places is accurate, preferring instead to pose the question: what work does this representation or imagery do in the world? Why do we believe it to have foundation? Poststructuralist human geographies support (indeed,

they often underpin) other postfoundational human geographies such as critical geog-
raphies, postcolonial geographies, feminist geographies, queer studies, children's geog-
raphies, anarchist geographies, geographies of disability, black geographies, relational
geographies, nonrepresentational geographies, more than representational geographies
and affective geographies.

While postfoundational, these geographies for the most part refuse the relativism that
postmodernism so readily celebrates. Precisely because we create understandings of the
world (they are our vistas on the world rather than the truth), we have a responsibility
to be conscious of the work these understandings do and for whom. As British geographer Doreen Massey (2005) declared emphatically, geographical perspectives may lack
foundation, but that does not absolve their authors from bearing responsibility for the
impacts they have once published, disseminated, and popularized. And so, the politics
and ethics of geographical representations continue to be important. Whose interests
are served by this or that way of seeing the world? How is power consolidated, or challenged, by the existence of particular imaginings about other peoples and places? How
can we adopt an ethics of care when trying to describe and explain other people's lives?
Can we ever transcend our deeply entrenched ways of knowing the world and court
alternative geographical traditions? (See Deep Dive Box 2.5.)

Deep Dive Box 2.5 On Decolonizing Human Geography

In his 1994 book by the same name, British-born but Canadian-based geographer
Derek Gregory coined the phrase *The Geographical Imagination.* Earlier in 1961, C.
Wright Mills had introduced the idea of "sociological imagination" to refer to the
ways in which sociologists look at the world by construing individual lives within
the larger social structures that characterize specific historical periods. According
to Gregory, the concept of the geographical imagination denotes the historical and
cultural frames that particular social groups – including human geographers – use
to imagine and render space. Having newly settled in British Columbia (Vancouver,
Canada), Gregory was struck with the juxtaposition of British imperial control over
space – the imperialism of the straight line – and First Canadians' orientations
toward space, place, nature, and landscape. Far from being superior by dint of their
logic and rationality, Western ways of seeing the world in terms of abstract, commodified, and bureaucratic space (enclosure, control, grids, etc.) were culturally
exceptional, unique, and aberrant. Indigenous ways of relating to space (spiritual,
holistic, embodied) were not to be compared unfavorably as more primitive or
backward; they were just different. For Gregory, the goal was to heighten awareness
of the provincial status of the deeply ingrained geographical imaginations that per-
vade human geography and to place under scrutiny the impact of these imaginations on the peoples and places that fall under their gaze.

Has human geography cleansed itself of its heritage as the European colonial
subject? According to some, the answer is a resounding NO! There remains an
ongoing need to "decolonize" human geography.

Postcolonial studies is customarily traced to the foundational works of both
Global North metropolitan scholars and anticolonial activists located in the Global
South, including French philosopher Jean-Paul Sartre, Martinique-born anticolonial
activist Frantz Fanon, French novelist Albert Memmi, US sociologist W.E.B. Du

Bois, Italian Marxist Antonio Gramsci, and Palestinian-American literary critic Edward Said; and it is translated and amplified by, among others, Indian-born cultural theorist Homi Bhabha, Indian-born feminist Gayatri Spivak, British historian Paul Gilroy, and American feminist bell hooks. Today, postcolonial human geography is being pioneered by British-, US-, Indian-, and Canadian-born geographers such as Alison Blunt, Catherine Nash, Derek Gregory, Cheryl McEwan, Sarah Radcliffe, Jennifer Robinson, Joanne Sharp, Ruth Craggs, Tariq Jazeel, Audrey Kobayashi, Patricia Noxolo, James Sidaway, Avril Maddrell, and Parvati Raghuram.

Because the discipline remains so steeped in its past, human geographers (these scholars say) continue to look at the world in a very particular way; one that is situated, ideological, Western, metrocentric, and masculinist. Even though European empires are dead and human geography is now more self-aware and committed to taking a critical stance toward imperial projects, its "ways of seeing" continue to bear the stamp of its colonial heritage. To decolonize human geography is to recognize that what many human geographers take to be an objective, true, and accurate "view from nowhere" is in fact a subjective, partial, and loaded "view from somewhere," and somewhere very specific. These scholars work to deconstruct European ways of imagining, demonizing, and belittling other, non-European, and past and present colonized societies. They question the primacy of European and Anglo-American worldviews. They critique mainstream geography for lapsing into an insufficiently reflexive commitment to the superiority of particularly European ways of looking and knowing. To decolonize human geography is to come to terms with and transcend the unwanted inheritances that lurk in the discipline's unconscious.

Human Geographies in Real Time: Geocomputation and Spatial Data Science

At precisely the same moment that postmodern human geographers are concluding that all science lacks foundation, human reason is not to be trusted, and the world is ultimately unknowable, AI, geocomputation/computerized data analytics, and big data are radically changing the ability of human geographers to make sense of the world and its workings. A second quantitative revolution is exploding in the discipline, giving rise to a new spatial data science.

Human geographers are no strangers to large data sets. In the past, they have made much use of huge data sets collected, for example, through remote sensing, by weather stations, and via national census. However, given the costs and difficulties of generating, processing, manipulating, analyzing, and representing large volumes of data, hitherto big data sets were collected only periodically and were mined only partially. But an era of rapid technological change is now upon us, and transformative possibilities for human geography are now being unlocked.

As long as we saturate the world with sensors capable of gathering data on all aspects of our lives (mobile phones have made a significant start), we can know the world as it actually is and in real time – every second and every inch. There exists an avalanche or deluge of data. big data comprises data sets that are huge in *volume*, consisting of terabytes or petabytes of data, and rapid in *velocity*, being produced in real time. Instead of collecting samples from a population, it is now possible to track the whole of a population under investigation (*n* = all) (see Kitchin 2014). But there exists a key hurdle. Data is of no value if it cannot be analyzed, digested, and put to use. The challenge of analyzing big data is coping with its sheer abundance and ubiquity, exhaustive coverage, and relentless production. Until recently, the analytical tools needed to crunch and render intelligible massive data sets were limited. To assist in this, Geographical Information Systems (GIS) has been devised. GIS provides tools that permit spatial data to be collected, stored, archived, manipulated, analyzed, and presented. But the kinds of analyses that have been possible have been limited by the GIS software adopted. Today the emerging field of Geographical Information Science (GIScience), or Geocomputation, is radically expanding what is possible (see Brunsdon and Singleton 2014). The term geocomputation was invented by British geographer Stan Openshaw in 1996. It represents a marriage between geography, statistical science, and computing science. Skilled in computer programming, geocomputation scholars write complex algorithms that scan and detect spatial patterns in even the most complex of spatial data sets.

There is no need for human geographers to invent new approaches, philosophies, and theories, then; we no longer need accept postmodernism's paralysis and needless hesitancy. We can move beyond situated knowledge and return to universal scientific truth. Technology has put the search for facts, truth, and objectivity back on the agenda, and it is now our job to make sense of the vast real-time data that is pouring into data centers and consider what it might tell us about how the world actually works (Plate 2.3).

Does spatial data science effectively resolve human geographers' reticence to use big ideas to render the world intelligible? It would be a mistake to rush to such a conclusion. While the data that is available to human geographers today is certainly bigger than at any point in human history, it is far from exhaustive. We remain some way off from

Plate 2.3　Data everywhere and everywhere data? Human geographies in real time. Source: Google Maps.

mapping every inch of the world at every second of the day. Choices are still being made about what data should be collected and why. It is often the powerful in society who are able to produce, store, interrogate, and visualize big data. The algorithms used to discern patterns in the data are often inherently biased, bearing the values of the coder. It is only some people's worlds that we might be mapping. A critical turn in geocomputation scholarship may help. But, once more, we may need to see new spatial data science not as a move beyond the postmodern challenge but as a situated social practice, producing particular and partial human geographical knowledge for particular and partial consumers of human geographical knowledge.

Conclusion

Geography has a long and complex premodern history, coming of age properly in the Greco-Roman period. But the subject as we know it today was effectively forged only in the modern era in conjunction with the rise of Western civilization from the fifteenth century. For much of the past 500 years, geography has been essentially a European branch of knowledge, reflecting European ways of making sense of the world. During the modern period, human geography developed a variety of explanatory frameworks through which to make sense of variations in human inhabitation of the surface of the earth. But in the absence of an agreed overarching framework, the discipline fragmented and a sense of mission was lost. As the West has faltered, a postmodern future has beckoned. Human geographers have become aware of the limits of the big stories that the West – and by implication they – have used to render the world legible. The era of the grand idea and the search for powerful explanatory frameworks to make the world understandable, it seems, is over.

An anxiety has haunted this chapter (and, indeed, the writing of this book). Human geography, as a Western intellectual project, is both indispensable and insufficient.

If geography is a child of the West, then one might think of human geography today as a young adult, tentatively leaving the family home, drawing strength from but also reflecting critically upon its childhood, and readying itself for life in the wider world. The West has made human geography, and for that we should be grateful. But human geography is recognizing that its personality and outlook have been overly shaped by its upbringing and that it has lots to learn about the world and its workings. Certainly, it has lots to learn, good and bad, about its parents' role in forging the world, and about its own complicity in their deeds. Human geography cannot and should not disavow its childhood; its rearing has bequeathed many positive and progressive developments. But certainly, if it is to mature into adulthood it needs to leave the family nest, face up to its parents' and its own impact on the world, and take stock of the strengths and weaknesses it has inherited as a consequence of its privileged upbringing. It needs to be open to unlearning things about the world and thinking about the world anew.

In short, if there is one conclusion to take from this chapter, it is that there is today a need to decolonize human geography.

But the risk is, of course, that if we decolonize human geography from a parochial location, this project itself is liable to be parochial. And this risk attends to writing histories of the discipline. Can a privileged group tell the story of their own history – even when explicitly setting out to be self-admonishing – without that story bearing the stamp of their own privilege? We must ask: who gets to tell the story of the history of geography?

Checklist of Key Ideas

- There are at least two ways to tell the story of the history of human geography: an internalist account and a contextualist account. This chapter adopts the latter. The rise of Western civilization was a pivotal moment in the emergence of geography and thereafter human geography. Accordingly, three phases can be identified in the historical development of geography: a premodern era (before the West), a modern era (when the West rose, peaked, and stumbled), and a postmodern era (when the West has shown signs of malaise and possible collapse).
- In the premodern era (before the fifteenth century), it was Greco-Roman and Islamic scholars who advanced the cause of geography most. Greco-Roman scholars were most interested in the shape, size, and physics of planet earth, and the cartographic techniques needed to depict its landmasses.
- From the fifteenth century, Western civilization climbed to the summit of world history. With the rise of the West came a revolution in intellectual life. The European Enlightenment marked the beginning of the modern era. The modern era stretched from the fifteenth century to the late twentieth century. During the modern era, human geography became in effect a European subject. Its fortunes fluctuated as the fortunes of the West waxed and waned. In the early modern period, from the fifteenth century to the mid-1800s, it was German scholars Alexander von Humboldt and Carl Ritter who advanced the cause of geography most. In the modern period, from 1874, geography became a university subject and cohered around the idea of environmental determinism. The late modern period, from the early twentieth century, witnessed a period of intense debate about geography's identity, purpose, and ends.
- In parallel with the faltering of the West as the world's leading civilization, in the postmodern era, human geography has entered a period of crisis. It now doubts its ability to fashion an overarching explanatory framework through which sense might be made of the world.
- But human geography has not given up on human reason! In the twenty-first century, it is likely that human geography will develop along two pathways. A first pathway will continue to strive for rigorous "situated reason," which is grasped as at once postfoundational but also anti-relativist. Not all ways of interpreting the world are equally valid objectively, morally, and practically. In producing human geographical knowledge, human geographers must be held radically accountable, and their ideas continually tested. A second pathway will witness the increased use of artificial intelligence, geocomputation and computerized data analytics, and real-time big data and the cultivation of a new spatial data science.
- The history of geography cannot be divorced from the history of European imperialism. Given this, critics claim that even today geography needs to decolonize its ways of knowing the world. Other critics observe a need to decolonize histories of geography. Western human geography continues to provide powerful intellectual resources with which to understand the world; it is progress that we now cherish these resources while constantly questioning their adequacy.

Chapter Essay Questions

1) Identify three key thinkers who have shaped the history of geography, and describe and comment upon the contributions these thinkers have made.
2) During the modern period, the discipline of geography was for all intents and purposes a European discipline. Discuss
3) Outline and comment upon the ways in which environmental determinism sought to explain variations in human culture from one place to another.
4) Postmodern human geography approaches the explanatory frameworks that modern geography invented with a degree of incredulity. Discuss.

References and Guidance for Further Reading

Scholars seeking to recount the history of geography have never been better served: in the past 40 years, there has emerged a whole slew of archival resources, textbooks, dictionaries, encyclopedias, "Geographers on Film" libraries, and key thinkers and biobibliographical studies, chronicling the people, ideas, and debates that have shaped the discipline's development.

The most popular "internalist" and accessible history of the Anglo-American geographical tradition remains:

Johnston, R. and Sidaway, J. (2016). *Geography and Geographers: Anglo-North American Human Geography Since 1945*, 7e. London: Routledge.

Other important books that focus on shifting geographical approaches over time include:

Cloke, P., Philo, C., and Sadler, D. (eds.) (1991). *Approaching Human Geography*. London: Chapman.

Couper, P. (2014). *A Student's Introduction to Geographical Thought: Theories, Philosophies, Methodologies*. London: Sage.

Cox, K. (2014). *Making Human Geography*. New York: Guilford Press.

Cresswell, T. (2013). *Geographic Thought: A Critical Introduction*. Oxford: Wiley Blackwell.

Holt Jensen, A. (2018). *Geography: History and Concepts*. London: Sage.

Martin, G.J. (2015). *American Geography and Geographers: Toward Geographical Science*. Oxford: Oxford University Press.

Martin, G.J. and James, P.A. (2005). *All Possible Worlds: A History of Geographical Ideas*, 4e. Hoboken, NJ: Wiley.

Nayak, A. and Jeffreys, A. (2011). *Geographical Thought: An Introduction to Key Ideas in Human Geography*. London: Routledge.

Peet, R. (1998). *Modern Geographic Thought*. Oxford: Blackwell.

Unwin, T. (1992). *The Place of Geography*. New York;: Longman.

A popular framework that human geographers use when writing "internalist" historiographies is provided in:

Kuhn, T.S. (1962). *The Structure of Scientific Revolutions*. Chicago: University of Chicago Press.

An important statement charting the intertwining and embroilment of human geography in the rise of the West from the fifteenth century is provided in Livingston's contextualist historical account:

Livingston, D. (1993). *The Geographical Tradition: Episodes in the History of a Contested Tradition*. Oxford: Blackwell.

Other books also make important contributions to this agenda:

Bell, M. (2005). *Geography and Imperialism 1820 to 1940*. Manchester: Manchester University Press.

Benton, L. (2009). *A Search for Sovereignty: Law and Geography in European Empires, 1400–1900*. Cambridge: Cambridge University Press.

Driver, F. (1999). *Geography Militant: Cultures of Exploration in the Age of Empire*. Oxford: Blackwell.

Edney, M.H. (1997). *Mapping an Empire: The Geographical Construction of British India, 1765–1843*. Chicago: University of Chicago Press.

Godlewska, A. and Smith, N. (1994). *Geography and Empire*. Oxford: Blackwell.

Gregory, D. (1994). *Geographical Imaginations*. Oxford: Blackwell.

Gregory, D. (2004). *The Colonial Present: Afghanistan, Palestine, Iraq*. Oxford: Blackwell.

Kearns, G. (2009). *Geography and Empire: The Legacy of Halford Mackinder*. New York;: Oxford University Press.

Philo, P. (ed.) (2008). *Theory and Methods: Critical Essays in Human Geography*. London: Routledge.

Rothenburg, T. (2007). *Presenting America's World: Strategies of Innocence in National Geographic Magazine 1888–1945*. London: Ashgate.

Smith, N. (2004). *American Empire: Roosevelt's Geographer and the Prelude to Globalization*. Berkeley: University of California Press.

Withers, C.W.J. (2001). *Geography, Science and National Identity: Scotland Since 1520*. Cambridge: Cambridge University Press.

An introduction to the ideas of key geographers can be found in:
Hubbard, P., Kitchin, R., and Valentine, G. (eds.) (2011). *Key Thinkers on Space and Place*, 2e. London: Sage.

Important books in historiographies of human geography include:
Churchill-Semple, E. (1911). *Influences of Geographic Environment: On the Basis of Ratzel's System of Anthropo-Geography*. New York;: Henry Holt.

Harvey, D. (1969). *Explanation in Geography*. London: Edward Arnold.

Hartshorne, R. (1939). *The Nature of Geography*. Washington, DC: Association of American Geographers.

Sauer, C.O. (1925). *The Morphology of Landscape*. Berkeley: University of California Publications.

A summation of a lifetime of thinking about "space" is provided by key thinker Doreen Massey in:
Massey, D. (2005). *For Space*. London: Sage.

A good introduction to behavioral geography can be found in:
Lynch, K. (1960). *The Image of the City*. Boston: MIT Press.

A good introduction to humanistic geography is provided by:
Buttimer, A. (1994). *Geography and the Human Spirit*. Baltimore: Johns Hopkins University Press.

Still the most sophisticated outline of Marxist geography can be found in:
Harvey, D. (1982). *The Limits to Capital*. Oxford: Blackwell.

Key introductions to postmodernism and its relevance to human geography are:
Harvey, D. (1989). *The Condition of Postmodernity*. Oxford: Blackwell.

Soja, E.W. (1989). *Postmodern Geographies: The Reassertion of Space in Critical Social Theory*. London: Verso.

Influential introductions to poststructuralist perspectives in human geography can be found at:
Duncan, J.S. and Ley, D. (2013). *Place/Culture/Representation*. London: Routledge.

Barnes, T.J. and Duncan, J.S. (2013). *Writing Worlds: Discourse, Text and Metaphor in the Representation of Landscape*. London: Routledge.

An introduction to the neglected contribution of women to the making of the history of geography can be found in:

Maddrell, A. (2011). *Complex Locations: Women's Geographical Work in the UK 1850–1970*. Oxford: Wiley.

Audrey Kobayashi has played a key role in the development of antiracist human geographies:

Kobayashi, A. (2014). The dialectic of race and the discipline of geography. *Annals of the Association of American Geographers* 104 (6): 1101–1115.

An introduction to anarchism in human geography is offered in:

Springer, S. (2016). *The Anarchist Roots of Geography: Toward Spatial Emancipation*. Minneapolis: University of Minnesota Press.

Important works in postcolonial human geography include:

Jazeel, T. (2019). *Postcolonialism (Key Ideas in Geography)*. London: Routledge.

Nash, C. (2002). Cultural geography: postcolonial cultural geographies. *Progress in Human Geography* 26 (2): 219–230.

Noxolo, P., Raghuram, P., and Madge, C. (2012). Unsettling responsibility: postcolonial interventions. *Transactions of the Institute of British Geographers* 37 (3): 418–429.

Radcliffe, S.A. (2017). Decolonising geographical knowledges. *Transactions of the Institute of British Geographers* 42 (3): 329–333.

A seminal text in the rise of nonrepresentational theory in human geography is:

Thrift, N. (2007). *Non-Representational Theory*. London: Routledge.

The recent revival of pragmatism in human geography is captured in:

Wills, J. and Lake, B. (eds.) (2020). *The Power of Pragmatism: Knowledge Production and Social Research*. Manchester: Manchester University Press.

Important works in applied human geography include:

Boyle, M., Hall, T., Lin, S. et al. (2020). Geography and public policy. In: *International Encyclopaedia of Human Geography*, 2e (ed. A. Kobayashi), 93–101. Boston: Elsevier.

Pacione, M. (ed.) (1999). *Applied Geography: Principles and Practice: An Introduction to Useful Research in Physical, Environmental and Human Geography*. London: Psychology Press.

Important works on computerized data analytics and human geography include:

Brunsdon, C. and Singleton, A. (2015). *Geocomputation: A Primer*. London: Sage.

Kitchin, R. (2014). *The Data Revolution: big data, Open Data, Data Infrastructures and Their Consequences*. London: Sage.

Longley, P.A., Goodchild, M.F., Maguire, D.J., and Rhind, D.W. (2015). *Geographic Information Systems and Science*, 3e. Chichester: Wiley.

Website Support Material

A range of useful resources to support your reading of this chapter are available from the Wiley *Human Geography: An Essential Introduction* Companion Site http://www.wiley.com/go/boyle.

Chapter 3

Big History: Watersheds in Human History

Chapter Table of Contents

Chapter Learning Objectives

By the end of this chapter you should be able to:
- Define "big history," and identify its key concerns.
- Describe Darwin's theory of evolution through natural selection, and comment specifically on how this theory explains the origins of the human species.
- Map and date the early migrations of humans from their common ancestral home in Central and East Africa, migrations that over time led to the colonization by humans of the entire planet.

Human Geography: An Essential Introduction, Second Edition. Mark Boyle.
© 2021 John Wiley & Sons Ltd. Published 2021 by John Wiley & Sons Ltd.
Companion website: www.wiley.com/go/boyle

- List key trends in the development of human culture, and locate and account for the times and places in which settled agriculture was first discovered and established.
- Identify the principal civilizations that have risen and fallen in world history, and compare and contrast environmental and societal explanations as to why civilizations succeed and fail.
- Describe and comment upon the idea that the West emerged as a leading civilization principally because it enjoyed a favorable environmental history, and that it is possible to date the rise of the West to as early as the tenth century BCE.

Introduction

Human history predates the ascendance of the West, and it is important to remind ourselves that many prior developments in human history – and, indeed, earth history – were needed to make the story of the West possible.

This chapter is inspired by recent developments in "big history": deep historical investigation that helps us to understand how we have arrived at where we are at today. Our focus is specifically upon one slice of big history: the epic story of the very recent arrival of human beings on planet earth and the key moments of innovation that have underpinned the incredible human journey. The chapter will provide you with a brief introduction to four principal watersheds that marked human history prior to the rise of the West: the origins of the human species, first migrations and the peopling of the planet, the development of human culture and birth of settled agriculture, and the rise and fall of civilizations. It will conclude with a reflection upon US anthropologist and geographer Jared Diamond's claim that the deep origins of the West's ascendance and dominance lie in the head start Europe secured following the Neolithic Revolution. According to Diamond, the rise of Europe, which we will understand here to constitute a fifth watershed in human history, can be dated not, as it is commonly supposed, to the fifteenth century CE, but to the first agricultural hearth in Mesopotamia as long ago as the tenth century BCE.

Introducing Big History: From the Big Bang to the Sixth Mass Extinction!

That human beings exist at all is a miracle – in many ways, an incredible accident of natural history starting with the Big Bang some 13.7 billion years before present (YBP), progressing with the formation of the earth 4.5 billion YBP, and concluding with the birth and evolution of living organisms through natural selection. Even more incredulous again has been the fate of modern humans since their very late arrival on this little blue planet some 200 000 YBP. From as early as 1991, US-born and Australian-based historian David Christian has called for historians to widen their lens and to consider the vast sweep of history that has enabled the present condition of human existence to be possible. Big history is an interdisciplinary and panoramic field of study that blends cosmology, astronomy, geology, biology, physics, chemistry, geography, anthropology, politics, sociology, and the humanities to probe into the deep existential concerns that weigh on humans today. These include humbling questions such as: from exactly where have we come, and how random is it that we are here at all? And, in the light of the fate of prior dominant species, to where might we be heading, and is our fate secure? Christian delivered a TED talk in Long Beach, California, in 2011, which so inspired philanthropist Bill Gates that he collaborated with him to establish *The Big History Project* – an initiative designed to encourage wider learning about the deep history of the universe, the natural history of the earth, and the genesis and meaning of human history.

First Watershed: The Origins of the Human Species

A timeline of human history can be quickly recounted: the earth formed approximately 4.5 billion years before the present (YBP); first evidence of life can be traced to 3.6 billion YBP; photosynthesis, which led to oxygen first entering the atmosphere, at 3.4 billion YBP; the first complex cells emerged circa 2 billion YBP; and the first multicellular life at 1 billion YBP. The first fish appeared 500 million YBP, land plants 475 million YBP, insects 400 million YBP, forests 375 million YBP, reptiles 300 million YBP, dinosaurs 230 million YBP, mammals 200 million YBP, and birds 155 million YBP. Primates first appeared just under 65 million YBP; hominin species first diverged from the Hominidae family (which includes chimpanzees) circa 7 million YBP; and through successive mutations, modern humans emerged in the form of *Homo sapiens* around 200 000 years ago (Figure 3.1). If this timeline is expressed in terms of a 24-hour period, with the first second announcing the formation of the earth, primates would emerge only at 21 minutes to midnight, hominins just over 2 minutes to midnight, and *Homo sapiens* a mere 4 seconds before midnight.

Until 1859, beyond attributing human existence to divine providence, very little was known about the origins of the human species. In that year, a great leap forward took place in human intellectual history and enlightenment.

The HMS *Beagle* set sail from Plymouth, England, on December 27, 1831. Following a circumnavigation of the globe, which incorporated visits to the Canary Islands, South America (and famously the Galapagos Islands), Australia, New Zealand, South Africa, and the Azores, it returned to Falmouth, England, on October 2, 1836 (Map 3.1). On board was English naturalist Charles Robert Darwin. Darwin's mission was to observe and collect samples of fossils, flora, and fauna encountered *en route*. This voyage was to change Darwin's life, and, in turn, Darwin was to revolutionize scientific and religious thought.

By 1838, Darwin had devised the central tenets of his theory of evolution through natural selection. For Darwin, all species display a tendency to propagate too many members for the given environment they occupy. This creates a competition for survival. Natural variation in member traits equips some species to prosper in certain environments better than others, and these constituencies survive and endure, passing on their competitive advantages to their offspring. The fittest survive, and the less well-adapted drift to extinction. Genetic lines of descent are determined by the continuous sifting and sorting of the strong from the weak, and only some gene pools are permitted to survive over the long haul. Living always on the edge of extinction, these survivors themselves are in constant interaction with the natural environment and mutate endlessly as they attempt to gain the upper hand (Deep Dive Box 3.1).

It was not until 1859, however, that Darwin published his theory in his famous book *On the Origin of Species by Means of Natural Selection, or the Preservation of Favoured Races in the Struggle for Life* (Darwin 1859). While in part a consequence of his laudable quest to garner sufficient evidence to put this theory beyond doubt, Darwin's reticence to

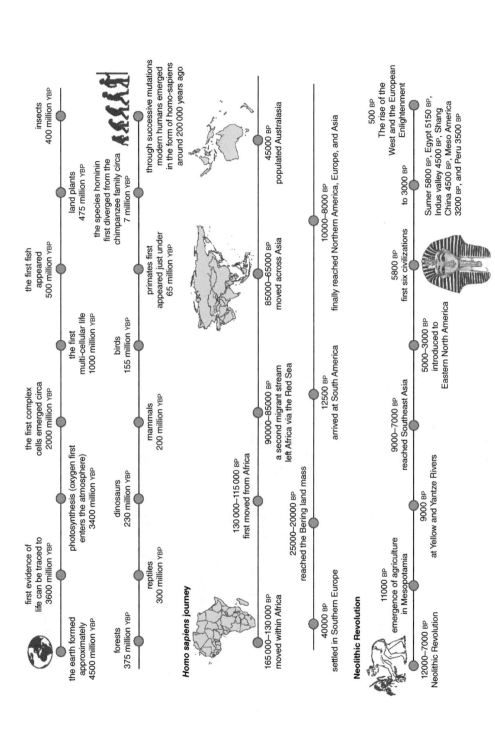

Figure 3.1 Timeline of world and human history.

first evidence of
life can be traced to
3600 million YBP

the earth formed
approximately
4500 million YBP

forests
375 million YBP

the first complex
cells emerged circa
2000 million YBP

photosynthesis (oxygen first
enters the atmosphere)
3400 million YBP

dinosaurs
230 million YBP

reptiles
300 million YBP

the first fish
appeared
500 million YBP

the first
multi-cellular life
1000 million YBP

birds
155 million YBP

mammals
200 million YBP

insects
400 million YBP

land plants
475 million YBP

the species hominin
first diverged from the
chimpanzee family circa
7 million YBP

through successive mutations
modern humans emerged
in the form of homo-sapiens
around 200 000 years ago

primates first
appeared just under
65 million YBP

Homo sapiens journey

130 000–115 000 BP
first moved from Africa

165 000–130 000 BP
moved within Africa

25 000–20 000 BP
reached the Bering land mass

90 000–85 000 BP
a second migrant stream
left Africa via the Red Sea

85 000–65 000 BP
moved across Asia

45 000 BP
populated Australasia

40 000 BP
settled in Southern Europe

12 500 BP
arrived at South America

10 000–8000 BP
finally reached Northern America, Europe, and Asia

Neolithic Revolution

11 000 BP
emergence of agriculture
in Mesopotamia

12 000–7000 BP
Neolithic Revolution

9000 BP
at Yellow and Yantize Rivers

9000–7000 BP
reached Southeast Asia

5000–3000 BP
introduced to
Eastern North America

5800 BP
first six civilizations

to 3000 BP

500 BP
The rise of the
West and the European
Enlightenment

Sumer 5800 BP, Egypt 5150 BP,
Indus valley 4500 BP, Shang
China 4500 BP, Meso America
3200 BP, and Peru 3500 BP

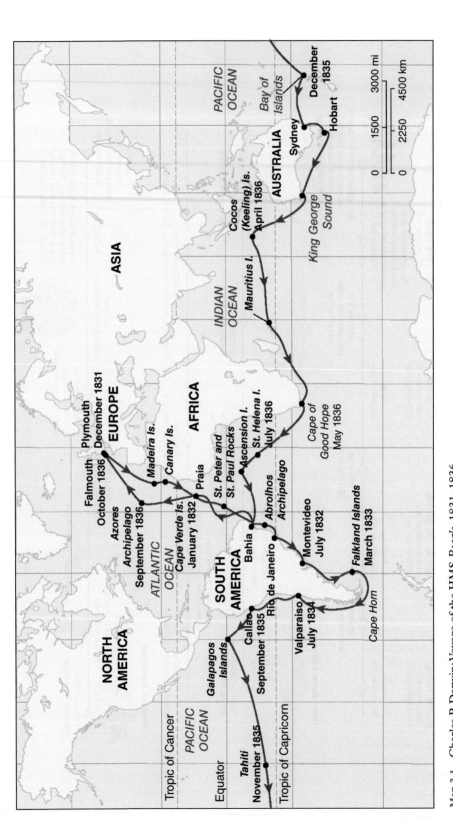

Map 3.1   Charles R Darwin: Voyage of the HMS *Beagle*, 1831–1836.

rush to publication stemmed principally from his awareness of the meaning of his theory for Christianity and biblical teachings. Contrary to the account of the origin of the human species contained in the Book of Genesis, if all species evolved through natural selection from a primitive ancestor, then this had to be true of the human species too. While triggering much furor, Darwin's *On the Origin of Species* in fact carefully avoided

---

Deep Dive Box 3.1 Mass Extinction Events: Natural Selection Through Revolution

The evolution of species has been a precarious business marked by periods of catastrophic setback. The geological record shows that there have been at least five major extinction events in the past: the first about 450 million YBP, and the last about 66 million YBP. For a period, life on earth was culled, and the evolutionary clock was interrupted and reset. Estimates conjecture that as much as 99% of the species that have ever lived (estimated at four to five billion) have become extinct. But life soon returned and new species emerged. A new evolutionary path was established. Natural selection would have looked dramatically different if these mass extinctions had not occurred; perhaps humans would not have evolved at all.

The five key mass extinction events are:

- The Ordovician–Silurian extinction (c. 450–440 million YBP), which wiped out over 70% of all species then in existence, the majority of which lived in the world's oceans.
- The Late Devonian extinction (c. 375–350 million YBP), which terminated nearly 70% of all species then present, most of them marine-based.
- The Permian–Triassic extinction (the Great Dying; c. 251 million YBP), which rendered extinct 90–96% of all species then in existence, including the majority of all marine species and nearly 70% of all land species.
- The Triassic–Jurassic extinction (c. 200 million YBP), during which 70–75% of all species then in existence were forced into extinction, including most of the earth's largest amphibians and reptiles.
- The Cretaceous–Paleogene extinction (c. 66 million YBP), during which 75% of all species then living became extinct, including dinosaurs.

A growing number of scientists believe that we are living today through a sixth mass extinction. According to some, in the process of plundering and polluting planet earth for their own ends, human beings have so modified nature that we now live in a new geological time period – the Anthropocene, or the "age of humans." The Anthropocene might well be a period in which the capacity of the planet to tolerate the human species reaches its limits and when planetary ecosystems finally collapse. Human-created environmental change, including climate change, will render our planet incapable of supporting the growing human population and other existing life forms.

Big history shows us that all dominant species eventually fall; have we any reason to believe that, in time, we humans will not go the way of the dinosaurs? The difference is that, on this occasion, humans are bringing about their own demise by their own careless neglect and abuse of the environment.

specific consideration of the lineage of the human species, merely noting that "the the-ory of natural selection might shed light on the origin of man [*sic*] and his history."

It was only in 1871, when Darwin published his later book *The Descent of Man, and Selection in Relation to Sex*, that he worked his theory to its logical conclusion. In *The Descent of Man*, Darwin questioned whether the human species was so distinctive in bodily structure, mental prowess, emotional capacity, and moral reasoning that the claim that humans could have evolved from lower life forms could be rendered intel-lectually untenable. Darwin argued that the human species in fact was not as excep-tional as might be surmised and that all human faculties could be shown to exist in some form among higher-order mammals. The most civilized human qualities could be interpreted as social instincts birthed at early stages in the evolutionary cycle to assist in the battle for survival. The human species had developed from a lower life form, now understood to be early primates, and probably for Darwin, this evolutionary process had African origins.

Darwin also recognized the political dangers of applying his theory to the human spe-cies, dangers well founded in the light of the subsequent eugenics movement, which advocated racial purity, believed that some races were superior to others, and promoted the pursuit of a master race. "Weaker" humans, eugenicists argued, should not be allowed to bear offspring, as the result would be a degenerate human community. For Darwin, racial differences were cosmetic, and all races could be shown to derive from a single parent. In any case, the search for a community of perfect human beings was an ill-conceived and dangerous one. Darwin argued that while it might buck evolutionary trends, society had a responsibility to protect and defend the reproductive rights of all communities, irrespective of their physical and mental capacities.

Although arguments between so-called "evolutionists" (who advocate evolution through natural selection) and "creationists" (who see in evolution the hidden hand of a higher being, God) continue to rage, few now doubt that the basic tenets of natural selection are substantially valid.

Second Watershed: First Migrations and the Peopling of the Planet

There remains debate over *where* human life first originated and when and how *Homo sapiens* learned to move across and dwell in a range of different environments, thereby populating the entire planet. *Pace* Darwin, the two most popular theories begin with the proposition that hominins first arose as a distinctive species in Central and East Africa.

The Multiregional Continuity Model suggests that from this point of origin, from as early as two million YBP, primitive hominin mutants managed to migrate and to popu-late new regions, for instance *Homo ergaster* elsewhere in Africa, *Homo erectus* and Denisovans in Asia, and *Homo neanderthalensis* in Europe. These early hominins sur-vived, and modern humans emerged from these separate lineages. The Recent African Origin Model, in contrast, posits that early migrations from Africa were ultimately fruit-less, with primitive hominin species gradually ebbing to extinction. The *Homo sapien* conquest of the world began in an enduring way only circa 100 000 years ago from a relatively small population (to emphasize how small, scientists often speak of all humans deriving ultimately from a single common mother dwelling in this group, often given the title "Mitochondrial Eve") and from a specific point of origin in Central and East Africa (modern-day Ethiopia). These later migrations replaced the degenerate and decaying early hominin communities.

For a long time now, commentators have favored the Recent African Origin Model. In 2003, British geneticist Stephen Oppenheimer proposed a detailed map of the earliest migrations through which *Homo sapiens* colonized the world. The timing and scale, waxing and waning of population migration into and out of certain world regions were affected by climate change; at different times it became more possible to breach natural barriers, while at other times the environment proved too inhospitable to enable traverse of certain territories (Oppenheimer 2003a,b). Starting from East Africa around 165 000 to 130 000 YBP, *Homo sapiens* moved first within Africa to the Cape of Good Hope, the Congo Basin, and the Ivory Coast. During a period of climate change, which created a more hospitable Sahara Desert, between 130 000 and 115 000 YBP modern humans first left Africa through modern-day Egypt, only to perish in the Levant at approximately 90 000 YBP during a period of climate cooling. Between 90 000 and 85 000 YBP, a further migrant stream left Africa by crossing the mouth of the Red Sea. All non-African populations today can be traced to this community. Between 85 000 and 65 000 YBP, these pioneers moved across Asia, traversing India, Southern China, Southeast Asia, and finally Australasia. By 45 000 YBP, Australia was substantially populated. Climate warming circa 50 000 YBP and the development of a more agreeable climate in northern regions of the world finally allowed northwest migration to Europe and a variety of northeast migrations through continental Asia. By 40 000 YBP, modern humans had settled in Southern Europe, and by 25 000 to 22 000 YBP they had reached the Bering landmass connecting Asia with the Americas (but see also Deep Dive Box 3.2). A variety of successful and unsuccessful migrant routes brought migrant streams down both the west and east coasts of the Americas, and by 12 500 YBP they arrived at the southern tip of South America. Finally, the period from 10 000 to 8000 YBP witnessed the penetration of *Homo sapiens* into the colder environments of North America, Northern Europe including Scandinavia, and Northern Asia (Map 3.2).

Since Oppenheimer's pioneering work, further genetic and haplogroup analyses of mitochondrial DNA (providing information on the maternal lineage of us all) and Y chromosomes (providing information on the paternal lineage of men) has bequeathed a more complicated and at times contested picture. In the past few years alone, a number of studies have afforded a richer set of insights.

- **The Simons Genome Diversity Project** analyzed DNA from 142 populations around the world and concluded that all modern humans living today can trace their ancestry back to a single population that emerged in Africa 200 000 years ago and that split 130 000 years ago with one faction exiting Africa. Meanwhile, within Africa, humans split into subpopulations, which evolved in isolation: the KhoeSan in South Africa, for example, separated from the Yoruba in Nigeria circa 87 000 YBP, while the Mbuti split from the Yoruba circa 56 000 YBP.
- **The Estonian Biocentre Human Genome Diversity Panel** examined 483 genomes from 148 populations and concluded that indigenous populations in modern Papua New Guinea owe 2% of their genomes to a now extinct group of *Homo sapiens*. The implication is that an earlier exodus of modern humans from Africa (c. 120 000 years ago) must have arrived in Papua New Guinea and interbred with much later human arrivals before withering to extinction.
- **The Cambridge Aboriginal Australian study** scrutinized genomes from 83 Aboriginal Australians and 25 Papuans from New Guinea and found that the ancestors of modern Aboriginal Australians and Papuans split from a single migration pulse out of

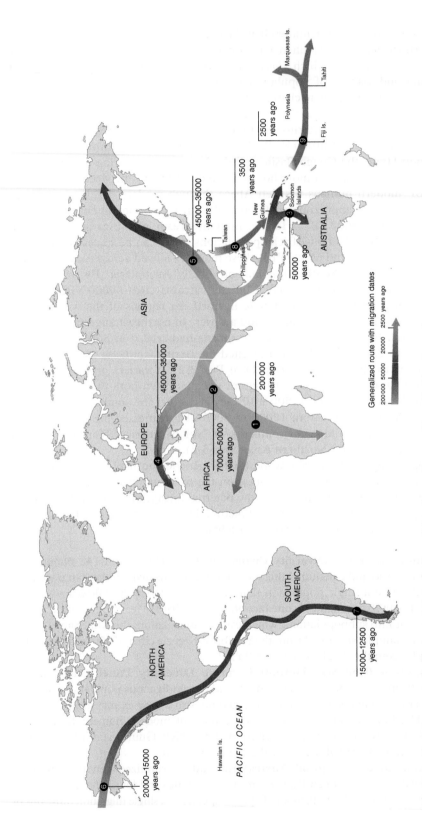

Map 3.2 Early human migrations and the peopling of the planet. Source: Compiled by the author

Africa into Eurasia some 58 000 YBP and that these two populations themselves later diverged around 37 000 YBP.

- **The International Pacific Research Center (IPRC) study** created an integrated climate–human migration computer model to re-create the spread of *Homo sapiens* over the past 125 000 years, and it established that modern humans first left Africa 100 000 YBP in a series of slow-paced migration waves and that *Homo sapiens* first arrived in Southern Europe around 80 000–90 000 YBP, much earlier than hitherto believed.
- **The Lipsom Model**, based upon a synthesis of previous studies, argues that present-day human diversity in Eurasia was created soon after modern humans left Africa. From about 45 000 YBP, a western migration pulse split from an eastern pulse, and thereafter different gene pools emerged in the east and in the west.
- **The Reich Study** analyzed genome data from the remains of 51 modern humans who lived between 45 000 and 7000 YBP. Focusing upon Europeans, this study argues that present-day Europeans can trace their ancestry back to a single population who lived in northwest Europe (likely Belgium) 35 000 years ago and who somehow survived the last Ice Age. This Aurignacian culture became displaced when another group of early humans (the Gravettians) arrived in Europe 33 000 years ago. But circa 19 000 years ago, a population related to the Aurignacian culture re-expanded across Europe and proceeded to repopulate Europe after the Ice Age.

We might conclude, then, that although the Recent African Origin Model's thesis remains dominant, there is much work to do if the details of this thesis are to be fleshed out and if consensus is to be reached.

Deep Dive Box 3.2   Peopling the Americas

The entry into and dispersal of the First Americans across North and South America continues to excite much controversy.

For much of the Pleistocene period (circa. 1.8 million YBP to 12000 YPB), Asia and North America were joined by a landmass called "Beringia." This landmass had the potential to facilitate human migration from Asia into the Americas. But during periods of global cooling, ice would form over Beringia, blocking passage. The last Ice Age occurred between 110 000 and 12000 YBP, with the last glacial maximum (LGM – period when glaciers were most advanced) at 22000 YBP. As the ice melted, the Holocene period began – a geological period marked by global warming – and melting and flooding caused sea levels to rise, and Asia and the Americas became separated by a sea.

The Beringian Incubation Model or Beringian Standstill Model was first proposed in the 1930s and suggests that the Americas were peopled by a single migrant stream, comprising peoples who reached Beringia from East Asia and Siberia between 35000 and 25000 YBP, who were marooned on Beringia for up to 15 000 years, and who entered the Americas only at the start of the Holocene period. The First Americans, then, were effectively trapped during the LGM as glaciers in the Verkhoyansk Range in Siberia and in the Mackenzie River valley in Alaska cut them off from returning to Asia or moving further on and into the Americas. Somehow, living as hunters and gatherers and eking out a precarious existence on

(Continued)

Box 3.2 *(Continued)*

barren tundra landscapes, they managed to survive in isolation. Pollen studies drawing samples from vegetation now submerged under the Bering Strait indicate that the area was indeed populated at the LGM by trees, such as spruce, birch, and alder, which all provided wood for fire.

More recently, however, it has been suggested that Native Americans derive from three different migration pulses after the LGM, which flowed through Beringia and later the Bering Strait (see Map 3.3). A first wave (c. 11000 YBP) brought Amerindians to the Americas, and they were later to disperse throughout North and South America. A second wave some 9000 YBP brought Athabascan-, Nadene-, and Chipewyan-speaking populations, who concentrated in Alaska and Northwest Canada but also made it to the Southwest United States. Paleo-Eskimos (from 3000 BCE to 1300 CE) stand as a third migration pulse into the Americas independent of these two Native American pulses. Paleo-Eskimos developed a series of cultures, including Saqqaq and three separate Dorset cultures. They settled the Arctic from Alaska to Greenland. The Paleo-Eskimo population survived in near isolation for almost 4000 years, only to vanish around 700 years ago. Later Thule people arrived (beginning around 1000 years ago) and became ancestors of the modern-day Inuit. There is evidence that movement was two-way, back and forth across the Bering Strait. And there is some evidence of limited interaction and interbreeding between Native Americans, Paleo-Eskimos, and Inuits.

Further complicating matters, not everyone believes that the first Americans arrived through Beringia. The principal rival is the Solutrean thesis, which holds that in fact Europeans were able to traverse the Atlantic using wood-frame and seal-skin boats and found their way to Canada and the Eastern United States, and from there dispersed into the continent. Support for the idea rests upon the claim that Solutrean technologies found in Solutrean cultures in France and Spain share similarities with Clovis technologies found in Central and North America, and that these technologies cannot be found in other East Asian, Siberian, or Beringian cultures. Genetic analysis has also subsequently found Western Eurasian genetic signatures in present-day Native Americans. But there is also evidence of West Eurasian genetic flows in Siberian populations at the time of the LGM, suggesting that Western Eurasians had managed to migrate toward Northeast Eurasia in spite of global cooling. This haplogroup passed through Siberia and through Beringia subsequently, and may account for the West Eurasian genetic signal in the Native American population.

Attention is increasingly being given to the spread of people within the Americas following their passage through Beringia. Emerging evidence points to the fact that human migration from North to South America fanned populations out in different directions, and this dispersal led to an early genetic diversification of the first Americans. The peopling of the South America involved multiple, autonomous, and geographically variegated migrations, which bequeathed in turn a diversity of genetically distinctive populations across the continent. There was no single or dominant migration pulse from Alaska to Argentina.

Other studies have found that populations in Amazonia and the Central Brazilian Plateau display traces of Australasian, Andamanese, and New Guinean lineage that are not found in East Asian and Siberian populations. No such trace is found in other Native American populations. Perhaps humans populated these parts of the Americas by boat, somehow successfully crossing the vast Pacific Ocean. Perhaps

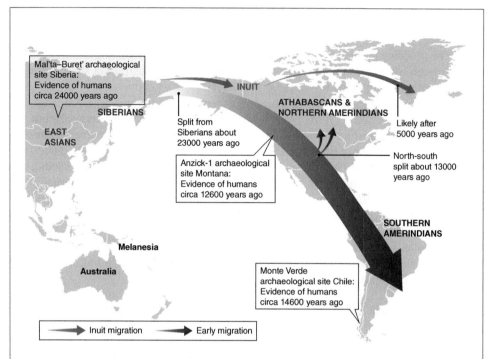

Map 3.3   Peopling the Americas. Source: Raghavan et al. (2015). © 2015, The American Association for the Advancement of Science.

there might have been two streams of migration, one through Beringia, the other by sea to Amazonia. Equally possible, however, there might have been a second migration pulse (perhaps a much later one and after the Beringian incubation period) that passed through Beringia and settled in Amazonia but whose distant origins are Australian. This pulse may have traveled around the Pacific Rim leaving no genetic trace, settling and surviving only in Amazonia.

Most recently, archeologists working at the late Pleistocene Cerutti Mastodon site near San Diego, California, claim to have found evidence of an unidentified species of *Homo* lineage that displayed manual dexterity and knowledge of tool making. Bone fragments from bodily remains found at the site can be dated to circa 130 000 YBP; if true, this finding would again challenge radically existing theories of when and how the Americas were peopled. How is it possible that a mentally advanced *Homo* species came to be in modern-day California 130 000 years ago when scientific convention holds that humans did not reach North America until circa 23 000 years ago?

Bringing some of these strands together, recent work on the origins and dispersal of the First Americans conducted by Raghaven et al. (2015 – see also Map 3.3) suggests:

1   Native Americans diverged from their East Asian source populations during the LGM and no earlier than 23000 YBP.
2   The Beringian Incubation Model has merit, but the population did not "stand still" for 15 000 years and instead remained trapped in Beringia for only circa 8000 years. Moreover, whether the standstill occurred in Beringia or Siberia remains to be determined.

(Continued)

Box 3.2   *(Continued)*

3   The peopling of the Americas occurred on the basis of a single migration pulse through Siberia. Later gene flows from East Asia and Australasia should not confuse or complicate the single-pulse theory.

4   Once in the Americas, there emerged a further split within the Native American population, between a northern branch (comprising Athabascans and northern Amerindians) and a southern branch of Amerindians, which occurred sometime around 13000 YBP. This was triggered by the opening up (from c. 16000 YBP to 14000 YBP) of habitable and navigable routes along the (now Pacific) coast southward and into a by-now deglaciating continental North America.

5   The (often assumed as pioneering) Clovis complex was not the first cultural complex created by new arrivals on the continent, and in fact developed much later and after other cultural groupings had emerged.

6   A smaller number of later migration streams, also passing across the Bering Strait, went on to populate North America and the Arctic, including Eskimos and the Inuit populations.

Our story ends here, but of course any study of the evolution of the Native American community needs to examine the post-Columbian period when Europeans "discovered" and colonized the Americas – dramatically affecting (in many cases exterminating, on occasions by design) many Native American cultures. The US National Geographic Genographic Project (now titled Geno 2.0) (2005–2020), directed by Spencer Wells and more recently Miguel Vilar, has distributed DNA Ancestry Testing Kits to over one million people across 140 countries. Whereas many studies draw a distinction between broad genetic groups, research by Dai et al. (2020) using data from the Genographic Project has enabled fine-scaled subgroup variations to be discerned. Their results for the United States ($n = 32000$) point to the impacts of admixture proportions, migration rates, haplotype sharing, and ancestral birth origins on the complex genetic histories of the present population. Key findings include:

1   Large variations in admixture proportions within and between Hispanic subpopulations have given rise to a US Latino population that is genetically diverse and admixed.

2   On gaining freedom, African Americans captured and traded as slaves to southern colonies journeyed North in a Great Migration and settled in northern and midwestern cities. Barriers to migration (the Appalachian Mountains and Mississippi River, and the New Mexico desert) restricted gene flow across the full landmass. A "Northern North American African" population resulted. Later, waves of migration of African descendants from the Caribbean have created a distinctively different, more diverse, and admixed "Southern North American" African population.

3   More recent migrant populations from East Asia, Southeast Asia, South Asia, and the Middle East present with heightened "homozygosity" (have genetic traces that are similar) reflecting enduring "consanguinity" (relatedness). Given the comparative lateness of their arrival, this relative genetic homogeneity and close relatedness are to be expected.

4   Genetic traces deposited by French Scandinavian, Jewish, Polish, Irish, and Scottish migrants to North America has also enabled migration scholars to better track the complex spread of the European diaspora across the continent.

Third Watershed: The Development of Human Culture and Invention of Settled Agriculture

For most of their history, humans have survived by adopting a hunter-gatherer mode of subsistence, foraging for food among wild plants and animals. While often thought of as primitive, in fact hunter-gatherers were skillful people who learned through bitter experience how to maximize the extraction of food from a given area of land. Hunter-gatherers developed complex social structures built around the idea of the "Band." Bands were groups united by kin that numbered no more than 100 people and normally fewer than 30. For the most part they practiced nomadism, traveling to avoid depleting food supplies to the point of exhaustion in any one place. Bands required extensive land areas to survive: in areas of abundance, Bands required up to 7 square miles of land per capita, but in regions of scarcity up to 500 square miles of land per capita were needed. Survival depended upon capturing wild game, fishing, and gathering plants and insects. Hunter-gatherers relied upon temporary huts, tents, lean-tos made of plant materials or animal skin, and caves for shelter.

Steadily mastering their environment, through time humans learned how to move beyond hunter-gatherer modes of subsistence. Around 10000 to 5000 BCE, humans began to pioneer settled agricultural hearths, and by 4000 to 2000 BCE the first civilizations began to appear. A variety of typologies have been proposed to characterize these developments in human culture. For some, it is appropriate to speak of the passage from the savage state to nomadism, agriculture, and civilization. For others, the movement from the Stone Age (with a further subdivision into Paleolithic, Mesolithic, and Neolithic) to the Bronze Age and, finally, the Iron Age provides a better description. Others again identify a development in the pattern of culture in terms of a chronological progression from Lithic to Archaic, to Formative, to Classic, and eventually to Post-Classic. Finally, yet other scholars prefer the sequence to be depicted in terms of the titles of savagery, barbarism, and civilization (see Deep Dive Box 3.3).

Deep Dive Box 3.3   Lewis H. Morgan's Seven Stages of Human Evolution

In his book *Ancient Society; or, Researches in the Lines of Human Progress from Savagery Through Barbarism to Civilization*, published in 1877, US anthropologist Lewis H. Morgan used the features of means of subsistence, forms of government, language (verbal and written), religion, house life and architecture, family form and function, and property entitlements to identify seven specific stages in human evolution – what he terms ethnical periods.

1 **Lower Status of Savagery**: From the emergence of the human race, through the reliance on foraging for fruits and nuts, to the invention of fishing and the use of fire. The dominant family form was the *Consanguine family*, and human communication occurred principally through gestures.
2 **Middle Status of Savagery**: From the discovery of fire to the invention of the bow and arrow. The *Punaluan family* form was now dominant, and humans communicated through monosyllabic languages.
3 **Upper Status of Savagery**: From the development of the bow and arrow to the invention of pottery, a key indicator of moving beyond savagery. The

(Continued)

Box 3.3 *(Continued)*

Syndyasmian family type was now preeminent, and the first multisyllabic languages were emerging. Social stratification developed, and religion was based on worship of the elements.

4 **Lower Status of Barbarism**: From the manufacture of pottery, through the domestication of animals and the cultivation of maize and plants by irrigation, to the use of adobe-brick and stone in house building. The *tribal confederacy* now emerged as the dominant form of government.

5 **Middle Status of Barbarism**: From the domestication of animals to the smelting of iron ore. Agriculture was enhanced by new techniques of irrigation. Communal tenement housing in defensive sites (for example, forts) emerged.

6 **Upper Status of Barbarism**: From the manufacture of iron to the invention of the phonetic alphabet, and the use of writing in literary composition, a key indicator of moving beyond barbarism. The *Monogamian family* now predominated, and the ideas of individual property, public life, poetry, and mythology emerged. Communities preferred to dwell in walled cities.

7 **Status of Civilization**: The period of the phonetic alphabet and the production of literary records; divided into *Ancient, Medieval,* and *Modern* civilizations depending upon their degree of sophistication.

Each ethnical period had a dominant pattern of inventions, discoveries, and institutions. Any single region could only occupy one ethnical stage at any point in time. Movement through each stage constituted human progress.

Evidently the Neolithic Revolution, the period when the human race discovered how to domesticate plants and animals and to practice settled agriculture, marks a watershed in human history. The Neolithic Revolution occurred "simultaneously" only in a specific number of locations – agricultural hearths – and at a specific moment in human history (between 10000 and 5000 BCE).

In his famous 1952 winter Bowman Lectures, convened by the American Geographical Society and given in the Harkness Theatre of Columbia University, New York geographer Carl Sauer addressed the topic of "Agricultural Origins and Dispersal" (Sauer 1952). Sauer's approach has been described as akin to Sherlock Holmes' method of solving a crime. A list of suspects are convened and one by one eliminated if they fail to reconcile with the evidence. Logical deduction steadily thins the herd down. Those suspects still standing become candidates for arrest. In Sauer's case different explanations for the rise of the first farmers were the suspects and those left standing after rigorous interrogation the perpetrators.

Sauer attempted to speculate upon where agriculture first began by postulating first six deductive principles:

1. "Agriculture did not originate from a growing or chronic shortage of food." Agriculture could only have developed in areas of food surplus where humans could experiment with plant and animal domestication knowing that if they failed food was still at hand. A comfortable standard of living was an essential prerequisite.

2. "The hearths of domestication are to be sought in areas of marked diversity of plants or animals, where there were varied and good raw materials to experiment with, or in other words, where there was a large reservoir of genes to be sorted out and recombined." Biodiversity was a prerequisite, since a wide gene pool was required to crossbreed species.

3. "Primitive cultivators could not establish themselves in large river valleys subject to lengthy floods and requiring protective dams, drainage, or irrigation." Because flood defenses were limited, agriculture could not develop in large river basins, a proposition that ran counter to the established Potamic thesis of the day, which held that agriculture first developed in the fertile valleys of the Near East.

4. "Agriculture began in wooded lands." Areas of forestry would be preferred to grasslands, as primitive technology made grasslands harder to till and cultivate. Forests could be cleared with fire.

5. "The inventors of agriculture had previously acquired special skills in other directions that predisposed them to agricultural experiments." Communities had to possess a number of skill sets that, while they might have derived from other needs, were of direct advantage in agriculture.

6. "Above all, the founders of agriculture were sedentary folk." Agriculture could only have been invented by populations who were already sedentary or who were inclined to cede their nomadic lifestyles.

Sauer then consulted four maps and, by testing places against his six "necessary conditions" for agriculture to blossom, identified areas he believed to be good candidates. A first map covering the "Old World of Asia, Africa and Europe" proposed the existence of an agricultural hearth area in Southeast Asia radiating out via established routes to places as distant as Southern Europe, West Africa, and New Guinea. Sauer's second map covered Central America, and hypothesized the existence of an early agricultural hearth in northern Andean valleys, and a later one in southern Mexico. Adopters stretched north into the Mississippi Valley, and south and east into the Amazon River basin. The third map centers upon the Middle East, and identifies Southwest Asia and Northeast Africa as potential hearths. Sauer argued that the Nile, Tigris-Euphrates, Indus, and Ganges Rivers were not conducive incubators of early agriculture. The fourth map depicts the world at 1500 CE, delimiting the extent to which agriculture had dispersed by the time Europe was ready to rise as a world historical power.

In the absence of factual evidence on the ground, for example sufficient surviving agricultural ruins and traces, and without the benefit of modern technology and dating tools, Sauer's method was both intriguing and ingenious. But it was not without its critics who argued that more was known about the first farmers than Sauer had alluded to and that he would have reached different conclusions had he checked his propositions against the then state of the art. Others argued that even if deductive logic was to be applied, Sauer's six propositions were inadequate, poorly formulated, and partial and from this point of departure he was destined to err.

Using updated evidence, US agronomist J.R. Harlan (1971) offered a centers and non-centers theory of the origins of agriculture and its dispersal. Centers were locations where agriculture first developed, while non-centers were secondary hearths whose existence relied upon the displacement of innovations from primary locations. For Harlan, distances between centers and non-centers could be as much as 5000 to 10 000 km.

Harlan identified three centers where agriculture first developed: a Near East center with an African non-center, a Northern China center with a Southeast Asia and South Pacific non-center, and a Mesoamerican center with a South American non-center. According to Harlan, centers and non-centers interacted with one another, and it was not necessarily the case that all crops or agricultural innovations were birthed in centers. Through time, non-centers led some innovations and transferred insights back to the center.

Australian archeologist Peter Bellwood (2004) has used the origins and diffusion of languages to better map the complex routes that agricultural dispersal took from the first agricultural hearths (see Map 3.4). According to Bellwood, agriculture began around 11000 YBP in the Near East Fertile Crescent of Mesopotamia (Plate 3.1); it appeared in the Yellow and Yangtze River basins around 9000 YBP; and between 9000 and 7000 YBP it surfaced in Southeast Asia, reaching as far as the New Guinean highlands. Around 6000 to 5000 YBP, agriculture was practiced in West Africa, Central Mexico, and the Peruvian Andes, and between 5000 and 3000 YBP it was first introduced into Eastern North America. Bellwood claims that farming spread from these hearths primarily because farmers migrated and took their knowledge with them to new locations. But how might one reconstruct geographies of migration that unfolded so many years ago? Bellwood's innovation was to note that migration patterns can be tracked by following the geographical diffusion of languages. He used the origins and dispersals of major language families such as Indo-European, Austronesian, Sino-Tibetan, Niger-Congo, and Uto-Aztecan to propose the pathways through which agriculture spread. Bellwood also made use of recent advances in genetic science to support his view that language diffusion was driven by migration and that the geographical dispersal of languages provides a useful proxy for the geographical dispersal of agriculture.

Why humans discovered agriculture at all – and why they did so in such a small number of places, apparently independently and relatively simultaneously – remains a hotly disputed topic. Initially, an "oases theory" was proposed, which held that the earth had become warmer and drier during this period and that, forced to live in a small number of hospitable places and in close proximity to wild animals, human beings were stimulated and motivated to bring nature under control simply to survive amidst other dangerous carnivores. But more recent evidence suggests that the earth became a more habitable rather than less habitable place in the years leading up to the Neolithic Revolution. Indeed, according to some commentators, the stabilization throughout the world of a more conducive climate was decisive; if the period from 22000 to 13500 YBP was marked by global cooling, a frozen planet, glacial advance, and erratic swings in climate, the Holocene period (from 12000 YBP to the present day) has been characterized by warmer, wetter, and more stable conditions, especially in temperate and tropical regions. But climate only enabled human beings to discover agriculture; it did not trigger this discovery. And given that the climate of many areas improved, it remains a puzzle as to why settled agriculture was pioneered in such a few select locations.

For some scholars, human progress alone is a sufficient explanation; agriculture is more technologically advanced than foraging, and it was inevitable that as humans evolved, they would achieve increased control over their environment. But human beings dwelled on planet earth for many tens of thousands of years before the advent of agriculture, and it is unclear why they did not achieve the breakthrough much earlier. Moreover, there is little reason to believe that humans who lived in the first agricultural hearths were any more intellectually, biologically, and genetically "evolved"

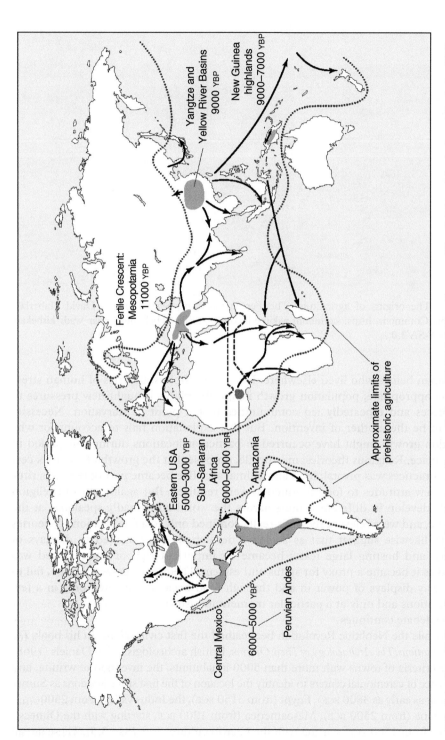

Map 3.4   Peter Bellwood's map of the first farmers. Source: Bellwood P (2004) First Farmers: The Origins of Agricultural Societies (Wiley Blackwell, Oxford). © 2004, John Wiley & Sons.

Plate 3.1 The origins of agriculture: The Fertile Crescent, Syria. Source: David Haberlah, Wikimedia Commons, https://commons.wikimedia.org/wiki/File:Gebel_musa_near_wadi_kirbekan. jpg. CC BY-SA 3.0..

than human beings who lived elsewhere. For other scholars, theories of human stress are more appropriate; population growth in specific places brought new pressures to those places and repeatedly led communities to the point of starvation. Necessity proved to be the mother of invention. But this explanation fails to account for why population growth might have occurred in particular locations during this period in the first place. Religious theories, meanwhile, propose that the growth of complex ceremonial practices was pivotal; food and animal sacrifice became part of religious rituals, and new attitudes to food production were required. But again, why did religion suddenly develop in different centers across the world all (broadly speaking) at the same time, and were relevant religions really confined only to a few locations? Theories of power likewise suggest that as societies developed, elites aspired to displays of authority, and hosting large feasts became an important sign of power. Food was required as it became a proxy for status and esteem. Theories of power, though, fail to explain why displays of power in and through food became venerated only in a few select locations and only at a particular moment in history.

So the debate continues.

Meanwhile the Neolithic Revolution bequeathed the first civilizations. In his book *The First Civilisations: The Archaeology of Their Origins*, British archeologist Glyn Daniels (1968) used the criteria of towns with more than 5000 inhabitants, the invention of writing, and the existence of ceremonial centers to identify the location of the first six civilizations as Sumer (beginning as early as 3800 BCE), Egypt (from 3150 BCE), the Indus Valley (from 2500 BCE), Shang China (from 2500 BCE), Mesoamerica (from 1200 BCE, starting with the Olmecs), and Peru (from 1500 BCE, starting with the Chavín culture) (see Map 3.5). These are all regions of the world that enjoyed a head start in the Neolithic Revolution; they grew as civilizations because food supplies permitted them to aspire to more than mere survival.

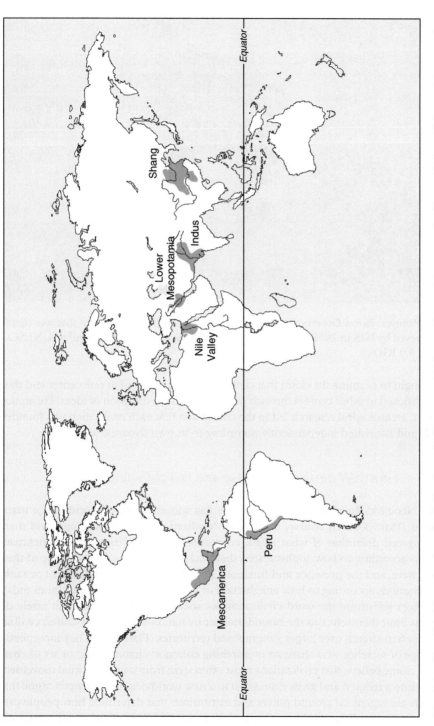

Map 3.5 Glyn Daniels – the first six civilizations, 3000 BCE to 1200 BCE: Sumer Mesopotamia, Egypt, Indus, Chinese, Mesoamerica (Aztecs, Maya), and Peru (Inca). Source: Daniels G (1968) The First Civilisations: The Archaeology of their Origins (Pelican, London).

Plate 3.2   Palmyra, Homs Governorate, Syria. A UNESCO World Heritage site that was significantly destroyed by ISIS in 2015–2016. It is now undergoing reconstruction. Source: UNESCO. CC BY-SA 3.0 IGO.

Daniels sought to examine the claim that civilizations first emerged in one center and then became trafficked to other centers through migration and the diffusion of ideas. He argued that, in fact, archeological research led to the conclusion that each civilization was founded separately and flourished independently according to its own dynamic and tempo.

Fourth Watershed: The Rise and Fall of Civilizations

Since the Neolithic Revolution, human history has witnessed the rise and fall of many civilizations (Plate 3.4). Of course, the idea of "civilization" is a slippery one, and there exists no agreed definition of what might constitute one. Scholars tend to benchmark civilizations according to how sophisticated their technologies were, how developed their languages were, and the presence and functions of their cities. This arrangement or ranking of civilizations according to how enlightened or backwards they were continues today. Some authors substitute the word civilization for society, and vice versa. But rarely do civilizations limit themselves to the boundaries set by nation-states or city-states; civilizations more often stretch over larger geographical territories (Plate 3.3). They incorporate a wide range of societies who share an overarching common characteristic or set of characteristics. Some believe that civilizations most often stem from a new spiritual movement that forms into a religion and gives expression to a new worldview. Others again argue that civilizations are organized around pillars and institutions that determine how people prefer to be governed (politics) and how they organize work (division of labor and economy).

Perhaps the most ambitious and most influential study of the rise and fall of civilizations ever to have been produced was that undertaken by British historian Arnold J.

Plate 3.3 Teotihuacan, Mexico (in 450 CE, perhaps the largest city in the world). Source: Angelo Hornak / Alamy Stock Photo.

Toynbee. Published in 12 volumes between 1934 and 1961, Toynbee's *A Study of History* sought to chart the rise and fall of all the principal civilizations ever to have visited the human stage. Toynbee's study of human history led him to identify the existence of 19 past civilizations, most of them related as parent or offspring to one or more of the others: the Western, the Orthodox, the Iranic, the Arabic, the Hindu, the Far Eastern, the Hellenic, the Syriac, the Indic, the Sinic, the Minoan, the Sumeric, the Hittite, the Babylonic, the Egyptaic, the Andean, the Mexic, the Yucatec, and the Mayan civilizations. Later he was to suggest a unification of the Iranic and the Arabic into a united Islamic civilization, and a division of Orthodox Christian civilization into Orthodox-Byzantine and Orthodox-Russian civilizations. In addition, he identified five additional "arrested civilizations": Polynesian, Eskimo, Nomadic, Ottoman, and Spartan. Toynbee argued that already by 1940, 16 civilizations had disappeared, leaving only five surviving: Western Christian, Orthodox Christian, Islamic, Hindu, and Far Eastern civilization.

More recently, in 2011 the US National Geographic Society attempted to map important empires in world history in its book *Great Empires: An Illustrated Atlas* (National Geographic 2012). This *Atlas* divided world empires into three categories, depending upon the era when they reigned: the Ancient World, the Middle Ages, and Modern Empires. While arguing that it is possible to identify as many as 150 great empires in world history, National Geographic focused upon only 31 empires that it considered to be especially noteworthy. The Ancient World (2600 BCE–500 CE) incorporated empires in Greece, Carthage, Rome, India, and China. The Middle Ages (500–1500 CE), meanwhile, included empires in Byzantium and the Arab World; medieval Asia and the Khmer, Mongols, and Ming; the Christian and Muslim dynasties of medieval Europe and Africa; and the Native American empires of the Toltec, Aztec, and Inca. Finally, Modern Empires (1500–1900 CE) focused on the Spanish-American and British Empires and the Imperial Tribes of North America; the Ottoman and Asante Empires in the Mediterranean and Africa; and the Mughals, Qing, and Romanovs in the East.

Why Do Civilizations Rise and Fall?

Environmental explanations emphasize the continuing importance of the natural environment in the sifting and sorting of winning and losing civilizations. Civilizations that dwell in more hospitable environments over the long run prosper when compared with counterparts that inhabit more extreme environments and/or environments of poorer quality and with more hazards and risks. Moreover, in his book *Collapse: How Societies Chose to Fail or Survive,* US evolutionary biologist and geographer Jared Diamond (2005) calls for more attention to be given to the role of environmental factors in the collapse of societies.

- First, societies can fail because they have destroyed or lost vital natural resources; natural habitats have been raided, wild food sources have been depleted, biodiversity has been reduced, and soils have been leached, bleached, and eroded.
- Second, societies can collapse because they come up against a number of environmental ceilings that prove difficult to raise; energy resources are mined to scarcity, demand for water resources exceeds supply, and exploitation of the earth's photosynthetic capacity approaches its peak.
- Third, societies can decline because human beings have polluted the environment through the mismanagement of toxic chemicals, the release of atmospheric gases, and the haphazard introduction of alien species into ecosystems.
- Finally, societies can falter because population growth and expanded consumption place unsustainable pressure on the available resource base.

Diamond illustrates the importance of these environmental threats in the collapse of past societies, including that on Easter Island, the Polynesians of Pitcairn and Henderson Islands, the Anasazi in North America, the Mayans in Central America, and the Greenland Norse. He also applies his framework to examine the role of environmental threats in present-day Rwanda, Haiti, China, and Australia.

It is often assumed, nevertheless, that as human beings learned to master planet earth, so too they loosened the shackles imposed upon them by nature and lessened the importance of the environment as a determinant of the location and success of civilizations. Some scholars seek to explain the rise and fall of civilizations principally in terms of the *social* organization of these civilizations. Those civilizations that organize themselves effectively take off, while those that do not falter, irrespective of the natural environment in which they dwell.

Toynbee's own theory of the rise and fall of civilizations provides a good example. This theory is predicated upon Toynbee's concept of the "creative minority." Civilizations emerge when a community is forced to solve a problem that threatens its existence but does not overwhelm it. This problem is solved by a creative minority within the community, who establish a pattern of life that is then copied by the majority and thereby comes to structure the community. There are no inherent reasons why civilizations are predestined to fail. Provided the creative minority continues to function, the civilization will continue to exist. Creative minorities, however, often falter and, instead of leading communities, become parasites. Recognizing that they are being exploited, the majority – what Toynbee calls the "internal proletariat" – stop copying the minority. Meanwhile, the "external proletariat," populations who lie beyond the civilization but who are also exploited by the creative minority, sense weakness and they revolt. Creative minorities can impose their authority by force through what Toynbee calls a "universal state": a body governing forcibly across an area. But the fate of the civilization is by that stage sealed. For Toynbee, civilizations commit suicide and are rarely murdered.

Since Toynbee, a number of other important works have followed suit in arguing that it is internal social advantages and flaws within civilizations themselves that lead to their rise and collapse. US anthropologist and historian Joseph Tainter's (1988) book *The Collapse of Complex Societies* argues that societies fail largely because they are unable to sustain the complex bureaucracies they invent to govern themselves. According to Tainter, as societies address and overcome the problems they encounter, they creep incrementally toward greater levels of complexity. To sustain this complexity, they require "energy subsidies," defined as resources procured from elsewhere in the system, which generate the capacity necessary to diagnose and address new problems (learning to be more efficient and/or unearthing new resources). Because energy subsidies are usually limited and finite, societies eventually struggle to create sufficient capacity to sustain their complex administrations; fresh problems then go unaddressed. As these problems accumulate, they start to attack the key institutions that are keeping societies afloat, and collapse results. Tainter notes that, although it might seem counterintuitive, often the best solution available to societies under stress is to simplify. Tainter applies his argument to the collapse of the Western Roman Empire, the Maya civilization, and the Chaco culture.

Meanwhile, US evolutionary anthropologist Peter Turchin has pioneered an approach to the study of the rise and fall of states that he has labeled *Cliodynamics* (from *Clio*, the muse of history, and *dynamics*, the study of why things change with time). Believing there to be a logic to long-term social and political change, Cliodynamics attempts to model long-term historical processes using mathematical and statistical equations and large-scale data sets. In *Historical Dynamics: Why States Rise and Fall*, Turchin (2003) offers what he terms a "metaethnic frontier theory" of societal growth and collapse. Turchin hypothesizes that the development of political and social cohesion within any given political community (what he calls the "Asabiyyah" of ethnic groups) is pivotal in determining whether that community develops into a leading civilization. Asabiyyah does not develop in a vacuum but is the product of a prolonged period of conflict and encounter between different ethnic groups. Successful empire-builders are formed in a cauldron of ethnic rivalry, but as they become dominant this rivalry becomes one-sided. Paradoxically however, now in the ascendancy, solidarity and mutuality in the group wanes, and eventually the capacity for collective action diminishes. These processes are mediated by population pressures too; population growth creates scarcities and antagonisms, and sabotages collective cohesion. As empires lose their capacity for political unity, their ability to act on the world is reduced.

Perhaps it is best to conclude, then, that civilizations collapse because of the complex relationships that exist between human beings, societal structures, and natural environments, or what some have termed the sustainability and resilience of the coupled human and natural systems that emerge in specific places and specific times. In their review of 12 societies under stress, Butzer and Endfield (2012, p. 3628) conclude: "Societal collapse represents transformation at a large social or spatial scale, with long-term impact on combinations of interdependent variables: (i) environmental change and resilience; (ii) demography or settlement; (iii) socioeconomic patterns; (iv) political or societal structures; and (v) ideology or cultural memory."

Fifth Watershed: The Rise of Western Civilization from the Tenth Century BCE?

Jared Diamond's 1997 book *Guns, Germs and Steel: A Short History of Everybody for the Last 13 000 Years* – hereinafter GGS – was an instant classic on publication. But the book has provoked the ire of many.

Diamond (1997) offers an environmental history explanation for the rise of Europe as the dominant imperial power of the past 500 years. His core claim is that while the West is often thought to have begun its climb to the summit of world history from the fifteenth century CE, in fact its deepest origins lie in the Neolithic Revolution and the head start from which Eurasia (the supercontinent combining the landmasses of Europe, the Middle East, North Africa, and Asia) benefited at the time of the first farmers. For Diamond, the West has been thousands of years in the making, and it is not inaccurate to say that the ascendance of the West can be traced as far back as even the tenth century BCE.

Diamond begins with the question: why was it that Europe and not other continents broke from the pack and became a global superpower? Why was it that Europe colonized Africa, Asia, Australasia, and the Americas, and not the other way around? Immediately, he rules out any suggestion that Europe succeeded because white Europeans were an innately superior race. He then turns to sociological explanations for the rise of the West. Certainly, Europe witnessed profound social, political, cultural, and economic change in the fifteenth century CE and gained advantages over the rest. Diamond acknowledges the role of these factors but argues they are at best proximate determinants of Europe's success. They still raise the question: why did these factors coalesce in Europe and not elsewhere? What stimulated and enabled social, political, cultural, and economic change? A deeper explanation is required. Diamond centers his attention on the rich and fertile bounty that the natural environment afforded Eurasia at the start of the Neolithic Revolution. The rise of Europe as a global superpower has deep origins in the environmental history of Eurasia and can be traced ultimately to Eurasians' uniquely successful encounter with the origins of agriculture.

According to Diamond, in the year 10 000 BCE there was very little difference in the development of different continents; hunter-gatherer societies dominated all regions. For this reason, Diamond refers to all history prior to 10 000 BCE as merely "a limp to the starting line." As we have seen, however, the Neolithic Revolution brought food production first to some areas and then only later to others, and flourished more fully in some locations than in others. The uneven geography of the Neolithic Revolution created a pattern of differential development and progress. Through time this pattern consolidated itself, and the gap between the more advanced civilizations and the less advanced backwaters steadily widened. Eurasia, and in particular the Fertile Crescent of Mesopotamia, was among the first to discover food production. Its Neolithic Revolution was enhanced by the abundant and rich diversity of wild plants and animals it hosted that were capable of domestication. The Fertile Crescent domesticated local plants earlier and domesticated a greater variety of species of plant than any other continent; this region founded eight crops of global significance: cereals (emmer wheat, einkorn wheat, and barley), pulses (lentil, pea, chickpea, and bitter vetch), and the fiber crop flax. Globally, of the 148 "big wild herbivorous mammals" who present themselves as candidates for domestication, only 14 in the end have been domesticated. The only region to house a large number of these mammals was Eurasia; Eurasia, in fact, was home to 13 of the 14 domesticable species, including the principal five: sheep, goats, cows, pigs, and horses.

Eurasia's early and comparatively successful encounter with the Neolithic Revolution generated four developments that cumulatively tipped the balance in favor of Eurasians, in particular European Eurasians: microbes, writing, weapons, and centralized political organization.

First, the large quantity of domesticated mammals in Eurasia proved an effective breeding ground for new infectious disease. European peoples developed immunity to these microbes over time. But as part of their mission to colonize large parts of Latin America, Asia, and Africa, European explorers and colonizers brought with them to far-flung colonies microbes and pathogens to which indigenous natives had no prior exposure (Plate 3.4). The infectious diseases that arrived with European colonists and imperialists were certainly more virulent than those that prevailed in the colonies. And so this lethal freight resulted in massacre. Germs became a more effective weapon than guns in subduing the indigenous population.

Second, freed from the immediate demands of survival, societies with food surpluses began to invent alphabets and writing systems. Writing furnished these societies with a sophisticated means of recording, storing, and transmitting knowledge, both from distant lands and from ancient wisdom. Writing allowed complex colonial endeavors to be administered and coordinated.

Plate 3.4 Spanish Conquistadors like Hernan Cortes de Monroy y Pizarro Altamirano brought diseases like smallpox to Mexico, which played a significant role in the demise of the Aztec Empire. Source: https://everipedia.org/wiki/lang_en/Smallpox.

Third, freed from the land, populations living in regions of food surplus could now develop complex divisions of labor and new skills. This facilitated the development of new technologies, including military technologies, which provided Europeans with a decisive edge in warfare.

Finally, agricultural surpluses facilitated the development of larger and more complex societies and more sophisticated and capable centralized political institutions. This provided Europeans with exceptional command-and-control capability.

For Diamond, then, it was the Neolithic Revolution that put Eurasians on a long trajectory that, in many ways, was destined to end in Europe's global supremacy.

But why did Western Europe subsequently outperform the rest of Eurasia? Why did the Fertile Crescent not go on to dominate the world? Iraq and Syria failed to benefit from their head start because they overexploited the environment and rendered it barren and exhausted. Critically, the diffusion of agriculture was also rapid throughout Eurasia on account of its west–east axis; since regions in this continental block shared similar latitudes and were not separated by any formidable physical barriers, diffusion was easier than in, say, the Americas or Africa (where greater north–south variations in physical environments constrained diffusion). Although agriculture first developed in the Fertile Crescent of Mesopotamia, it quickly spread to Western Europe – through the influence of Greek civilization and then Roman civilization. After its arrival, and over time, it combined with a number of additional comparative advantages to render Western Europe the most promising of all of Eurasia's regions. Of these other advantages, perhaps the most important was Europe's political fragmentation. In contrast to other regions – and in particular China, where political authority was controlled by geographically expansive and monolithic all-powerful political dynasties – Europe was characterized by a number of comparatively weak and warring monarchies. In these conditions, competition between political and economic organizations for access to and control over scarce resources prospered and spurred the development of capitalism.

Deep Dive Box 3.4 Critical Encounters with Jared Diamond's *Guns, Germs and Steel (GGS)*

Many admire the originality and the breadth of knowledge displayed in Diamond's environmental history explanation for the rise of the West. GGS is a work of tremendous confidence and reach; it could only have been written by a profoundly knowledgeable polymath. But the book has attracted a lot of criticism, at times quite hostile criticism.

1 GGS presents itself as predicated upon scientific arguments – human history can be treated as a science. But some question just how scientific the book is and whether the big sweeping story Diamond proposes is in fact little more than the speculation of an imaginative scholar.

2 GGS ignores the fact that agriculture developed in many other centers and gives undue significance to the Fertile Crescent in Mesopotamia as the point of origin of the rise of the West. There is evidence to suggest that agriculture was practiced almost as early in other places (in China and tropical New Guinea, for example) as it was in Mesopotamia. But Diamond suggests that the Neolithic Revolution was less significant in these regions because they

domesticated less nutritious crops (root crops, maize, and rice), which were inferior to the wheat and barley found in the Fertile Crescent. But does this claim underestimate the quality of domesticates in other centers of agricultural innovation? These were areas of food surplus and abundance. Were they not just as likely to have spawned a world-domineering civilization; and if they were, why did they not do so?

3 GGS overlooks the significance of transmissions of agricultural innovations between north and south in all continents, and overstates the primacy and ease of the diffusion of innovations along east-to-west axis. It overplays the importance of the shape of continents (Eurasia, west to east; Africa and North America, north to south). It also fails to account for why the innovations pioneered in the Fertile Crescent spread to the west (to Europe) more effectively than they did to the east. Diamond underestimates the extent to which it was possible to adapt crops to new climatic conditions and therefore to spread them from north to south (he ignores, for example, the fact that maize was cultivated in both Peru and North America). Moreover, even west-to-east transmission is problematic, as there are different climates by longitude even in regions that share the same latitude.

4 GGS, at least according to some, creates the impression that Europeans were "inadvertent colonizers," destined by history to colonize other peoples. But just because one has guns, germs, and steel does not mean one has to use them toward certain ends. The pathway to colonization was not "locked in" with the onset of the Neolithic Revolution; this would be to let the colonizers off the hook. The social, political, and economic "proximate" factors that Diamond talks about are often more fundamental than the factors he considers to be fundamental. Some scholars prefer to explain the rise of the West in terms of more recent social, political, cultural, and economic factors, and refuse to see these factors as merely piggybacking on favorable environmental conditions inherited from the past. Diamond is said to downplay to too great an extent such factors as Western culture, the transformation from feudalism to capitalism in Europe, and the rise of nation-states and the rapacious global empires these states built.

5 Most significantly, GGS has been criticized for placing too much emphasis upon environmental factors in the deep past. Diamond, it is said, practices a kind of geographical determinism – the environment is the ultimate cause of the rise of superior civilizations. Of course, Diamond steers clear of the sins of environmental determinism – the racist claim that hospitable temperate climates bred a superior species of humanity who colonized the world to civilize more barbarous peoples. But he does suggest the environment is critical in furnishing some peoples with a decisive advantage. In the light of the uses and abuses of environmental determinism in human geography's past, even this lighter proposition is a potentially dangerous one to make.

Conclusion

Big history takes seriously the long and incredible human journey – that has made possible the human geographies we witness today. *Homo sapiens* secured their status in the natural order only following a long process of evolution through natural selection. Recent developments in population genetics and migration suggest that all modern humans can be traced to a common ancestor in Central and East Africa. From this location, modern humans began the long process of migration, and over the past 100 000 years they steadily colonized first the African continent and then the entire planet. The timing and scale, waxing and waning of population migration into and out of certain world regions were affected by climate change; at different times it became more possible to breach natural barriers, while at other times the environment proved too inhospitable for certain territories to be navigated through or inhabited. They survived as hunter-gatherers for much of this time, and significant developments in human culture and civilization occurred only over the past 12 000 years, beginning with the Neolithic Revolution and featuring the rise and fall of different world civilizations.

When telling the incredible story of how human beings gradually made planet earth their home, it is tempting to marvel at the many achievements of the human race. In an important sense, human history is the history of how one species emerged from, struggled with, secured victories over, and now to a certain extent lives free from limitations imposed by the natural environment. Although emerging from the animal world as a dominant species only recently, and for much of its history ravaged by natural hazards, the human race has steadily imposed itself on planet earth and liberated itself from nature's most binding constraints. With the emergence of the West, and the dazzling new technologies now at our disposal, humanity's triumph over nature is today complete.

Certainly, for much of its history the human race merely eked out an often-precarious existence by foraging for fruits and nuts in a very limited number of more hospitable environments. But by innovating new technologies, human beings worked to exert control over the natural environment. They have learned to exploit nature's abundant resources to the full. They have incubated themselves from many of nature's harshest extremes. As the human race has become less preoccupied with mere survival, human culture has prospered, and many important and exalted civilizations have emerged. However, while such an account speaks to an essential truth, it also risks encouraging an exaggerated sense of the capacities of the human species. We are vastly more vulnerable than we suppose. Accordingly, the extent to which this story might be said to be anthropocentric will be the subject of discussion in later chapters in the book.

Checklist of Key Ideas

- "big history" sets human history into much longer natural and earth history and provides humans with a more humbling understanding of their location in time and place.
- The human species emerged and secured its status in the natural order only following a long process (over at least 3.6 billion years) of evolution through natural selection. Charles Robert Darwin was the first to understand this process. Recent developments in population genetics suggest that all modern humans first appeared some 200 000 years ago and can be traced to a common ancestor in Central and East Africa.
- From this location, modern humans began the long process of migration around 100 000 BCE, and over the next 90 000 years learned how to traverse and dwell in all

of the world's environments, steadily colonizing the entire planet. By 10 000 BCE, human beings inhabited all the main regions of the world.

- After surviving as hunter-gatherers for much of this time, significant developments in human culture and civilization occurred only over the past 12 000 years. The Neolithic Revolution (the domestication of plants and animals) marked a key moment in people's capacities to exert control over the physical environment. Why the Neolithic Revolution began when it did and where it did remains the subject of debate.
- In the past 6000 years, human history has been marked by the rise and fall of different civilizations. British historian Arnold J. Toynbee has identified as many as 19 historically significant past civilizations. There exist both environmental and social explanations of the rise and fall of civilizations. The concept of the sustainability and resilience of coupled natural and human systems provides a useful way to study civilizational collapse.
- Jared Diamond proposes that the rise of Western civilization has deep roots in the environmental history of Eurasia, and in particular that supercontinent's bounty-filled Neolithic Revolution. According to Diamond, the rise of the West can be dated not to the fifteenth century CE but to as early as the tenth century BCE.

Chapter Essay Questions

1) Provide a synopsis of the origins and dispersal of the human species across the planet.
2) The first agricultural hearths and early civilizations appeared at a specific moment in human history and only in a selective number of locations. Discuss.
3) Describe and comment upon *natural environmental,* and *societal* explanations for the rise and fall of civilizations.
4) Outline and comment on Jared Diamond's claim that the emergence of the West as a leading world civilization, which is commonly dated to the fifteenth century CE, in fact has much deeper roots in the favorable natural environmental conditions enjoyed by Eurasia at the time of the Neolithic Revolution.

References and Guidance for Further Reading

The seminal article that announced the birth of big history is:
Christian, D. (1991). The case for big history. *Journal of World History* 2 (2): 223–238.

Popular books explaining and applying big history frameworks include:
Ansary, T. (2019). *The Invention of Yesterday: A 50,000 Year History of Human Culture, Conflict and Connection.* London: Public Affairs.
Brown, C.S. (2007). *Big History: From the Big Bang to the Present.* London: New Press.
Christian, D. (2011). *Maps of Time: An Introduction to Big History.* Berkeley: University of California Press.
Christian, D. (2019). *Origin Story: A Big History of Everything.* New York: Penguin.
Harari Yuval, N. (2014). *Sapiens: A Brief History of Humankind.* London: Random House.
Kolbert, E. (2014). *The Sixth Extinction: An Unnatural History.* Edinburgh: A&C Black.
Mann, C. (2005). *1491: New Revelations of the Americas Before Columbus.* London: Alfred A. Knopf.

Morris, I. (2010). *Why the West Rules – for Now: The Patterns of History and What They Reveal About the Future*. London: Profile.

Scott, J.C. (2017). *Against the Grain: A Deep History of the Earliest States*. New Haven, CT: Yale University Press.

Spier, F. (2015). *Big History and the Future of Humanity*. Oxford: Wiley.

Important statements on the theory of evolution through natural selection are provided in:

Darwin, C.R. (1859). *On the Origin of Species by Means of Natural Selection, or the Preservation of Favoured Races in the Struggle for Life*. London: John Murray.

Darwin, C.R. (1871). *The Descent of Man and Selection in Relation to Sex*. London: John Murray.

For excellent and accessible accounts of the origins and dispersal of the human species around the planet:

Higham, T. (2021) *The World Before Us: How Science is Revealing a New Story of Our Human Origins*. New York: Penguin.

Olson, S. (2002). *Mapping Human History: Genes, Race, and Our Common Origins*. New York: Houghton Mifflin Harcourt.

Oppenheimer, S. (2003a). *Out of Eden: The Peopling of the World*. London: Constable and Robinson.

Oppenheimer, S. (2003b). *The Real Eve: Modern Man's Journey out of Africa*. New York: Carroll & Graff.

Roberts, A. (2009). *The Incredible Human Journey*. London: Bloomsbury.

Roberts, A. (2011). *Evolution: The Human Story*. London: DK/Penguin.

Sykes, B. (2002). *The Seven Daughters of Eve: The Science That Reveals Our Genetic Ancestry*. New York: Norton.

Wells, S. (2006). *Deep Ancestry: Inside the Genographic Project*. Washington, DC: National Geographic Society.

Wells, S. (2017). *The Journey of Man: A Genetic Odyssey*. Princeton, NJ: Princeton University Press.

Good overviews of the state of the debate over the peopling of the Americas can be found in:

Dai, C.L., Vazifeh, M.M., Yeang, C.H. et al. (2020). Population histories of the United States revealed through fine-scale migration and haplotype analysis. *American Journal of Human Genetics* 106 (3): 371–388.

Moreno-Mayar, J., Víctor, L.V., de Barros Damgaard, P. et al. (2018). Early human dispersals within the Americas. *Science* 362 (6419): eaav2621.

Raghavan, M., Steinrücken, M., Harris, K. et al. (2015). Genomic evidence for the Pleistocene and recent population history of Native Americans. *Science* 349 (6250): aab3884–aab3881.

The emergence of complex societies from hunter-gatherer societies is introduced in:

Morgan, L.H. (1877). *Ancient Society; or, Researches in the Lines of Human Progress from Savagery Through Barbarism to Civilization*. London: MacMillan & Company.

Carl Sauer's deductive logical approach to the geography of the birth of agriculture can be found at:

Sauer, C. (1952). *Agricultural Origins and Dispersal*. New York: The American Geographical Society.

An excellent account of the history and geography of the first farmers can be found in:

Bellwood, P. (2004). *First Farmers: The Origins of Agricultural Societies*. Oxford: Wiley Blackwell.

Harlan, J.R. (1971). Agricultural origins: centres and noncentres. *Science* 174: 468–474.

Important studies charting the rise and fall of civilizations can be found in:

Butzer, K.W. and Endfield, G.H. (2012). Critical perspectives on historical collapse. *Proceedings of the National Academy of Sciences* 109 (10): 3628–3631.

Daniels, G. (1968). *The First Civilisations: The Archaeology of Their Origins*. London: Pelican.

Diamond, J. (2005). *Collapse: How Societies Chose to Fail or Survive*. New York: Penguin.

Diamond, J. (2019). *Upheaval: Turning Points for Nations in Crisis*. New York: Little, Brown.

National Geographic (2012). *Great Empires: An Illustrated Atlas*. Washington, DC: National Geographic Society.

Tainter, J.A. (1988). *The Collapse of Complex Societies*. Cambridge: Cambridge University Press.

Toynbee, A. (1934–1961). *A Study of History*, vol. 12. Oxford: Oxford University Press.

Turchin, P. (2003). *Historical Dynamics: Why States Rise and Fall*. Princeton, NJ: Princeton University Press.

The full citation for Jared Diamond's most prominent book is:

Diamond, J. (1997). *Guns, Germs and Steel: A Short History of Everybody for the Last 13,000 Years*. New York: Norton.

An account of the Spanish impact on Latin America, including the smallpox pandemic and its impacts on the Aztecs, can be found in:

Mann, C. (2011). *1493: Uncovering the New World Columbus Created*. New York: Vintage.

Website Support Material

A range of useful resources to support your reading of this chapter are available from the Wiley *Human Geography: An Essential Introduction* Companion Site http://www.wiley.com/go/boyle.

Chapter 4

The Commanding Heights: A Brief History of the European World Capitalist Economy from 1450

Chapter Table of Contents

Chapter Learning Objectives

By the end of this chapter you should be able to:
- Define and explain the significance in the field of economic geography of the concepts of Global Commodity Chains (GCCs), Global Value Chains (GVCs), and Global Production Networks (GPNs).
- Outline the tenets of Wallerstein's world-systems analysis, and define and explain the rise from 1450 of the European world economy and the Old International Division of Labor (OIDL).

Human Geography: An Essential Introduction, Second Edition. Mark Boyle.
© 2021 John Wiley & Sons Ltd. Published 2021 by John Wiley & Sons Ltd.
Companion website: www.wiley.com/go/boyle

- With reference to Parisian regulation theory, explain what is meant by the concepts of "regime of accumulation" and "mode of regulation," and why these concepts help us to make sense of capitalist economic growth.
- Identify the conditions that gave rise to the thirty glory years of capitalism (*Les Trente Glorieuses*, 1945–1975), and explain why core countries in the world economy were plunged into crisis in the 1970s.
- Define the New International Division of Labor (NIDL), and explain why it provided a solution to the crisis in profitability that firms encountered in the 1970s.
- Outline the tenets of the neoliberal economic model, and explain why this model has shown itself to be especially vulnerable to cycles of boom and bust.
- Identify and comment on the geographies of post-Fordist production systems.
- Comment on the claim that we exist on the precipice of a Fourth industrial revolution (IR4) based upon artificial intelligence and data-driven economies – and speculate upon the potential economic geographies this revolution might create by 2050.

Introduction

The rise and reign of the West from the fifteenth century have left in their wake a grossly unequal and socially differentiated world. For nearly five centuries, countries in the Global North have industrialized, developed, and accumulated vast riches. For much of the same period, countries in the Global South have been left to languish in poverty and underdevelopment.

While our unequal world has complex roots, undoubtedly its etiology can be traced in part to the rise of the European-led world capitalist economy from the fifteenth century and the mutation and evolution of this economy over 500 years. The purpose of this chapter will be to reveal the ways in which the emergence of European capitalism conspired to create uneven geographical development across the surface of the earth and an Old International Division of Labor (OIDL). Since 1979, Western core economies have sought to assert their prowess in the face of an emerging battle for control over the world economy. In a bid to retain world economic leadership, they have increasingly turned to neoliberal economic orthodoxies. But the neoliberal model of capitalist economic growth has shown itself vulnerable to boom-and-bust cycles and today itself is in crisis. The hegemony of the European world economy has never looked less secure.

In this chapter, we will encounter the story of the rise of the European world economy from 1450, and will examine the travails of the twentieth- and early twenty-first-century political-economic models that have sought to sustain the primacy of Western core economies as the sun has set on European empires.

Key Concepts: Global Commodity Chains, Value Chains, and Production Networks

British geographer Peter Dicken's classic book *Global Shift: Industrial Change in a Turbulent World* was first published in 1986; in 2015 Dicken published the seventh edition, now entitled *Global Shift: Mapping the Changing Contours of the World Economy*.

Global Shift provides readers with an introduction to key concepts within the field of economic geography. As successive editions were published, Dicken increasingly engaged the ideas of the global commodity chain (GCC) and the global value chain (GVC), and most recently the concept of the global production network (GPN).

The ideas of the GCC and the GVC are used to capture the stages that need to be passed through if a product is to be made available to a customer, from the extraction of raw materials from the earth to the manufacturing of the good and its transportation to the market. At each stage in this sequence or chain, more value is added to the product, building to its final price at the point of sale. Stages that add significant value tend to rely on high technology and are referred to as high-value-added. Stages that are necessary but add little value are normally labor-intensive, use little technology, and are referred to as low-value-added. We might say, then, that GCCs refer to the integrated bundle of activities (from sourcing raw materials to processing those materials to producing products and to the delivery of those products to final markets) that occur when producing commodities, while GVCs refer to the extent of the value that is added by companies at each stage. When companies couple with or acquire other industries upstream or downstream of their own level on the value chain, this might be referred to as vertical integration. When industries or companies couple with or acquire other industries on the same level of the value chain as themselves, this might be referred to as horizontal integration.

Economic geographers find these ideas useful when thinking of the tangled webs that connect continents, countries, regions, and cities to the world economy. For regions to be incorporated into the global economic system, they need to contribute to the aggregate product that is traded and consumed within this system. In this way, they can play a role in the overall division of labor. Because different regions specialize in different stages of the chain, it is possible to think of GCCs and GVCs as geographically expressed; there are *spatial* divisions of labor and now today *international* divisions of labor. Increasingly, it is transnational corporations (TNCs) that are allocating different functions to different regions of the globe and in consequence structuring the geographical manifestations of GCCs and GVCs and shaping the economic trajectory of places. Regions that perform high-value-added functions tend to be high technology and prosperous ones. Regions that perform low-value-added functions tend to be low technology in economic structure and comparatively underdeveloped. The concept of "strategic coupling" has been used to refer to the dynamic processes through which local actors work to "coordinate, mediate, and arbitrage" their entanglements with global economic actors for strategic gain and value capture (that is, position themselves more advantageously within the global division of labor).

While inspired by these concepts, Dicken, and more recently economic geographers based at the National University of Singapore Neil Coe and Henry Yeung, have developed the related concept of the GPN. The "chain" approaches are limited, it is argued, because they tend to view economic activity from the perspective of vertical and linear journeys of particular commodities from farm to fork. Economic processes are messier with chains being imbricated, through horizontal and vertical linkages, to a wider variety of supporting actors, infrastructures, institutions, and resources. The focus, then, is upon all the interconnected functions, operations, and transactions (social, political, cultural, institutional, and ecological) that are embroiled in the production, distribution, and consumption of a specific product or service. The idea of the GPN is now a central one in economic geography; economic geographers seek to understand the genesis of and map the configuration of particular GPNs. They are concerned with the ways in which the development strategies of cities and regions lead them to couple with, and uncouple from, these GPNs.

The Rise of the European World Economy, 1450–1945: Wallerstein's World-Systems Analysis

The geography of the world economy that presents today is substantially a product of the historical emergence from 1450 of what might be termed the European world economy.

In her book *Before European Hegemony: The World System A.D. 1250–1350*, US sociologist Janet Abu-Lughod argues that to understand the rise from 1450 of Europe as a global economic hegemon, it is necessary to understand the pre-modern world system that emerged in the thirteenth and early fourteenth centuries and that spanned Eurasia (from Northwest Europe to China). Starting with the French "Champagne trade fairs," and expanding to the Flanders markets (Bruges and Ghent) and then the "compradorial" (brokerage) city-states of Venice and Genoa, Europe began to trade with the Mongol Empire, Baghdad (Persian Gulf), Cairo (Mameluke Empire), India, the Malacca Strait, and Yuan China. In the complex regional trading networks that arose, Europe was but one of a number of economic hearths. From these trading links, there sprung nascent capitalist economic practices – entitled proto-capitalism. This system, however, decayed and withered with the arrival of the mid-fourteenth-century bubonic plague, a period of cold temperatures that reduced harvests, and constant wars between Venice and Genoa. Somehow, Europe rose to fill the vacuum.

US sociologist Immanuel Wallerstein has provided a cogent rendition of this world historical event, its causes, and its consequences (Wallerstein 1974, 1980, 1988). According to Wallerstein, there have existed three types of economic system in the history of the human species. *Mini systems* were the predominant economic systems in pre-agricultural societies. Small in scale and culturally homogeneous, these systems were self-contained and largely based upon the principle of subsistence. *World empires* were the predominant economic systems in the period from the earliest civilizations to the fifteenth century. They emerged when leading civilizations rose to govern over culturally and geographically heterogeneous societies. These civilizations worked to incorporate other societies into an overarching and coordinated division of labor. World empires ruled over large regions of the globe, the Roman Empire being the iconic example. Finally, a single *world economy* has dominated from the fifteenth century. World regions have been drawn into a single division of labor and a single economic system based upon a globally integrated network of production, circulation, and consumption.

Wallerstein's primary focus was upon the third economic system, the world economy. According to Wallerstein, around 1450 a fundamental shift occurred in the social structure of Europe. Feudalism, the dominant social structure prior to 1450, began to suffer from a number of debilitating failings and limitations and steadily gave way to a new capitalist order. This transition created a European capitalist system. But this system was inherently predatory and expansive. It needed to exploit the resources of other world regions to function. Steadily the European capitalist economy junked, swept away, appropriated, and absorbed into its fold other prior mini systems and world empires. A European-led global capitalist economy emerged. Europe was to be the primary beneficiary. Europe prospered only because other world regions submitted, by choice or through coercion and compulsion, to its directives. Our unequal world is at root an outcome of this history.

According to Wallerstein, the European-led global capitalist economy never developed into a world empire, but this political failure was to be an integral part of its successful march to the four corners of the earth. Certainly, European nation-states established global empires during this period, and these empires played a central role in the building of the European-led global capitalist economy. Nevertheless, each country presided over its own empire, and no single European-wide or pan-European imperial power emerged. The European-led world capitalist economy was never dominated by one single government, and Europe's many empires were birthed, grew, nested, matured, and perished inside this wider world economic system. Unencumbered by the crippling and expensive bureaucracies that often marked world empires, this lack of a single political center allowed the European capitalist system to remain nimble and flexible and to conquer the world more efficiently and rapidly.

For Wallerstein, the rise of the European-led global capitalist economy brought with it four types of global region, each playing a specific role in the global division of labor: core, peripheral, semi-peripheral, and external. This system has come to be labeled, the Old International Division of Labor (OIDL). It has been possible, though difficult, for some regions to move between these categories (see Deep Dive Box 4.1 and Map 4.1).

- Core regions, of which Europe (in particular France, England, and the Netherlands) was preeminent, enjoyed command-and-control authority over the world economy. Deploying their more advanced technologies, these regions transitioned first to more capitalist, intensive-agricultural economies, and then to industrial regions and manufacturing hearths. Core regions exploited the raw materials of peripheral areas and sold manufactured goods back to these areas, and through this unequal relationship accumulated wealth and riches.
- Subordinated and marginalized by the core, peripheral regions were marked by low-technology, labor-intensive agriculture, and primary and extractive industries. Their role was to supply the core and the semi-periphery with cheap raw materials and labor.
- Semi-peripheral regions lay between these two extremes and included former core regions that had fallen from prosperity (like Spain and Portugal) and former peripheral nations on the rise (such as China, India, Brazil, and Mexico). Semi-peripheral regions were often exploited by core regions but in turn themselves exploited truly peripheral regions.
- Finally, a number of external areas managed to steer clear of the European world capitalist economy (such as Russia) and continued to exist as separate world empires.

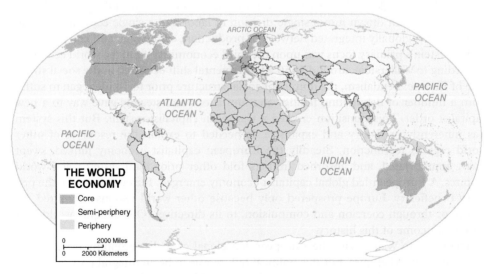

Map 4.1   Legacies of the Old International Division of Labor (OIDL): Stubborn geographies of the world economy. Source: Based on http://aphug.wikispaces.com/.

Deep Dive Box 4.1   The Old International Division of Labor: The Case of Rubber

The geographical processes at work in the production of rubber for use in the production of tires for use in the production of automobiles serve to illustrate the machinations of the OIDL.

In 2019, total global rubber consumption amounted to 28.4 million metric tons (mmt), 13.2 mmt of which was worldwide natural rubber consumption and 15.2 mmt of which synthetic rubber consumption. Worldwide, in 2020 global revenues from rubber production amounted to US$56 billion (Figure 4.1).

Historically, tires for use in the production of automobiles were manufactured using natural rubber tapped directly by cutting into the bark of rubber trees. Four stages were involved in the GPN. In Stage 1, natural rubber was extracted from rubber trees and, through treatment using water and acid, converted into milky-white latex (a low-value-added process). In Stage 2 (using low- to medium-value-added technology), this latex was then subjected to more sophisticated treatments, during which chemicals were added and heating and cooling processes applied to toughen up the latex. In Stage 3 (through medium- to high-value-added manufacturing), this fortified and treated latex then became molded into tires. In Stage 4 (very high-value-added stage), tires would be bolted to automobile wheels, and these automobiles would then be offered for sale to the final market.

Leading exporters of natural rubber by revenue (US$) 2019 or nearest year	Leading tire companies in the world by revenue (US$) 2019 or nearest year	Leading car producers in the world by revenue (US$) 2019 or nearest year
Thailand 4.6 billion	Michelin (France) 25.4 billion	Volkswagen (German) 280 billion
Indonesia 4 billion	Bridgestone Corporation (Japan) 27.2 billion	Toyota (Japan) 272 billion
Vietnam 984 million	Goodyear (United States) 15.4 billion	Daimler (Germany) 198 billion
Malaysia 936.5 million	Continental (Germany) 13.1 billion	Ford (USA) 160 billion
Ivory Coast 752.6 million	Sumitomo Rubber Industries (Japan) 6.8 billion	General Motors (USA) 147 billion
Belgium 202 million	Pirelli (Italy) 6.2 billion	Honda Motors (Japan) 143 billion
Myanmar (Burma) 183.8 million	Hankook Tire (South Korea) 6.1 billion	SAIC (China)135 billion
Laos 168.3 million	Zhongce Rubber Group Co. (China) 4.8 billion	Fiat Chrysler (Italy/USA/UK/Netherlands/USA)133 billion
Guatemala 152.7 million	Yokohama Rubber (Japan) 4.3 billion	BMW (Germany) 115 billion
Liberia: 126.2 million	Cheng Sin Rubber (Taiwan) 2.7 billion	Nissan Motor (Japan) 108.7 billion

Figure 4.1   Top 10 companies embroiled in the global rubber industry, by stage in the division of labor. Sources: Composed from: Association of Natural Rubber Producing Countries (ANRPC); Market Research Reports - Tyres Industry and Organisation Internationale des Constructeurs d'Automobiles (OICA).

(Continued)

Box 4.1   *(Continued)*

This division of labor was very clearly etched unevenly onto the face of the earth. The natural rubber used to produce tires was tapped almost exclusively from a tall softwood tree called *Hevea brasiliensis*. Indigenous to the Brazilian Amazon, this tree was replanted and cultivated by European colonists in the hot and humid tropical regions that they colonized – especially in Asia. For example, in 1876, British explorers and scientists appropriated rubber-tree seeds and experimented with them at the Botanical Gardens in London. Through trial and error, they pioneered more sophisticated rubber trees and sent seeds from these trees to colonies in Asia such as Malaysia, Ceylon (modern-day Sri Lanka), and Singapore. Densely packed rubber plantations proved immensely profitable.

To this day, Stage 1 and 2 processes – the extraction of latex from rubber trees – remain concentrated in Thailand, Indonesia, Malaysia, India, Vietnam, China, Sri Lanka, and Cambodia (Plate 4.1). These regions perform low- to medium-value-added activities, and they are peripheral and semi-peripheral regions in the global economy. Meanwhile, Stages 3 and 4, which require higher levels of technology, remain concentrated in the Global North. These regions are high-value-added regions and remain economic cores. Efforts have been made by some peripheral regions to attract investment from core regions or to build higher-value-added indigenous industry themselves (vertically integrate

Plate 4.1   Harvesting natural rubber from the Havea Brasiliensis tree in Vietnam. Source: Tatsiana Volskaya/Moment Open/Getty Images.

up the value chain). At most, this has helped some peripheral regions become middle-income, semi-peripheral regions in the global economy. Stage 4 (car manufacturing), however, remains principally concentrated in the Global North.

Following World War II, the production of synthetic rubber from petrochemical compounds (oil, shale, coal, and even natural gas) became more important. The acquisition by Japan of former European colonies in Asia brought shortages of rubber to the Allied Western forces. The United States was forced to pioneer new forms of synthetic rubber. Today synthetic rubber production remains a relatively high-value-added and high-technology activity, and the industry is centered in the United States, Europe, and Russia. The consequence is that efforts by peripheral and semi-peripheral countries to climb the value chain and to become higher-technology regions, making latex into usable rubber, making tires for global vehicle production, and even manufacturing automobiles themselves, are facing stiffer challenges. This example shows how, with its high-technology and innovative capacities, core regions can consolidate their dominance of global industries, frustrating the development of peripheral and semi-peripheral regions.

The implication of Wallerstein's work for the future of the West is clear. The degree to which the West will continue to be at the heart of the global capitalist economy depends upon its ongoing capacity to subordinate and appropriate resources from other world regions. That other regions are no longer prepared to accept the existing global economic order and are themselves seeking to change this order suggests that a fractious and uncertain future beckons.

The Fate of the European World Economy After the Age of Empire

Notwithstanding the loss of its remaining colonies throughout the twentieth century, the OIDL has proven stubborn and ingrained; indeed, up until the late 1960s and early 1970s, Western core economies continued to pull ahead of the rest. The economic growth of these countries was driven by new developments in industrial production processes and systems, and the emergence from 1945 of social partnerships that promoted a virtuous relationship between firms, workers, and governments. This was called the Fordist-Keynesian economic model. From the late 1970s, as Fordist-Keynesianism faltered, there emerged a neoliberal economic model that has generated growth of sorts (albeit less impressive growth than that recorded under Fordist-Keynesianism). But this model has shown itself to be crisis-prone and has failed to secure a social license. Today, neoliberalism is showing evidence of exhaustion. There is every reason to suppose that we reside at another hinge point in history, when neoliberalism will be replaced by a new social democratic economic model, much in the same way as it itself replaced the failing Fordist-Keynesian model in the mid-1970s. Whether the West can innovate itself out of trouble and continue to power the world economy remains to be seen: the

Plate 4.2   Three influential twentieth-century economists: from left to right Friedrich August von Hayek, John Maynard Keynes and Milton Friedman. Source: The University of Chicago and Wikipedia.

alternative is a "global shift" in the world economy and the rise of new economic powerhouses elsewhere.

Stabilizing Capitalism: Parisian Regulation Theory

In this section, we will draw upon what has been termed Parisian regulation theory to make sense of periods of growth and crisis within the global capitalist economy.

Published in 1976 as a book entitled *A Theory of Capitalist Regulation: The US Experience*, Michel Aglietta's monumental study of the trajectory of the US economy from the Civil War of the early 1860s to Jimmy Carter's election as US president in 1976 (he took office in 1977) has come to be viewed as the seminal text in the Parisian Regulation School (*l'école de la regulation*). Aglietta asks: given that capitalism has an innate tendency toward crisis, why has it survived over time and managed to endure? After all, if companies of all varieties are allowed to compete according to strict laissez-faire principles, anarchy and chaos would surely quickly engulf. What makes capitalism so resilient given its manifest capacity to go awry?

According to Aglietta, at any given point in time there exists a dominant "regime of accumulation": a particular structure, pattern, and order to the production and consumption of commodities. Left to their own devices, regimes of accumulation are unlikely to survive very long. What allows them to be sustained, at least for a period, is the existence of parallel "modes of regulation": formal and informal institutions that prop up given systems of production and consumption and afford them technical efficacy and social legitimacy. But contradictions in the system can only be suspended for a window of time. When regulation fails, the prevailing politico-economic model defaults and a crisis arises. Crisis could, of course, persist endlessly as the new norm; the alternative is that a new regime of accumulation and mode of regulation emerge from the struggle to "build back better," enabling capitalism to prosper for another epoch. The history of capitalism, then, can be construed as a cycle with periods of stability and growth punctuated by periods of crisis – or, more accurately, periods of crisis interspersed with moments of deferred crisis.

The Thirty Glory Years of Capitalism in the Core: The Fordist-Keynesian Compromise (1945–1975)

From the 1880s onward, and guided by US mechanical engineer Frederick Taylor, there emerged in the Western world a tradition of applied scholarship called scientific management. Scientific management placed under scrutiny the working practices of industries and factories with a view to making those practices more reliable, efficient, and productive. It was in this climate that US industrialist Henry Ford began to examine critically the methods through which automobiles were being produced. Ford was convinced that the prevailing approach to car production, which involved craftspeople undertaking a range of tasks, was inefficient and could be improved. He studied how workers were being deployed and concluded that by simply applying labor more efficiently and effectively, it ought to be possible to produce more vehicles per person per year.

Ford established a car production plant in Detroit, Michigan, and in this plant pioneered two notable innovations. He recognized that it was costly and inefficient for a single craftsperson to be involved in various stages of a production process. A division of labor ought to be introduced in which that production process was broken down into its component parts and workers dedicated their labor only to finite and discrete tasks. Of course, this strategy had the potential to be deskilling, monotonous, and dehumanizing for the worker. But it did allow far more work to be accomplished per worker in a given period of time. In addition, Ford recognized that time was being lost as workers moved around factories completing this or that task. It was better, he concluded, if tasks were made mobile and presented to stationary workers rather than workers circulating from one part of the factory to another. Accordingly, Ford introduced into his factory flow lines or assembly lines.

The result of these changes was a quantum increase in worker productivity and the advent of the mass production of automobiles. Ford produced his first mass-produced Model T car in 1908 and went on to build a world-class automobile-manufacturing company that persists to this day.

Ford recognized that the sustainability of his firm was dependent upon worker loyalty and the existence of a market for his cars. He solved both in part by offering workers pay increases in line with improvements in productivity. This had the benefit of persuading workers that the monotony of undertaking largely unskilled tasks and being pinned to an assembly line was worthwhile. Employees had a direct stake in improving productivity. Wages improved as productivity increased. Although the work was mind-numbingly dreary, application and endeavor were sufficiently rewarded, and standards of living were improving. Moreover, improving salaries enabled many of Ford's workers to purchase one of his cars themselves. Higher wages meant greater purchasing power, and greater purchasing power meant more business for Ford.

Ford's approach to production came to be copied by manufacturers in many industries, in the United States and subsequently throughout the industrial world. By World War II, Fordist mass production was assumed to hold the key to economic growth and prosperity. But Ford's interest in building employee loyalty, securing peaceable labor relations, and ensuring the existence of a market for mass-produced goods presented a challenge to governments. Firms could not solve this problem by themselves; a structural solution across the entire economy was needed.

In searching for such a solution, many governments turned to the ideas of British economist John Maynard Keynes. Keynes wrote his classic textbook *The General Theory of Employment, Interest and Money* in 1936 during the Great Depression, when unemployment grew to as much as 30% in many countries (unemployment was as high as

25% in the United States). The market, Keynes famously declared, could remain irrational longer than people could stay solvent; markets were moved by "animal spirits" and not "reason." Capitalism was based on "the extraordinary belief that the nastiest of men for the nastiest of motives will somehow work together for the benefit of all." State regulation was essential. But what type of regulation was required? For Keynes, aggregate demand in the economy was key. As long as there were buyers, there would be sellers; as long as there were sellers, there would be employment; and as long as there were workers, there would be consumer demand. In conditions of market failure, it was the job of government to inject demand into the economy; to spend more even when spending less appears to be more logical.

The extent to which Keynes' ideas had any impact upon President Franklin D. Roosevelt's New Deal (1933–1939) and massive public work program remains the subject of debate. But certainly after World War II, inspired by Keynes' theory of demand-led economic growth, governments entered into social partnerships with Fordist firms and labor and trade union movements. Collective agreements were forged. Workers would receive pay rises pegged to productivity improvements. The higher the productivity of the labor force, the greater the renumeration. Also, governments established welfare states. Through general taxation, workers would be provided with social protection such as unemployment benefits and pensions and items of collective consumption, such as health care, education, and housing. Productivity increases would guarantee better protection and more services.

This marriage between the political system and the economy came to be known as the Fordist-Keynesian compromise. Using the language of the regulation school, we may say that Fordism emerged as the principal regime of accumulation, while the Keynesian welfare state emerged as its mode of regulation. Between 1945 and 1975, this partnership led to historically unrivaled economic growth, prosperity, and improved standards of living throughout the Western world. This period is now referred to as the thirty glory years of capitalism (per regulation theory, *Les Trente Glorieuses*). But by the mid-1970s, the Fordist-Keynesian compromise began to come unstuck. In part, problems were caused by a number of external shocks to the capitalist world, including a series of wars and oil crises. In particular, the decision in 1973 by the Organization of Arab Petroleum Exporting Countries (OPEC) to quadruple the price of oil in protest of support for Israel during the Yom Kippur War was followed by a second oil crisis in 1979, when the price of oil doubled as the Iranian Revolution unfolded. Combined with the economic impacts of the Vietnam War (1955–1975), these hikes caused a profit squeeze in the West's oil-fired economies and led to economic turbulence.

But in fact, by the late 1960s and early 1970s, the system itself was in crisis (Harvey 1989). External shocks merely toppled a regime of accumulation that was already structurally compromised. Scientific management and Fordism had squeezed out of production processes and systems all of the productivity gains that were possible. Annual increases in production across the economy began to wane, and the law of diminishing returns set in. But social partnerships remained strong, and the labor movement continued to demand more wages and governments more taxation. Having enjoyed thirty years of expansion, neither were minded to enter into a new relationship with the capitalist economy. There were to be no new deals between firms, workers, and governments. It proved politically impossible to curb demands for wage inflation and continuous expansion of welfare states. Firms simply had to subsidize both. With costs increasing as before but productivity increases stalling and output stabilizing, firms found themselves unable to maintain their profit margins, and their rate of profit decreased. Thrashing to reverse this squeeze upon their profitability and capacity to pay dividends to shareholders, they conspired to plunge the entire system into crisis.

An Early Response: The New International Division of Labor (NIDL)

In an effort to restore profitability, capitalist firms devised a variety of strategies, many of which have carved out some interesting new economic geographies. Toward the late 1970s and early 1980s, scholars began to notice the development of a new international division of labour (NIDL). The NIDL, it was asserted, was systematically reworking the geography of the world economy laid down by the OIDL. While the OIDL conceived of the world in terms of the role of different regions in GVCs, the NIDL conceived of the world in terms of which function (within firms occupying any given tier in GVCs) regions are most able to attract, and how significant the added value of that function is (see Deep Dive Box 4.2).

Deep Dive Box 4.2   Mapping Intel's Worldwide Operations

The semiconductor industry is a pivotal industry because microchips are the foundation for the entire electronics industry. Virtually every electronic good that is now manufactured, from televisions to computers, mobile phones to fridges, game consoles to radios, and washing machines to electric heaters, depends upon a silicon chip to operate. Many of the world's leading semiconductor firms base their HQs in Silicon Valley in California. One such company is the **Int**egrated **El**ectronic Corporation, or Intel. Founded in 1968, Intel is the largest semiconductor or microchip producer in the world (Plate 4.3). It manufactures the semiconductors that are used in the majority of laptop computers and makes many other devices that relate to computing and communication. In 2019, Intel generated a net revenue of $72.2 billion and employed 107 600 workers worldwide, 45% of whom resided outside the United States.

Plate 4.3   Intel microchip. Source: Photo by Slejven Djurakovic on Unsplash.

(Continued)

Box 4.2   *(Continued)*

Intel has created a worldwide division of labor predicated upon allocating six types of operation or function to different locations (see Map 4.2):

Global headquarters: Intel has its global HQ in Santa Clara, California. This HQ functions as a central decision-making nerve center. By drawing upon world-leading technology innovators and very high-skilled executive managers, it provides the company with strategic global command and control.

Wafer fabrication (Fab) plants: Intel locates its prized Fab plants in Oregon, Arizona, and New Mexico in the United States, as well as Ireland, Israel, and Dalian, China. Requiring high-skilled professional and technical labor, Fab plants use a photolithographic printing process to etch microchips layer by layer onto silicon masks.

Assembly and testing facilities: Intel has three major assembly and testing facilities, requiring both high- and lower-skilled technical labor; in Chengdu, China, Malaysia, and Vietnam. Finished wafers are sent to these facilities, where they are cut into individual chips with a diamond saw, tested for their functionality, and packaged.

General manufacturing functions (like mask making): Intel locates its global Mask Operations in Santa Clara, California, and in Oregon. Mask making is a medium-skilled manufacturing process requiring forensic precision. Six-inch films of quartz, a quarter-inch thick, are produced, upon which wafer circuitry is later imprinted.

Research and development (R&D) and product development units: Intel's principal R&D centers are located in Silicon Valley, Arizona, New Mexico, Massachusetts, Ireland, Israel, Moscow, Dubai, Swindon (UK), Munich, Istanbul, Stockholm, Bangalore, Shanghai, Beijing, and Kuala Lumpur.

Sales marketing and services support offices: Intel has significant sales, marketing, and services support offices across 46 countries worldwide.

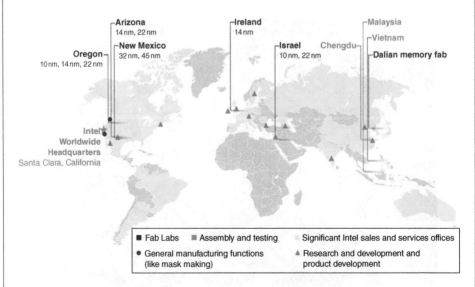

Map 4.2 Intel's global division of labor. Source: Compiled by the author from https://www.intel.com/content/www/us/en/location/worldwide.html

In her 1984 book *Spatial Divisions of Labour: Social Structures and the Geography of Production*, British Marxist geographer Doreen Massey sought to investigate changes that were then occurring in the geography of industrial production across the United Kingdom. Massey recognized that many companies were exploiting their Fordist division of labor by developing spatial divisions of labor. Production systems were so fragmented that it was now possible for companies to physically separate the various tasks they were engaged in. These tasks could then be located in regions of the country that had the cheapest wage rates for any given skill level. Companies needed to keep their headquarter (HQ) functions in large cities like London; although labor was costly, these cities were home to pools of high-quality graduates, white-collared professionals, and senior managers. But research and development (R&D) functions and high-technology tasks could be delegated to areas such as Cambridge, Oxford, and Bristol, where there existed pools of technical and professional workers. Labor cost less in these regions than it did in London. Finally, companies could locate their low-technology activities and assembly and packaging functions to areas such as the Northeast of England, South Wales, or the central belt of Scotland. These regions had pools of unskilled and unemployed workers that could be exploited. By adopting spatial divisions of labor, firms could drive down their labor costs and maximize profitability.

While Massey's focus was upon spatial divisions of labor within a single country, already by the early 1980s it was becoming clear that firms were stretching their spatial divisions of labor across the entire globe in a bid to exploit variable wage rates. In their 1980 book *The New International Division of Labour: Structural Unemployment in Industrialized Countries and Industrialization in Developing Countries*, German political economists Folker Fröbel, Jürgen Heinrichs, and Otto Kreye argued that German firms were now building an NIDL. Although company HQ functions were being left behind in German cities, R&D and high-technology functions were being relocated to regions of the world with cheaper but still technically proficient pools of labor, while low-technology manufacturing and assembly functions were being located among the world's poor.

Fröbel, Heinrichs, and Kreye argued that this trend was being driven by four factors. First, the profit squeeze of the 1970s had stimulated German companies to search for ways of reducing labor costs and readied them to scour the world in search of the right workers at the lowest cost. Second, Fordism had run so deep in German companies that their divisions of labor were highly fragmented. It was entirely possible to physically separate bits of a company and to relocate those bits to locations around the world. Third, developments in telecommunications (then computer, phone, and fax technologies) and air travel made it possible for HQs to manage branch plants from a distance. Finally, the modernization of agriculture in the developing world had triggered mass migration from rural areas to Global South cities and had created in these cities a vast army of desperate and powerless workers. These workers were ripe for super-exploitation, could be paid very little, and could easily be replaced if they found themselves injured, ill, or simply exhausted.

Today, TNCs create international divisions of labor certainly to exploit cheaper wages elsewhere, but they also do so to access markets, avail of weaker environmental regulations, escape tariffs on trade, access specialized workers, source protected and scarce raw materials, and so on.

In a seminal academic article published in 1996, US urban planner John Friedmann (1986) argued that the NIDL was reworking urban fortunes in core countries. Although the exodus of manufacturing from the core to lower-wage regions of the world economy has devastated industrial cities and left former manufacturing heartlands in ruins (giving birth to the label "the rustbelt"), TNCs have left their HQs in core countries and there has emerged a small group of highly powerful and prosperous *world cities*. One of the impacts of the NIDL in core countries has been the rise of these world cities or global cities – which have become magnets for the HQs of the largest companies in the world. As TNC HQs have gravitated to world cities, they have become powerful command-and-control centers orchestrating the entire global economy. It is because of these twin processes of deindustrialization and world city formation that core economies today suffer from stark regional inequalities in wealth and income. The NIDL has left in its wake an economic core characterized by both declining smokestack cities such as Detroit and Liverpool, and booming, all-conquering world cities like New York and London.

Why have only a small number of core cities become home to the HQs of the leading companies in the world?

US sociologist Saskia Sassen (2001, 2018) has sought to advance the world city thesis through the development of the concept of "global command capability." The practice of global command is fraught with difficulties. When these problems become too complex for TNCs, HQs find it useful to outsource solutions to international-oriented producer service firms. Support is sought from information and communications technology (ICT) consultants, language training firms, marketing and public relations (PR) companies, recruitment agencies and executive search specialists, international legal experts, academics and political analysts, accountancy firms, and management consultants. To the extent that an army of international-oriented producer service firms operate in any city, they can be said to furnish multinational headquarters with a capability to manage internationally. But only a few cities have sufficient concentrations of these support services. A geography of international-oriented producer services exists, and, therefore, so too does the global command-and-control infrastructure available to firms vary over space. For this reason, the geography of the tools through which global command capability is made possible becomes an important factor in determining the geography of company headquarters. Given this, it is not surprising to find that only a small number of so-called "alpha world cities" exist, and that these cities house most of the HQ functions of the largest TNCs and therefore perform as key command-and-control nerve centers in the new world economy (see Deep Dive Box 4.3).

Deep Dive Box 4.3   Where Are World Cities?

Created by the Department of Geography at the University of Loughborough in the United Kingdom, the Globalization and World Cities Research Network (GaWC) produces biannual rankings of cities according to their global command-and-control power and reach. Three categories (with subcategories) of world cities are recognized.

- Alpha-level world cities are leading command-and-control crucibles of power from which the world economy is orchestrated.
- Beta-level world cities link their region or state into the world economy.
- Gamma-level world cities link smaller regions or states into the world economy, or cities with unusual global capacity.

Map 4.3 profiles alpha world cities with subclassifications showing the most powerful cities within this already-powerful group of cities.

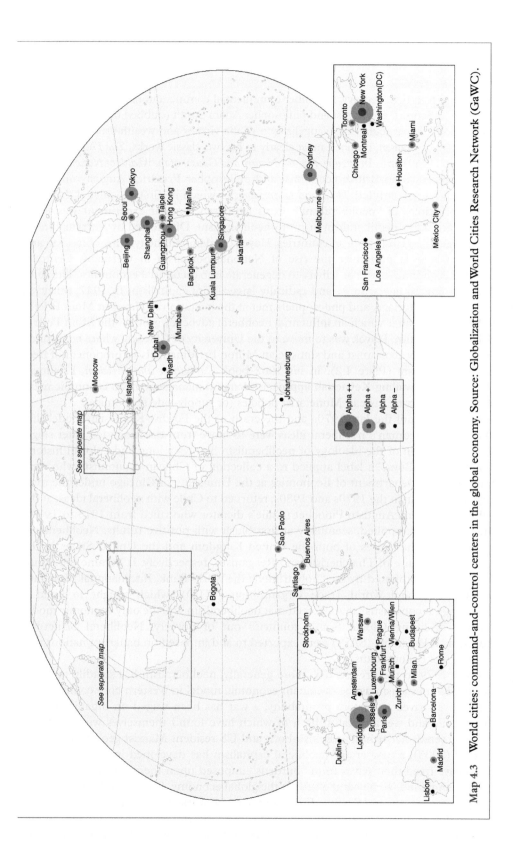

Map 4.3   World cities: command-and-control centers in the global economy. Source: Globalization and World Cities Research Network (GaWC).

Boom and Carnage in the Core: The Neoliberal
Juggernaut (1979–Present)

In the mid-twentieth century, a number of influential commentators sought to rediscover the ideas of Scottish economist and philosopher Adam Smith (dubbed the "father of capitalism") and in particular his evangelization of the virtues and wealth-creating power of free market capitalism, outlined most clearly in his two classic works, *The Theory of Moral Sentiments* (1759) and *An Inquiry into the Nature and Causes of the Wealth of Nations* (1776).

In 1944, Austrian-British economist and philosopher Friedrich August von Hayek published a book entitled *The Road to Serfdom*, in which he first articulated the theoretical principles of "neoliberalism" – this lab-based thought experiment was to be labeled prototype neoliberalism or proto-neoliberalism. Disaffected by worrying political developments in European countries, Hayek condemned the tyranny of fascism and socialism, and in particular the rise of National Socialism in Germany. Centrally planned economies had a proclivity to degenerate into dictatorial economies. The solution was limited government and radically laissez-faire capitalism. In 1947, thirty six economists, historians, and philosophers met at the invitation of Hayek at Mont Pelerin in Switzerland, after which an influential neoliberal advocacy group – the Mont Pelerin Society – was born. Hayek was to move to the University of Chicago, where he inspired American fellow economist and statistician Milton Friedman, also a proponent of free-market capitalism (Plate 4.2). In his 1962 book *Capitalism and Freedom*, Friedman rejected the prevailing Keynesian approach (demand management) to economic management, preferring instead Monetarist policies (or supply-side economics).

It was not until the late 1970s and the crash of the Fordist-Keynesian model that Hayek and Friedman's neoliberal ideas were to move from the fringe to center stage. The propagation of the ideology of neoliberalism betrays a long and colorful history. The "Chicago Boys," a label applied to a collection of Chilean economists who were trained at the Department of Economics at the University of Chicago under the tutelage of Friedman in the 1970s and 1980s, returned to Chile with neoliberal ideas; at the behest of General Augusto Pinochet (Chile's dictator who ruled from 1973 to 1990), these ideas were applied, seemingly at first blush with positive results. Neoliberalism crashed onto the shores of both the United Kingdom and the United States in the 1980s in the guise of Thatcherism and Reaganism, respectively. It became the dominant ideology proffered by the "holy trinity": the World Bank, International Monetary Fund (IMF), and World Trade Organization (WTO), which sought to build a "Washington consensus" that decreed that neoliberalism was the only economic model capable of lifting Global South countries out of poverty. Neoliberal Structural Adjustment Programs (SAPs) were exported to and imposed on bankrupt nations as a condition of bailout loans.

From 1979 to the present day, more generally, neoliberalism has steadily replaced Fordist-Keynesianism as the dominant economic model in Western core countries. As firms have strived to restore profitability, a war has been waged on organized labor movements and welfare states, both of which have found themselves weakened and disempowered. According to British-born and US resident Marxist geographer David Harvey (2005), a new era of neoliberal capitalism has developed in which capitalist firms have freed themselves from "burdens" imposed upon them during the Fordist-Keynesian period. To be competitive in the global economy, firms need to be entrepreneurial, lean, mean, and flexible. For this to happen, workers, it is argued, need to offer their labor at cheaper rates and be prepared to accept flexible terms and conditions. It also requires, it is contended, that governments reduce taxation and downsize

Plate 4.4 A symbol of aggressive neoliberal capitalism: The Bull of Wall Street, New York, NY. Source: Photo by Alec Favale on Unsplash.

welfare states. But neoliberal capitalism has created a volatile and unstable world and has yet to discover a formula as successful as the Fordist-Keynesian compromise. The law of the jungle has set in; in the struggle for survival, only the nastiest of the nastiest – to invoke Keynesian – will make it (Plate 4.4).

According to Harvey (2005), the recent history of neoliberalism can be read as a decisive development in the class war over the division of the national product – an overt elite ideological project. Through what Harvey (2005) calls "accumulation by dispossession," capital has reversed the gains to labor and rising wages that were ushered in with the Fordist-Keynesian state, and restored inequalities in wealth back to Victorian levels. Rereading German economist, philosopher, and revolutionary Karl Marx's *Capital Volume 1: A Critique of Political Economy* (first published in 1867), Harvey drews parallels between neoliberalism and what Marx called "original or primitive accumulation," the dispossession by Europe of capital in the colonies through imperialism. Only now this dispossession was creating economic injustices for workers living in core countries, who were being divested by four processes:

- Public assets, resources, property, and infrastructure have been privatized and commodified; this loss of public goods has benefitted a small number of private interests.
- Financial services (banks, insurance companies, pension and investment funds, and so on) have been deregulated, inducing practices that, while having the potential of yielding huge profits, have also proven risky and at times reckless. But as witnessed during the global financial crisis in 2008, whilst private actors claimed full ownership of high returns in times of plenty, they were prepared to pass all the responsibility and pain of loss to governments and the public when things went awry.
- Through what has been termed "disaster capitalism," powerful private interests have created and exploited economic crises to secure "vulture" profits; meanwhile, ordinary citizens have had to live with the carnage.
- The fiscal policies of neoliberal governments have persistently favored actions that lubricate the market over those designed to redistribute wealth to those most in need.

Although the power of the "unleashed" market has led to impressive growth in some places, in fact generally economic growth rates under neoliberalism have been inferior to those achieved under the Fordist-Keynesian model. But returns to societies' elites, by contrast, have been spectacular. For Harvey, neoliberalism is a political-economic project orchestrated by the top 1% for the top 1% (or, at best, top 10%) at the expense of the comparative wealth and income accrued by the bottom 50% and especially the bottom 10%.

Meanwhile, having lived through half a century of improving living standards and expanding government expenditure, both organized labor movements and welfare states have found it difficult to accept the collapse of social partnerships and attacks on wage increases, social entitlements, workers' rights, and public services. According to Harvey (2010), working-class people have sought to weather the vagaries and vicissitudes of neoliberal capitalism by borrowing from financial institutions and entering into debt. Expectations of continuous rises in standards of living were to be serviced only through borrowing and indebtedness. Keen to raise capital to lend to this growing market of borrowers, financial institutions have borrowed from other financial institutions, creating in turn a deeply interconnected global financial system. But the debt being accumulated is vast, and servicing it is going to be a challenge. In 2008, financial institutions woke up to the realization of just how vulnerable they were to debt default. A global meltdown of the financial system resulted, in turn bringing the global economy to its knees.

Alas the global financial crash of 2008 has not deterred neoliberals. Across the West, a distinctive feature of the post-crash regulatory environment has been neoliberalism's "Houdini-like" ability to appropriate a crisis it was centrally implicated in causing to gain further momentum and entrenchment. But neoliberalism redux remains a work in progress, and its future remains uncertain. In this, Covid-19 is likely to have a say.

In the early 1970s, following thirty glory years of capitalist growth in the advanced capitalist world, few would have predicted the collapse of the Fordist-Keynesian compromise. And yet, it rapidly unraveled and quickly became obsolete. Like then, the system now is structurally exhausted and limping. There is no reason to think that neoliberalism will have a longer life expectancy, and there is every reason to believe that we may be living at yet another fulcrum point in political-economic history. Both the technical and social performance of the neoliberal model now lack legitimacy. Hayek and Friedman believed in trickle-down economics: a rising tide would lift all boats. This belief has shown itself to be erroneous; a rising tide, it transpires, lifts only luxury yachts. Inequalities have deepened, leading to an explosion of populist political movements and a revolt of the left-behind rustbelt. The climate and ecological crisis has become an acute emergency. Like in the 1970s, exogenous shocks are taking an already-precarious system to the brink. Then, it was the OPEC oil crisis, Vietnam War, and Iranian Revolution; now it is the Covid-19 pandemic. Perhaps a new social democratic project is just around the corner, even if its contours are hard to imagine. The mantra "there is no alternative" has never looked less shaky (see Deep Dive Box 4.4).

In their 1984 book *The Second Industrial Divide: Possibilities for Prosperity*, US political economists Michael Piore and Charles Sabel examined why some firms seemed to have thrived and prospered throughout the 1970s and early 1980s, while others struggled to survive. Their eye was drawn to a region in Italy called the "Third Italy." While Italy is normally thought to comprise a rich North and poor South, in fact a third region exists

Deep Dive Box 4.4   Ireland's Adoption of the Neoliberal Economic Model

The Republic of Ireland's adoption of a neoliberal model of economic development bears witness to the Janus face of neoliberalism; alongside boom and prosperity, there have been bust and carnage.

From 1922 when it became the Irish Free State, to 1948 when it became a full republic, the Republic of Ireland suffered from an impoverished and insular colonial economy that was further impaired by the economic policy of protectionism and import substitution. Following decades of economic stagnation, in the 1950s and 1960s Irish Taoiseach (prime minister) Seán Lemass and state technocrat TK Whitaker sought to open up Ireland to the global economy as a hyper-liberalized entrepôt for global capital. Economic and fiscal crises in the 1980s only led to a redoubling of the mission. In the 1990s, Ireland embraced yet deeper deregulation, entrepreneurial freedoms, and free-market principles and aggressively courted high-value-added export-oriented foreign direct investment (FDI).

Steadily, the world's largest TNCs began to open plants in Ireland, not least to benefit from the country's very low corporate tax rate of 12.5%, access to the European Union market, stable social compact with unionized labor, lax planning, and technical training colleges. The opening by Intel of a wafer fabrication plant near Dublin in 1987 catalyzed an influx of other leading US companies; Ireland became known as the 51st state. The "Celtic Tiger" years (1993-2007) saw a dramatic transformation in the social and economic life of a country that had, until the start of the 1990s, been a relatively poor and peripheral state, perched on the edge of Europe. Ireland was reinvented, from a former colony with a weak industrial base and only 70% of the average EU Gross Domestic Product (GDP), into a high-tech economy housing the world's leading TNCs and boasting a GDP of 140% of the European average.

However, from 2000, unchecked financialization of real estate, especially housing developments in cities, was increasingly used to sustain economic growth. In 2008, as a number of worldly dramas unfolded, so too did the Celtic Tiger model unravel. The domino effect of the global financial crisis unearthed fragilities, overextensions, and tenuous solidarities within international financial markets. As a small open economy, Ireland was always going to be exposed to the collapse of the global financial system, but the extent of this exposure was significantly exacerbated by reckless banks, maverick property developers, pro-development fiscal policy, liberal planning, and in consequence, a home-grown property bubble. What started most explicitly as a financial crisis would inevitably become a wide-reaching economic crisis. And so, from 2008, Ireland's particularly epic property bubble burst, unmasking gargantuan levels of debt and bringing the economy to the edge. Housing, hotel, and retail developments were abandoned overnight, and the country was blighted with so-called "ghost estates" and "zombie developments."

Remarkably, the response of the Irish government amounted to "more of the same": to patch up, rather than transform, the political economic system. There was little appetite for any radical departure from neoliberal orthodoxies. Given its perilous economic state, the government accepted an €85 billion Troika (European Commission [EC], European Central Bank [ECB], and IMF) bailout in 2010,

(Continued)

Box 4.4 *(Continued)*

using this money to rescue Irish banks and property developers by nationalizing toxic loans. To service this debt and notwithstanding its origins in the excesses of politicians, banks, and developers, there has followed a period of savage austerity cuts and economic pain for the wider population, in particular for the bottom 50% and especially the bottom 10%. But, in 2013 the country exited its bailout compact with the Troika and regained economic sovereignty. Quickly, the economy returned to growth. Between 2014 and 2020, economic growth continued apace, to the extent that fears began to arise that the economy might at some point overheat again.

And so Ireland has been held up by the Troika as a model pupil of bailout and austerity-led recovery. Ireland is alleged "proof" that deep neoliberalism is the only effective solution to economic crisis, even neoliberalism-induced crisis.

But steadily, critical voices emerged questioning the idea of a pain-free "Celtic comeback." In 2015 and 2016, grassroots anti-austerity movements galvanized into national protests over the imposition of water charges, forcing the Irish state to abort the policy. A bankrupt property sector has led to a decade of zero house building. Opposition has grown around the ongoing housing and homelessness crisis, comprising a deficit of affordable housing, rising property and rent prices, and the growing problem of (family) homelessness. The housing crisis stands as the most visible echo effect of Ireland's most recent encounter with boom and bust economics, but it is likely to be joined by other unwanted inheritances as the deferred impacts of austerity work their course.

Ireland's dominant right-of-center political parties (Fianna Fail and Fine Gael) have continued to rule since 2008 but in steadily more fragile coalitions. But the center-right has been variously held accountable in successive elections, and in a 2020 "earthquake" election, left-of-center parties (Sinn Féin and independent social parties) protesting Ireland's ongoing adherence to the neoliberal economic model and calling for "people before profit," came close to a historical victory. Ireland's political establishment know that if their neoliberal Celtic comeback model is to survive, it will need to do better.

---

in the northeast of the country, centered upon the administrative region of Emilia Romagna and its capital city of Bologna.

Already suppressed by Benito Mussolini's fascist movement, World War II impacted the Emilia Romagna region perhaps more so than any other Italian region. In response, local leaders embarked upon an economic development strategy that centered upon small businesses, employee ownership, consumer cooperatives, and agricultural cooperatives. These organizations worked closely together so that through cooperation, the sum would be greater than the individual parts. The region had a significant volume of small enterprises that were extraordinarily innovative and entrepreneurial. These firms differentiated themselves from Fordist factories, which mass-produced standardized goods for the global market. Their specialty was to produce small batches of fashionable goods for differentiated and niche markets. When tastes shifted in these niche markets, these firms were flexible and nimble enough to adjust their production processes and product ranges.

Piore and Sabel concluded that industrial districts comprising specialist agglomerations of small firms with high levels of flexibility and innovation would increasingly win out over traditional, rigid, and stale Fordist manufacturing heartlands. Consequently, they believed that the Third Italy contained lessons for the entire world economy. The world no longer needed large-batch production of mass-produced and standardized goods. Instead, it required firms that were capable of catering to the whims of an ever more volatile and differentiated marketplace. Fordist production systems were too inflexible and rigid. What was needed was a revolution in the ways in which firms operated. *Flexible specialism* was required; firms needed to be able to adjust their systems and products at a moment's notice. They needed to be capable of producing small-batch runs of specialized, trend- and fashion-led, customized products.

Across the world, post-Fordist manufacturing practices have now replaced their Fordist counterparts. Post-Fordist firms are characterized by their flexibility:

- Fixed assembly lines have been replaced with robotic technology that can be reprogrammed to perform different tasks as manufacturing processes change to accommodate consumer trends (Plate 4.5).
- Part-time, casual, zero-hours and temporary workers are favored over full-time and permanent employees. Workers are hired only as and when they are needed.
- Companies are increasingly subcontracting core duties to other companies, thereby freeing themselves from obligations to workers (tax payments, pension rights, holiday entitlements, health and safety insurance, and so on).
- Traditional just-in-case supply systems (where firms would order and stock a vast inventory of supplies based upon predictable future production targets) have been replaced by just-in-time supply systems (where supplies are delivered only as and when needed).

Plate 4.5 Post-Fordism at Ford! A post-Fordist assembly line in Ford's car production factory in Chennai, India. Source: Creative Commons, https://www.flickr.com/photos/fordapa/albums/72157622110280353.

- Marketing (producing products based upon market research) is now prioritized over advertisement (simply selling products better). Marketing teams now work directly with R&D offices and engineers to ensure that customer needs drive manufacturing processes and not vice versa.

Although Fordist production strategies gave birth to the NIDL and a spreading of economic functions to low-wage regions, post-Fordist production systems are presumed to result in geographical reconcentration and clustering of flexible firms in high-technology industrial hearths in core economies.

- Flexible production requires firms to cluster their branches and offices in a single location and to locate close to suppliers (vertical integration). Firms that have scattered their functions to various locations around the world have seen merit in the reintegration and (re)consolidation of activity in particular locations.
- Firms that produce goods flexibly are finding it beneficial to locate next to firms doing likewise (horizontal integration). The geographical clustering of separate firms is being recognized to confer onto post-Fordist firms a decisive competitive advantage.

As early as 1890, British economist Alfred Marshall coined the term "Industrial Districts" to refer to these pioneering regions of innovation (Lancashire and Sheffield, both in England, were cases in point). Marshall identified pooled labor markets, specialized suppliers, and knowledge spillovers as the three main reasons why groups of firms locate in close proximity (known as the Marshallian Trinity). In his seminal book *The Competitive Advantage of Nations* (Porter 1990), US business scholar Michael E. Porter likewise conceives of industrial clusters as geographical concentrations of interconnected companies, specialized suppliers, service providers, and associated institutions in a particular product sector. For Porter, clusters deliver competitive advantage to firms by enhancing their capacity to source inputs and access information, technology, and other cognate institutions; coordinate more effectively with related companies; and measure and motivate improvement.

2050: Toward a Fourth Industrial Revolution (IR4)?

In the early twentieth century, the Soviet economist Nikolai Dimitrievich Kondratiev sought to identify in the capitalist economy surges of technological disruption and economic advancement. In contrast to other economists, who claimed to discern the existence of short business cycles (anything from 3 years to 25 years), Kondratiev believed that technology paradigms, which shifted every 45–65 years, created longer business cycles (indeed, he thought these cycles lasted on average for 54 years). A number of generative technologies rose and fell across this time period, rebooting the entire economic system and propelling (for a time, at least) the rise of whole new industries. Kondratiev waves might be understood in terms of the four seasons: spring (improvement), summer (acceleration and prosperity), fall (plateau), and winter (decline and depression). In his 1925 book *The Major Economic Cycles*, Kondratiev applied his theory to nineteenth-century economic development and proposed the existence of two cycles (one between 1790 and 1849 powered by steam engines, and the other between 1850 and 1896 powered by railways and steel). He argued that a new cycle had commenced in 1896 (powered by electrification and chemicals). Later, elaborating upon

Kondratiev's theory in his own 1939 book *Business Cycles* – by nesting models of shorter business cycles with longer cycles to understand their cumulative impacts – Austrian-American political economist Joseph Schumpeter referred to Kondratiev's economic cycles as "K-waves."

In the spirit of Kondratiev, today there is emerging an increasingly influential claim that a fourth industrial revolution (also known as a cyber-physical revolution, Industry 4.0, IR4, and/or Society 5.0) is imminent, predicated upon computerized data analytics (see Deep Dive Box 4.5). A particular variant of K-wave theory is proposed. Whereas the first industrial revolution used water and steam to power production, the second used electricity to create mass production, and the third used electronics and information technology to automate production, the future prosperity of Western core economies will depend on networked computing power and computerized data analytics.

Increasingly, computing power will become embedded in foundational operational systems and in the fabric of everyday life, to enable institutions, infrastructure, services (including municipal administration, education, health care, public safety, real estate,

---

Deep Dive Box 4.5   IR4: Key Terms and Definitions

According to US technology firm IBM, by 2050 the presence, scale, quality, and alignment of instrumentation, interconnection, and intelligence in any given economy will determine its prosperity and fortunes. But what does this mean? Some definitions are necessary.

- **Instrumentation** is the capability to generate and capture live real-time data through the use of sensors, meters, appliances, and personal devices.
- **Big Data** is a label applied variously to data that is: huge in volume, consisting of terabytes or petabytes; high in velocity, being created in or near real time; diverse in variety, being structured and unstructured in nature; exhaustive in scope, striving to capture entire populations or systems (n = all); fine-grained in resolution and uniquely indexical in identification; relational in nature, containing common fields that enable the conjoining of different data sets; and flexible, holding the traits of extensionality (can add new fields easily) and scalability (can expand in size rapidly).
- **Interconnection** refers to the capacity to circulate, pool, exchange, integrate, and communicate this data, through digitalization and technologies such as the Internet of Things (IoT), 5G, radiofrequency identification (RFID), data platforms, and interfaces.
- **Intelligent** refers to the application of data science (statistics, modeling, optimization, and visualization) and computing power (artificial intelligence/autonomous systems/machine learning, and cognitive, quantum, and ubiquitous computing) to enable big data to make decisions and/or assist humans in doing so.
- **Artificial intelligence (AI)** refers to intelligence demonstrated by machines, in contrast to the natural intelligence displayed by humans.
- An **algorithm** is a set of software instructions, written in code, typically to solve a class of problems or perform a computation.

transportation, and utilities), and individual citizens to benefit from increased instrumentation, interconnection, and intelligence (sometimes abbreviated to "In3"). Data will rise to become the primary raw material of the new economy – the new oil. Just as oil shortages constitute a serious threat to an oil-fired economy, any scarcity of data, particularly well-curated data, represents an existential threat to a computer-fired economy. To catalyze IR4, governments will need to unlock the full value of "big data" by encouraging significant data owners to radically increase accessibility for public, private, and third-sector stakeholders – and in particular their software coders – to the data they hold.

Today, neoliberalism has been the dominant engine behind IR4. Parisian regulation theory instructs that if IR4 is to prosper into a mutant neoliberal regime of accumulation, it will need to create a parallel mode of regulation that affords it social legitimacy and technical efficacy. This will be difficult, because IR4 presents profound logistical, ethical, legal, commercial, and political challenges. Critical market failures indicate that IR4 experiments are accelerating ahead of ethics, law, public policy, purpose, regulation, and governance, and there is legitimate concern that burgeoning computerized data analytic solutions are lacking in social responsibility and failing to deliver for the public good. In fact, the origins and development of IR4 within a framework of what business scholar Shoshana Zuboff (2019) calls "surveillance capitalism" have given rise to technology that is not only configured primarily to serve the interests of commercial data harvesters, but also substantially – and manifestly – underregulated. For Zuboff, surveillance capitalism is driving computerized data analytics, and overdetermining the terms of its application and the trajectory of its development. Many now worry about the power of companies like Facebook, Google, and Twitter to spy on our everyday life, collecting data about us without our permission, and selling it to third parties with commercial and surveillance agendas. Challenges include:

- The loss of privacy and sovereignty over our personal data (for example, data is routinely collected from our mobile phones)
- Threats from cyber-criminals (for example, hacking our bank accounts)
- Bias in algorithms and prejudicial software (for example, software that uses facial recognition to predict criminal activity but is loaded against "ethnic" faces)
- Corporate abuses of personal data (for example, insurance companies gaining access to our medical records to set premiums)
- The loss of freedom of speech (for example, the use by the "deep state" of personal data to monitor those with alternative political views)
- Threats to democracy (for example, the use of social media to interfere in elections)
- Digital exclusion and poverty (for example, the exclusion of, say, the elderly community from participating in the digital revolution owing to lack of computer skills)
- Digital addition (for example, the inability of some teenagers to disengage from toxic social media platforms)
- Job displacement and mass unemployment (once machines replace human beings: "No Humans Need Apply!")

But innovation and regulation are not incommensurable. Whether IR4 becomes a new neoliberal long wave or collapses under the weight of its problems remains to be seen. It could, of course, be claimed and put to use by a radically alternative regime of accumulation. Whether or not computerized data analytics will lead to a better tomorrow will depend upon social and political choices we make today. Everything depends upon the political constitution of data markets and other data sharing platforms: specifically how data sharing arrangements are designed, regulated, and governed and whether they command a social license (see Deep Dive Box 4.6).

Deep Dive Box 4.6   Who Will Lead IR4? Silicon Valley? Cloned Silicon Valleys?

It is likely that the geography of IR4 will consolidate economic power in a small number of high-tech core cities. New AI start-up companies continue to favor only a small number of Western countries (Map 4.4). The largest AI companies in the world by HQ location at present are: Baidu (Beijing), Tesla (San Francisco Bay Area), Alphabet (San Francisco Bay Area), NVidia (Delaware), Enlitic (San Francisco Bay Area), Facebook (San Francisco Bay Area), Didi Chuxing (Beijing), Microsoft (Seattle), Fanuc Fujitsu (Tokyo), Improbable (London), Bosch (Stuttgart), Line (Tokyo), and IBM (New York) (see Plate 4.6).

In 2050, will Silicon Valley in California continue to lead the IR4 world economy as the primordial "lighthouse" high-tech city-region that others aspire to replicate? Or will Silicon Valley be copied and bettered by rivals? Emulating Silicon Valley has proven difficult. Many countries have sought to copy and clone Silicon Valley, and today there exists a variety of rival "silicon valleys," "silicon glens," "silicon canals," "silicon docks," "silicon beaches," "silicon corridors," and "silicon oases," to name but a few. In truth, very few pretenders to the throne have been able to live up to the billing.

Perhaps the most ambitious attempt to replicate Silicon Valley today can be witnessed in ongoing efforts in Russia to build a "Skolkovo Innovation Center" from scratch on a greenfield site located in the suburbs of Moscow (some 40 minutes by rail from the center of the city).

Skolkovo is a flagship project of the Russian government. It was first announced in 2009 under the presidency of Dmitry Medvedev and signed into law in 2010. Medvedev was concerned with the speed with which Russia was modernizing. In particular, he was concerned with Russia's overdependency on natural resource exports, including oil and gas, and advocated the need for a more diverse economic base that was more creative, Western, innovative, and technologically advanced. Although to a degree casting off its centrally planned past and embracing democracy and market rule, in fact Russia continues to combine elements of both state autocracy and capitalist entrepreneurialism. Unlike liberal Western states, the Russian state remains centrally embedded in the Russian economy (through, for example, state-owned enterprises [SOEs]) and has capacity to steer and guide development. Under the tight grip of President Vladimir Putin, Russia can be said to be presiding over a form of "authoritarian modernization": modernizing engineered by the muscle and strong arm of the Russian state. It is in this context that the Skolkovo project is to be understood.

(Continued)

Box 4.6 (*Continued*)

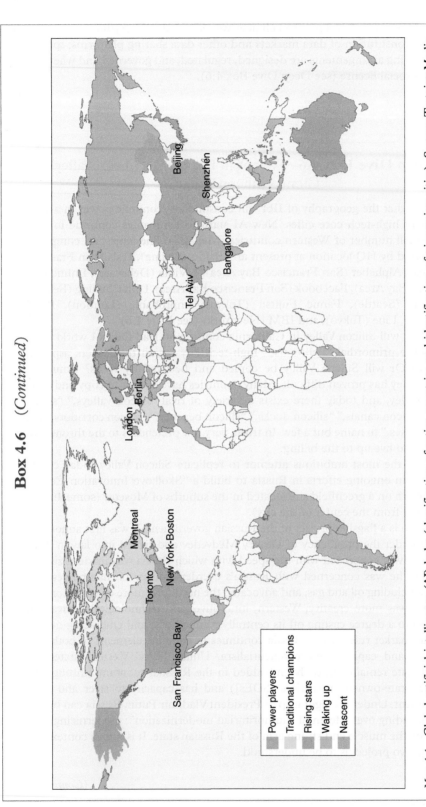

Map 4.4 Global artificial intelligence (AI) hubs (classification of active countries and identification of pioneering cities). Source: Tortoise Media 2020. Retrieved from https://www.tortoisemedia.com/intelligence/ai/Tortoise Media 2020.

Plate 4.6   Map of the Silicon Valley area of California: from San Jose to San Francisco. Source: zimmytws/iStock/Getty Images Plus/Getty Images.

The project is being overseen by a not-for-profit company called the Skolkovo Foundation. This foundation is seeking to bring together a number of stakeholders to build a world-class technology ecosystem (see Figure 4.2). These include higher education institutions (centered upon the Skolkovo Institute of Science and Technology – Skoltech), start-up Russian technology companies, the R&D functions of larger Russian and global technology companies, a technology park, and venture capitalists. The focus is upon fortifying five key technology sectors: ICT, Energy-Efficient Technologies, Nuclear Technologies, Biomedical Technologies, and Space Technologies and Communication. Partnering with other Russian universities and the Massachusetts Institute of Technology (MIT), Skoltech is a private research-led university that was opened in 2011. Meanwhile, Skolkovo Technopark now has facilities spread over 800 000 square meters and hosts around 500 start-ups, while supporting an additional 1500 enterprises beyond its campus. Skolkovo hosts around 50 research centers employing more than 15 000 people. Thirty leading TNCs, including Boeing, Cisco Systems, EADS, GE, Johnson & Johnson, IBM, Intel, Microsoft, Siemens, Nokia, and Samsung, have signed R&D partnership agreements with the Skolkovo Foundation.

These early achievements notwithstanding, the extent to which Skolkovo will ever become a rival to Silicon Valley remains an open question. Some critics think that Skolkovo has been overhyped and lament the extent to which anything meaningful is happening in reality on the ground. To try to replicate something as organic and spontaneous as Silicon Valley, critics say – under the Russian model of authoritarian modernization – is likely to fail. Meanwhile, Russia has consistently scored badly in global rankings of corruption. Indeed, questions have been

(Continued)

Box 4.6 *(Continued)*

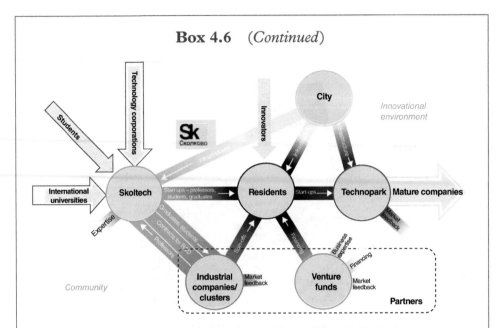

Figure 4.2   The Skolkovo innovation ecosystem, Moscow. Source: Skoltech.

leveled at the Skolkovo Foundation itself. Some believe that Skolkovo has become a victim of political disagreement within the Duma and that allegations of fraud are as much political as they are factual. Moreover, there remains tensions within the Russian state between building regional development policies around the principle of spatial polarization (based on growth poles and mega-projects in a few privileged centers) on the one hand, and spatial equalization (spreading development to all regions in a more dispersed way) between regions on the other. The success of Skolkovo will depend upon the extent to which the former triumphs over the latter; that is, if glamorous mega-projects like Skolkovo remain in vogue. And, of course, aggravated by tensions in Ukraine and Syria, relations between Russia and the West are poor.

Given the political clout and finances that have been plowed into the project, if Skolkovo fails, it is unlikely other regions with less political will and fewer resources will ever be able to clone Silicon Valley.

Conclusion

A product of the rise of the European world capitalist economy from 1450, our world is marked by uneven geographical development, denoted crudely by the terms Global North and Global South. But it is a restless world in which inherited patterns of uneven development (established by an OIDL and comprising core, semi-peripheral, and peripheral regions) are being reworked and revised. Up until the mid-1970s and notwithstanding the end of the age of empire, Global North core countries thriving under a Fordist-Keynesian economic model continued to grow economically faster than the countries of the Global South. The crisis of the global capitalist system that has been working its course since the 1970s has created a new era of neoliberal capitalism. Amid the restructuring that is taking place, a range of complex new economic spaces are emerging, reshuffling relationships between core, semi-peripheral, and peripheral regions. It is certainly fair to conclude that neoliberal capitalism has not solved the crisis in the capitalist system. Indeed, the economic recession that devastated the global economy following the 2007 financial crash indicates the extent to which neoliberal capitalism is a poor political and economic system, destined to create further crises. But it continues to endure. What the recession could not break, perhaps Covid-19 and its economic aftershocks will. If Western countries are to retain their roles as leaders of the world economy, they will need to secure competitive advantage in the rising AI- and data-driven economy. Silicon Valley in California will continue to set the standard, but it is a model that has proven difficult to replicate.

Checklist of Key Ideas

- Economic geographers make use of the key concepts of Global Commodity Chains (GCCs), Global Value Chains (GVCs), and Global Production Networks (GPNs), understanding local and regional economies in terms of the ways in which they couple and decouple from these chains and networks.
- Our unequal world was formed by the rise from the fifteenth century of a European-led world capitalist economy, which gave birth to an Old International Division of Labor (OIDL) comprising core regions, semi-peripheral regions, and peripheral regions.
- Parisian regulation theory uses the concepts of "regime of accumulation" and "mode of regulation" to understand cycles of growth and crisis in the capitalist economy.
- Although the Fordist-Keynesian compromise led to thirty glory years of capitalist growth (1945–1975) in the core, this compromise failed in the 1970s, plunging the capitalist world into a crisis from which it is still to recover. A period of neoliberal capitalism has emerged. As the global economic crisis that has burdened the world since 2007 testifies, it is doubtful if neoliberal capitalism will be able to deliver sustained economic growth over time.
- Faced with a profit squeeze in the 1970s, capitalist firms have created both a New International Division of Labor (NIDL) and a set of new flexible production practices that can be termed post-Fordist. In core countries, the NIDL has both deindustrialized older industrial cities and towns and created a small number of powerful world cities. Post-Fordist industrial production systems are marked by clusters of firms embracing flexible specialization. Post-Fordist firms tend to reconcentrate branches and functions around one location in these clusters.
- To maintain economic hegemony in the twenty-first century, Western core countries will need to secure competitive advantage in the emerging computerized data analytics and data-driven economy. To achieve economic growth, they will need to establish a regulatory system that subordinates technology to public oversight, champions 'tech for good', and secures for computerized data analytics a social license and technical efficacy.

Chapter Essay Questions

1) Explain what is meant by both the OIDL and the NIDL. In what ways and to what extent has the NIDL restructured the geography of the world economy?
2) The "Third Italy" provides a template that other regions ought to copy. Discuss.
3) Using concepts from Parisian regulation theory, write an essay on the fate of Western core economies from 1945 to the present.
4) Silicon Valley in California provides a template that other regions ought to copy. Discuss.

References and Guidance for Further Reading

Excellent introductions to economic geography are provided in:
Barnes, T. and Brett, C. (2018). *Economic Geography: A Critical Introduction*. Oxford: Wiley.
Clark, G.L., Feldman, M.P., Gertler, M.S., and Wójcik, D. (eds.) (2018). *The New Oxford Handbook of Economic Geography*. Oxford: Oxford University Press.
Coe, N.M., Kelly, P.F., and Yeung, H.W.C. (2020). *Economic Geography: A Contemporary Introduction*. Oxford: Wiley Blackwell.
Dicken, P. (2015). *Global Shift: Mapping the Changing Contours of the World Economy*, 7e. London: Guilford Press.
Knox, P., Agnew, J.A., and McCarthy, L. (2014). *The Geography of the World Economy*. London: Routledge.
Gilpin, R. (2018). *The Challenge of Global Capitalism: The World Economy in the 21st Century*. Princeton, NJ: Princeton University Press.
Hudson, R. (2016). *Approaches to Economic Geography: Towards a Geographical Political Economy*. London: Routledge.
MacKinnon, D. and Cumbers, A. (2020). *Introduction to Economic Geography: Globalization, Uneven Development and Place*. London: Routledge.
Yeung, H.W.C. (2016). *Strategic Coupling: East Asian Industrial Transformation in the New Global Economy*. Ithaca, NY: Cornell University Press.

An important book that charts the rise of the European world economy that began in 1450 to an earlier thirteenth- and fourteenth-century Eurasian economy is:
Abu-Lughod, J.L. (1991). *Before European Hegemony: The World System A.D. 1250–1350*. Oxford: Oxford University Press.

World-systems analysis was first pioneered by Immanuel Wallerstein:
Wallerstein, I. (1974). *The Modern World System: Capitalist Agriculture and the Origins of the European World Economy*. London: Academic Press.
Wallerstein, I. (1980). *Mercantilism and the Consolidation of the European World Economy, 1600–1750*. New York: Academic Press.
Wallerstein, I. (1988). *The Second Era of Great Expansion of the Capitalist World Economy, 1730 to 1840s*. San Diego, CA: Academic Press.

The pioneering book that established the French Regulation School was published in French in 1976, and in English in 1979 as:
Aglietta, M. (1979). *A Theory of Capitalist Regulation*. London: New Left Books.

Other important books in the Parisian regulation tradition are:
Boyer, R. and Saillard, Y. (2001). *Regulation Theory: The State of the Art*. London: Routledge.
Lipietz, A. (1987). *Mirages et Miracles: The Crises of Global Fordism*. London: Verso.

The important works of key twentieth-century economists John Maynard Keynes, Friedrich August von Hayek, and Milton Friedman include:

Hayek, F.A. (1944). *The Road to Serfdom*. Chicago: University of Chicago Press.

Keynes, J.M. (1936). *The General Theory of Employment, Interest and Money*. London: MacMillan.

Friedman, M. (1962). *Capitalism and Freedom*. Chicago: University of Chicago Press.

The following books written by David Harvey provide an excellent overview of the Fordist-Keynesian compromise, the thirty glory years of capitalism, the 1970s crises in capitalism, and the rise of neoliberal capitalism:

Harvey, D. (1989). *The Condition of Postmodernity: An Enquiry into the Origins of Cultural Change*. Oxford: Wiley Blackwell.

Harvey, D. (2005). *A Brief History of Neoliberalism*. Oxford: Oxford University Press.

Harvey, D. (2010). *The Enigma of Capital: And the Crises of Capitalism*. Oxford: Oxford University Press.

Harvey, D. (2014). *Seventeen Contradictions and the End of Capitalism*. New York: Oxford University Press.

Harvey, D. (2017). *Marx, Capital, and the Madness of Economic Reason*. Oxford: Oxford University Press.

Original introductions to the ideas of the spatial division of labor and the New International Division of Labor (NIDL), respectively, are provided in:

Massey, D. (1984). *Spatial Divisions of Labour: Social Structures and the Geography of Production*. London: Methuen.

Fröbel, F., Heinrichs, J., and Kreye, O. (1980). *The New International Division of Labour: Structural Unemployment in Industrialised Countries and Industrialisation in Developing Countries*. Cambridge: Cambridge University Press.

The best introductions to world cities remain:

Friedmann, J. (1986). The world city hypothesis. *Development and Change* 17: 69–83.

Sassen, S. (2001). *The Global City: New York, London, and Tokyo*, 2e. Princeton, NJ: Princeton University Press.

Sassen, S. (2018). *Cities in a World Economy*. London: Sage.

Important writings on Ireland's neoliberal economic model can be found in:

McCabe, C. (2013). *The Sins of the Father: Tracing the Decisions That Shaped the Irish Economy*. Dublin: The History Press.

Riain, S.Ó. (2014). *The Rise and Fall of Ireland's Celtic Tiger: Liberalism, Boom and Bust*. Cambridge: Cambridge University Press.

Roche, W.K., O'Connell, P.J., and Prothero, A. (eds.) (2017). *Austerity and Recovery in Ireland: Europe's Poster Child and the Great Recession*. Oxford: Oxford University Press.

The growing importance of post-Fordist production systems was first charted by:

Piore, M.J. and Sabel, C.F. (1984). *The Second Industrial Divide: Possibilities for Prosperity*. New York: Basic Books.

Important works examining post-Fordist clustering include:

Barnes, T.J. and Gertler, M.S. (eds.) (1999). *The New Industrial Geography: Regions, Regulation and Institutions*. London: Routledge.

Cooke, P. and Morgan, K. (1998). *The Associational Economy: Firms, Regions, and Innovation*. Oxford: Oxford University Press.

Gertler, M.S. (1988). The limits to flexibility – comments on the post-Fordist vision of production and its geography. *Transactions of the Institute of British Geographers* 13 (4): 419–432.

Martin, R. and Sunley, P. (2003). Deconstructing clusters: chaotic concept or policy panacea? *Journal of Economic Geography* 3: 5–35.

Porter, M.E. (1990). *The Competitive Advantage of Nations*. New York: Free Press.

Scott, A.J. (1988). *New Industrial Spaces: Flexible Production, Organisation and Regional Development in North America and Western Europe*. London: Pion.

Influential books announcing the arrival of Industry 4.0 (IR4) and its limitations are:

Lee, K.F. (2018). *AI Superpowers: China, Silicon Valley, and the New World Order*. Boston: Houghton Mifflin Harcourt.

Schwab, K. (2017). *The Fourth Industrial Revolution*. New York: Currency Books.

Zuboff, S. (2019). *The Age of Surveillance Capitalism: The Fight for a Human Future at the New Frontier of Power*. New York: Profile Books.

Good introductions to Silicon Valley as a high-technology industrial cluster can be found in:

Kenney, M. (2000). *Understanding Silicon Valley: The Anatomy of an Entrepreneurial Region*. Stanford, CA: Stanford University Press.

O'Mara, M. (2019). *The Code: Silicon Valley and the Remaking of America*. New York: Penguin Press.

Saxenian, A.L. (1994). *Regional Advantage: Culture and Competition in Silicon Valley and Route 128*. Cambridge, MA: Harvard University Press.

Saxenian, A.L. (2007). *The New Argonauts: Regional Advantage in a Global Economy*. Cambridge, MA: Harvard University Press.

Storper, M., Kemeny, T., Makarem, N., and Osman, T. (2015). *The Rise and Fall of Urban Economies: Lessons from San Francisco and Los Angeles*. Stanford, CA: Stanford University Press.

Website Support Material

A range of useful resources to support your reading of this chapter are available from the *Wiley Human Geography: An Essential Introduction* Companion Site http://www.wiley.com/go/boyle.

Chapter 5

Power: The Governmental Machine of the West

Chapter Table of Contents

Chapter Learning Objectives

By the end of this chapter you should be able to:
- Discern the classical (Greek and Roman) origins of many of the West's central political constructs. Using the ideas of sovereign "power over" and governmental "power to," unpack the various concepts of power that underpin the governmental machine of the West.
- Approach human territorialization as historical and processual rather than an innate species condition, and appreciate the historical novelty and provincial (Europe, Westphalia, 1648) origins of the idea of the sovereign nation-state.

Human Geography: An Essential Introduction, Second Edition. Mark Boyle.
© 2021 John Wiley & Sons Ltd. Published 2021 by John Wiley & Sons Ltd.
Companion website: www.wiley.com/go/boyle

- Comment on the extent to which the sovereign nation-state is now ceding to a new medievalism characterized by a return to multiscalar governance.
- Recognize the meaning and implications of the rise of the West as the first global hegemon, and explain how a number of more rapacious and expansive European nation-states have played a central role in shaping the world in which we now live. Apprehend anticolonialism, and document the collapse of Europe's empires. Present a detailed case study of Europe's frenetic late nineteenth-century and early twentieth-century "scramble for Africa."
- Comment upon the extent to which there exists a US empire. With reference to late twentieth- and early twentieth-first-century geopolitics, comment on changing relations between the United States, Russia, militant Islam, and China. Reach a conclusion on whether we are heading towards a multipolar world.

Introduction

Human territorialization – claims made by people to exclusive ownership of and sovereign rule over discrete patches of turf – took a decisive turn with the rise of Europe's nation–states from the seventeenth century. Had Europe's nation–states not existed and had these states not sought to project their power and impose their wills and agendas, first on European peoples and then on other peoples in other places, undoubtedly we would live in a very different world today.

To explore the story of the rise of the West as the world's first truly global hegemon, this chapter will first ruminate on the classical (Greek and Roman) origins of the concepts of power that have defined the West's governmental machine. It will then examine the origins in Europe (in Westphalia in 1648 and with the 1848 spring time of nations) of the idea of the sovereign nation-state and the claim that the nation-state enjoys only a fragile existence. It will examine the current hollowing out of nation–states and the possible return to models of multiscalar governance more commonly associated with the medieval period. Although in the context of Big History it is likely that the appearance on the historical stage of the sovereign nation-state will be fleeting, these states have enjoyed hugely consequential lives. This chapter will explore the ways in which Europe's unique polity has left an indelible imprint not only on European societies but also on other peoples in other places, including by globalizing, through empire and in other ways, their own very particular ideas about statehood and statecraft. It will note the role the United States now plays in defending the West's hegemony and will question whether there exists such a thing as a US empire. It will conclude by examining rival powers that have threatened and continue to threaten the dominance of the United States, in particular Russian authoritarian capitalism, Islamic fundamentalist theocracies, and Chinese command capitalism.

Power: The Governmental Machine of the West

In the encounter between the world and the West that has been going on now for 400 or 500 years, the world, not the West, is the party that, up to now, has had the significant experience. It has not been the West that has been hit by the world; it is the world that has been hit – and hit hard – by the West.

British historian Arnold J. Toynbee, Reith Lectures, BBC, 1952.

By what means have Western states hit, and hit hard, distant others? Scholarship on Western power has ruminated over the status of two competing conceptions of power: a sovereign paradigm and a governmental paradigm. Two ideas of power then follow.

- **Power over**: Supreme political authority is invested in a sovereign (such as a monarchy or parliament), which then rules and governs absolutely. Sovereigns have the ultimate power over life and death; subjects have little option but to be subservient. Power is discharged through the rule of law. Tyrants and dictators reveal sovereign power's darkest possibilities.
- **Power to**: Institutions rule only indirectly by inculcating in subjects certain expectations about what constitutes acceptable and unacceptable behavior. Power is hidden in the language we use in everyday life; words create worlds, do work in the world, and enable power to be exercised. Power is discharged at a remove, by responsibilizing citizens and building their ability to self-regulate and self-calibrate according to prescribed ideas of "common sense" (See Deep Dive Box 5.1).

Perhaps in their earliest days, Europe's monarchical nation–states ruled principally through sovereign power – "power over." Understanding themselves to be divine, omnipotent, and free from any earthly authority, kings and queens ruled with impunity. Subjects cowered under autocratic leaders who had it in their gift to confiscate land and destroy and take lives at a whim.

According to French philosopher Michel Foucault, the maxim of monarchical rule is: *the king both rules and governs.*

Deep Dive Box 5.1   Key Thinkers on the Governmental Machine of the West: Michel Foucault and Giorgio Agamben

French philosopher Michel Foucault (1926–1984) and Italian philosopher Giorgio Agamben (1942–) have made decisive contributions to our understanding of the exercise of power in the governmental machine of the West. Foucault, a twentieth-century intellectual giant, wrote extensively on power and its implications for the workings of modern institutions such as prisons, asylums, clinics, and education establishments. Power was the center of attention in his famous *College de France Lectures* delivered in Paris between 1970 and 1984. In these lectures, Foucault claimed to discern a historical shift from "power over" to "power to" from the mid-eighteenth century, at least in advanced liberal states. Meanwhile, Agamben's contribution has been to excavate the classical (Greek and Roman) origins of many of the West's central political constructs. In his acclaimed *Homo Sacer* series of books (1995–2015), Agamben offers a "theological genealogy" of Western political constructs, noting the formative influence of classical, Christian, and medieval political, economic, legal, and linguistic history on contemporary forms of Western power and political theory. For Agamben, although seemingly "antinomical" (contradictory), sovereign power and governmental power are by necessity "functionally related"; it is a mistake therefore to consider either in isolation (Figure 5.1).

(Continued)

Box 5.1   *(Continued)*

Foucault's principal books (1962–1984) and *College de France Lectures* delivered in Paris (1970–1984)	Giorgio Agamben's *Homo Sacer* series of books (1995–2015)
Books (in order) Mental illness and psychology Madness and civilization: A history of insanity in the age of reason The birth of the clinic: an archaeology of medical perception The order of things: an archaeology of the human sciences The archaeology of knowledge Discipline and punish: the birth of the prison The history of sexuality, vol. 1: an introduction The history of sexuality, vol. 2: the use of pleasure The history of sexuality, vol. 3: the care of the self Lectures (in order) 1970–1971 The Will To Know 1971–1972 Penal Theories And Institutions 1972–1973 The Punitive Society 1973–1974 Psychiatric Power 1974–1975 Abnormal 1975–1976 'Society Must Be Defended' 1976–1977 No Lectures 1977–1978 Security, Territory, Population 1978–1979 The Birth Of Biopolitics 1979–1980 On The Government Of The Living 1980–1981 Subjectivity And Truth 1981–1982 The Hermeneutics Of The Subject 1982–1983 The Government Of Self And Others 1983–1984 The Courage Of Truth	Books Although published out of sequence and in no way linear, Agamben arranges the books into four categories: Category 1 (1) Homo Sacer: Sovereign Power And Bare Life Category 2 (2.1) State Of Exception (2.2) Stasis: Civil War As A Political Paradigm (2.3) The Sacrament Of Language: An Archaeology Of The Oath (2.4) The Kingdom And The Glory: For A Theological Genealogy Of Economy And Glory (2.5) Opus Dei: An Archaeology Of Duty Category 3 (3) Remnants Of Auschwitz: The Witness And The Archive Category 4 (4.1) The Highest Poverty: Monastic Rules And Form-Of-Life (4.2) The Use Of Bodies

Figure 5.1   Key works by Michel Foucault and Giorgio Agamben. Source: Inspired by Philo (2012).

Foucault's genius was to identify that power was not so much wielded by an omnipotent centralized autocrat but instead was cultivated through what he called *dispositifs* (a dispersed battery of techniques of control, discourse, and regimes of truth). For Foucault, at least in Western liberal polities, power understood as absolute domination over territory and people has ceded to new machinations centered upon the management of populations through the promotion of new modes of subjectification – socializing human beings so that they internalize values and regulate their behavior in preferred ways. Governmentality can be understood as the "art of government," or the calculated means through which governments condition our mentalities and rationalities and work to "normalize" us. Power and knowledge act together; institutions gain the "power to" do things to human beings only if prevailing thought structures make it appear that they are acting on the basis of truth. *Dispositifs* enable institutions to pathologize human behavior that falls outside the mainstream, construing it as deficient, deviant, and even a sign of insanity, and licenses them to impose corrective and curative therapies. Power does not emanate from a sovereign; it is achieved as citizens yield to technologies of subject formation and regulate their own behavior. Power is everywhere – embedded in the wide range of everyday rituals that enable common sense to be reproduced as common sense.

Foucault's own maxim, then, is: *we must cut off the king's head.*

In the seminal opening book in the series, *Homo Sacer: Sovereign Power and Bare Life*, Italian philosopher Giorgio Agamben (2005) argues that sovereign power is very much alive and well. But, he asks, what does sovereign rule look like today?

For Agamben, sovereignty now expresses itself in the form of denying citizenship and human rights rather than enforcing them. In Roman law, the phrase *homo sacer* was used to denote a person who could be killed (although not sacrificed in a religious ritual) by any citizen without legal implications for the perpetrator. The sovereign is powerful because it has the ability to create what Agamben terms a "state of exception." Sovereigns have the authority to strip citizens of political status (bio) and degrade them to "bare life" (zoë) by placing them outside of law and juridical protection. This is power in its rawest form. Examples include denying human rights to prisoners, extrajudicial state killings and torture, illegal wars, and holding indefinitely refugees in temporary camps.

While insisting upon the ongoing importance of sovereign power ("power over"), like Foucault, Agamben recognizes also the central importance in liberal societies of governmental "power to." His point of departure is the claim that genealogies of the concepts of power that underpin the governmental machine of the West often stop prematurely in the mid-eighteenth century. As a consequence, we wrongly suppose that both conceptions of power are best figured in isolation. For Agamben, the modern dualism between sovereign power and governmentality is but a secularized reworking of the early Christian dualism between God as transcendent (God as a supernatural being, a Godhead) and God as immanent (God as an active agent in the world). As such, per Saint Augustine of Hippo's (354–430 CE) Trinitarian theology, they must be understood as they were originally intended: as consubstantial.

St Augustine's theology reminds us that God created humans as autonomous beings with freedom of choice and understood the potential for this to create satanic anarchy. God retains sovereign power: a plague of locusts, meteorite storm, or flood could be visited on the human race at any time should it wander too far off course. But God also loves us and recognizes the weaknesses that make his (*sic*) gift of free will a burden. He intercedes in the world indirectly to support us, sending angels as his messengers (intermediaries who help us to regulate and mainstream our behavior so that it aligns with his providential plan). What is most important is that both interventions issue from a single being, God.

To sustain the integrity of their origins, the same logic must be applied to the ideas of power that underpin the governmental machine of the West. The only difference is that today, it is governments and not God who put into play sovereign power, and this power takes the form of "states of exception" and not floods, meteor showers, or locusts. And it is no longer angels who serve as intermediaries between the sovereign and the governed, but the disciplining rationalities, technologies, and modes of subjectification that align Western subjects with Western understandings of providential design. Still, "power over" and "power to" remain as they were initially conceived in Trinitarian theology: consubstantial. Without the threat of "power over," "power to" would not be as effective; and without "power to," "power over" would be too dictatorial and fascistic to be sustainable.

Agamben's maxim, then, is: *the king rules but (mostly) does not govern.*

There has arisen a debate over whether these ideas of power apply equally to the actions of states in the international arena; that is, whether Western states have interiorized these ideas so that they frame the way governance is accomplished not only domestically but also internationally. The ideas of hard power and soft power, first introduced by US political scientist Joseph Nye (2004), provide a way of forging a link:

- **Hard power** can be construed as the deployment by one country of aggressive and direct military and economic force so as to control the direction of another country. For Nye, hard power uses guns (compulsion), sticks (coercive diplomacy), and carrots (inducements to conform) to force another to act in particular ways. Brute force, the use of colonial violence, imperialist law, the imposition of neocolonial puppet regimes, and heavily conditioned financial loans are all examples of hard power.
- **Soft power**, in contrast, can be understood as the ability to shape the preferences of others through appeal and attraction. In part, soft power derives from propaganda. A defining feature of soft power, however, is that it is noncoercive; instead, it seeks to persuade distant others to inculcate belief and value systems that serve "our interests." The manipulation of others by creating and disciplining "safe" subjectivities is key. People from beyond the West come to serve the West by becoming Western subjects.

Perhaps in light of the work of Giorgio Agamben, it is wise to conclude that **hard power/power over** and **soft power/power to** be considered together. Indeed, according to Nye (2004), states are most effective in imposing their agendas on the world when they exercise both hard power and soft power judiciously: states that find the correct balance between both practice overall "smart power."

The Rise (and Fall?) of the "European" Nation-State

Beyond Medieval Polities: The Rise of the "European" Nation-State from 1648 and 1848

We tend to take the political map of the world that presents itself today as a given. The world is divided into political units called countries. Unless the international community sees fit to intervene, the rulers of these countries enjoy sovereign authority to command as they please. It seems that it has been this way forever. But it has to be remembered that the idea of the sovereign nation-state is of recent origin and is a peculiar invention when set into historical context.

As early as 1986, United States geographer Robert D Sack challenged the claim that human territoriality is a species condition arguing instead that it be theorized and historicized. In his book *The Birth of Territory*, British political geographer Stuart Elden (2013) likewise argues that human beings are not by biological design territorial creatures. They are not hard-wired or compelled by their DNA to claim ownership of turf. According to Elden, it is important to approach the idea of human territoriality conceptually and historically. Political geographers often date the notion that every community has the right to exercise exclusive sovereignty over its territory to the Peace of Westphalia (1648). But for Elden, human territorialization has been much longer in germination; key supporting concepts were developed from as early as the Classical period, evolved throughout the dark Middle Ages, and came to fruition in the West during the Renaissance and Age of Reason. Using the tools of etymology, semantics, philology, and hermeneutics, it is possible to trace core Western political concepts to political speeches, texts, thought, and debates dating from at least antiquity. The fore-bears of the signature political ideas that now undergird the governmental machine of the West first appeared in Greek and Roman public life and journeyed to their present form and meaning over the past 2000 years. Elden's assertion, then, is that the proclivity and competency of human beings to lay claims to territory emerged processually and contingently *within* history.

Still, while boiling and frothing in the wider atmosphere, in Europe in the medieval period the concept of the sovereign nation-state would have appeared a strange and foreign one to many. No single authority presided over any particular territory. Instead, territories were administered by a range of overlapping governing regimes, including emperors, the papacy, kings and queens, feudal knights, barons and baronesses, counts and countesses, dukes and duchesses, and so on. Increasingly, this arrangement created conflict and tension. Polities made competing claims over space, and quarrels, spats, and wars became the norm. In the seventeenth century, the most important of these conflicts came to a head. Europe's disastrous Thirty Years' War (1618–1648) was triggered by efforts by the Holy Roman Empire to reassert Catholic hegemony over adjacent reformist Protestant regions. Warring parties agreed to resolve their differences through diplomacy and negotiation. The Peace of Westphalia (Germany) was signed in 1648, and Europe's landmass was effectively divided into discrete parcels and allocated to particular polities. The idea of the sovereign nation-state was born. Thereafter, the Westphalian system became more widely propagated throughout Europe in the eighteenth century and came to dominate the political organization of the world in the nineteenth and twentieth centuries.

But how was the map of Europe to be carved up? Who had a legitimate claim to sovereign self-rule? How were borders to be decided?

The revolutions across Europe in 1848, known as the "Springtime of Nations," held that there existed nations and that these nations deserved to be self-governing (to be nation–states). But how did the substantive entities we now recognize as *nations* materialize? Challenging a tendency to reify nations – to ascribe to them some natural and matter-of-fact ontological existence – there is now a growing body of scholarship that conceives of nations as relatively recent social constructions that grew out of conditions prevalent from the seventeenth century. Far from being natural or organic entities that have existed from the dawn of time, "nations" are best conceived as historical inventions (see Deep Dive Box 5.2). Of course, to say that nations are relatively recent constructs that exist only in the realms of the imagination is to go too far (Smith 2004). But nations are in an important sense "narratives" (Bhabha 1990), 'invented traditions' (Hobsbawn and Ranger 1983), and "imagined communities" (Anderson 1983); they depend for their existence on an irrational affinity between perfect strangers! When nation builders invoke common ethnicity and race, ethnic nationalisms arise; alternatively, when they mobilize common political ideals and economic and social values, civic nationalisms emerge. Of course, not all nations have statehood, and not all states are built around the idea of a unified nation. For this reason, nation and nation–state building continues to be an active and contested project to this day (Plate 5.1).

Plate 5.1   The Israeli-Palestinian interface: peace wall or apartheid? Source: © DARREN WHITESIDE/Reuters.

Deep Dive Box 5.2   The Invented Traditions of the Irish Nation

Although the Irish nation likes to think of itself as a nation that has existed as long as time itself, in fact it is a relatively recent creation. The most potent narratives of the Irish nation were those created by Ireland's cultural nationalist movements, which came of age in the late eighteenth and the nineteenth centuries. Encouraged by the European revolutions of 1848, cultural nationalism played an integral part in Ireland leaving the British Empire and becoming A free state in 1921, and finally a full republic in effect in 1937 and in title in 1948. These narratives were propagated in the oral tradition, and through folklore, literature, journalism, poetry, theater, film, music, monuments, street and place naming, cultural festivals, religious devotions (for the most part Roman Catholic), sport (through the Gaelic Athletic Association, or GAA), political cartoons, and so on.

The Irish state was founded upon five key narrative traditions.

Narratives of origin: Comprising myths of ethnogenesis, homeland and foundation myths, and myths of descent. These recall the arrival onto the island of Ireland of the Goidelic Celts (the Gaels) between 500 and 300 BCE, the legitimacy of the land claims made by the Gaels over their new homeland, the foundation of Gaelic society as a distinct polity, and the continuous and harmonious descent of the Gaels over the first millennium.

Narratives of a golden age: These seek to recall the greatest achievements of Gaelic society at its pinnacle before (subsequent) foreign intervention. Invariably, these tend to focus upon the period from the sixth to the eighth century CE, when Ireland became a European center of religious and secular learning, and a leading guardian of European civilization.

Narratives of British colonization: Myths that track British involvement in Ireland, representing the British as driven by imperial greed, capable of acts of evil and at times cowardly aggression in pursuit of cultural and economic supremacy, and immune from any appreciation of the rights of other peoples. British rule is invariably traced back to the landing of Strongbow in 1169 CE, and thus occupation is represented as occurring over an 800-year period.

Narratives of the Irish diaspora: These record the ways in which the British repression of Ireland proved to be midwife to massive out-migration, peaking in the period 1845 to around 1920, and leading to the formation of one of the world's largest diasporic communities (estimated today to be around 70 million in strength). The Great Irish Famine of 1848 to 1851 (Plate 5.2) and associated large-scale emigrations to the United Kingdom, United States, Canada, Australia, New Zealand, and Argentina are taken to be the chief exemplars.

Narratives of rebellion and uprising: These seek to portray the stoic suffering the Irish have endured under British colonization and colonial rule, the repeated and heroic acts of resistance and rebellion they have put up to this rule leading to independence, and the assertion that Ireland's Gaelic past will never be extinguished and that British rule in Northern Ireland will eventually be broken, leading to a United Ireland.

Plate 5.2   The Great Irish Famine (1847–1851) Memorial IFSC, Dublin, Ireland.

Is the Sovereign Nation-State Obsolete? Back to Medieval Polities?

There are few grounds to assume that the idea of the sovereign nation–state will last over the *longue durée*. Indeed, according to some, we are already in the throes of great change, and the contours of a new political system are emerging. Nation–states are currently being "hollowed out," and a new era of multilevel governance is appearing. We are witnessing, they say, a reversal back to a medieval system of governance based upon a series of overlapping and sliding layers of authority.

What might be driving such a shift? The power of the world economy over weakened and vulnerable nation–states means that these states are becoming ever more toothless and impotent when confronted by gigantic, bruising, and bullying transnational corporations (TNCs). Accordingly, nation–states are ceding power to bigger political entities operating at larger geographical scales. Continental or even global political institutions, it is contended, have more muscle and command over global processes and can shape these processes more effectively. Equally, while nation–states are often too small to tackle the big forces at work in the world today, they are often too large to address the everyday concerns of people. Accordingly, there has been a surge in interest in many countries in devolving more powers from central states down to existing and newly created regional and local authorities.

Evidence of the hollowing out of the nation-state is plentiful. Increasingly, national states are turning to global political institutions (international organizations [IOs]) such as the United Nations (UN), the International Monetary Fund (IMF), the World Bank, the Organisation for Economic Co-operation and Development (OECD), the World Health Organization (WHO), and the North Atlantic Treaty Organization (NATO) to help them solve their problems. Moreover, they are increasingly clubbing together into supranational regional alliances, trading blocs, economic communities, and even political unions. Perhaps the European Union is the foremost example of a continent-wide political union in the world today, knitting together as it does no fewer than 27 individual member states into ever more binding social, economic, environmental, cultural, and political agreements (see Map 5.1). At the same time, in order to sharpen and refine their capacity to solve subregional and local problems, nation–states are, through both choice and coercion, ceding ever more authority to smaller political units within their borders. Some states are already federal in structure (such as Australia, Canada, the United States, India, Belgium, Spain, and Germany), that is, they already enshrine within their constitutions the rights of subnational and local authorities to govern autonomously in certain policy areas. Others, however, remain highly centralized (such as Greece, Russia, Bulgaria, Croatia, Norway, Ireland, Denmark, France, and the United Kingdom) and are facing pressures to devolve further power and executive functions to regional, city, and local institutions (Deep Dive Box 5.3). In some cases, subnational separatist movements are placing the very future of the unitary nation–state at risk.

Perhaps, then, the twenty-first century will see a return to medieval-style polities, what Scottish geographer Ronan Paddison once called *The Fragmented State* (1983), in which there exists no single sovereign but only sliding layers of authority, each vying to rule over this or that strip of the earth's surface.

However, the nation-state might just be a more stubborn and enduring form of polity than presumed. Recently an opposing set of forces have come to the fore.

Key facts:

Founded: 1952

Member states (2021): 27 states

Population: 445 million

Candidate countries: Albania, the Republic of North Macedonia, Montenegro, Serbia and Turkey

Function: Began as a customs union, developed as a common market, progressed to an economic and monetary union, and now stands on the threshold of becoming a full-blown political union.

Map 5.1   Map of the European Union at 2021. (Source: European Commission)

Globalization, neoliberalism, inequalities, the global financial crash, and austerity have combined, it seems, to effect a growing dissonance between representative democracy and popular sovereignty. Multiscalar governance appears to be an inadequate response. The variety of nationalist populist movements that are currently springing up around the world (for example, led by Trump [United States], Johnston [United Kingdom], Wilders [the Netherlands], Hoffer/Kurz [Austria], Orbán [Hungary], Le Pen [France], Bolsonaro [Brazil], Syriza [Greece], and Podemos [Spain]) appear to be signaling a revival of patriotism and nationalism and a reassertion of strong centralized governments.

Deep Dive Box 5.3   Brexit Britain: A Fragmenting *and* Defragmenting State?

With the Westminster Parliament at its political epicenter, the United Kingdom remains one of the most centralized nation-states in Europe. But, as elsewhere, the British state is now subject to processes of "hollowing out."

The fate of Britain is increasingly entangled in the work of the World Bank, UN, WHO, OECD, NATO, and IMF. More significantly, the United Kingdom joined the European Economic Community (EEC) on 1 January 1973, and remained a committed member of this community for 47 subsequent years as the EEC broadened (incrementally to 28 member states) and deepened (morphed into the world's most sophisticated supranational political union, the European Union [EU]). Meanwhile, devolution of powers and resources from Westminster has commenced, albeit unevenly across the country: the Scottish Parliament, the National Assembly for Wales, the Northern Ireland Assembly, and the Greater London Authority are all examples. Moreover, English devolution is much discussed today, and across England new combined city–region authorities have been created, and directly elected *Metro-Mayors* have been installed (Map 5.2).

But has this rescaling of the British state weakened the power of Westminster? Arguably no.

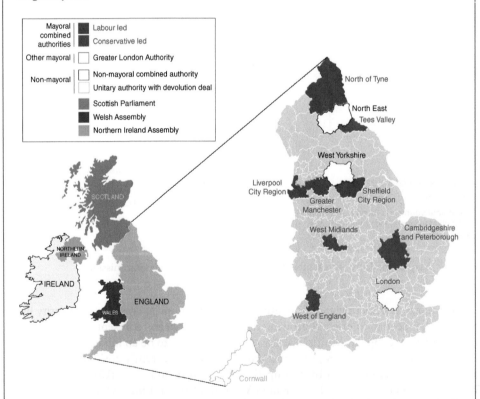

Map 5.2   Devolution in the United Kingdom at 2020: A mosaic of sub-national authorities. Source: Institute for Government June 2019.

On 23 June 2016, the United Kingdom held a plebiscite on the status of its ongoing membership in the EU: with a turnout of 72.2%, 51.9% voted to "Leave the EU." In December 2019, the United Kingdom concluded a trade deal with the EU to ensure an orderly "Brexit" (British exit from the EU) and left the EU orbit officially on January 1st 2020. Meanwhile, in spite of promises of further devolution ("devo-max"), there seems little appetite in government to transfer additional and significant powers and resources to UK regions. English devolution has stalled, and a new referendum on Scottish independence has been blocked. With Brexit and now Covid-19 dominating political attention, devolution looks set to be relegated as a priority for now.

Why is multiscalar governance not accelerating at pace in the United Kingdom?

Globalization, neoliberalism, and deindustrialization have created a *North–South* divide; the North of England comprises Northern England, Wales, Scotland, and Northern Ireland, and the South, London and the South East. According to critics, the hollowing out and fragmented British state has been unable to protect declining industrial cities and left-behind places. Brexit has come to be associated with the politics of the urban "left behinds." Unfairly, Europe has been blamed for promoting fiscal austerity and failing to step up to help declining regions. In this sense, the vote to leave the EU has been characterized as a "revolt of the rustbelt." Meanwhile, although devolution is seen as a solution to uneven geographical development, it is recognized that a strong Westminster-led redistribution of the national product is key. Westminster needs to take back control, level up the space economy, and rescue declining blue-collar cities and towns. Brexit, moreover, is a nostalgic reflex that betrays an imperial yearning to restore Britain's lost place in the world: in the words of British geographers Danny Dorling and Sally Tomlinson (2018), "the last vestiges of empire working their way out of the British psyche."

And so two competing logics – and sociospatial processes – are currently at work:

A fragmenting state: Westminster is feeling helpless and hapless, unable to shape events that are unfolding at larger scales and over which it has little control, and to strengthen its muscle it is leveraging supranational institutions. It is also being recognized that insofar as they operate at finer geographical scales, more localized state tiers know better the challenges facing particular localities, and therefore there are grounds to scale devolution. A break-up of Britain is afoot in which Parliament is ceding power to institutions operating at a wide range of overlapping scales and better suited to remediate different kinds of problems. Centrifugal forces are overwhelming centripetal forces.

A defragmenting state: British nationalism is on the ascendancy, and Westminster is consolidating its standing as a crucible of power. Multiscalar states have proven themselves unable to deal with problems created by globalization, neoliberalism, and uneven geographical development. Supranational organizations are too big to tackle the problems nations are encountering and insensitive to nations' histories, but subnational regional and local authorities are too small to do anything meaningful about local challenges even if they understand these challenges more intimately. Centripetal forces are overwhelming centrifugal forces.

Europe's Nation–States and Empires: Europe's Scramble for the World

The Age of European Empires

While all the time jostling for control of lands in Europe held by their European neighbors, from the start of the sixteenth century Europe's polities (and, after 1648 and 1848, European nation–states) steadily turned their attention to the colonization and annexation of lands further afield. Power at a distance was exercised through the imposition of both colonial (direct rule) and neocolonial (quasi-direct rule but through third parties, including puppet governments) regimes (see Deep Dive Box 5.4). Colonization and control of territory, especially lands rich in resources and endowed with a number of strategic advantages, came to be viewed as essential if European nation–states were to prosper and avoid being outflanked by near rivals. Following a period of expedition and exploration, successive waves of colonization steadily brought vast areas of the world under the control of European powers. Cover was provided in the form of the idea of colonization as a *mission civilisatrice* (civilizing mission) (see Deep Dive Box 5.5). Among the countries of Europe to establish empires were Portugal, Spain, the Netherlands, France, the United Kingdom, Russia, Sweden, Germany, Belgium, Denmark, Italy, and Norway.

Deep Dive Box 5.4   Halford J. Mackinder's Heartland Thesis

At the Royal Geographical Society in London in 1904, British geographer Halford J. Mackinder delivered a lecture entitled "The Geographical Pivot of History," during which he outlined his famous Heartland thesis (Map 5.3). This was followed with a fuller exposition of the Heartland thesis in 1919 in his book *Democratic Ideals and Reality: A Study in the Politics of Reconstruction.*

Mackinder conceived of the earth's surface as being organized into a number of broad regions. At the center was the World-Island. Incorporating Europe, Asia, and Africa, the World-Island was the world's richest and most heavily populated, geographically expansive, and powerful region. It was further divided between the pivot area (the heartland stretching from the Volga to the Yangtze Rivers and from the Himalayas to the Arctic), the inner or marginal crescent (from Europe through North Africa to South Asia and Southeast Asia), and the outer islands of Britain and Japan. The further one moved from the heartland, the more one ventured into the "lands of the outer or insular crescent," which for Mackinder centered upon the continents of North America, South America, and Australia.

Mackinder believed that any country that controlled the World-Island also controlled over half of the world's known resources. To control the World-Island it was necessary, in turn, to control the heartland. For geographical reasons, it had proved difficult for any country to dominate the heartland historically. Polar conditions to the north, desert conditions to the south, and lack of communication routes from east to west militated against comprehensive colonization of the region. Although the heartland was governed substantially by the Russian Empire, Mackinder believed that it was possible that a European power or a power from Asia (China or Japan) might attempt to seize it. He warned the British government that Britain could not afford to lose the heartland to another power if it

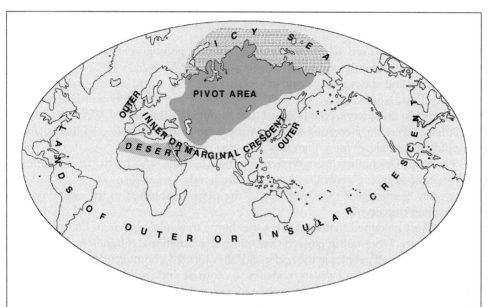

Map 5.3   Halford J. Mackinder's Heartland thesis. Source: Mackinder (1919).

wished to continue to be the most dominant country on earth. For Mackinder, "Who rules East Europe commands the heartland; who rules the heartland commands the World-Island; who rules the World-Island controls the world."

Mackinder's thesis was heavily studied by, and proved influential in shaping the thinking of, political strategists in countries as far and wide as Japan, China, Russia, Germany, Britain, and the United States. Perhaps the greatest test of the validity of the thesis came during the Cold War period from 1945 to 1989. In spite of Russian dominance of the pivot area, the Soviet Union was unable to compete over time with the United States and Europe, and from the 1970s onward began to fail as a global superpower and to crumble. For some, this indicates that Mackinder was wrong to assume that control of the heartland inevitably led to control of the World-Island and therefore the world. For others, however, it merely proved Mackinder's point that ownership by a single power of the pivot area would only confer competitive advantage to that power if it had the ability to use the resources of the region wisely. Based upon communist ideals, Russia had created a type of society that was singularly unable to make the most of its natural advantage.

Deep Dive Box 5.5   Edward Said's Orientalism: Colonization as a *Mission Civilisatrice*?

The publication, in 1978, by Palestinian and US literary scholar and activist Edward Said, of *Orientalism* marked a watershed in the understanding of the ways in which the West sought to cultivate a story about the world that emphasized its superiority over other world cultures, a story that later paved the way for its colonial project.

(Continued)

Box 5.5   (Continued)

Orientalism focused upon the writings, diaries, photographs, paintings, and speeches of nineteenth- and early twentieth-century "Western" scholars, travelers, journalists, and statespeople, with respect to the Arab world (incorporating the Middle East, North Africa, and Asia). Appalled by the simplicity and ignorance displayed by these works, Said argued that throughout this period, Western commentators were busy creating images of the Arab world that reflected their own prejudices and biases more than they reflected reality.

For Said, Orientalists were Western opinion formers who participated in the creation of sweeping generalizations and falsehoods, Orientalism was the worldview they created, and the Oriental was the person subjected to the crude and misleading stereotypes.

The West imagined that the Orient was home to a single coherent and monolithic culture. Orientalism glossed over cultural and national differences and collected together all peoples in the region as if they formed a homogeneous grouping. A feature of Orientalism was its tendency to exoticize and eroticize the Orient and the Oriental, often in ways that demeaned, patronized, and belittled. The Orient was imagined to be a dusky, sultry, and irrational place, home to eccentric, primitive, deranged, and villainous populations. These populations were prepared to succumb to dictators and shied away from harnessing human reason in the service of progress. The Orientalist often imagined the Orient in feminine terms, an object of allure waiting and willing to be directed by the master. The Oriental man was often characterized as being feminine and weak but nevertheless with unpredictable sexual appetites that might threaten white women. Meanwhile, the Oriental woman was considered to be exceptionally beautiful while being prepared to be subordinate to, and eager to please, men.

Said recognized the importance of Orientalism in the European colonial adventure. Orientalism served to legitimate Western colonization of North Africa, the Middle East, and Asia by painting the region as vulnerable to domination, in need of direction and a strong master, and capable of being improved by the introduction of European reason, modernization, and progress.

The phasing of the European colonial adventure is complex and defies simple description. We might note, though, that Portugal and Spain were first to embark on sustained imperial conquest, that the Netherlands too subsequently emerged as a leading pioneer, and that by the start of the twentieth century France and especially the United Kingdom presided over the largest empires on earth. In part reflecting this sequence, the lands of the Americas, and in particular Latin America, were first to be colonized, from the middle of the sixteenth century, and first to secure independence. Although Europe had gained a toehold in coastal Africa, it was not until the turn of the twentieth century that European powers dared to venture into the interior of the continent and to colonize it in a substantial way. Last to be colonized, African territories were among the last to be decolonized. Colonization of Asia remained an ongoing process from the sixteenth century, and both annexation and decolonization occurred at different periods in different regions of that continent.

It was Portugal and Spain who were first to establish global empires. The Iberian origins of the European imperial adventure are not surprising. It was largely thanks to navigational and seafaring techniques pioneered by both nations that European voyages of discovery and exploration were made possible in the first instance. Portugal's early empire extended into Latin America, Asia, and Africa. Among its first and principal acquisitions were Mozambique (which came under Portuguese control from 1498 until 1975, when it secured independence), Brazil (colonized in 1500 and held until 1822), Goa (1510–1946), Angola (1575–1975), and East Timor (1769–1975).

The Spanish Empire too stretched into the four corners of the world, but its principal focus was central and southern America. Spain organized its colonial administration around four viceroyalties (see Map 5.4). The first to be established, the Viceroyalty of New Spain (1535–1821), had Ciudad de México (Mexico City) as its capital, and covered what is known today as the southern United States (California and Florida), Mexico, the Spanish West Indies (including Cuba, Haiti, Jamaica, and Trinidad), and the Spanish East Indies (the Philippines and Taiwan). The Viceroyalty of Peru (1542–1824) was governed from Lima, and included present-day Peru, Chile, Colombia, Panama, Ecuador, Bolivia, Paraguay, Uruguay, Argentina, Venezuela, and parts of Brazil. The Viceroyalty of New Granada was established later (1717–1819), and from its capital city of Bogotá it governed over present-day Colombia, Ecuador, Panama, and Venezuela. Spain's final viceroyalty, the Viceroyalty of Río de la Plata (1776–1810), had as its base Buenos Aires and ruled over present-day Argentina, Bolivia, Paraguay, and Uruguay.

Having itself served as a colony of Spain, the Netherlands gained effective independence in 1581 and became Europe's next great imperial power. Throughout the 1600s, Holland became the world's principal seafaring power. It established two companies, the Dutch East India and Dutch West India companies, and together these companies expanded the reach of the Dutch state throughout the world. By the mid-seventeenth century, the Netherlands was the most powerful country on earth. Its colonies included the eastern seaboard of the present-day United States (New York and New Jersey, which it held from 1614 to 1664), Guyana (1616–1815), Suriname (1664–1975), Ghana (1598–1872), South Africa (1652–1806), Sri Lanka (1640–1796), Indonesia (1603–1945), and Taiwan (1624–1662).

Although Spain, Portugal, and the Netherlands remained global superpowers, their empires were to be eclipsed by those held by other European countries, especially France and Britain. France became one of the first European countries to substantially colonize parts of present-day North America. The Viceroyalty of New France (held from 1534 to 1763) was commanded from Quebec and at its height included parts of Canada (including Newfoundland, the Hudson Bay, the Great Lakes, and Quebec) and parts of the present-day United States (a belt extending through the present-day Midwest and reaching as far south as Texas and Louisiana). In Asia, France's colonies included Indochina (present-day Laos, Thailand, and Vietnam, which it held from 1887 to 1954). Meanwhile, France also colonized large sections of North and West Africa, including present-day Niger (1890–1960), the Ivory Coast (1840–1960), Chad (1900–1960), Cameroon (1916–1960), Algeria (1830–1962), and Tunisia (1881–1956). It also brought Madagascar (1896–1960) under its control.

Notwithstanding the expansion of the French empire, by the turn of the twentieth century it was the United Kingdom that presided over the largest empire in the world. In 1922 nearly 460 million people belonged to the British empire (approximately 20% of the entire population of the world at that time) and Britain claimed ownership over nearly 25% of the entire surface of the earth. Britain began its rapacious march around

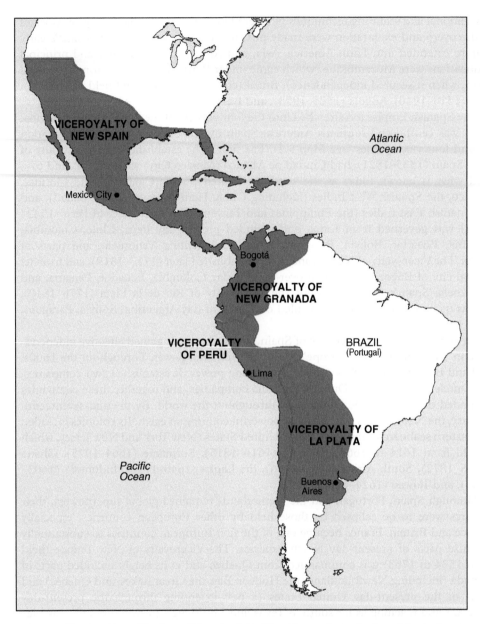

Map 5.4 Viceroyalties of Spain at the peak of the Spanish empire in the late eighteenth century.
Source: Map composed by the author.

the world following the Act of Union in 1707, which brought England, Scotland, Wales, and Ireland together under one rule. Throughout the eighteenth century, it sought to colonize North America and acquired land in the present-day United States and Canada. Following the American War of Independence (1775–1782), in which 13 of Britain's colonies broke free to form the United States, Britain consolidated its holdings in British North America (this covered much of present-day Canada, and this entity existed in changing forms between 1783 and 1907). Principal British colonies in the Caribbean included Barbados (1624–1966), Jamaica (1655–1962), and Trinidad and Tobago (1762–1962). In Africa the British incorporated Uganda (1890–1962), Nigeria (1900–1960), Kenya (1920–1963), and Egypt (1801–1922). In the Middle East they annexed Palestine (1920–1948) and Aden (1839–1967). In Asia they acquired Afghanistan (1839–1919), Burma (1824–1948), India (including today's Pakistan and Bangladesh) (1757–1947), and Malaya (1824–1963). Meanwhile, British settlers established the colonies of Australia (1788–1942) and New Zealand (1769–1947).

Case Study: Europe's Scramble for Africa

In spite of its geographical proximity to Europe, and notwithstanding settlement in its coastal regions, Africa was the last continent to be substantially colonized by European nations. Among others, early Scottish, English, German, and French explorers such as James Bruce (1730–1794), Mungo Park (1771–1806), René-Auguste Caillié (1799–1838), Heinrich Barth (1821–1865), Samuel Baker (1821–1893), Richard Burton (1821–1890), John Hanning Speke (1827–1864), David Livingstone (1813–1873), Henry Morton Stanley (1841–1904), Carl Peters (1856–1918), and Mary Kingsley (1862–1900) had all painted a picture of a barbarous and dangerous interior, dominated by fierce and harsh environments and lethal tropical diseases. "Deepest darkest Africa" was perceived to provide a formidable obstacle for would-be colonizers (Map 5.5a).

Between 1870 and 1914, European states effectively carved up Africa between them, imposing new political boundaries upon a continent hitherto marked by complex tribal units and other types of polities. Recognizing the folly of fighting among themselves for the spoils of Africa, European leaders met at the so-called Berlin Conference of 1884–1885 (also known as the Congo Conference) to agree on a framework through which the colonization of the continent might proceed more peaceably. Within a few decades, almost all of Africa was subsumed under European rule – the exceptions being Liberia (territory reclaimed by former African American slaves) and Ethiopia (which had fought off Italian imperial advances). Britain and France, and to a lesser extent Germany, Spain, Italy, Belgium, and Portugal, were now effectively ruling Africa from afar. Anglo-Irish historian Thomas Pakenham (1991) famously described this period as "the scramble for Africa" (Map 5.5b).

European powers used four principal strategies to govern over Africa.

Private companies: Particularly at the onset of the European colonial (mis)adventure, some European nations chartered private companies to run their colonies. Examples include the British East Africa Company (which colonized Kenya on Britain's behalf) and the British South Africa Company (led by British imperialist John Cecil Rhodes, which colonized present-day Malawi, Zambia, and Zimbabwe; see Plate 5.3). These companies generated disappointing returns for investors and provoked hostility from indigenous populations. By 1924, they were wound down and replaced with other forms of rule.

Direct rule: Pioneered by France, Belgium, Germany, and Portugal, the strategy of direct rule consisted of European governments establishing colonial administrations, normally based in the largest city in the colony, and using these administrations to impose their will directly on colonized populations. This model normally refused to countenance dialogue and collaboration with African populations on the basis that

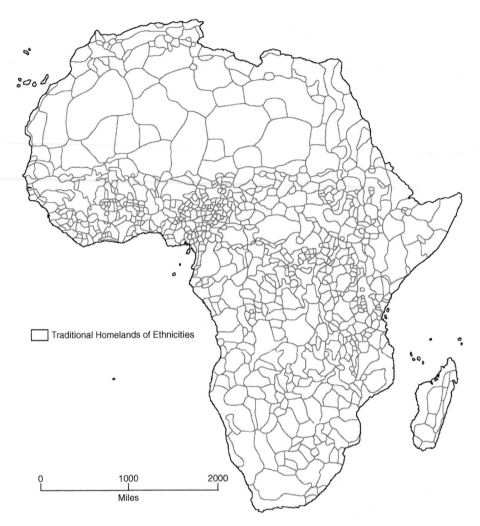

Map 5.5a Political geography of Africa before European colonization. Source: Data from Murdock (1959) and Fage (1978).

these populations were inferior and lacked European standard organizational and technical skills.

Indirect rule: Indirect rule was the preferred strategy of the United Kingdom. It comprised identifying local indigenous leaders and enlisting them in the governance of the country. Although more cooperative on the surface, by inviting local power brokers to collude in the European colonial experiment, this model might also be considered more Machiavellian. Often, it meant rule through indigenous elites as opposed to rule in collaboration with native populations.

Settler rule: Settler rule was rule by European migrants who settled in colonies in large numbers and who governed these colonies by themselves and for themselves. Settlers abused their privileged positions to suppress the indigenous population and to extract benefits from colonial lands for personal self enhancement. Settlers from the Netherlands, France, Britain, Germany, and Portugal practiced settler rule in countries

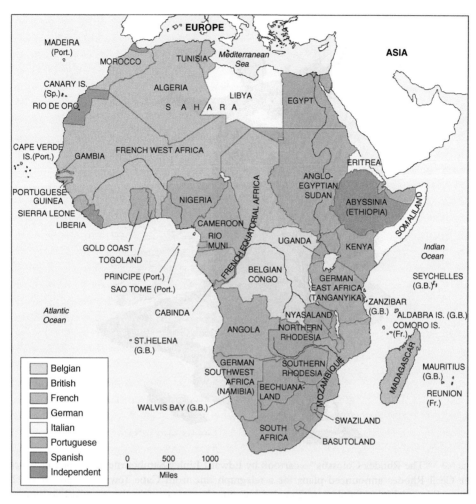

Map 5.5b　Africa, 1914: Political geography of Africa after the European scramble for the continent. Source: Data from Murdock (1959) and Fage (1978).

such as present-day South Africa, Zimbabwe, Zambia, Angola, Mozambique, Namibia, Kenya, and Algeria.

Although opinion today generally holds that colonialism was morally indefensible and largely detrimental to development on the African continent, there remains considerable debate over the exact social, economic, and political consequences of European rule. Critics point to the economic exploitation of the natural resources found in the colonies and the enforced impoverishment of local economies, the political fascism of some colonial states, the damage European plantations inflicted on local ecosystems, the suppression of traditional cultures, the spreading of lethal "European" infectious disease and mass epidemics, and the social fragmentation of communities and families wrought by colonial labor markets. Those more sympathetic to the "civilizing mission" of European colonialism suggest, in contrast, that it left in its wake better education systems, improved health care, and better technologies, infrastructure, and political capacity.

Plate 5.3   "The Rhodes Colossus" – cartoon by Edward Linley Sambourne, published in *Punch* after Cecil Rhodes announced plans for a telegraph line from Cape Town to Cairo in 1892. Source: Edward Linley Sambourne/Punch Magazine, https://en.wikipedia.org/wiki/File:Punch_ Rhodes_Colossus.png. Public Domain.

The rapidity of European colonization of Africa was matched by the rapidity of European withdrawal and decolonization. Opposition in Africa to European colonization existed from the outset. Even during the period of the "scramble" for the continent, indigenous populations in some parts of Africa (in particular Zimbabwe, Ethiopia, Libya, Isandlwana [present-day South Africa], Ghana, Tanganyika, and those commanded by the warrior Samori Ture from his base in present-day Guinea) sought to resist European aggression. These rebellions were largely futile. Forced to accept the reality of being ruled from afar, African resistance then took the form of demands for equality of opportunity, religious freedom, and labor rights. Europe's grip on Africa, however, began to weaken only after World War II. Having enlisted and fought for a European parent country during the war, many Africans returned home with a renewed expectation that they too deserved to be free from oppression. Throughout the 1940s–1960s, mass anticolonial and nationalist movements swelled throughout the continent (see Deep Dive Box 5.6). European imperial powers were forced to reflect upon the morality and *realpolitik* of their imperial (mis)adventures, both in Africa and elsewhere. In response, some ceded their colonies relatively peaceably

Deep Dive Box 5.6 Frantz Fanon's *The Wretched of the Earth*

The Caribbean island of Martinique was effectively claimed as a French colony in 1815 and to this day remains one of France's overseas departments. There, Frantz Fanon was born in 1925. Fanon fled Martinique for France and enlisted in the French army. He served during World War II in Morocco, Algeria, and France itself. He returned to Martinique in 1945 and worked for, and was influenced by, Aimé Césaire, a leading communist on the island, advocate of black freedom movements, and ardent critic of French colonization. Fanon subsequently returned to France, where he studied medicine and psychiatry and qualified as a psychiatrist in 1951.

France was present in Algeria for 124 years before the Algerian nationalist uprising of 1954. In the ensuing eight-year War of Independence, 250 000 Algerians and 25 000 French soldiers were killed. Between 1953 and 1957, Fanon served at the Blida-Joinville psychiatric hospital in Algeria. Based upon his experiences with patients, both French and Algerian, Fanon became a supporter of Algerian independence from France, and was a member of the Algerian National Liberation Front (Front de Libération Nationale [FLN]). Fearing his fiery and seditious writings and speeches, France disbanded the Blida-Joinville psychiatric hospital in 1957 and deported Fanon. Fanon quickly worked his way back to Tunisia, from where he actively supported the FLN.

Fanon wrote his famous book *The Wretched of the Earth* in 1961, at a moment when the Algerian War of Independence was reaching its crescendo. It rapidly became a bible for anticolonial liberation movements.

Fanon's focus was upon the ways in which colonizers depicted colonized populations as inherently inferior so that they might legitimate acts of criminality and violence toward them. France repressed any conscience it had about plundering Algeria by imagining the Algerian population as illiterate, uncivilized, primitive, and barbaric. There was a development lag between Europe and the rest of the world, and it was a duty of France to spread the message of progress.

Fanon believed that the treatment of indigenous peoples as less than human was taxing on the mental health of both the colonizer and the colonized. Colonized peoples, and especially peoples with skin colors other than white, were especially vulnerable to internalizing European stereotypes, with two possible consequences: they either displayed psychiatric scars and became ill, or they sought to self-correct by imitating their European "superiors." Fanon excoriated those who chose this second response (normally local middle-class natives who were given a minimal stake in colonial profits) and coined the term "black skin, white masks" to underscore the extent of his scorn.

Fanon argued that colonized peoples needed to rise up in arms and defeat European oppressors through acts of violence. *The Wretched of the Earth* continues to stand as the most brutal and controversial statement on the cultural value of anticolonial violence. Violence was not only needed to oust foreign powers who themselves were committing acts of violence in their attempts to suppress uprising. Violence was also needed to restore the dignity of those hitherto forced to live in fear and burdened with feelings of cowardice.

(Continued)

Box 5.6 *(Continued)*

Fanon concluded with a rally cry to all colonized peoples to reject the idea that they were culturally inferior to white Europeans and needed to imitate their political masters and catch up with them. Following decolonization, there should be no building of a new Europe in the Third World. Working together, people of color from the three continents of Latin America, Asia, and Africa needed to draw upon their experiences as colonized peoples and build a new world.

(Italy and the United Kingdom, for instance). In other cases, independence was won only after violent struggles (in French, Belgian, and Portuguese colonies in particular).

Decolonization of Africa began first in Libya (1951) and Egypt (1952). British-French humiliation by Egypt during the Suez Canal crisis of 1956 shattered the confidence of imperialists. In 1960, an additional 14 countries gained independence. By 1966, only six African countries remained under European control: Angola, Guinea Bissau and Cape Verde, and Mozambique (Portuguese colonies), Namibia (which remained under the rule of colonial settlers in South Africa), South Africa (principally ruled by Dutch settlers), and Zimbabwe (British-ruled). Five of these were settler colonies, governed directly by migrant settlers from the European parent country. Settlers had a lot to lose and clung onto their riches. Nationalist leaders in these countries initially sought to secure improved constitutional rights (such as the right to vote) through peaceable means. In time, they came to agitate for outright freedom from Europe's orbit and for the right of self-determination. Although settler governments sought to quell insurrection, often in violent ways, all six settler colonies were to secure their independence in time (Map 5.6).

Although welcomed, decolonization created a sense of political instability in former African colonies. This instability derived in part from the contrived countries and artificial borders that colonists left behind. The political map of Africa today is an imposition by Europe and pays scant attention to precolonial indigenous polities and to the complex ethnic and linguistic communities that prevailed prior to colonization. The consequence is that some newly created states contain many complex groupings. In this context, African countries have found it challenging to build properly functioning political institutions that practice good governance. In many cases, despotic regimes have emerged taking the form of corrupt socialist dictatorships and/or right-wing family dictatorships. These regimes have on occasion presided over the violent political suppression of their opponents, genocide, mass refugee movements, famines, and civil wars. The most recent political legacy of European decolonization of Africa is the so-called Arab Spring, a series of popular uprisings and revolutionary wars seeking to replace postcolonial despotic regimes, which began in Tunisia in 2010, swept across Egypt and Libya (and Yemen, Bahrain, Iraq, and Syria), and has also reverberated in Algeria, Sudan, and Morocco (and Jordan and Kuwait). Notwithstanding promising beginnings, the Arab Spring has led to an Arab Winter, not least in Syria.

It is clear that decolonization has failed to bequeath to Africa and elsewhere a settled political geography and that the process of building nation–states in former colonies is continuing to run its course.

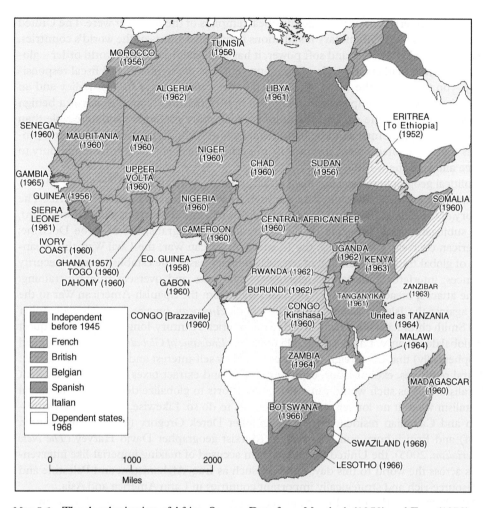

Map 5.6   The decolonization of Africa. Source: Data from Murdock (1959) and Fage (1978).

American Empire: The Eagle, the Bear, the Theocrat, and the Dragon

Is the United States an empire? This idea has long seemed preposterous to Americans; American foreign policy, after all, does not seek to invade and annex other countries. Throughout its history, the United States has concentrated on its own internal affairs, only venturing overseas when morally obliged to by world historical events such as checking Japanese, Nazi, Soviet, and militant Islamic expansionism. With the demise of the Cold War and the transition from a bipolar to a unipolar world order, it is true that the United States has emerged as the world's sole superpower and hegemon. But terminology should not be confused; there is a world of difference between the ideas of unipolarity and hegemony on the one hand and empire on the other. Or so argue some.

In his 2005 book *Colossus*, Scottish-born but US-based historian Niall Ferguson contends that in the way it uses both its military and economic might, the United States is more of an imperial power than the European empires of the past ever were. The United States presides over 750 military installations in over two-thirds of the world's countries. Through both hard power and soft power, it has imposed its vision of world order – globalization, free markets, the rule of law, representative government, and fiscal responsibility – on the rest of the world. But a friend of neoconservative foreign policy and an apologist for Western imperialism, Ferguson argues that the United States is a benign imperial power and the American empire is much more preferable and agreeable than alternative empires. Instead of being in denial about its imperial stature, Ferguson proposes that the United States embrace its imperial prowess and accept its moral duty to police and lead the world.

Political geographers have not been afraid to refer to the idea of an American empire, but more often they have sought to question its virtues than evangelize its merits. The history of the United States is a history of European white settlement, a march westward, and subjugation and genocide of First Americans; land grabs; the Monroe Doctrine; American overseas missionaries; the Mexican-American War; the Civil War; the expansion of global trade; the building of overseas military bases; the establishment of security alliances; interference in foreign elections; arms sales and overseas military training; drone attacks; breaches of sovereignty; and wars from the Spanish-American War to the war against the Islamic State. In his 2001 book *American Empire*, Scottish geographer Neil Smith chronicled what he believed to be America's century-long and rapacious quest for global dominance. Later, in his 2005 book *The Endgame of Globalization*, Smith argued (prophetically) that US foreign policy was fueled by self-interest and nationalism (extract natural resources, exploit labor, capture markets, and extract taxes) and not benevolent liberalism, and as such would put a halt to its efforts to globalize democracy and liberal capitalism when it no longer served its interests to do so. Likewise, in books by British-born and Canadian resident critical geographer Derek Gregory (*The Colonial Present*, 2004) and British-born and US resident Marxist geographer David Harvey (*The New Imperialism*, 2005), the United States has been accused of making imperial-like interventions across the world to this day: in places such as Iraq, Afghanistan, and Palestine and in resource-rich and strategically important countries in Latin America and Asia.

After World War II and as European decolonization proceeded, world politics came to be structured around an ideological conflict between the Western world, led by the now-dominant United States, and the communist world, led by the Soviet Union (the Union of Soviet Socialist Republics, or USSR). The Bolshevik Revolution in Russia (1917) had given birth to a communist state, which the Bolsheviks claimed was built around principles first articulated by German philosopher Karl Marx. The Bolsheviks viewed capitalism as an iniquitous and unjust political system and were bent on sweeping their alternative ideology across the world. Russia, of course, had already shown itself to be an imperial power and was not afraid to assert its will by violent means if necessary. At the close of World War II, the Soviet Empire had grown to include, largely by coercion and compulsion, the countries of Eastern Europe. In contrast, the United States was bent on promoting a world in which liberal capitalist democracies prospered. The United States viewed the USSR's revolutionary agenda and imperial tendencies with suspicion. The greater the territory ceded to the USSR, the more dangerous communism was to US interests. As they jostled for supremacy, these two superpowers, both with nuclear capabilities, sponsored a dangerous bipolar world. (See Deep Dive Box 5.7.) Other countries were called upon (or forced) to take sides, and many found it difficult to maintain their neutrality.

By the late 1970s, it was becoming obvious that the USSR was an empire on the brink of collapse. The communist model, or at least the model of communism developed by the Soviet Union, had proved undemocratic, oppressive, inefficient, and unsustainable. As the USSR struggled to maintain its economic and military strength, it came to recognize that it could no longer finance a standoff with the United States. The West, it seemed, had won the Cold War. Its political and economic institutions had proved superior. Sensing the weakness of the Russian Empire, toward the late 1980s a number of nations subsumed within the Russian fold began to plot a new course. By 1989, many

---

Deep Dive Box 5.7   The Vietnam/Indochina War (1955–1975)

The Vietnam War (1955–1975) captures well the kinds of geopolitical struggles that developed during the Cold War period.

The roots of US military intervention in Vietnam, Cambodia, and Laos lie in the rise and fall of the French colonial adventure in Indochina. In 1954, following a protracted war between France and its allies and the Soviet- and Chinese-backed Vietnam national independence movement led by Ho Chi Minh (the Viet Minh), France granted independence to Vietnam, Cambodia, and Laos. Recognizing political differences between north Vietnam and south Vietnam, the Geneva Conference of 1954 divided Vietnam at the 17th Parallel. North Vietnam, renamed the Democratic Republic of Vietnam, was to be governed by the Viet Minh from Hanoi; south Vietnam, renamed simply the State of Vietnam, was to be governed by the pro-Western Ngo Dinh Diem from Saigon. Reunification was to come following national elections in 1956. These elections were never held.

The Vietnam War emerged out of this vacuum. The Vietnam People's Army (the north Vietnamese army), in partnership with the Viet Cong (National Liberation Front, a south Vietnamese guerrilla force directed by the North) and communist allies (particularly the Soviet Union, China, North Korea, and Cuba), fought for a unified Vietnam under communist rule. These groups viewed partition as a neo-colonial strategy pursued by the West against the Indochina region. Meanwhile, the South Vietnamese army, in collaboration with Western allies (in particular the United States, but also Australia, Japan, Thailand, and the Philippines), fought for an independent south Vietnam organized around liberal capitalist and democratic principles. These groups viewed partition as at once a necessary defense of a minority population and an essential brake on the global march of communism.

The Vietnam War ran its course between 1955 and 1975, inflicting as it unfolded collateral damage on both Cambodia and Laos. It is estimated that between 800 000 and 3.1 million Vietnamese soldiers and civilians perished, between 200 000 and 300 000 Cambodians died, between 20 000 and 200 000 Laotians were killed, and nearly 60 000 members of the US military died in battle. Faced with the horror of such a loss of life, and unable to defeat enemy combatants with brute military force, enthusiasm for the war waned in the United States (Plate 5.4). By 1973, US military personnel withdrew from the conflict; by 1975, the Vietnam People's Army gained control of Saigon; and in 1976, a unified Socialist Republic of Vietnam was declared (Map 5.7).

(Continued)

Box 5.7   *(Continued)*

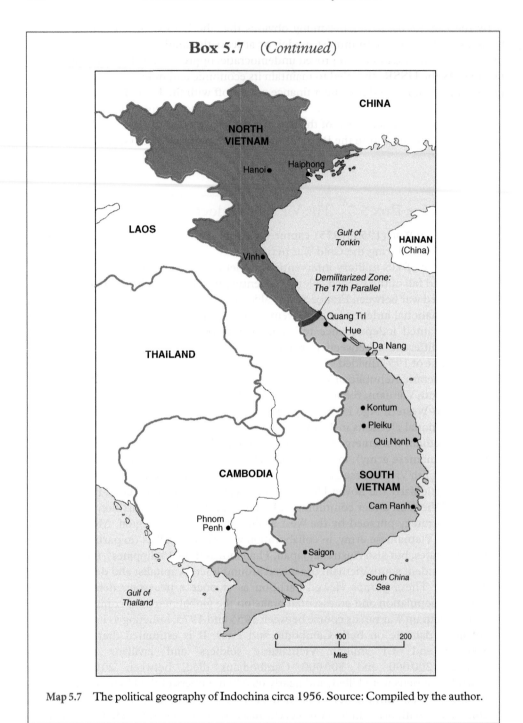

Map 5.7   The political geography of Indochina circa 1956. Source: Compiled by the author.

states had secured liberation, and by 1991, the USSR was officially wound down. With the collapse of the Soviet Empire came a transformation in the political map of Eastern Europe and Asia (Map 5.8). The state of Russia (the Russian Federation) was created. In Eastern Europe, six new (or revived) countries were formed: Estonia, Latvia,

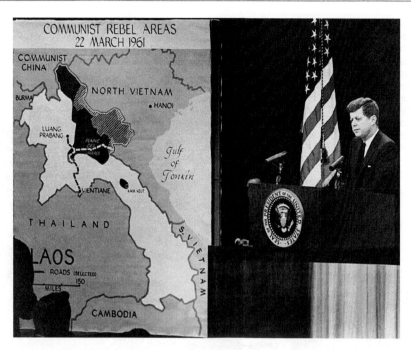

Plate 5.4   President John F. Kennedy at a press conference on the Vietnam War, 23 March 1961. Source: Abbie Rowe. White House Photographs. John F. Kennedy Presidential Library and Museum, Boston.

Lithuania, Belarus, Ukraine, and Moldova. In Asia, eight new (or newly created) countries emerged: Kazakhstan, Uzbekistan, Kyrgyzstan, Turkmenistan, Tajikistan, Georgia, Armenia, and Azerbaijan. A number of other Eastern bloc countries secured full autonomy over their affairs: Poland, East Germany, Czechoslovakia, Romania, Bulgaria, Hungary, Albania, and Yugoslavia. Thereafter, East Germany reunited with West Germany, and Czechoslovakia separated into the Czech Republic and Slovakia. Meanwhile, Yugoslavia fell into a civil war, and from this war there emerged the countries of Serbia, Slovenia, Croatia, Bosnia, Montenegro, Macedonia, and Kosovo. All the while, separatist movements in Moldova, Abkhazia, and South Ossetia continue to lobby for independence.

Although many believe that the story of the Cold War has now run its course, recent developments in Ukraine suggest that tensions between the West and Russia remain. Encouraged by the Russian president, Vladimir Putin, the pro-Russian region of Crimea and Sevastopol in southeast Ukraine declared independence from pro-European Kiev in 2014 and signed a treaty of accession into the Russian Federation. Meanwhile, allegations abound concerning Russian cyberattacks and interference in Western elections through manipulation of social media platforms, leading in 2017–2019 to the Mueller Special Counsel investigation into Russian interference in the 2016 US elections and related matters. The bitterness of the Cold War, it seems, continues to reverberate in today's world, and it is not beyond the bounds of possibility that Western relations with Russia will re-emerge as a significant battleground in the twenty-first century.

1	Estonia	12	Montenegro
2	Latvia	13	Albania
3	Lithuania	14	Macedonia
4	East Germany	15	Bulgaria
5	Czech Republic	16	Moldova
6	Slovakia	17	Georgia
7	Hungary	18	Armenia
8	Slovenia	19	Azerbaijan
9	Croatia	20	Tajikistan
10	Bosnia	21	Kyrgyzstan
11	Serbia		

Map 5.8   The political Geography of the former USSR. Source: Compiled by the author.

Meanwhile, in 1996, US political scientist Samuel P. Huntington published a book entitled *The Clash of Civilizations and the Remaking of World Order* (Huntington 1996). He sought to draw attention to the shape of future global conflicts in the post–Cold War era. Huntington's contention was that if, in the Cold War period, conflict was most likely to occur between the Western free world and the communist bloc around questions of ideology and economy, it was now most likely to take place between the world's major civilizations and religions, which were (re)emerging with new potency. Huntington identified eight (re)emerging civilizations, with a possible ninth: Western, Latin American, Islamic, Sinic, Hindu, Orthodox, Buddhist, and Japanese, and the possible ninth, African (see Map 5.9). For Huntington, the hegemony of the Christian West would be most threatened by the Sinic civilization (spurred on by Chinese economic growth), Islamic (fueled by a youthful population bulge and age structure), and Latin American (with Mexican and other migrants transforming the culture of cities like Los Angeles) civilisations.

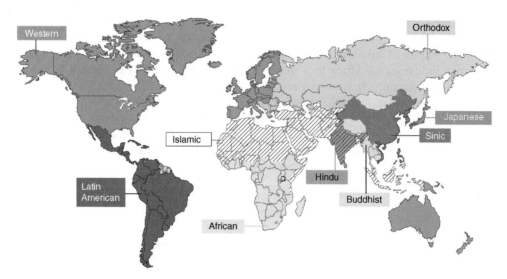

Map 5.9 Samuel P. Huntington's clashing civilisations mapped. Source: Based on https: //upload.wikimedia.org/wikipedia/commons/2/2f/Clash_of_Civilizations_map.png and The clash of civilizations

Huntington's thesis has been viewed by some as prophetic of subsequent events. He offered a way to understand wars, such as the one that followed the breakup of Yugoslavia; those in Chechnya, Burma, and Israel and Palestine; and that between India and Pakistan. And the attacks on the Twin Towers in New York, on 11 September 2001, and the events that followed were taken by many to be proof that Huntington's view of the world was correct. The rise of militant Islamic fundamentalism and threats posed by terrorist organizations such al Qaeda, Boko Haram, the Taliban, Egyptian Islamic Jihad, Al-Shabaab, and of course the ISIS Caliphate in Syria and Iraq, among many other cases, also appeared to validate the idea of a civilizational confrontation between Western secularists and Islamic theocrats.

But close examination of recent and current wars suggests that conflicts most often occur within civilizations and not between civilizations; it is more accurate to speak today about the "crash of civilizations" (through internal feuding) than a clash of civilizations. Moreover, according to Palestinian-born and US resident activist and literary scholar Edward Said (2001), Huntington's clash of civilizations thesis is nothing more than "a gimmick which panders to unwarranted and irrational fear and insecurity in the United States and furnishes terrorists throughout the Islamic world with a warped justification for heinous crimes." Huntington's thesis trades in "vast abstractions that may give momentary satisfaction but little self-knowledge or informed analysis" (Said 2001).

Perhaps in the twenty-first century, China will rise to become the most serious threat to US hegemony. Enjoying remarkable economic growth from the 1980s onward, China is already signposting its credentials as a new world superpower and for some global empire. The ambitious Belt and Road Initiative is seeking to build out economic infrastructures along ancient trade routes between China and Europe, including Marco Polo's Silk Road in the north and the maritime expedition routes of Admiral Zheng He in the south. China is lending significant sums to developing countries to construct flagship infrastructure projects (ports, airports, and rail infrastructure), many of which are failing, encumbering these countries in a Chinese debt trap. The price of debt forgiveness has been increased Chinese influence in their internal affairs. Meanwhile, many of

China's *state-owned enterprises* (SOEs) service contracts, looping wealth back to China. Inviting parallels with the Old International Division of Labor (OIDL) orchestrated by European empires, China's model of command capitalism has also brought SOEs in the extractive industries to Africa in a quest to source raw materials to service China's expanding production. It is not an accident, then, that China presented as a central policy concern during Donald Trump's presidency and is likely to be at the top of the agenda of the Biden administration. Whether it be with respect to questions of currency manipulation, political freedoms in Hong Kong, the autonomy of Taiwan, terms of trade, the intentions of the Chinese company Huawei, or the origins of Covid-19, it is clear that Sino-US relations are deteriorating.

Conclusion

The Classical (Greek and Roman) period has bequeathed many of the political constructs that underpin the governmental machine of the West, including ideas regarding power and territorialization. By dint of the rise of the West, these ideas have been hugely consequential for all peoples. Were it not for the rise of Europe's nation–states from the seventeenth century onward, it is likely that the political map of the world we take for granted today would be significantly different. We might wonder, however, whether the sun is now setting on the idea of the sovereign nation-state to be replaced by a new variant of medieval multilevel governance. During its life and times, and through colonization and then later decolonization, Europe's nation–states carved new political geographies – new polities and political boundaries – onto the lands of Latin America, Asia, and Africa. Moreover, the colonization and decolonization of these continents, in turn, have created political, social, economic, and cultural legacies that continue to work themselves through today. Across the twentieth century, the United States emerged as the most important superpower in the West, and it too has actively forged the political map of the world. Whether the twenty-first century is a unipolar American century, a bipolar century in which the United States and China compete for hegemony, or a multipolar world comprising geopolitical alignments and fractiousness between the American eagle, the Russian bear, Islamic (and other) theocracies, and the Chinese dragon remains to be seen. And of course the role of countries such as Iran, North Korea, Pakistan, India and Saudi Arabia cannot be underestimated too.

Checklist of Key Ideas

- Many of the political constructs that underpin the governmental machine of the West are Classical (Greek and Roman) in origin. Rooted in early Christian theology, Western power is predicated on two competing conceptions of power: a sovereign paradigm ("power over" or, crudely, hard power) and a governmental paradigm ("power to" or, crudely, soft power).
- Although often given ontological status – that is, treated as if they have always existed – the ideas of the sovereign state, nations, and nation-states are recent inventions, given first life at Westphalia in 1648 and energised by the Spring of Nations in 1848.
- The term "multilevel governance" is used to refer to the idea that the sovereign nation-state is being replaced today by political entities operating at large scales

(supranational organizations) and smaller scales (local and regional authorities). But resurgent populism and nationalization suggest that it is premature to announce the death of the nation-state as a distinctive polity.

- Europe's nation-states began the process of building empires and, at different speeds and through different means, colonized the Americ, Asia, Africa, Oceania, and the Polar Regions. Although the process of decolonization began in the eighteenth century (especially in Latin America), it was only in the twentieth century that most colonized states secured independence. European decolonization bequeathed to these continents a volatile set of polities (and societies, cultures, and economies) that remain contested and in transformation to this day.
- The extent to which the United States is an imperial nation-state is a matter of debate. Ongoing American hegemony will depend upon the outcome of global power struggles between the United States, Russia, militant Islam, and China. Iran and North Korea too will be influential.

Chapter Essay Questions

1) Identify and comment upon the Classical (Greek and Roman) origins of key political constructs in the governmental machine of the West.
2) Outline what is meant by multilevel governance. Discuss the claim that we are currently entering a period in which sovereign nation-states will be replaced by multilevel governance.
3) Describe the ways in which European colonization and decolonization of Latin America, Asia, and Africa transformed the political geographies of these continents.
4) Is the United States an empire? Is China building an empire? Why do these questions matter?

References and Guidance for Further Reading

Good general introductions to political geography and geopolitics can be found in:

Agnew, J.A. (2013). *Geopolitics: Re-Visioning World Politics*. London: Routledge.

Agnew, J.A. and Muscarà, L. (2012). *Making Political Geography*. Lanham, MD: Rowman and Littlefield Publishers.

Agnew, J., Mitchell, K., and Toal, G. (2017). *A Companion to Political Geography*. London: Wiley Blackwell.

Agnew, J. and Shin, M. (2019). *Mapping Populism: Taking Politics to the People*. Lanham, MD: Rowman and Littlefield Publishers.

Demko, G.J. (2018). *Reordering the World: Geopolitical Perspectives on the 21st Century*. New York: Routledge.

Dittmer, J. and Sharp, J. (eds.) (2014). *Geopolitics: An Introductory Reader*. London: Routledge.

Dodds, K.J. (2014). *Global Geopolitics: A Critical Introduction*. London: Routledge.

Flint, C. (2016). *Introduction to Geopolitics*, 3e. New York: Routledge.

Flint, C. and Taylor, P. (2018). *Political Geography: World-Economy, Nation-State and Locality*, 7e. New York: Routledge.

Painter, J. and Jeffrey, A. (2009). *Political Geography: An Introduction to Space and Power*, 2e. London: Sage.

Jones, M., Jones, R., Woods, M. et al. (2014). *An Introduction to Political Geography: Space, Place and Politics*. London: Routledge.

Muir, R. (2015). *Modern Political Geography*. London: Macmillan International Higher Education.

O'Loughlin, J. (2018). Thirty-five years of political geography and political geography: the good, the bad and the ugly. *Political Geography* 65: 143–151.

Paddison, R. (1983). *The Fragmented State: The Political Geography of Power*. Oxford: Blackwell.

Slowe, P.M. (2015). *Geography and Political Power: The Geography of Nations and States*. New York: Routledge.

Smith, S. (2020). *Political Geography: A Critical Introduction*. Oxford: Wiley.

Two important works in Foucault's oeuvre are:
Foucault, M. (1991). *Governmentality. The Foucault Effect: Studies in Governmentality*. London: Harvester Wheatsheaf.

Foucault, M. (1995). *Discipline and Punish: The Birth of the Prison*. London: Vintage.

Two important works in Agamben's Homo Sacer series are:
Agamben, G. (2005). *State of Exception*. Chicago: University of Chicago Press.

Agamben, G. (2011). *The Kingdom and the Glory: For a Theological Genealogy of Economy and Government*. Stanford, CA: Stanford University Press.

The ideas of hard power, soft power, and smart power were first introduced at length in:
Nye, J. (2004). *Soft Power: The Means to Success in World Politics*. New York: Public Affairs.

An excellent summary of recent political philosophies of Western power can be found in:
Dean, M. (2013). *The Signature of Power: Sovereignty, Governmentality and Biopolitics*. London: Sage.

Important works on Foucault and Agamben by geographers are:
Elden, S. (2017). *Foucault: The Birth of Power*. London: Wiley.

Philo, C. (2012). A 'new Foucault' with lively implications–or 'the crawfish advances sideways'. *Transactions of the Institute of British Geographers* 37: 496–514.

Katz, I., Martín, D., and Minca, C. (eds.) (2018). *Camps Revisited: Multifaceted Spatialities of a Modern Political Technology*. London: Rowman and Littlefield Publishers.

An excellent introduction to the ways in which human geographers think about the idea of territoriality can be found in:
Storey, D. (2012). *Territories: The Claiming of Space*. London: Routledge.

Important studies of the rise of European modernity and the birth of new ideas about territoriality include:
Elden, S. (2013). *The Birth of Territory*. Chicago: University of Chicago Press.

Maier, C.S. (2016). *Once Within Borders: Territories of Power, Wealth and Belonging Since 1500*. Cambridge, MA: Harvard University Press.

Sack, R.D. (1986) Human territoriality: Its theory and Practice (Cambridge University Press, Cambridge).

Key texts in the study of the nation-state as a historical invention are:
Anderson, B. (1983). *Imagined Communities: Reflections on the Origin and Spread of Nationalism* (rev. ed.). London: Verso.

Bhabha, H. (1990). *Nation and Narration*. London: Routledge.

Hobsbawn, E. and Ranger, T. (eds.) (1983). *The Invention of Tradition*. Cambridge: Cambridge University Press.

Smith, D. (2004). *The Antiquity of Nations*. Cambridge: Polity Press.

The best introduction to the hollowing out of the nation-state remains:
Jessop, B. (2013). Hollowing out of the 'nation state' and multi-level governance. In: *A Handbook of Comparative Social Policy*, 2ee (ed. P. Kennett), 11–26. Cheltenham: Edward Elgar.

Examination of the hollowing out of the UK state and Brexit can be found in:
Dorling, D. and Tomlinson, S. (2018). *Rule Britannia: BREXIT and the End of Empire*. London: Biteback.
Jones, M. (2019). *Cities and Regions in Crisis: The Political Economy of Sub-National Economic Development*. London: Edward Elgar Publishing.

Perhaps the most important book written by Halford Mackinder was:
Mackinder, H.J. (1919). *Democratic Ideals and Reality*. New York: Holt.

An authoritative guide to the work of Halford MacKinder is:
Kearns, G. (2009). *Geopolitics and Empire: The Legacy of Halford MacKinder*. Oxford: Oxford University Press.

Important works on European imperialism and the decolonization of Africa, Asia, and Latin America include:
Abernethy, D. (2000). *The Dynamics of Global Dominance: European Empires 1414–1980*. New Haven, CT: Yale University Press.
Baly, C.A. (2014). *The Birth of the Modern World 1780–1914. Global Connections and Comparisons*. Oxford: Blackwell.
Fanon, F. (1961). *The Wretched of the Earth*. London: Grove Press.
Jazeel, T. (2013). *Sacred Modernity: Nature, Environment and the Postcolonial Geographies of Sri Lankan Nationhood*. Liverpool: Liverpool University Press.
Manjapra, K. (2020). *Colonialism in Global Perspective*. Cambridge: Cambridge University Press.
Said, E.W. (1978). *Orientalism*. New York: Pantheon Books.
Thomas, M. and Thompson, A. (2019). *The Oxford Handbook of the Ends of Empire*. Oxford: Oxford Handbooks.

The most impressive study of European imperial expansion into Africa is provided in:
Pakenham, T. (1991). *The Scramble for Africa*. London: Abacus.

Also of importance in the history of knowledge of imperialism in Africa are:
Fage, J.D. (1978). *History of Africa*. New York: Knopf.
Murdock, G.P. (1959). *Africa: Its Peoples and Their Culture History*. New York: McGraw-Hill.

Important works on American empire include:
Agnew, J.A. (2005). *Hegemony: The New Shape of Global Power*. Philadelphia: Temple University Press.
Ferguson, N. (2005). *Colossus: The Rise and Fall of the American Empire*. New York: Penguin.
Gregory, D. (2004). *The Colonial Present: Afghanistan, Palestine, Iraq*. Oxford: Wiley Blackwell.
Gregory, D. and Pred, A. (eds.) (2007). *Violent Geographies: Fear, Terror, and Political Violence*. London: Taylor & Francis.
Harvey, D. (2004). *The New Imperialism*. Oxford: Oxford University Press.
Maçães, B. (2020). *History Has Begun: The Birth of a New America*. Oxford: Oxford University Press.
Morrissey, J. (2017). *The Long War: CENTCOM, Grand Strategy, and Global Security*. Athens: University of Georgia Press.
Smith, N. (2004). *American Empire: Roosevelt's Geographer and the Prelude to Globalization*. Berkeley: University of California Press.
Smith, N. (2004). *The Endgame of Globalization*. Berkeley: University of California Press.

Important political geographies exploring the meaning and collapse of the USSR are:
Smith, G. (1999). *The Post-Soviet States: Mapping the Politics of Transition*. London: Routledge.
Toal, G. (2017). *Near Abroad: Putin, the West, and the Contest over Ukraine and the Caucasus*. Oxford: Oxford University Press.

The Clash of Civilizations thesis and its limitations are explored in:

Huntington, S.P. (1996). *The Clash of Civilizations and the Remaking of World Order*. New York: Simon & Schuster.
Said, E.W. (2001) The clash of ignorance. The Nation, 4 October.

The extent to which China is behaving like an imperial power is discussed in:

Carmody, P. (2017). *The New Scramble for Africa*. Hoboken, NJ: Wiley.
Maçães, B. (2019). *Belt and Road: A Chinese World Order*. Oxford: Oxford University Press.

Website Support Material

A range of useful resources to support your reading of this chapter are available from the Wiley *Human Geography: An Essential Introduction* Companion Site http://www.wiley.com/go/boyle.

Chapter 6

Worlds of Meaning: Power, Landscape, and Place

Chapter Table of Contents

Chapter Learning Objectives

By the end of this chapter you should be able to:
- Provide a brief history of the rise of Western culture, and identify the central tenets of Western culture; outline and comment upon the claim that the West's

Human Geography: An Essential Introduction, Second Edition. Mark Boyle.
© 2021 John Wiley & Sons Ltd. Published 2021 by John Wiley & Sons Ltd.
Companion website: www.wiley.com/go/boyle

central cultural institutions lie behind its ascent to the summit of world history and that the West is currently faltering because it has allowed these cultural institutions to decay.

- Outline the ways in which cultural geographers might approach the study of culture, and describe and reflect upon the ways in which Sauerian cultural geography and "new cultural geography" might frame Western culture.
- Explain in what ways it makes sense to speak about both the idea of landscape in Western culture and the idea of the West in the cultural landscape; explain what is meant by the concept of "the imperialism of the straight line."
- With reference to the work of seminal thinkers, discuss the ways in which human geographers have sought to understand the interplay between landscapes of power and living landscapes.
- Identify and document the thinking of some of the influential scholars who studied the impact of the rapid rise of nineteenth- and early twentieth-century industrial cities on human personality and culture.
- Examine the impact of European urban planning on living landscapes in colonial cities; explain the phrase "worlding cities," and identify non-Western mega-development projects that seek to project global power in Asian city-landscapes.
- Discuss the ways in which urban planning in North America has frustrated the capacity of the African-American community to humanize their living environments.

Introduction

When asked what he thought of the idea of Western civilization, India's great leader Mahatma Gandhi famously declared, "I think it would be a good idea"!

Evangelists of the West like to tell their own story about the contributions of Western culture to the world. Western culture, they hold, is humankind's crowning achievement; superior in every way to all past and present world cultures. It was the destiny of the West to become the preeminent civilization in the world. A time lag existed between Western culture and other cultures. Civilized, advanced, technologically sophisticated, rational, and scientific, Western culture was millennia ahead of other cultures, which remained by comparison primitive, savage, barbaric, and backward. It was the moral duty of the West to share its culture with the world by persuasion, coercion, and, if necessary, compulsion. The West was entitled to sweep away, destroy, erase, and neuter alternative cultures with impunity. According to supporters of the West, it was the responsibility of non-Western cultures to awaken from the medieval stupor they were mired in and to begin the task of catching up by opening themselves to Westernization. But not everyone has been seduced by this account. Through occupation, being, becoming, and making meaning in, human beings have humanized the West's symbolic landscapes, birthing complex local human ecologies and living landscapes with complex relations with landscapes of power.

The purpose of this chapter is to study the idea of landscape in the culture of the West and the idea of the West in the cultural landscape, and to consider the dialectical interplay between what might be loosely termed "the imperialism of the straight line" and human natality and flourishing.

Enlightenment(s): The West's Culture(s)

Evangelists of the West believe that the steady emergence and blossoming of Western culture announced a decisive new moment in human history. Some even argue that it was developments in culture that lay behind the West's ascendance. The rise of the West as an economic and political force was made possible only because Western culture enabled the peoples of Europe to harness human reason like never before and to build new institutions around which the perfect society might be created. Western culture was rational, scientific, objective, and analytical. It believed in laws and, in particular, in its capacity to unearth the laws that shape natural worlds and societies. It placed confidence in the ability of enlightened human beings to use reason to build new and better worlds. It was at root Western culture that enabled Western nation–states to build successful social, economic, and political institutions. The latter would not have been possible without the former.

Western civilization is commonly assumed to be the product of four historical pillars: Greek philosophy, Roman law, Judeo-Christian theology, and modern science. Less appreciated is the fact that it was also influenced by the fruits of the Islamic enlightenment from the ninth century to the thirteenth century CE and the Chinese enlightenment from the tenth century to the twelfth century CE. The rise of the West announced a moment of great awakening from centuries of intellectual slumber. Superstitions, legends, folk cultures, quasi-religions, and supernatural cults had piled up so as to confuse and mystify the human mind. According to evangelists of the West, the genius of Western culture was that it recognized that human beings had become enslaved, oppressed, and infantilized by these false and irrational belief systems. The human mind needed to free itself from such debris, clutter, and ignorance. The power of human reason alone, and not blind faith in tradition or in lazy and untested beliefs, needed to be prioritized.

The greatest minds in Europe became convinced that both the natural world and the social world were systematically arranged into regular patterns. This order reflected physical and social processes and was not, as had been believed up to that point, the work of the divine hand of God. Human reason could figure out the physical and social laws that governed the natural environment and the operation of society. By developing powerful explanatory frameworks that laid bare the workings of the world, humans no longer needed to be passive recipients of judgments handed down by an almighty creator and could control their own fate.

The West's intellectual revolution was manifest at the time of the European Renaissance (fifteenth to seventeenth centuries), found encouragement in the Protestant Reformation (sixteenth century), and came of age with the European Enlightenment (seventeenth century onward).

The Renaissance was a period of great learning and creativity in Europe. Europeans sought to recover and consider anew the wisdom, literature, art, architecture, and culture of ancient and, in particular, classical civilizations, which had been buried and neglected during the medieval period. Finding life first in Florence and Venice in Italy, and trending next through cities in the Netherlands like Amsterdam, Rotterdam, and The Hague, the Renaissance was to invigorate cultural life throughout the whole of Europe. It was followed by the Protestant Reformation, a moment of schism within Christianity. Reformers lamented the financial, sexual, and political abuses and indulgences of the then-dominant Roman Catholic Church. The Catholic Church deprived people of an authentic relationship with God. Reformers believed that Christians needed to be liberated from the vices of earthly Catholicism, so that they might build their own personal relationships with God, guided by their own intellect and the Bible alone.

The European Enlightenment from the seventeenth century built upon the intellectual creativity of the Renaissance and the new intellectual freedoms secured by the Protestant Reformation, and called for a new culture to be built that championed the power of human reason alone. In a famous essay entitled "What is Enlightenment?" published in 1784, German philosopher Immanuel Kant declared:

> The Enlightenment is man's [*sic*] leaving his self-caused immaturity. Immaturity is the incapacity to use one's intelligence without the guidance of another. Such immaturity is self-caused if it is not caused by lack of intelligence, but by lack of determination and courage to use one's intelligence without being guided by another. Sapere Aude! Have the courage to use your own intelligence! is therefore the motto of the enlightenment   (Kant 1784).

According to German sociologist and philosopher Jürgen Habermas, the work of the European Enlightenment was to search for and establish universal truths in the spheres of science, morality, and esthetics (Habermas 1985).

- **Science**: Truth had to be more than prejudice parading itself as fact. Using human reason, human beings could cut through irrational and false belief systems and lay bare the truth about the processes that created the world.
- **Morality**: Judgments as to what was right and wrong had to be based upon more solid ground than individual conscience or religious dogma. Through human reason, human beings could arrive at a universal set of morals and ethics.
- **Esthetics**: Beauty had to be more than mere beauty in the eye of the beholder. By using human reason, human beings could arrive at a universal consensus on the aesthetic value of art, music, poetry, literature, sculpture, the human body, natural landscapes, and so on.

Once science had figured out how the world worked, morality had established how the world ought to work, and esthetics had gleaned insights into how the world should look, it was then possible to build a better world. It was the duty, then, of the intellectual to harness human reason to build what English philosopher Thomas More in 1516 labeled "Utopia," and thereafter to maximize human happiness.

In calling for greater interest in the geography of the Enlightenment, Scottish historical geographer Charles Withers (2007) has challenged conventional geographical imaginaries that consider the Enlightenment to have originated in France or the Netherlands and then diffused intact to peripheral countries such as Portugal, Spain, the Greek-speaking countries of southeast Europe, or the Spanish-speaking colonies in Latin America. To speak of the "enlightenment in France," "in America," "in Spain," and so on is to understand it to be an essential "thing" that expresses itself variously. It makes more sense, Withers argues, to think in terms of the "Dutch Enlightenment," the "Scottish Enlightenment," the "French Enlightenment," and so on and thus to understand the Enlightenment as a locally produced, multiple, and geographically variegated cultural movement. It may be necessary, then, to understand the Enlightenment not only historically as a coherent cultural movement arising at a particular historical moment, but also spatially as a series of connected but distinctive cultural movements that need to be *placed.*

In his book *Civilization: The West and the Rest*, Scottish-born and US resident historian Niall Ferguson (2011) argues that Western culture bequeathed the extraordinary cultural institutions of the market, science, property rights, medicine, consumerism, and the work ethic. For Ferguson, the West's culture remains unrivaled. Why then, he muses, is it losing ground to the rest and faltering today? The West, it seems, has been overcome by a number of cultural pathologies. Apathy, confusion, hesitancy, doubt, and relativism (the belief that all ideas are of equal value) have all replaced confidence and faith in the capacity of human beings to master their own destiny. Successive attempts to shape the world have ended with mixed results. Instead of celebrating its successes, the West has become fixated with its failures. It has lost confidence in its ability to understand the laws through which the social and natural worlds work, and, worse, has begun to doubt that there existed any such laws in the first instance. Ferguson concludes that the West might be being dethroned in the twenty-first century (by, for example, China) not because other civilizations are proving superior, but because it is less good at being the West than other rival civilizations (for example, witness China's seemingly formidable imitation of the West). The West and its youth need to be re-educated about the unique virtues of Western culture.

Cultural Geography and the Study of Western Culture

How might human geographers approach the study of Western culture? To address this question, it is first necessary to examine how human geographers approach the study of culture itself.

Although ultimately overly simplistic, it is customary to distinguish between two traditions of cultural geography: that which was championed by the Berkeley School of Geography in California from the 1920s, and new cultural geography, which was pioneered by a number of US and British human geographers from the 1980s onward. While traditional cultural geography treated culture as a "superorganism" and focused upon the "cultural landscape," new cultural geography has explored the concept that culture is a social construction that both is formed from and plays a role in sustaining and/or interrupting power relations and wider social, political, and economic processes.

Sauerian Cultural Geography: Culture as a Superorganism

According to the Berkeley School, cultures are best thought of as superorganic entities. Superorganic concepts of culture treat cultures as:

- Guided by their own internal laws and workings and beyond the control of any particular individual or social group. People are passive bearers of culture, not creators of culture.
- Homogeneous groupings; everyone belonging to a culture shares a common worldview, has a similar set of traits, and conforms to a singular set of traits.
- Causal agents in their own right, working to make the world alongside social, political, and economic processes.

Cultures have a life of their own. They are born, grow, live, and die according to their own life cycle. Human geographers should not try to explain this lifecycle. Instead, working alongside anthropologists and archeologists, they should simply attempt to map traces of culture on the earth's surface – their material deposits and etchings onto the landscape (see Deep Dive Box 6.1).

Can Western culture be viewed as a superorganism? Certainly, this conceptual framework consciously and unconsciously pervades populist writings on the West. The rise, reign, and faltering of the West as a social, political, and economic entity might therein be explained with reference to the birth, adult life, and death of Western culture. Western culture somehow emerged from the amalgamation of a myriad of prior cultures. No individual or social group consciously willed it into existence or deliberately crafted its emergence. Once in the world it has done a tremendous amount of work, building new social, political, and economic institutions in the pursuit of tomorrow's ideal world. It has left its imprint on the face of the earth in the form of cultural landscapes, which, as we will see, are ordered, rational, planned, and sanitized. But for

---

Deep Dive Box 6.1   Carl O. Sauer's (1925) "The Morphology of Landscape"

In his famous 1925 essay "The Morphology of Landscape," US geographer and leader of the Berkeley School of Geography Carl Sauer sought to make the case that the field of geography ought to be construed as the study of the cultural landscape (Sauer 1925). For Sauer, the terms 'culture', 'cultural geography' and 'cultural landscape' assumed particular meanings.

In 1925, the doctrine of environmental determinism prevailed in human geography. Human culture was considered to be a product of the natural environment. Coastal regions gave birth to coastal cultures, tropical regions tropical cultures, mountainous regions mountainous cultures, desert regions desert cultures, and so on. Sauer questioned the claim that the geography of human culture was determined by the geography of the natural environment. Different cultures had developed on either side of the US -Mexico border even though in both cases a semi desert ecology prevailed. Both Mexican and US culture must have emerged then by dint of non- environmental factors. Culture had to have a life of its own beyond climate and biogeography. Sauer's motif was "Culture is the agent, nature is the medium, the cultural landscape is the result."

Sauer believed that culture could be conceived of as a superorganic entity. Cultures were real existing entities in and of themselves, and were guided by their own internal laws and workings. Sauer dismissed the idea that conscious actions by individuals and/or social groups were responsible for the emergence, maintenance, and withering of cultures. Cultures may form from gatherings of individuals, but as people fuse together a group dynamic emerges, and it is this dynamic that guides the trajectory of the collective. The whole is greater than the sum of the individual parts; cultures were constituted through collective behavior and perhaps even the invisible hand of crowd behavior.

From the 1920s to his death in 1975, Sauer mapped cultural landscapes in the Midwestern United States and those created from European interactions with native cultures in the United States, the Caribbean, Mexico, and Latin America. Every human modification of the natural environment was of interest, from field patterns to irrigation channels, spiritual centers to burial grounds, castles to ruins, buildings to barns, and roads to canals.

reasons internal to itself, it is currently "committing suicide." The West has lost confidence in the certainties that hitherto underpinned its culture. Its boldness and assertiveness have given way to a sense of apathy, procrastination, and self-doubt. Western culture is suffering a degree of psychosis. And as its culture ebbs, the central pillars of Western culture are crumbling, at least in the West itself.

New Directions in Cultural Geography from the 1980s

From the 1980s, many came to believe that for cultural geography to prosper, it needed to move away from its concerns with traditional anthropology and archeology and to begin a conversation with cultural studies, sociology, economics, and political science. In his classic critique of the superorganic idea in American cultural geography in 1980, US geographer James S. Duncan criticized the Berkeley School for failing to understand that people are not just passive bearers of all-powerful cultures that mysteriously hypnotize them and over which they have no control (Duncan 1980). Moreover, cultures do not just spring to life and endure in a vacuum. Cultures are, in fact, actively made by social groups embroiled in power relations at particular moments in particular places and to serve particular ends. A new cultural geography was needed that placed at its heart the origins and functions of culture in society. Cultures were formed in the context of asymmetric power geographies and wider social, political, and economic dynamics.

It is through this lens that the idea of the cultural landscape should be understood. Duncan (1980, p. 11) argued that landscapes are less products of superorganic cultures and more "constituent elements in socio-political processes of cultural reproduction and change." One of the principal ways in which power is sought, claimed, and entrenched is through the control and manipulation of landscapes. Elite actors (state, capital, or dominant group by class, nationality, ethnicity, gender, religion, and so on) materialize their ideologies and project their power through the landscape. Landscapes naturalize and affirm their authority and views of reality. Although landscapes are more often complicit in the production and reproduction of power relations, they are also foci for contestation and resistance to political authority. The cultural landscape, then, is not only a political construction; it is also a provocation that invites cultural conflict with alternative (subcultural) groups.

Central to the new cultural geography are the ideas of the *social construction of cultural practices* and the *social construction of ideas about culture*.

In his 1989 book *Maps of Meaning: Introduction to Cultural Geography*, British cultural geographer Peter Jackson captured and defined for a new generation the idea that culture is a social construction. Jackson called upon cultural geographers to pay more attention to the ways in which culture plays a role in shaping power relationships and social inequalities, for example on the bases of class, race, sexuality, gender, age, disability, and so on. For Jackson, cultures work in the same way as ideologies do. Elite or dominant groups construct and police the dominant culture to maintain their position of power – or, as he puts it, to preserve their hegemony. They act as if their way of seeing the world is true, a given. It is the job of subcultures and countercultures to challenge the dominant culture and to provide new ways of seeing reality. Cultural politics or culture wars emerge when competing understandings of reality grate up against one another.

For Jackson, the objective of cultural geography is to examine and challenge the ways in which dominant groups (for example, white, middle-class, heterosexual, able-bodied men living in the developed world) attempt to produce and normalize particular images of their cultures to retain their dominance.

While sympathetic to Jackson's call for a new cultural geography, US and Swedish based cultural geographer Don Mitchell has cautioned cultural geographers not to treat culture as something that actually exists. In his 2000 book *Cultural Geography: A Critical Introduction*, Mitchell in fact went as far as to say "there is no such thing as culture." Instead, he argues, there is only *the idea of culture.*

The idea of culture is used in a multiplicity of ways by different people at different times. For instance, culture variously refers to: the dominant ways of life and institutions around which societies are built; a set of traits, habits, and customs that develop around particular activities such as workplaces, professions, hobbies, and interests; the level of civilization, sophistication, and refinement of a society; a set of artistic practices, including painting, theater, literature, poetry, film-making, sculpting, and so on; and communities that are not part of mainstream society. According to Mitchell, the idea of culture itself should be studied in relation to power dynamics and wider and contested social, economic, and political praxis. For example, dominant groups often use and abuse the idea of culture to control and define "others" and thereby secure their own power base. Instead of treating these meanings literally, that is, assuming that they refer to something that actually exists in the world, it is more useful, Mitchell argues, to examine how the concept is used by powerful groups in particular settings and toward purposeful ends.

For Mitchell, then, cultural geography should be less concerned with studying cultures *per se* and more concerned with the social construction of the idea of culture: its deployment in particular places and situations to serve specific interests in particular social, economic, and political contexts.

Inspired by the idea that society is culturally constructed and that ideas about cultures are as powerful as cultures themselves, cultural geographers are interested in how the West seeks to represent, imagine, depict, and project its superiority, including through the cultural landscape. As British political and cultural geographer Alastair Bonnett has shown, cultural representations, images, and projections of Western supremacy continue to exist because they accomplish important work in the world for powerful Western (and sometimes non-Western) social, economic, and political elites (Bonnett 2017). Cultural geographers are less interested in testing the veracity of the story that West is best and are more interested in probing into who invented this story, who put it into the world, what work it does in the world, who benefits from it, and how it survives and prospers over time.

Social Formations and Symbolic Landscapes

British cultural and historical geographer Denis E. Cosgrove's 1984 book *Social Formation and Symbolic Landscape* and British feminist geographer Gillian Rose's 1993 book *Feminism and Geography: The Limits of Geographical Knowledge* provide pathbreaking insights into the West's projection of power in and through the cultural landscape. Each argues that landscape is a "way of seeing" that emerged only with the rise of European capitalism, European nation-states, and European patriarchal society and that has worked principally to consolidate the dominance of elite and powerful groups in society.

According to Cosgrove, every social formation (type of society) has a dominant mode of production (economic system) from which particular models of symbolic production (mode of producing culture) arise. In primitive societies, where reciprocity within extended family networks dominated, symbolic production was rooted in kinship ties, and landscapes expressed common ancestry and the culture of kindred groups. In archaic societies where religion defined prevailing models of economic production and consumption, including symbolic production, landscapes were freighted with religious messages and organized around a central sacred site. Finally, in capitalist societies, symbolic production is rooted in the market economy and expresses itself in the form of private property, abstract grids and geometries, and the parceling of land and property. It is here that landscape emerges as a particularly important symbolic product.

Landscape, for Cosgrove, was a way of seeing the world that was ideologically loaded; it was a product of the ascendance of the West and the social, political, and economic changes the rise of the West ushered in. The West looks at earth with an all-seeing eye: it adopts an omnipotent and visual perspective, an eye of providence. Undoubtedly, the West's preference to view from the sky stems from its desire to control and engineer space. Specifically, the idea of landscape emerged first during the Italian (Venetian) and Dutch Renaissances as colonists, merchants, and industrialists commissioned paintings to venerate their status and holdings. Cosgrove had a particular interest in Italian architect Andrea Palladio (1508–1580) and his projection of Western power on the landscape of Venice (Plate 6.1). Representations of landscapes served to undergird the idea that landowners were all-powerful, commanding over vast territories and subduing and ordering nature. As political and economic elites in France, Prussia, and the United Kingdom accumulated vast riches from their colonial adventures, they too began to sponsor the idea of landscape as a peculiar and useful perspective.

Rose's contribution was to extend Cosgrove's argument by noting that gender, too, played an important role in the emergence of landscape as a way of seeing. Purveying the landscape from a perch from afar was a male-centered practice; it constituted a masculinist gaze. It was not an accident that landowners and political and economic aristocrats were also by and large men. These men enjoyed puffing their chest out and

Plate 6.1 A Palladian villa near Venice, Italy. Source: Bruno Perousse/agefotostock/Alamy Stock Photo.

viewing in a self-satisfied way the land they owned and commanded. Landscape paintings portraying their property affirmed to the world their superiority over serfs and landless laborers. According to Rose, women were rarely included in landscape frames, and when they were, they were depicted as sexualized mistresses beholden to the landowner. Meanwhile, landscape paintings were often organized so as to depict heroic men striving to bring a fertile but unruly mother nature under control. In any case, landscape as a perspective was inconsistent with feminine ways of experiencing the world. Women invariably preferred getting off the helicopter and were more comfortable seeing themselves in and of the world than detached and above the world.

The West in the Cultural Landscape: On the Imperialism of the Straight Line

A central feature of Western society is its preference for rational, ordered, and civilized spaces over anarchic, chaotic, and unruly places. US historian John Merriman coined the phrase "the imperialism of the straight line" to underscore the power of the design of the built environment and its architecture to denote omnipotence and authority. To explore the uses of the concept of the imperialism of the straight line, it is instructive to look at projections of Western power in the cityscapes of Paris and Washington, DC, arguably the first and the latest major capitals of Western modernity.

Haussmann's Paris: The Capital City of Modernity

In his book *Paris: The Capital of Modernity*, David Harvey (2003) recounts the epic story of the Haussmanization of Paris in the period of 1848–1870. Haussmann's ambition was to rebuild Paris after the 1848 workers' socialist rebellion against the French Monarchy. With the coup d'état against King Louis Philippe in February 1848 and institution of the Second French Republic, Napoléon III soon became French president. By 1851, however, Napoléon III consolidated power, acting himself as a de facto French monarch and establishing a Second French Empire. In 1853, Bonaparte appointed Prefect of Seine Georges-Eugène Haussmann to plan and renew Paris's urban form. His instruction was to build Paris so that it might wear the same clothes as imperial Rome once did, and to serve as the crucible of power and capital city for Europe's surging civilization. Haussmann carried out this mission between 1853 and 1870 with brutal efficiency, ripping up urban fabric and building Paris' streetscapes anew.

Harvey begins with images of Paris in the period of Restoration of the monarchy (1815–1830) painted by French novelist and playwright Honoré de Balzac. Although self-identifying as a conservative monarchist and a Catholic, in realist novels such as *La Comédie humaine* (*The Human Comedy*) Balzac developed in forensic detail a social commentary on the conditions of the working class and the inner-city slums in which they lived. He was to become an unwitting apostle for the revolutionary movement. Stung by the insurrection of 1848, Haussmann came to view his role as building social order through physical planning. By 1870, the urban form of Paris was unrecognizable from that which prevailed in 1850. While insurrection returned with the Paris Commune of 1871, Haussmann's reconstruction of the physical fabric of the city altered the tenor of rebellion. And by the time Haussmann was dismissed from his role, the momentum he had garnered ensured that his vision for urban transformation lived on.

Haussmann was a forceful, energetic, and skillful planner and quickly stamped his authority on the city. He was famed for viewing Paris as an integrated whole (and not a collection of disparate districts) but equally for attending to minute detail and precision. Hausmann set out to create a streetscape that projected French cultural superiority and imperial authority, which paved the way for modern large-scale capitalism, and which removed spatial conditions that bred insurrection. His goal was to reclaim the narrow chaotic slums and streets, ridding them of vice, overcrowding, and disease and imposing on them a rational organization of space. The conception of urban space that Haussmann championed was historically novel. Wide, light-filled, and grandiose boulevards were to radiate from the center, interspersed with iconic buildings and architectural jewels, "variously sepulchral and ethereal and dominated by haughty grandeur" (Plates 6.2 and 6.3). A new sanitary infrastructure was built, predicated upon underground sewage tunnels. New transport connections and railway stations were constructed. The city was to be divided into 20 *arrondissements*, centrally governed but each with a Mairie (city hall). Parks were built, like the Bois de Boulogne, to enhance health, urban nature, and public sociality. The urban fabric was designed to illuminate significant symbols of imperial power; public spaces were built to host theatrical spectacle-events that venerated imperial munificence. The Bon Marché (opened in 1852) and Louvre (completed in 1855) announced a new age of opulence and consumerism.

L'Enfant and McMillan's Washington, DC, and National Mall

In her 2016 book *The National Mall: No Ordinary Public Space*, US geographer Lisa Benton-Short examines the planning of Washington, DC, and in particular the National

Plate 6.2   Haussmann's City of Light: The Arc de Triomphe at the center of the Place Charles de Gaulle, also known as the "Place de l'Etoile." Source: DigitalGlobe, Inc. CC BY 4.0.

Plate 6.3   Street view, Arc de Triomphe, from Avenue des Champs-Élysées, Paris. Source: Photo by Matt Boitor on Unsplash.

Mall, aptly entitled "the nation's front yard" and in many senses today the "world's front yard." Benton draws attention to the successive layering of the National Mall with added symbolic strata.

The book begins with the story of President George Washington's decision to build a capital city for the new United States alongside the Potomac River. In 1791 Washington commissioned French-born US based engineer, architect, and urban designer Pierre Charles L'Enfant to create a master plan. L'Enfant envisaged Washington, DC, to be built to a gridiron of irregular rectangular blocks upon which broad diagonal avenues were to be superimposed. The creation of public squares, circles, and triangles at street intersections would enable monuments to be displayed and convey a city of political gravity. At the heart of L'Enfant's plan was the National Mall, a "grand promenade" that radiated westward for two miles from Capitol Hill to the Potomac River. This National Mall was to be a public national commemorative space emphasizing democratic openness and patriotism. L'Enfant conceived of Capitol Hill, the White House, and the Washington Monument to align along two axes, symbolizing the importance of both the executive and legislative arms of government, and the calibrating watchful eye of the Washington Monument at the fulcrum, reminding all of both of the nation's origin story and Constitution.

Alas, L'Enfant was to be dismissed before fully enacting his plan, and for most of the nineteenth century Washington, DC, remained a partially completed building project. But his vision was revived by the McMillan Commission in 1901, and the McMillan plan

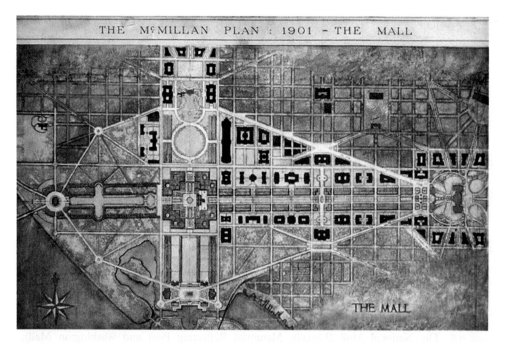

Plate 6.4 The McMillan Plan of 1901: Laying out the National Mall in Washington, DC. Source: National Capital Planning Commission, https://commons.wikimedia.org/wiki/File:McMillan_ Plan.jpg. Public Domain.

advanced its implementation (Plate 6.4). This plan reclaimed land from the Potomac River, extending the Mall westward, closer to Arlington Cemetery (an iconic resting place for US servicewomen and servicemen) and with enough space to enable the construction of the present-day Lincoln Memorial. The plan injected fresh symbolic significance into the east-to-west axis: to the east was Capitol Hill, a "monument to democracy" at the center was the Washington Monument, a "monument to freedom and independence"; and to the west was Lincoln Memorial, a "monument to unity and equality." Meanwhile, the north-south axis was strengthened by the building of the Jefferson Memorial.

The iconography of the National Mall is tightly policed. Conservatism preaches against additions to the mall; not only would added buildings diminish the availability of public space and indeed spaces of protest, but also it would interfere with the "text" of the mall by adding fresh historiographical narratives. Lamenting "sprawl on the mall" and asking "how many memorials are enough?" some defenders of the original mall have waged a series of planning battles. Nevertheless, since 1980 eight new memorials have added: the Vietnam Memorial, the Korean War Memorial, the Franklin Delano Roosevelt Memorial, the Museum of the American Indian, the George Mason Memorial, the World War II Memorial, the Martin Luther King Memorial, and the Black Revolutionary War Patriots Memorial. A ninth, the John Adams Memorial, has been approved and is awaiting construction. New commemorative builds serve to complexify and enrich the mall as a site of national memory, but the culture wars surrounding interventions designed to layer new features on top of the older symbolic landscape point to ongoing controversy over who owns the American story (Plate 6.5). The insurrection mounted by the far right on January 6th 2021 and reckless and violent highjacking of the Capitol building in a bid to overturn the election of Joseph Biden as US president provides a case in point.

Plate 6.5   The National Mall (Lincoln Memorial Reflecting Pool and Washington Mall), Washington, DC. Source: Photo by Jacob Creswick on Unsplash.

Worlds of Meaning: Landscapes of Power and Living Landscapes

The Ville and the Cité: Building and Dwelling in Western Spaces

One of the key concerns within contemporary critical human geography is the super-imposition and hyperextension of what might be termed abstract space onto the concrete spaces of everyday life and the implications of the empires of abstract space for fragile human ecologies, local cultures, and communities who live space sensuously, emotionally, and affectively. Are landscapes of power also capable of serving as living landscapes? Human geographers have answered this question in various ways.

In his 1977 book *Space and Place: The Perspective of Experience* (Tuan 1977), Chinese-born and US resident humanistic geographer Yi Fu Tuan distinguished between the concepts of space and place. Space can be thought of as abstract, impersonal, and pre-fabricated. Abstract space is the product of instrumental logic and acts of cognition rather than human creativity, festivity, and acts of affection; it is the craft of profession-alized spatial architects, investors, bureaucrats, planners, and engineers, and is most often manufactured, governed, and regulated to serve hegemonic social, political, and economic interests. Spaces are largely empty of meaning. Place, by contrast, refers to the meanings human beings attach to spaces, their senses of belonging to and estrangement from spaces. Topophilia, literally the love of place, can often turn even the most barren, arid, and inhumane environment into something fit for human existence. Spaces become places when human beings give them an identity and personality.

In his book *Place and Placelessness*, Canadian geographer Edward Relph (1976) argues that in Western societies, placelessness has waged a war on place; to a significant degree, abstract space has colonized everyday life and uprooted settled attachments to place. The globalization of the West's symbolic landscapes means that we are now finding it difficult to invest meaning in the synthetic and cloned environments in which we live. Identikit cities are being built that look and feel the same. The global march and penetration into our lifeworld of capitalist space mean that differences between places are being smoothed over to the extent that differentiating one place from another is becoming increasingly difficult. For Relph, place is central to human flourishing; that we are being ripped apart from that which provides us with ontological security and finding it difficult to humanize the placeless spaces and concrete jungles in which we now live is existentially threatening.

In his 1974 book *The Production of Space*, French sociologist and Marxist Henry Lefebvre identified three kinds of spaces: the "perceived space" (*le percu*) of everyday life (people's subjective experiences of space); the "conceived space" (*le concu*) of cartographers, government projects and urban planners, and property speculators and architects (for example, the shopping mall); and the "lived space" (*le vecu*) of the imagination (Gotham City, for example). Conceived space had now colonized everyday life, there was a need to strip the West's abstract space of its fetishisms and reifications and, by redistributing "rights to the city," to restore control over the production of space to those who immediately inhabit it. He judged the concept of "alienation" to have "enjoyed a brilliant career as a truly enlightening notion" (Lefebvre 1991, p. 343) and aspired to recreate spaces that truly reflect "the full inventory of what the body has to give" (Lefebvre 1991, p. 197). Using the method of "spatial architectonics," he sought to peel back the successive strata of capitalist spatiality to recover, once more, human beings' primal and biomorphic relationships with space.

Inspired by the French tradition of existential Marxism, in 1989, Canadian geographers Audrey Kobayashi and Suzanne Mackenzie published a book outlining a case for *Remaking Human Geography* by bringing into conversation humanistic geography and historical materialism, and considering human psychology and culture within the context of wider social, political, and economic forces. There was a need to move beyond the parochialism of humanistic geography that tended to focus upon the existential dramas that mark the day-to-day lives of particular people in particular places. Equally, however, there was a need to move beyond grander historical theories of social change and economic determinism and to recognize that history has to be actively made by flesh-and-blood human beings and the phenomenological sensory tools they have at their disposal. A new integrated humanistic-Marxist geography was required. Marx's dictum that people make history but not in conditions of their own choosing was now to read "people make spaces and places but not in conditions of their own choosing."

Geographical Imaginations, a 1994 book by British-born but Canadian-based geographer Derek Gregory, stands as perhaps the most sophisticated human geographical statement adopting this perspective. Gregory argued that human geographers ought to have as their central concern the impacts of the hyperextension into people's everyday places and lifeworlds of the abstract grids and geometries imprinted on the earth's surface by Western capitalism and Western states. According to Gregory, abstract space takes one of two forms:

The commodification of space: Whereby private ownership of land results in the etching onto the face of the earth of capitalist circuits of production, circulation, and consumption. Examples would include the industrial parks, airports, suburban landscapes, and out-of-town shopping malls that now adorn virtually every major city. Hotel chains like the Hilton, Radisson, Sheraton, Best Western, Ibis, and Marriott pride themselves on creating identikit layouts – the lobby of any one feels and looks the same, no matter where in the world it is located. Some refer to this standardization of space as the McDonaldization of the world.

The bureaucratization of space: Whereby Western states claim authority over spaces and inscribe their will onto space in the form of highly regulated and planned public infrastructure and administrative systems. An example would be the sprawling public housing estates that quickly become destitute "sink estates" and that brutalize residents to the extent that they cannot humanize their own living environments creatively. The spaces of hospitals, schools, and prisons provide other examples. These often monotonous and homogenized spaces are designed by states to perform certain tasks, and citizens are expected to use and navigate through them in ways that submit to state-defined rules (See Deep Dive Box 6.2).

Deep Dive Box 6.2 Richard Sennett: Reconciling Ville and Cité

US sociologist Richard Sennett's (2002) book *Respect* takes him on a journey back to his roots in Cabrini Green. A huge public housing estate only eight blocks west of Chicago's Gold Coast, Cabrini Green stands as an exemplar of the rise and fall of the grandiose modernist housing experiments of the twentieth century. A futuristic utopian landscape when it first opened in 1942, the estate has steadily fallen into decline and became for a time widely stigmatized as one of the United States' most dangerous ghettoes. Sennett's story of the origins and evolution of Cabrini Green is structured around an examination of the interplay between the brutalized and mutilated spaces of the estate and the existential condition of the human beings who inhabit these spaces. In Cabrini Green, one witnesses a truly synthetic space that has wounded residents and significantly impaired their capacity to humanize their own living environment creatively.

In his later books, *Building and Dwelling: Ethics for the City* (2018) and *Designing Disorder: Experiments and Disruptions in the City* (2020), Sennett invokes the distinction between Ville and Cité. If the Ville is the built environment, empty, ordered, and engineered, the Cité is the populated and lived landscape, fleshy, colorful, and humanized by people with all their creativity, quirkiness, and perversions. In too many cities, the Ville has colonized the everyday spaces of the Cité, leading to sterile and alienating living environments. Building has overwhelmed dwelling.

For Sennett, the aim of urban planning should be to effect a reconciliation between Ville and Cité by creating an improvisational and "open city" that embraces complexity, ambiguity, and difference. Urban development needs to be authentically co-created, co-governed, and co-implemented by planners, policy makers and practitioners and the communities they serve. Instead of fixing problems and focusing upon what is absent, urban renewal needs to begin by mapping comprehensively the skills and assets that communities already have and work to build upon areas of strength and vitality. Communities need to steadily own the neighborhoods they inhabit.

Only by changing the organizing principles around which our cities work will it be possible to create pro-social spaces and ecologies of care that are structurally therapeutic and life-affirming. To fix ourselves, we need first to fix our cities.

A crucial caveat here is required. Not all living and breathing humanised places merit nurture. For British Marxist and feminist geographer Doreen Massey, the idea of "place" comes freighted with certain assumptions that can prove unhelpful. Places are often considered to be coterminous with coherent communities and are assumed to be predicated upon monolithic cultural identities, cherished customs, and valued traditions. Threatened by globalization and fearing the loss of local culture, self-appointed custodians of local tradition often take it upon themselves to pursue a series of hostile, reactionary, and regressive countertactics. Massey (1991) argues that in the globalized world of today, a more progressive concept of place is both necessary and possible. Places are no longer isolated from the world but instead are deeply intermeshed with and imbricated in the lives of peoples and places in other parts of the world. Places always have permeable borders, multiple identities, and complex relations with other places. Massey calls for a rethinking of the idea of place and for the introduction of a less defensive and reactionary and more progressive and relational concept of place, which she titles a "local sense of the global" or a "global sense of place."

Let us now put these ideas into practice and examine some of the ways in which human geographers have sought to study intersections between landscapes of power and living landscapes at particular times in particular places.

Case Study 1: The Metropolis and the Mind: Early Conjectures

Give the rise of great cities around them, it is unsurprising that a number of leading thinkers in nineteenth-century and early twentieth-century European and US cities tried to figure out the great puzzle of the impact city life on human culture. Would living in cities change us as a species? Although they ciphered it in different ways, these scholars observed a dichotomy at the heart of urban life: *Homo urbanus* was a liberated but tortured creature – at once free from iron-clad rural cultures and stifling traditions and exposed to the phantasmagorical technology, hyper-stimulating sensory bombardment, and exhilarating possibilities of the new metropolis, the urbanite was also prone to anonymity, grime, dirt, pollution, poverty, loneliness, alienation, and an erosion of personal identity (Deep Dive Box 6.3). Early conjectures on the psychological impact of living in rapidly growing cities include:

- French poet Charles Baudelaire's renditions of life in mid-nineteenth-century Paris. In his 1857 *Fleurs du mal* (*Flowers of Evil*), Baudelaire portrayed the city as a harbinger of a new urban aesthetic described as *modernité*: a condition of being in the world and perceiving the world marked by liberating transience, fleeting encounters, fluid identities, and ephemeral churn.
- Baudelaire was to influence German literary critic Walter Benjamin, who in his famous *Arcades Project* (1927–1940), a multifaceted reading of human encounters

Deep Dive Box 6.3 Gotham City: Utopia and Dystopia

At the start of the twentieth century, American cities were in the throes of great expansion, carrying as they grew the hopes and aspirations of teeming masses of immigrants. Skyscrapers adorned the skyline of many cities, beckoning migrants to come join a modern society on the march to greatness. Americans came to be in awe of landscapes of power but anxious that beneath the veneer, there lurked all that was bad in the human condition.

The DC Comics *Batman* series began in 1939. From as early as 1940, Gotham City was identified as the city Batman lived in and was destined to save. Although the Batman series has presented a number of histories of Gotham City, many have taken Gotham to be an exaggerated representation of any standard twentieth-century city in the United States. Moreover, whilst the character, atmosphere, architecture, and mood of Gotham vary from series to series and from film to film, the city can be thought of as a powerful image of dystopia. *Batman* writer and editor Dennis O'Neil states, "Batman's Gotham City is Manhattan below Fourteenth Street at eleven minutes past midnight on the coldest night in November." Often depicted only in the light of a full moon, Gotham is a dark and murky city, overwhelmed by gargoyles and graveyards, gothic architecture, and bombastic, gargantuan angular buildings, and drowning in vermin and crime. Personified most in the grotesque figure of the Joker, it is a city in which law and order has broken down and in which gangsters, lunatics, and violent psychopaths reign. No human law and order enforcement agency can arrest the degeneration of the city into anarchy; its only hope is to turn to a superhero with extra-human powers, a Batman, to fight the evil that lurks within.

In Todd Phillips' 2019 dark, grim, and tense film *Joker*, starring Joaquin Phoenix, an origin story of the Joker is told. Set in 1981, and against the backdrop of the turn to neoliberal policies and austerity, the film follows the mental degeneration of socially incompetent clown and comedian Arthur Fleck. Relentlessly mocked for being different, dis-anchored by revelations about this childhood, and worn down by an uncaring society, Fleck becomes increasingly nihilistic and unhinged and descends into grotesque and pointless violence. Although the movie ends with the Joker becoming a cult leader of clown goons also abused by the system and bent with rage on "killing the rich," in fact it is clear that he lacks any sociopolitical motivation. An existential threshold is breached, and life is lived as a useless passion; order and rationality make no more sense than chaos and insanity. The Joker exclaims: "What do you get when you cross a mentally ill loner with a society that abandons him and treats him like trash? You get what you f**kin' deserve!"

with spaces of consumption in pre-Haussmann Paris, sought to characterize the modern city in terms of an interplay between *Erlebnis* (overwhelming sensory bombardment) and *Erfahrung* (liberating wanderings of the *flâneur*). Benjamin excoriated Haussmann's Paris (Haussmann had bulldozed down the Arcades in the name of progress), believing that capitalism was creating through cities a visual façade that was destroying the soul of Paris while pretending to be expressing it. False utopianism would lead to "the hell of unfulfillment."

- For German sociologist Ferdinand Tönnies, the psychological implications of the shift from traditional forms of community to modern forms of social life were felt most keenly in the cities. In his 1887 book *Gemeinschaft und Gesellschaft*, Tonnies argued that the maintenance of a sense of personal significance and ontological security presented an ongoing challenge to city dwellers, who were increasingly embroiled in abstract and impersonal social networks.
- Reacting to the grime of the late nineteenth-century British industrial city in his 1898 book *To-Morrow: A Peaceful Path to Real Reform*, for urban planner Ebenezer Howard there was a need for a "Garden City Movement" to heighten urbanites access to urban greenery and the countryside and to improve their mental and physical health.
- The impact of urbanization on the psychology of urbanites was a source of sustained fascination for German sociologist Georg Simmel, who in 1903 in a book entitled *The Metropolis and Mental Life* worried that the city might create sensory overload for human beings. Life in the twentieth century would be marked by existential struggle; adapt to the neon bustle and neural blitz of the modern city – or die.
- In his classic essay *Urbanism as a Way of Life* (1938), Chicago School of Urban Sociology scholar Louis Wirth argued that cities are at once harbingers of alienation, anomie, isolation, and loneliness and citadels of liberalism, tolerance, freedom, and creative expression There was such a thing as urban culture; this culture became more pronounced as cities grew in size.
- Urban scholar Kevin Lynch's 1962 book Image of the City proved pathbreaking in arguing that human psychology led human beings to cope with the chaos of the city by constructing mental maps that helped them navigate and orient safely. Lynch argued that these cognitive maps are constructed around the mental constructs of paths, edges, districts, nodes, and landmarks.

Case Study 2: Beyond the Pale – Clean Lines and Crooked Colonies

The European colonization of countries such as the United States, Canada, Australia, New Zealand, and other settler colonies was made possible only by the violent dispossession by European peoples of lands held for millennia by indigenous, native, and First Nation peoples. Notwithstanding apologies for the actions of their forebears and gestures toward truth and reconciliation by descendants of European colonizers, many indigenous peoples feel that retribution and compensation for past deeds remain due. Native American Indian and aboriginal attitudes to land, of course, differed significantly from those of the European settlers. While Europeans believed in private ownership of land and traded land through the open market, many indigenes believed that, because land was a source of life, it was to be respected, not owned. Land that sustained life was considered to be imbued with spiritual significance. Accordingly, European appropriation of land was considered by First Nation peoples not only as theft but also as an abhorrent attack on indigenous spirituality. The American frontier progressed west on the basis that natives were less than enlightened and that their way of life was so primitive it deserved no recognition.

But the superimposition of Western space extended well beyond settler colonies (see Plates 6.6a through 6.6e). In his book *Of Planting and Planning: The Making of British Colonial Cities*, British planner Robert Home (2013) provides a magisterial survey of the application of British town-planning principles to the building and design of colonial cities. Home argues that colonial ruling elites sought to express their political authority, economic hegemony, and intellectual superiority by shaping the cultural landscape. The British town-planning tradition – and, in time, profession – became the "chief exporter of municipalities," forging first cities in settler colonies in the New World and in the Antipodes (1600-1850), then in colonial port cities (1850–1900), and thereafter across Asia and Africa and to a degree parts of Latin America. From around 1910, the formal profession of town planning in quick order swept through the colonies, and new legislative frameworks were created to support the adoption of new planning functions by colonial governments. The 1932 English Town and Country Planning Act provided an off-the-shelf legislative framework for colonial administers to deploy when creating their domestic national, regional, and town planning approaches.

The British model of colonial town planning – what Home terms the "Grand Modell" – comprised the following:

- A policy of deliberate urbanization, or town planning, in preference to dispersed settlement.
- Control of land markets and land rights allocated in a combination of town, suburban, and country lots.
- The town planned and laid out in advance of settlement.
- Classical ideals of esthetics, symmetry, and order; Baroque grandeur and surface decoration.
- Public squares (providing a market, church, arsenal, and courthouse).
- Wide streets laid out in geometric, usually gridiron form, usually on an area of one square mile.
- Standard-sized, rectangular plots, spacious in comparison with those in British towns of the time.
- Land reservations and some plots reserved for public purposes.
- A physical distinction between the central city and the countryside usually via the creation of common land or an encircling greenbelt.

While the legacy of Europe's symbolic landscapes persists in the cities of the Global South, in the postcolonial period it is clear that a different dynamic has emerged. No longer is world-class planning, design, and architecture defined and conceived in the metropole and exported to the colonies to project Western power in colonial cityscapes. Increasingly, it is cities in the Global South that are exporting globally iconic cityscapes to the West. (see Plates 6.6a through 6.6e).

In their 2011 book *Worlding Cities: Asian Experiments and the Art of Being Global*, postcolonial theorists and urban planning scholars Ananya Roy and Aihwa Ong examine world-aspiring cityscape experiments in Asian cities. They question whether the West is any longer the benchmark of urban modernity. It is no longer appropriate to study Asian cities only in terms of the legacies of colonialism or with respect to registers of progress inherited from the West. It is now Singapore, Shanghai, Hong Kong, Delhi, Dubai, and Kuala-Lumpur and not New York;, London, Paris, Berlin, Rome, and Brussels that produce urban planning models that must be "invoked, envied, and emulated" as paradigmatic instances of a "new urban normativity." There are now many more sites in the Global South where claims to world-class urban design are being globalized by the "staging of showy architecture, cutting-edge industry, and homegrown urban aesthetics." An Asian modernity, quite apart from Western canons and standards, is now pioneering new ways of designing city spaces as world-class spectacles.

Plate 6.6a Exporting the imperialism of the straight line. India Gate War Memorial at the Rajpath and the "ceremonial axis" of central New Delhi, India (ex-colony of Britain). Source: Shalender Kumar from Pixabay.

Plate 6.6b Exporting the imperialism of the straight line. Avenida 9 de Julio, Buenos Aires, Argentina (ex-colony of Spain). Source: a r c a n g e l t, https://commons.wikimedia.org/wiki/File:La_Avenida_m%C3%A1s_ancha_del_Mundo...._-_panoramio.jpg. CC BY-SA 3.0

Plate 6.6c   Exporting the imperialism of the straight line. The Australian Parliament, Canberra, Australia (ex-colony of Britain). Source: Photo by Aditya Joshi on Unsplash.

Plate 6.6d   Exporting the imperialism of the straight line. Brasilia Esplanada dos Ministérios, Brasilia, Brazil (ex-colony of Portugal). Source: Arturdiasr / CC BY-SA 4.0.

Plate 6.6e   Exporting the imperialism of the straight line. Nir Sultan (Astana), Kazakhstan (ex-colony of USSR but designed by Japanese architect Kisho Kurokawa). Source: flickr.com.

Two examples serve to illustrate the ways in which Singapore has transitioned from being a colony to a global city and an importer of the West's symbolic spaces to being an exporter of world-class Asian urban landscape ideas.

The 1822/1828 Raffles Town Plan/Jackson Plan

In 1819, British statesman Sir Stamford Raffles founded Singapore as a trading post of the British East India Company and appointed Major-General William Farquhar as commandant of colonial Singapore. Returning to the colony in 1822, Raffles encountered chaos and disarray, dismissed Farquhar, and recruited Lieutenant Philip Jackson to create a town plan to bring authority and order. The 1822/1828 Raffles Town Plan/Jackson Plan established a vision for the future development of what has become Singapore's downtown. Drawing learning he had acquired in other colonial towns in which he had served, including Georgetown in Penang (Prince of Wales Island), Raffles envisaged Singapore becoming a "place of considerable magnitude and importance" and fashioned a layout to symbolize this aspiration (Plate 6.7). To project political power, Raffles located government functions and buildings and cultural institutions at "Singapore Hill," Fort Canning. To express civility and affluence, he emphasized greenery, gardens, and parks (such as Padang and the Botanic Gardens). To convey order, he established a grid street pattern. To address British anxieties over sanitation and disease, Raffles developed a plan to segregate the population by race. The European Town was to be the most affluent; separate *kampongs* (urban villages) were to be built for the Chinese, Indian, Malay, and Arab communities. Buildings were to have common facades, and streets to have standardized pavements.

The Urban Redevelopment Authority (from 1974) and Marina Bay Master Plan (from 1983)

In 1974, in anticipation of the further growth of Singapore as a tiger economy and the likely scarcity of land that would arise, the Singapore government established the Urban Redevelopment Authority (URA). The authority has worked to reclaim 360 ha of land for development at Marina Bay (adjacent to the downtown core and at the waterfront). In 1983 the URA published their vision for Marina Bay, and in 1998 translated this vision into a plan. Marina Bay was to become a world-class focal point for the city-state, announcing Singapore's transition from colony to global city. The Singaporean state has long fretted over the impact of globalization on Singaporean culture and identity. Marina

Plate 6.7   British Singapore: Raffles Place, 1966. Source: www.flickr.com/photos/argentern/4000469334/ (David Ayres).

Plate 6.8   Asian experiments in worlding cities: Marina Bay, Singapore. Source: Photo by Meriç Dağlı on Unsplash.

Bay was to be a dazzling world-class space but one with a Singaporean signature. Over time, this master plan has been executed patiently and diligently. Marina Bay has developed into a leading financial services center, a majestic civic and public space, a spectacular skyline dotted with a range of awe-inspiring and iconic skyscrapers, a twenty-first-century living space, a pioneer of urban greening, a model of cosmopolitanism, and a hub for conspicuous consumption and leisure (Plate 6.8). Marina South (adjacent to the Bay) is now being developed as a world-class model of sustainable urban living.

Case Study 3: Slums and Projects: The African-American Search for a Sense of Place

In 1955, Robert Moses, New York's then "master-builder" and proponent of modernist urban planning (often referred to as New York's Haussmann) famously declared, "when you operate in an overbuilt metropolis, you have to hack your way with a meat ax." In the early to mid-twentieth century, clearing overcrowded and unsanitary urban slums and building social housing estates and high-rise buildings became a central preoccupation for civic leaders in North American cities. Little account was taken of the needs and desires of the communities for whom these places, as bad as they were, were homes. US urban scholar Jane Jacobs took to task urban renewal programs that failed to respect the needs of city-dwellers and called for grassroots resistance to large-scale slum clearance. Moreover, what were intended to be (by the standards of the time) futuristic and utopian housing projects for those displaced by slum clearance too often degenerated to become inhumane, soulless, monotonous, concrete gray, and dystopian landscapes, beset by concentrated poverty and social malaise.

In her 2004 book *Root Shock: How Tearing Up City Neighborhoods Hurts America and What You Can Do About It*, US urban policy scholar and psychiatrist Mindy Thompson Fullilove examines the impacts of US urban renewal programs between 1949 and 1973 on the mental and physical health of residents, and underscores the disproportionate wounding that slum clearance has inflicted upon poor African–American communities. Fullilove invokes the concept of mazeway to summarize her argument. Mazeway can be thought of as the aggregate totality of individual lifeways in a community. The African–American mazeway was healthy and upward social mobility was well underway by 1950. Urban renewal conspired to rupture that mazeway and encouraged a disintegration of the black community. The consequence has been the dismantling of a healthy and vibrant community and the social engineering of community pathologies – all in the name of progress.

Against the backdrop of the US civil rights movement, urban renewal conspired to undermine the collective power and wealth of African-Americans. Renamed by some as the "Negro removal" Act, the Housing Act of 1949 authorized cities to clear "blighted land." During the 14 years of the urban renewal program, 993 cities participated, carrying out more than 2500 "clearances." Of those displaced, 63% were African-Americans. The process of urban renewal tore communities apart, destroying their accumulated social, cultural, political, and economic capital, as well as undermining their competitive position. Fullilove argues that place identity and place attachment are fundamental dimensions of the human condition; good mental health arises from strong, dense tissues of supportive relationships, enabled by access to society's resources and opportunities. As a corollary, mental and physical health deteriorate as societal programs destroy communities. It is little surprise, then, that the Kerner Commission's study (1968) of civil disorder in 1967 included urban renewal in the list of factors that triggered race-related rioting.

Many displaced black communities found their way to the US social housing "projects." From the outset, it was clear that the projects could not serve as effective incubators of human flourishing and as life-affirming places. By ring-fencing people who were already in social need into one area, these projects outed marginal groups for greater public scrutiny, and the names of estates themselves quickly became a signifier of all that is bad. Pioneering urban planners and thought leaders such as Oscar Newman sought to help. Newman's defensible "space theory" led to housing developments that attempted to "design out" crime and integrate African-Americans into established

Plate 6.9 President Jimmy Carter tours the South Bronx, October 5, 1977: "The presidential motorcade passed block after block of burned-out and abandoned buildings, rubble-strewn lots, and open fire hydrants, and people shouting 'give us money!' and 'we want jobs!'" *New York; Times.* Source: U.S. National Archives and Records Administration, https://catalog.archives.gov/id/176391.

neighborhoods. But estates such as the Magnolia Projects in New Orleans; Techwood Homes in Atlanta; Queensbridge Houses in Queens, New York; Robert Taylor Homes and Cabrini Green in Chicago; Marcy Houses and Louis Heaton Pink Houses in Brooklyn, New York; Pruitt-Igoe in St. Louis; Watts in Los Angeles; and Johnston Square Apartments in Baltimore, became besmirched as national failures. As US cities increasingly embraced neoliberal economic policies in the late 1970s and 1980s, these estates fell into ever deeper despair, debilitated by unemployment, precarious and low-paid work, corrosive inequality, poor housing, air pollution, violence and vice, and social exclusion.

US cultural geographer Rashad Shabazz has undertaken a variety of studies of the settlement experiences of the black community in northern US cities (particularly Chicago and New York) from the onset of the twentieth century "Great Migration" from the southern states. Shabazz argues that white majorities consistently exercised carceral power as an anti-black process, containing the black community spatially. Of course, the disproportionate incarceration of African-Americans in prisons is the most visible expression of containment, but through policing and surveillance, racist restrictive rules, and disciplinary measures extend "prison-like" spatial isolation in neighborhoods, streets, and single-room kitchenettes. The idea of "prison" is best understood not as discrete institutional space of confinement and punishment, but as a wide set of techniques of social control that limited the mobility of the black community and created at a distance black spaces. These techniques thrived off cultural representations of black men as dangerous and sexually promiscuous and were given further license by the emergence of HIV/AIDS in the 1980s and 1990s and the US war of drugs.

Deprived of rights to the city, black men developed and expressed their masculinities in alternative and not always productive ways. Blocked from accessing acceptable codes of masculinity by spatial confinement, overly compensatory models of masculinity were embraced that led to heightened efforts by black men to access and to control what public space did exist in black communities. Abandonment of family and home and resort to gang violence were poor choices. Their subtext was a desperate search to reclaim respect in a society that afforded black men little. But the hegemonic black masculinity that emerged cultivated performances of masculinity that had serious consequences for black women already marginalized by class, race, and white patriarchy.

In the South Bronx, New York, Shabazz traces one particularly fruitful subculture to emerge from racial and class marginalization. In the wake of economic recession, there arose in parks, on street corners, in train stations, and at viaducts a vibrant youth culture that developed into the hip-hop movement. Lore holds that hip-hop was instituted in 1973 by Cindy Campbell and her brother Clive Campbell, aka Kool DJ Herc, at the Mitchell Lama housing complex in Morris Heights, the Bronx. The Bronx was remade from an abandoned, segregated, and deprived project to a space of art, creativity, and experimentation. According to Shabazz, "amid the rubble of deindustrialization and the wholesale abandonment of the inner core of the city – not idleness or a lack of a work ethic – hip-hop was born. Hip-hop comprised four distinct artistic practices: graffiti, b-boying10 (sometimes called breakdancing), DJing and MCing (or rapping)." While breakdancing was the dominant artistic practice in the 1970s and 1980s, it has steadily been replaced by rap, which globalized the culture.

Plate 6.10 Street dance, or hip-hop dance cipher on the street corner 1970s style, the Bronx, New York. Source: Joe Conzo/Cornell university Library.

Hip-hop challenged the spatial containment of the black community by enacting transgressive performances that repurposed the built environment. Street corners became sites of cipher, informal gatherings of rappers, beatboxers, and/or breakdancers in a circle, in order to jam musically and bodily. In the early 1990s, MCs from all the boroughs convened regularly at Washington Square Park in Greenwich Village, New York City, in ciphers that protested their exclusion from public space. More generally, rap music constituted a stance against deindustrialization, unemployment, and incarceration and opponents of black politics and cultivated for itself radical spaces.

More recently, a sexual and gender politics over the rights to cultural production has sought to break the assumption that "masculinity and the mic" are synonymous. Women MCs and performers such as Lauren Hill, Salt N Pepa, Missy Elliot, Lil Kim, Nicki Minaj, Ms. Melody, Queen Latifah, Yo-Yo, MC Lyte, Medusa, Jean Gray, Mystic, and Monie Love had made significant inroads into the industry. These developments have prompted black men to rethink hegemonic models of masculinity and to temper therapeutically a proclivity to assert overly compensatory but toxic masculinities.

Conclusion

Evangelists of the West believe that the West's superior culture, and in particular its key cultural institutions, played a significant role in the rise of Western civilization as a global economic and imperial power from the fifteenth century. But the mantra of "West is best" has been met with resistance and has been forced to confront and participate in a number of culture wars in which it has not always triumphed. Cultural geographers are particularly interested in studying the ways in which the West has sought to project its cultural superiority onto the cultural landscape, to depict some spaces as civilized and other places as unruly. They examine the West's fascination with the idea of "landscape" and the projection of Western power in and through symbolic landscapes. Mobilizing the concepts of landscapes of power and living landscapes, they also examine the ways in which human beings have occupied, encountered, and humanized these landscapes, often creating subcultures that have challenged hegemonic assumptions. Representations of the West play a crucial role in undergirding the ongoing hegemony of the West in world affairs. Whoever controls the myth that "West is best" in no small way controls the destiny of the West.

Checklist of Key Ideas

- An amalgam of cultural inheritances and borrowings, Western culture emerged first in Europe around the fifteenth century. Based upon Enlightenment ideals, at the core of Western culture is a profound faith in human reason and the belief that by harnessing human reason alone, it is possible to build a perfect world. Geographers believe that the Enlightenment was produced in different ways in different places, and therefore it might be more accurate to speak in terms of the Enlightenment(s) cultures.
- Rejecting the idea that culture is a superorganism, cultural geographers now approach culture as a social construction, inextricably embroiled and ensnared in wider social, political, and economic processes.

- Each social formation (form of society) has a core mode of production (system of production) that leads to particular modalities of symbolic production (cultural representations, including landscapes of power).
- The idea of landscape occupies an important position within Western culture. Accordingly, Western culture has attempted to project its superiority onto the landscape by crafting a series of symbolic civilized spaces (sometimes termed the imperialism of the straight line) and distinguishing these from crooked non-Western places.
- The dynamic between Ville and Cité (landscapes of power and living landscapes) can be witnessed in migrant encounters with the early nineteenth-century industrial city, colonial cities, and the African-American search for a sense of place.

Chapter Essay Questions

1) How can perspectives from cultural geography help us to understand Western culture?
2) Write an essay entitled "The West in the cultural landscape."
3) With reference to examples, explain what cultural geographers mean when they speak about the relationship between landscapes of power and living landscapes.
4) What is meant by the phrase "global sense of place," and why does it matter in this age of political populism?

References and Guidance for Further Reading

Books that explore the pillars of Western culture include:

Ferguson, N. (2011). *Civilisation: The West and the Rest*. London: Penguin Books.

Ferguson, N. (2013). *The Great Degeneration: How Institutions Decay and Economies Die*. London: Penguin Books.

Habermas, J. (1985). Modernity – an incomplete project. In: *Postmodern Culture* (ed. H. Foster), 3–15. London: Pluto Press.

Kant, I. (1774). *Answering the Question: What Is Enlightenment?* Berlin: Berlinische Monatsschrift.

Withers, C.W.J. (2007). *Placing the Enlightenment: Thinking Geographically About the Age of Reason*. Chicago: University of Chicago Press.

A critical introduction to the social production of the idea of the West can be found in:

Bonnett, A. (2017). *The Idea of the West: Culture, Politics and History*. London: Macmillan International.

Carl O. Sauer's approach to cultural geography is articulated most clearly in:

Sauer, C.O. (1925). *The morphology of landscape*. University of California Publications in Geography 2 (2): 19–53.

Good introductions to Carl O. Sauer's approach to cultural geography can be found in:

Denevan, W.D. and Mathewson, K. (eds.) (2009). *Carl Sauer on Culture and Landscape: Readings and Commentaries*. Baton Rouge: Louisiana State University Press.

Williams, M., Lowenthal, D., and Denevan, W.M. (2014). *To Pass on a Good Earth: The Life and Work of Carl O. Sauer*. Charlottesville: University of Virginia Press.

General introductions to what has become known as "new cultural geography" are provided in:
Cosgrove, D. and Jackson, P. (1987). New directions in cultural geography. *Area* 19: 95–101.
Duncan, J.S. (1980). The superorganic in American cultural geography. *Annals of the Association of American Geographers* 70: 181–198.
Jackson, P. (1989). *Maps of Meaning: Introduction to Cultural Geography*. London: Unwin Hyman.
Mitchell, D. (2000). *Cultural Geography: A Critical Introduction*. Oxford: Blackwell.
Naylor, S., Ryan, J., Cook, I., and Crouch, D. (2018). *Cultural Turns/Geographical Turns: Perspectives on Cultural Geography*. London: Routledge.
Philo, C. (1991) *New words, new worlds: reconceptualising social and cultural geography*. Edinburgh: Conference Proceedings.

Important works by Denis Cosgrove include:
Cosgrove, D. (1983). Towards a radical cultural geography: problems of theory. *Antipode* 15: 1–11.
Cosgrove, D. (1984). *Social Formation and Symbolic Landscape*. London: Croom Helm.
Cosgrove, D. and Daniels, S. (eds.) (1988). *The Iconography of Landscape*. Cambridge: Cambridge University Press.
Cosgrove, D. (1993). *The Palladian Landscape: Geographical Change and Its Cultural Representations in Sixteenth-Century Italy*. University Park: Pennsylvania State University Press.
Cosgrove, D. (2003). *Apollo's Eye: A Cartographic Genealogy of the Earth in the Western Imagination*. Baltimore: JHU Press.

The legacy of Denis Cosgrove in geography is commemorated in:
Vallerani, F. (ed.) (2018). *Everyday Geographies and Hidden Memories: Remembering Denis Cosgrove*. London: Centre for the GeoHumanities, Royal Holloway, University of London.

Other general introductions to cultural geography can be found in:
Anderson, J. (2015). *Understanding Cultural Geography: Places and Traces*. London: Routledge.
Horton, J. and Krafti, P. (2013). *Cultural Geographies: An Introduction*. London: Routledge.
Johnson, N.C. (2018). *Culture and Society: Critical Essays in Human Geography*. London: Routledge.
Johnson, N.C. and Schein, R.H. (2016). *The Wiley-Blackwell Companion to Cultural Geography*. Oxford: Wiley Blackwell.

Power and the cultural landscape are the subjects of books such as:
Barnes, T.J. and Duncan, J.S. (2013). *Writing Worlds: Discourse, Text and Metaphor in the Representation of Landscape*. London: Routledge.
Daniels, S. (1993). *Fields of Vision: Landscape Imagery and National Identity in England and the United States*. Cambridge: Polity Press.
Daniels, S., DeLyser, D., Entrikin, J.N., and Richardson, D. (eds.) (2012). *Envisioning Landscapes, Making Worlds: Geography and the Humanities*. London: Routledge.
Duncan, J.S. and Ley, D. (2013). *Place/Culture/Representation*. London: Routledge.
Lowenthal, D. (2015). *The Past is a Foreign Country-Revisited*. Cambridge: Cambridge University Press.
Matless, D. (2014). *In the Nature of Landscape: Cultural Geography on the Norfolk Broads*. Oxford: Wiley.
Mitchell, D. (1996). *Lie of the Land: Migrant Workers and the California Landscape*. Minneapolis: University of Minnesota Press.
Olwig, K.R. (2002). *Landscape, Nature, and the Body Politic: From Britain's Renaissance to America's New World*. Madison: University of Wisconsin Press.
Olwig, K. and Mitchell, D. (eds.) (2019). *Justice, Power and the Political Landscape*. London: Routledge.

Tyner, J.A. (2016). *Landscape, Memory, and Post-violence in Cambodia.* New York: Rowman & Littlefield.
Wylie, J. (2021). *Landscape: Key Ideas in Geography.* London: Routledge.

A brilliant study of Haussmann's Paris is found in:
Harvey, D. (2003). *Paris, Capital of Modernity.* London: Psychology Press.

An excellent study of power, urban planning, and design in Washington, DC (specifically the National Mall), is:
Benton-Short, L. (2016). *The National Mall: No Ordinary Public Space.* Toronto: University of Toronto Press.

Studies that explore the dialectical relationship between landscapes of power and living landscapes (also cast as space and place, place and placelessness, and Ville and Cité) are:
Cresswell, T. (2014). *Place: An Introduction.* Oxford: Wiley.
Cresswell, T. (2019). *Maxwell Street: Writing and Thinking Place.* Chicago: University of Chicago Press.
Gregory, D. (1994). *Geographical Imaginations.* Oxford: Blackwell.
Hawkins, H. (2013). *For Creative Geographies: Geography, Visual Arts and the Making of Worlds.* London: Routledge.
Kobayashi, A. and Mackenzie, S. (2014). *Remaking Human Geography.* Toronto: Routledge.
Lefebvre, H. (1991). *The Production of Space.* Oxford: Blackwell.
Massey, D. (1991). A global sense of place. *Marxism Today* 38: 24–29.
McCormack, D.P. (2014). *Refrains for Moving Bodies: Experience and Experiment in Affective Spaces.* Durham, NC: Duke University Press.
Price, L. and Hawkins, H. (2016). *Geographies of Making, Craft, and Creativity.* London: Routledge.
Rose, G. (1993). *Feminism and Geography: The Limits of Geographical Knowledge.* Minneapolis: University of Minnesota Press.
Sennett, R. (2002). *Respect in a World of Inequality.* New York: W. W. Norton & Company.
Sennett, R. (2018). *Building and Dwelling: Ethics for the City.* New York: Farrar, Straus and Giroux.
Sennett, R. and Sendra, P. (2020). *Designing Disorder: Experiments and Disruptions in the City.* New York;: Verso Books.
Till, K.E. (2005). *The New Berlin: Memory, Politics, Place.* Minneapolis: University of Minnesota Press.

A tradition of humanistic geographies that does similar work includes:
Buttimer, A. (1983). *The Practice of Geography.* London: Addison-Wesley Longman Ltd.
Casey, E. (2013). *The Fate of Place: A Philosophical History.* Berkeley: University of California Press.
Entrikin, N. (1991). *The Betweeness of Place.* Baltimore: The John Hopkins University Press.
Malpas, J. (1999). *Place and Experience.* Cambridge: Cambridge University Press.
Relph, E.C. (1976). *Place and Placelessness.* Los Angeles: Pion.
Relph, E.C. (1981). *Rational Landscapes and Humanistic Geography.* New York: Barnes and Noble.
Tuan, Y.F. (1977). *Space and Place: The Perspective of Experience.* Minneapolis: University of Minnesota Press.

A select flavor of early studies of the impact of nineteenth-century industrialization and urbanization on human culture and psychology is provided in:
Benjamin, W. (1999). *The Arcades Project.* Cambridge, MA: Harvard University Press.
Simmel, G. (1976). *The Metropolis and Mental Life.* New York: Free Press.
Wirth, L. (1938). Urbanism as a way of life. *American Journal of Sociology* 44: 1–24.

Studies that examine the hyperextension of Europe's symbolic landscapes into colonial cities (including Singapore) include:

Duffy, P. (2007). *Exploring the History and Heritage of Ireland's Landscapes*. Dublin: Four Courts Press.

Estes, N. (2019). *Our History Is the Future: Standing Rock Versus the Dakota Access Pipeline, and the Long Tradition of Indigenous Resistance*. London: Verso.

Ho, E.L.E., Woon, C.Y., Ramdas, K. et al. (2013). *Changing Landscapes of Singapore: Old Tensions, New Discoveries*. Singapore: NUS Press.

Home, R. (2013). *Of Planting and Planning: The Making of British Colonial Cities*. Abingdon, UK: Routledge.

King, A.D. (2012). *Colonial Urban Development: Culture, Social Power and Environment*. London: Routledge.

Kong, L. and Yeoh, B.S. (eds.) (2003). *The Politics of Landscapes in Singapore: Constructions of Nation*. New York: Syracuse University Press.

Roy, A. and Ong, A. (eds.) (2011). *Worlding Cities: Asian Experiments and the Art of Being Global*. Oxford: John Wiley & Sons.

Waziyatawin, A.W. and Yellow Bird, M. (2007). *For Indigenous Eyes Only: A Decolonization Handbook*. Santa Fe, NM: School of American Research Press.

Yeoh, B.S. (2003). *Contesting Space in Colonial Singapore*. Honolulu: University of Hawaii Press.

Studies of urban planning and the African-American search for place are:

Cooper, H.L.F. and Fullilove, M.T. (2020). *Ending Police Violence: A Public Health Primer*. Baltimore: Johns Hopkins University Press.

Fullilove, M.T. (2016). *Root Shock: How Tearing Up City Neighborhoods Hurts America, and What We Can Do About It*. New York: New Village Press.

Newman, O. (1972). *Defensible Space*. New York: Macmillan.

Shabazz, R. (2014). Masculinity and the mic: confronting the uneven geography of hip-hop *Gender, Place and Culture* 21: 370–386.

Shabazz, R. (2015). *Spatializing Blackness: Architectures of Confinement and Black Masculinity in Chicago*. Urbana: University of Illinois Press.

Website Support Material

A range of useful resources to support your reading of this chapter are available from the Wiley Human Geography: An Essential Introduction Companion Site http://www.wiley.com/go/boyle.

Chapter 7

(Under)Development: Challenging Inequalities Globally

Chapter Table of Contents

Chapter Learning Objectives

By the end of this chapter you should be able to:

- With reference to classical, neoclassical, and neoliberal economic doctrines, explain why free market fundamentalists believe that liberalized market economies are best placed to (i) radically transform human development and lift the world's population out of poverty, and (ii) reduce spatial and social inequalities in wealth and income.

Human Geography: An Essential Introduction, Second Edition. Mark Boyle.
© 2021 John Wiley & Sons Ltd. Published 2021 by John Wiley & Sons Ltd.
Companion website: www.wiley.com/go/boyle

- Drawing upon measures of human development, poverty, and inequality, comment on the claim that since 1800, the world has witnessed an unevenly distributed and incomplete revolution in human flourishing. Comment also on the extent to which twenty-first-century globalizing capitalism is now reorganizing inherited geographies of (under)development and human welfare at all spatial scales.
- Provide a brief critical history of development theory and practice. Note the significance of the Western origins of development theory and practice. Discuss the extent to which the Western laissez-faire market model and in particular "rollout" neoliberalism has influenced development agendas at all geographical scales – providing neoliberal remedies for spatial inequalities that it itself has been centrally implicated in creating.
- Describe and reflect upon the efficacy and politics of the capability and human rights approaches to development that are currently garnering support.
- Reflect upon the progress that countries in the Global South and the Global North are making with respect to the United Nations Sustainable Development Goals (SDGs) 2015–2030.
- Explain why some human geographers think it is important to talk in terms of both *development alternatives* and *alternatives to development*.

Introduction

Contemplate the radically different lives you would be leading had you been, by simple accident of birth, born in, say, Beverly Hills, Los Angeles, or Dharavi, Mumbai.

With a population of 33 792 stretching across 14.7 km² (population density of 2300 people per square kilometer), Beverly Hills is a city (in reality, a small neighborhood) within Los Angeles County, California (Plate 7.1). Home to the stars of Hollywood and playground to the rich and famous, Beverly Hills is adorned with majestic palm tree–lined avenues that lead to opulent homes, fashionable hotels, boutique shops, and Michelin star restaurants. Annual per capita income is $92 185, and life expectancy is 86.2 years of age. The average house has 2.2 inhabitants and is over 3100 square feet in size. In contrast, Dharavi, located in central Mumbai, is contained within an area of just 2.1 km² and is home to an "official" population of 600 000 (a population density of 285 714 people per km²). Described as Asia's most infamous slum, Dharavi is built on a former landfill site, remains contaminated with heavy metals, is stewing in open sewage and garbage, and is chronically rat infested (Plate 7.2). On average 6–8 people live in each squatter shack (rarely more than 250 square feet in size), and 50–60 people share a common toilet. Per capita income is $927, and life expectancy is 60. Living in Dharavi would mean having to work 10 years to earn the equivalent of a single month's salary in Beverly Hills. You would be living in a place 124 times more crowded with access to only 2% of the household space. You would be facing the prospect of dying 26 years younger.

How did it get to be this way? Can anything be done to help residents living in Dharavi improve their standard of living and quality of life? Who will bring about change, and how?

Plate 7.1 Aerial of Palm Tree Lined Street Beverly Hill Los Angeles. Source: Ryan Herron/E+/ Getty Images.

Plate 7.2 Dharavi slum, central Mumbai. Source: A. Savin, Wikimedia Commons, https://commons. wikimedia.org/wiki/File:Mumbai_03-2016_108_Bandra_station_surroundings.jpg

Market Fundamentalism and the Promise of Convergence

In the late eighteenth century, classical economists fashioned a powerful and consequential claim: liberalized market economies would: (i) increase the wealth of nations and transform the quality of human life, lifting the world's poor out of poverty; and (ii) correct over time diverging levels of economic prosperity and geographical inequalities in wealth and income at all geographical scales.

What Was the Reasoning Behind This Claim?

Scottish economist Adam Smith is widely regarded as the founding father of classical economics, an approach that came to replace mercantilism, which reigned in Europe from the sixteenth century to the eighteenth century. Mercantilism was based on the assumption that the world's wealth was fixed, and consequently the mission of any nation was to accumulate the largest possible share of that wealth (to hold the largest quantity of gold and silver) by maximizing exports and limiting imports. In his famous 1776 book *An Inquiry into the Nature and Causes of the Wealth of Nations*, Smith argued that self-interest and laissez-faire open and free markets operate to grow the wealth of the world and serve the public good by increasing the living standards of all peoples in all places. He introduced the idea of the "invisible hand" of the market; although private owners of businesses competing in an open marketplace might appear chaotic and anarchic, in fact the market embeds a "hidden" logic in the economy that leads to greater efficiency, optimum resource use, and more output.

Classical and later neoclassical economists (those inspired by and who have elaborated upon the basic tenets of classical economics) believed that a liberalized world capitalist economy would be a panacea for global poverty; indeed, free markets worked best for the poor and would lift humanity out of poverty and enable human beings to live long, dignified, and comfortable lives, free from want, pain, natural hazards, and necessity. Would it matter if inequalities were to widen? Even if the rich are getting richer at a faster pace, so long as overall wealth is growing and the poor are getting richer too, perhaps the free market model can be said to be succeeding. But no, market fundamentalists go further. For classical and neoclassical economists, although valid, this observation is moot because liberalized capitalism would remediate not only poverty but also in time social and spatial inequalities in wealth and income. To equalize the economy over space, it is best to leave markets to their own devices. By ruthlessly seeking out and purging inefficiencies, only the unfettered market could fix suboptimal economic geographies.

US economist Simon Kuznets (1955), for example, famously predicted that as any market economy developed, inequality would initially rise but at a certain stage of development would move into reverse gear, leading to greater convergence and creating a more equal society. Capitalism would birth industrialization and urbanization, and inequalities would grow as rural migrants flocked to cities and wages were depressed. But greater wealth would trickle down to everyone and wages would increase as growth bequeathed increasing employment opportunities and rural in-migration ran its course. A Kuznets curve – an inverted-U or bell-shaped curve – would result. Meanwhile, at the global scale, although market economies may initially produce hot growth cores and lagging weaker peripheries, in the long term market forces would act to correct imbalances and level up the space economy. Regions that grow fast will eventually overheat, causing diseconomies of scale and negative externalities (such as rising land and labor costs and pollution), and when productivity gains are exhausted and diminishing returns set in, returns on capital investment will be squeezed. At this point, capital will flow to less developed areas to recoup lost competitive advantage and to restore profitability.

Much then is claimed for – and expected of – globalizing liberalized free market capitalism. But are these claims misplaced, and are expectations proportionate and realistic?

Given that markets are historically, politically, and socially constructed, always striated and mottled and never pure, clean, and uncontaminated, there has never been a laboratory hygienic enough to put market fundamentalism to the test. But for most of the period up until World War I, including and in particular the Victorian era (1830–1870), and thereafter the 1920s or "roaring '20s," laissez-faire free market capitalism became the central pillar of Western economic policy. German philosopher Karl Marx referred to this period as the age of "vulgar capitalism": Darwinian survival of the fittest reigned. An interesting counterfactual is provided by periods in the twentieth century (WW1, WW2, and a Keynesian moment between 1945 and 1975) when the application of market rule was less disciplined and social democratic logic and state economic management prevailed. Nonetheless, the past 40 years have witnessed the rise once again of neoliberal economic orthodoxies and a rekindling of interest in small low-tax states, liberalized free markets, privatism, deregulation, and entrepreneurial freedoms.

Classical, neoclassical, and neoliberal philosophers and economists, then, have played a central role in shaping the West's approach to designing and managing the (space) economy, and evidence accumulated over the past two centuries provides some basis upon which to reach qualified conclusions about the efficacy and politics of market rule.

Tracking and Mapping Development and Human Welfare from 1800

In 1980, former German Chancellor Willy Brandt published *The Brandt Report*, a seminal investigation into uneven geographical development, in which he introduced the idea of the Brandt Line – a line dividing the rich developed Global North from the poor underdeveloped Global South (Map 7.1). Inherited geographies of development, poverty, and inequality continue to weigh on the present, and this lexicon continues to speak to an important reality. But given the complex ways in which globalizing capitalism is reterritorializing human development, poverty, and inequality today, the Brandt Line is becoming less and less fruitful as a heuristic. Clearly, mechanisms of uneven geographical development play out differently at different scales and are refracted through specific conditions in each locale. But scholars have found analytical, policy, and political value in bringing together in a single conversation a diverse range of spaces of poverty and inequality. The distinction between the Global South and Global North is giving way to cartographies of (under)development that assume that every country has a development problem and that conventional scales of analysis are less relevant (see Deep Dive Box 7.1). Taking as its cue the turn within development theory, practice, and activism – from an exclusive focus upon the Global South to a wider concern with planetary uneven geographical development or global (under)development – the remainder of this chapter will adopt an expansive view of the idea of geographies of development and human flourishing (Horner and Hulme 2019).

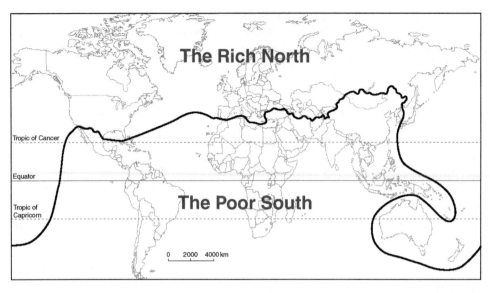

Map 7.1   A Global North and a Global South? The Brandt Line (1980). Source: Brandt (1980).

Deep Dive Box 7.1   Mapping a Moving Target? Hardt and Negri on Capital's *Empire*

Today, globalizing capitalism is refashioning inherited economic geographies, making it ever more difficult to conceptualize the task of tracking and mapping human development, poverty, and inequality.

   In their book *Empire*, published in 2000, US and Italian political philosophers Michael Hardt and Antonio Negri present a particularly cogent rendering of twenty-first-century capitalism and its emergent geographies of development and inequality. Colonialism and European capitalist modernity, they argue, understood as a state project dedicated to annexing territories occupied by others, are dead. Nonetheless, imperialism is still alive in the form of the hyper-globalizing neoliberal capitalism that now stretches its tentacles deeply and widely into the four corners of the world and every domain of our lifeworld. But this is a new breed of empire; there exists no single force orchestrating the global economy; there is no metropole and colony, no world system comprising cores, semi-peripheries, and peripheries. Hardt and Negri recall and invoke the Roman political constitution, which blended monarchy, aristocracy, and democracy to powerful effect, to argue that Capital's empire adopts an analogous distributed leadership structure, with the United States' nuclear supremacy exercising the monarchical function, the economic power of transnational corporations and the G7 performing the aristocratic function, and the internet providing the democratic pillar – rule through *Bomb, Money and Ether*. Capital accumulates across many sites, and it expands its reach by dint of the interests and toil of a multiplicity of stakeholders active across both conventional and unconventional geographies. The categories of the First, Second, and Third Worlds are now redundant; globalizing capitalism is reordering geographies of uneven development, and the First World can now be found in the Third and the Third World in the First.

Development, poverty, and inequality therefore are literally now everyone's problem. Our politics is being shaped by an exciting new "multitude" of alliances; new conversations are possible across seemingly disparate spaces, scholars, development practitioners, and activist groups, because universal, necessary, and generic mechanisms of uneven development appear to coexist with local, contingent, and particular forces of impoverishment.

Geographies of Human Development and Poverty from 1800

In their seminal article published in the *American Economic Review* in 2002, French economists François Bourguignon and Christian Morrisson provide global estimates of the long-running evolution of global poverty and income inequalities – both within and between countries – for the period 1820–1992.

- In 1750, the majority of the world's population lived in extreme poverty just as severe as that experienced by the poorest people living in the poorest countries today. In 1820 no less than 84% of the global population was living in extreme poverty. But by 1950 this had fallen to 55%, and by 1992 24%. European countries and their Anglo-Saxon settler colony offshoots reduced poverty first, fastest, and in time universally; Latin America too made tangible, if more limited, progress. But Asia and Africa were left behind, and as late as 1992 both were home to the largest concentrations of the world's poor (Figure 7.1).
- World income inequality worsened dramatically over the period 1820–1992: the Gini coefficient increased by 30%, and the Theil index by 60%. In 1820 the top 10%

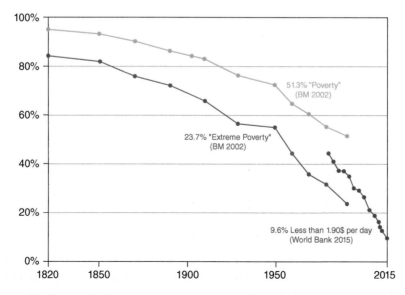

Figure 7.1 Declining global poverty: share of people living in extreme poverty, 1820–2015. Source: Bourguignon and Morrisson (2002) and World Bank (PovcalNet) (2015).

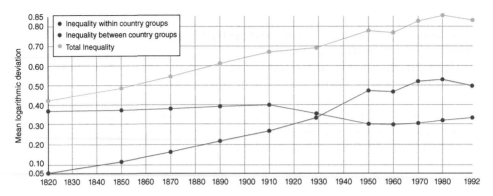

Figure 7.2   The relative importance of within- and between-country income inequality, 1820–1992. Source: Bourguignon and Morrisson (2002) and World Bank (PovcalNet) (2015).

of income earners in the world captured 42% of total income; by 1950 this figure grew to 51.3%, and by 1992 it stood at 54.4%. Across the same period, the bottom 40% captured only 13.5, 6.8, and 6.4%, respectively. In the period 1820 to 1950, the driver of divergence was inequality within countries; from 1950 to 1992, between-country variations appear to dominate (Figure 7.2).

Each will now be explored in greater detail, and we will note the further important developments that have emerged since 1992 (see Deep Dive Box 7.2).

Deep Dive Box 7.2   Measuring Geographical Inequalities in Income and Wealth

Human geographers focus primarily upon two kinds of inequality:

- **Income inequality**: The extent to which income (wages, bonuses, salaries, dividends, interest on savings, state benefits, pensions, rent, etc.) is distributed equally across a given population.
- **Wealth inequality**: The extent to which capital or financial assets (savings, property, bonds, gold, stocks, pensions, inheritances, etc.) are distributed equally across a given population.

Therein, they pay attention to each of the following:

- Within-country (or intra-country) inequality, which addresses income and wealth inequalities within a country.
- Between-country (or international, cross-country, or inter-country) inequality, which compares income and wealth differences between countries.
- Global inequality, the sum of both within- and between-country inequalities.

They measure inequality using a variety of metrics.

- Share of national income/wealth accruing to those at the top (top 0.1%, 1%, or 10%) and the bottom (bottom 50% or 10%) of the overall income/wealth distribution at a given time.

- The Palma ratio, which calculates the ratio of the share of the income/wealth earned by the richest 10% (or 0.1%, 1%, or 20%) to that earned by the poorest 40% (or 50% or 20%).
- The Gini coefficient, which uses the Lorenz curve to measure how far an income distribution varies from perfect equality; this coefficient ranges from 0 (perfect equality) to 1 (complete inequality).
- The Theil index, which allows scholars to decompose inequality into its subparts – in our case, to gauge the part that is due to inequality within areas (e.g., domestic inequalities within countries) and the part that is due to differences between areas (e.g., international inequalities between countries). Most often, the Theil measure is recalibrated to yield scores ranging from 0 (perfect equality) to 1 (complete inequality).

The United Nations' Human Development Index (HDI)

Devised in 1990, the United Nations Development Program's (UNDP) Human Development Index (HDI) provides one mechanism for mapping and tracking global variations in the quality of people's living standards. HDI scores range from a maximum of 1 (most developed) to a minimum of 0 (least developed). The UNDP has provided HDI scores by country from 1980, broken down into four tiers: very high human development (0.8–1.0), high human development (0.7–0.79), medium human development (0.55–0.70), and low human development (below 0.55) (see Figure 7.3).

The UNDP HDI has inspired a number of additional and related metrics that supplement the core index. These include:

- **Historical Index of Human Development** (HIHD), which profiles global variations in improvements in human development by country from 1870 to 2015.
- **Subnational Human Development Database** (SHDI), which tracks subnational HDI scores for 1865 regions within 187 countries for the period 1990–2018.
- **Inequality-adjusted Human Development Index** (IHDI), which measures losses in human development wrought by intra-country inequality or the human development cost of inequality.
- **Global Multidimensional Poverty Index** (MPI), which offers a more rounded measure of poverty based upon combining 10 separate indicators.
- **Capabilities Measurement Project** (CMP), which measures the state of human capital in a country on the basis of a number of "basic" and "enhanced" capabilities.

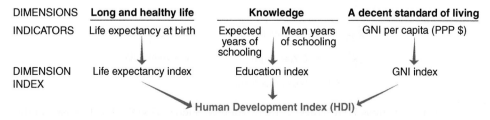

Figure 7.3   Methodology for calculating the UNDP Human Development Index. Source: UNDP New York.

In combination, what do these data sets tell us?

- The HIHD confirms that the period from 1800 has witnessed a historically unprecedented improvement in human welfare, nothing short of a revolution. The pace of improvement appears to have been particularly intense in the period from the First World War to the oil crisis in the mid-1970s. But this improvement was not universally shared; up until the mid-1970s, a gap between the Global North and the Global South opened and thereafter widened rapidly. Europe and its offshoot settler economies prospered most, Latin American countries nurtured meaningful improvements, while Asian and African countries for the most part languished in poverty.

- The HIHD and HDI confirm that since the mid-1970s, there has been an incomplete catching up, at least for some Global South countries (see Map 7.2). The growth of China and India has led and dominated this progress. Sub-Saharan Africa and parts of Asia continue to suffer from low levels of development.

- The Global MPI 2020 report notes that across the past decade, significant progress has been made in reducing world poverty, and at least 271 million people have been lifted out of multidimensional poverty. But still, at least 1.3 billion (22% of the population) out of the 5.9 billion people covered by the Global MPI (across 107 developing countries) are still living in multidimensional poverty – the vast majority (84.3%) in sub-Saharan Africa (558 million) and South Asia (530 million). Half (644 million) are children.

- Across the world in 2019, the IHDI suggests that inequalities within countries cost these countries a 20.2% reduction in their HDI score; that is, their HDI score was 20.2% below that which might be expected if everyone in the jurisdiction enjoyed

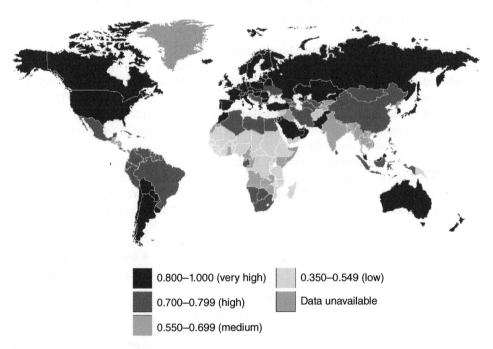

■ 0.800–1.000 (very high)	▨ 0.350–0.549 (low)
■ 0.700–0.799 (high)	▨ Data unavailable
▨ 0.550–0.699 (medium)	

Map 7.2 United Nations HDI scores, 2018, by country. Source: UNDP HDI Database of the Global Data Lab.

the average standard of living. At 10.7%, this loss was lowest for the very high human development group; it increased to 17.9% for the high human development cohort, 25.9% for the medium human development group, and peaked in the low human development cohort at 31.1%.

- The SHDI indicates significant inequality in HDI scores within countries themselves. When IHDI scores are calculated using subnational regions, countries suffer greater losses in human welfare due to inequality. This implies that national HDI scores are painting only a general picture, beneath which other more complex processes of uneven geographical development are at work.
- Capability measurements (CMP) point to a convergence since 1980 in basic capabilities (for example, early childhood survival, primary education, entry-level technology, and resilience to recurrent shocks) but a new "great divergence" in enhanced capabilities (for example, access to quality health at all levels, high-quality education at all levels, effective access to present-day technologies, and resilience to unknown new shocks).

Deep Dive Box 7.3   The World Bank's International Poverty Line (IPL)

In 1990, the World Bank introduced their International Poverty Line (IPL) – a measure of poverty that takes into account variations in prices between countries (based upon purchasing power parity [PPP] exchange rates that standardize what can be purchased for a given $1 at a standard date) and changes in prices over time (inflation). The IPL enables better comparisons to be made between countries.

The IPL is arbitrary. Initially, the World Bank set the extreme poverty level at $1 per person per day; anyone earning less could be said to be living in absolute poverty. In 2005 this was raised to $1.25 per person per day, and in 2015 to $1.90 per person per day. The threshold of US$1.90 does not derive from any rigorous assessment of what is actually required to meet basic needs. It was settled upon by taking the national poverty lines of the 15th poorest countries in the world (which tend to be artificially low to enable governments to exaggerate progress in eradicating poverty), converting these thresholds using PPP exchange rates to the value of the 2011 US dollar, and calculating the average.

The World Bank IPL data set suggests:

- Between 1981 and 2020 the percentage of the world's population living in absolute poverty fell significantly; at the start of the period, 42.3% or 1.9 billion people were estimated to be living on less than $1.90 a day; by 1999 this figure stood at 28.6% (1.8 billion people), and today it stands at 9.2% (705 million people).
- This decrease has to be understood against the backdrop of an increase between these dates in world population from 4.5 to 7.7 billion.
- Nevertheless, there exists an uneven geography of mass exit from extreme poverty. Since 1990, it has been the poor in South Asia and East Asia and the Pacific whom have benefitted most. Indeed, the vast bulk of the improvement registered can be attributed to growing income and wealth in China and India alone (Figure 7.4).

(Continued)

Box 7.3   *(Continued)*

- Although absolute poverty in sub-Saharan Africa has fallen, the majority of the world's poor now reside on the African continent; in particular, Nigeria with a poverty headcount of 50% has the largest population of poor people in the world (102 million); the Democratic Republic of Congo comes second with a headcount of 75%, or 66 million people living below the poverty line. By 2023, it is estimated that Africa will house over 80% of the world's poor.
- United Nations Sustainable Development Goal (SDG) 1 aims to bring extreme poverty down to 3% of the world's population by 2030. Current trends suggest that by 2030, at best 6% of the global population will still earn less than US$1.90, most of whom will live in Africa (Table 7.1).

Critics of the World Bank's IPL argue that the $1.90 limit is flawed; even as a benchmark of extreme poverty, it is absurdly low and gives the erroneous appearance that significant progress in the war against poverty is being made. In a report entitled *The Parlous State of Poverty Eradication*, United Nations Special Rapporteur on "extreme poverty and human rights" Philip Alston (2020) condemned the World Bank for benchmarking extreme poverty against a "standard of miserable subsistence rather than an even minimally adequate standard of living." In contrast, in his book *Factfulness: Ten Reasons We're Wrong About the World – and Why Things Are Better Than You Think*, Swedish physician Hans

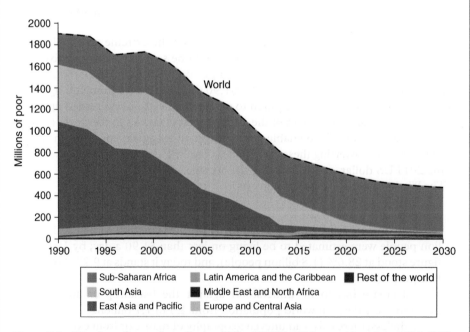

Figure 7.4   Number of poor (based upon IPL of US$1.90) by world region, 1990-2030. Source: World Bank (PovcalNet) (2020), World Development Indicators; World Economic Outlook; Global Economic Prospects; Economist Intelligence Unit. http://iresearch. worldbank.org//PovcalNet.

Table 7.1 International trends in poverty as measured by the international poverty line (IPL), 1981–2015/2020.

World Bank IPL, PPP, US$/day (standardized to 2011)	Headcount, number of poor, and poverty gap for various IPL thresholds				
	1981	*1990*	*1999*	*2010*	*2015–2020*
Global trends					
World population (billions)	4.5	5.3	60.4	6.9	7.6
US$1.90 Headcount (%)	42.3	35.9	28.6	15.7	9.2
US$1.90 Number of poor (billions)	1.9	1.9	1.8	1.1	0.7
US$1.90 Poverty gap (%)	17.9	12.8	9.5	4.7	3.1
US$3.20 Headcount (%)	57.1	55.3	50.7	35.2	26.4
US$3.20 Number of poor (billions)	2.6	2.9	3.1	2.3	1.9
US$3.20 Poverty gap (%)	31.4	26.7	22.3	13.4	9.3
US$5.50 Headcount (%)	66.4	67.1	66.9	53.9	46.2
US$5.50 Number of poor (billions)	2.9	3.5	4.0	3.7	3.4
US$5.50 Poverty gap (%)	47.58	41.60	38.10	26.91	20.9
Trends: China and India					
China population (billions)	1.0	1.1	1.3	1.3	1.4
US$1.90 Headcount (%)	88.0	66.2	40.2	11.8	0.52
US$1.90 Number of poor (billions)	0.9	0.8	0.5	0.2	0.007
US$1.90 Poverty gap (%)	42.7	24.1	1.3	2.7	0.1
India population (billions)	0.7	0.8	1.1	1.3	1.3
US$1.90 Headcount (%)	54.8	48.9	38.2	21.2	4.0
US$1.90 Number of poor (billions)	0.41	0.41	0.43	0.29	0.05
US$1.90 Poverty gap (%)	17.1	13.8	9.3	4.3	4.3

[a] The poverty headcount ratio is the number of people in a country living below a set poverty line divided by the population of that country. The number of poor refers to the absolute number of people living below a set poverty line. The poverty gap index captures the intensity of poverty as measured by the quantity of money required to bring everyone up above the poverty line or the average percentage of shortfall from the poverty line per capita.

[b] For the US$1.90 data, the headcount and number of poor calculated for the final column are for 2020, and the measure of the poverty gap is for 2015. For the US$3.20 data, the final column references 2015 data. For the US$5.50 data, the final column references data for 2018. For all three thresholds, total population is calculated for 2018. For the US$1.90 data for China, the final column references 2016 data. Accessing a data series for India is especially difficult. The 1980 column for India references data for 1983; the 1990 column, data from 1993; the 2000 column, data from 2004; the 2010 column, data from 2011; and the final column, data for 2020.

Source: World Bank (PovcalNet) (2020) data set.

Rosling et al. (2019) argue against using the categories of the "developed" and "developing" world. It is better to think, he contends, in terms of four categories of income per person (again based upon PPP 2011 exchange rates in US dollars): Level 1, less than $2 a day; Level 2, $2–8 a day; Level 3, $8–32 a day; and Level 4, $32+ a day. Assuming a population of 7 billion (the world's population in 2011), Rosling estimates that the breakdown of people living at each level

(Continued)

Box 7.3   *(Continued)*

would be 1, 3, 2, and 1 billion respectively. Given that six-sevenths of the world's population earn more than US$2 per day on this basis, and three-sevenths over US$8 per day, Rosling concludes that things are better than we assume. These figures are not especially reassuring, however, and it is difficult to see why Rosling would think them so.

While defending the US$1.90 yardstick as a measure of "extreme poverty," the World Bank also now calculates the number of people living below the poverty lines of US$3.20 a day and US$5.50 a day. A less promising picture emerges when using these more realistic (if still critically low) thresholds. Today, 26.4% of the world's population live on US$3.20 a day or less (1.9 billion people), and 46.1% on US$5.50 a day or less (3.4 billion). When judged against these higher benchmarks, alongside African countries Asian countries reappear as harbinger's of the world's poor.

Geographies of Income and Wealth Inequality from 1800

In 2011, led by the Paris School of Economics, over 100 researchers from across the world joined forces to create the World Top Incomes Database (WTID). In 2015, the WTID was subsumed into the World Inequality Database (WID). Later a World Inequalities Lab was charged with the responsibility of producing a regular *World Inequality Report* (the first report was published in 2018). The WID has had an enormous impact on the global inequality debate. It provides data on income inequalities for 70 countries and wealth inequalities for 30 countries.

French economist Thomas Piketty has played a central role in the development and mining of the WID. He catapulted to international fame in 2014 following publication of his groundbreaking book *Capital in the Twenty-First Century*. In 2019 Piketty published an equally ambitious sequel entitled *Capital and Ideology*. Across these two books, Piketty:

1) Charts inequalities in incomes (mainly in Global North OECD countries but also in a limited number of important Global South non-OECD countries) over time, demonstrating that since the 1980s, growing within-country inequalities have re-emerged as the dominant driver of global income inequality.
2) On the basis of the ideas of r (the rate of return on capital) and g (the rate of economic growth) and the formula r > g, argues that wealth inequalities are important progenitors of income inequalities.
3) Contending that inequalities are not inevitable and require the propagation of free market ideologies to endure, calls for a new era of "participatory socialism" underpinned by a progressive fiscal (taxes and state expenditure) regime.

Income Inequality: Is Income Distributed Equally Across a Given Population?
According to Piketty, in the period from 1800 to World War I, income inequalities within countries were persistently high across the present OECD world. From World War I to the mid-1970s, however, they began to reduce, with poorer people capturing larger shares of national income. Although tempered by the lavish lifestyles led by some elites

in the Great Gatsby "roaring '20s," between the First and Second World Wars and across the period of the 1930s depression, the income of the top earners stopped pulling away from the pack and the bottom 50% started to catch up. This continued between 1945 and 1975 during the thirty glory years of capitalism and the period when Fordist-Keynesian economic policy was in the ascendency, when income inequalities declined more rapidly. From the early 1980s, however, although forces of convergence have reduced the scale of inequalities between countries, inequalities within countries have displayed strong divergence. The top 1% of the population has captured 27% of the product of economic growth, leaving the bottom 50% with only 12% of that growth (Figure 7.5). Divergence has occurred more rapidly in Anglo-Saxon countries (the United States, the United Kingdom, Canada, Ireland, and New Zealand) than in the rest of Western Europe, and more rapidly in countries such as South Africa, India, Argentina, and China than in sub-Saharan Africa, Japan, the Middle East, and Brazil (although with respect to Latin America, although inequalities have stabilized they remain at a very high level) (Figure 7.6).

In consequence, in 2018, the share of income accruing to the top decile (the 10% highest income earners) was 34% in Europe, 41% in China, 46% in Russia, 48% in the United States, 54% in sub-Saharan Africa, 55% in India, 56% in Brazil, and 64% in the Middle East (see also Figure 7.7).

What Might Explain These Historical and Geographical Trends In Income Inequality?

Piketty draws a crucial distinction between what he calls r (rate of return on wealth in a given economy) and g (overall national income). The rate of return on capital refers to the value ("returns") that is added to a country's stock of wealth, normally on an annualized basis. Meanwhile, g refers to the rate of economic growth, that is a country's total income. Piketty makes the case that r has remained largely constant throughout history, at around 5%. He also argues that economic growth has most often fallen below 5%. Because capitalism is structurally wired to ensure that wealth grows faster than the economy, it is normal that its default position is r > g. Income returns to the owners of capital at a higher rate than it does to the rest of society, and in consequence inequalities grow over time.

Piketty reaches the following conclusions:

- *The rise of "vulgar capitalism" (1800–WW1 and the "roaring '20s")*. During this period, patrimonial capitalism reigned, and a comparatively limited number of wealthy people owned an increasingly larger portion of the wealth of nations and, through inheritance, passed on this privilege to their heirs. r > g and income accrued to the wealthy disproportionately, making them incrementally wealthier over time. Within-country inequalities widened dramatically.
- *Rewarding work, not wealth (WW1, the 1930s Great Depression, WW2, and 1945–1975)*. Only a burst of rapid growth (from technological progress or rising population) or government intervention can lead to g > r. These epochs are rare. But WWI, WW2, and the welfare Keynesianism era of 1945–1975 combined to serve as one such period. Damage inflicted by war reduced the share of total wealth held by elites. Meanwhile, state redistributive policies and rising wages meant that the rate of return accruing to labor exceeded the rate of return to capital. A convergence in income inequalities and in consequence wealth inequalities, was the result.

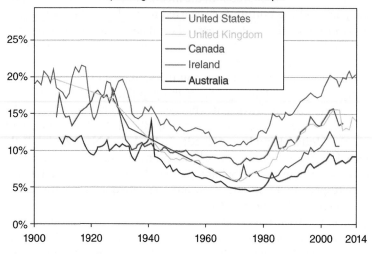

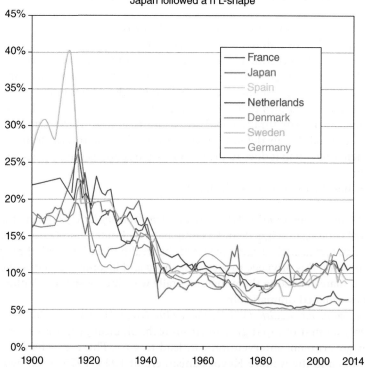

Figure 7.5 Share of total income going to the top 1% since 1900, before taxes and transfers. Source: World Wealth and Income Database. This data visualization is available at OurWorldinData. org, which has the raw data and more visualizations on inequality and how the world is changing. Licensed under CC-BY-SA by the author, Max Roser.

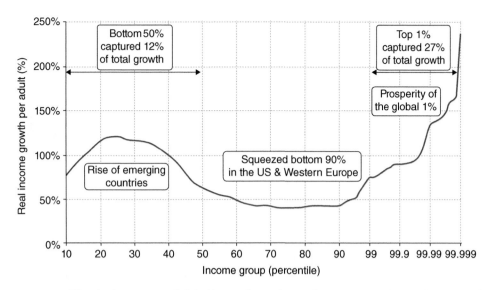

Figure 7.6 The elephant curve of global inequality and growth, 1980–2016. Source: WID.world, http://wir2018.wid.world.

- *Wealth is back! The rise of neoliberalism (from the early 1980s)*. With the ascent of the dominant neoliberal economic model, the elite has mounted a full frontal attack to recover their lost advantage; wealth and not work has become again a key progenitor of income, r > g and catching up has ceded to falling behind. Wealth is back! Although inequalities in income and wealth have not yet returned to that which prevailed during the age of "vulgar capitalism," they are growing again.

Of course, it is necessary to qualify and temper this chronology by attending to variations in the existence and extent of these trends across capitalism's geographies. In 1990 Danish sociologist Gøsta Esping-Andersen claimed to identify three worlds of welfare capitalism: liberal (personified by the United States), corporatist-statist (whose paradigmatic exemplar was Germany), and social democratic (embodied by Sweden). In a similar vein in 2001, US European studies scholar Peter Hall and British economist David Soskice called for more attention to be given to varieties of capitalism (VoC) and distinguished therein between liberal market economies (the Anglo-Saxon countries of the United States, the United Kingdom, Canada, Australia, New Zealand, and Ireland) and coordinated market economies (for example, the Nordic and European countries such as Germany, Belgium, the Netherlands, and Austria). Meanwhile, in a 2019 book, Serbian-American economist Branko Milanovic identified the existence of two types of capitalism: the liberal meritocratic capitalism of the West, and the state-led political, or authoritarian, capitalism that has fueled the rise of Asia. Perhaps the most significant conclusion to arise from this work is that social democratic Nordic capitalism (Sweden, Denmark, Finland, and Norway), based upon strong partnerships between government, business, and unions and relatively high taxes but world-class universal basic services, is best placed to deliver high development, little or no poverty, and low inequality while nurturing a vigorous democracy and strong citizenship rights.

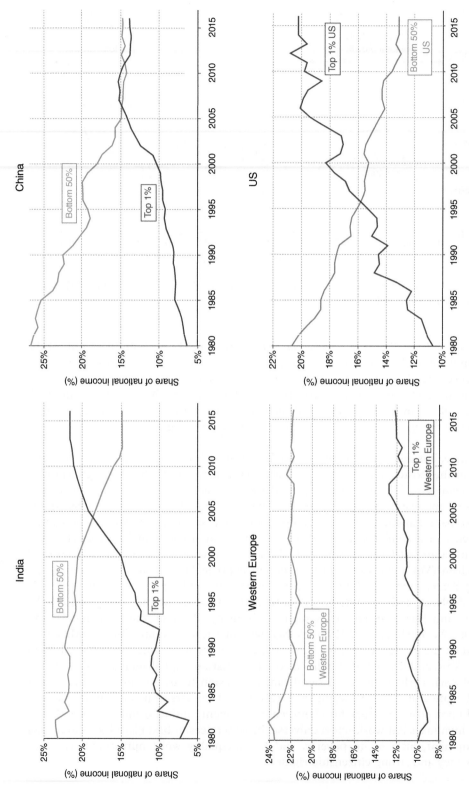

Figure 7.7 Income inequality for a select number of important economies, 1980–2015. Source: WID.world, http://wir2018.wid.world.

What Fiscal Measures Can Be Taken to Make Capitalism Fairer?

According to Piketty, there is nothing inevitable or innocent about inequality; it is not innate to human beings or a species condition. It is a structural feature of capitalism but moderated by political choices and institutional arrangements. Piketty coins the phrase "inequality regime" to understand the justification used for the structure of inequality and also the institutions – the legal system, the educational system, and the fiscal system – that help sustain a certain level of equality or inequality in a given society. Piketty accepts the need for and the advantages of private wealth and the value of a certain degree of inequality in society. But he argues that inequality has reached levels that are intolerable; they not only burden people living in both absolute and relative poverty, but they are bad for economic growth and sow the seeds of political disquiet. Growing inequalities have been centrally implicated in the rise of political populisms and nationalist political movements around the world, and the emergence of a politics of anger and resentment among those who feel that they have been "left behind" and languish in a dying "rustbelt" condemned to "managed decline."

Piketty's solution is a new *participatory socialism* and he argues that a global tax on wealth of up to 2% could free up capital to be redistributed with the goal of reducing inequality at all scales. By 2020, in his book *Capital and Ideology,* he goes further and proposes that governments raise their taxes more substantially (now a 15% tax on capital and 80% tax on high incomes). These "confiscatory taxes" could then be spent on a mixture of Universal Basic Services (for example, universal health care and access to higher education for all) and/or a Universal Basic Income (guaranteeing everyone a basic income of no less than 60% of average after-tax income) and bequeathing to all an inheritance cash payout worth no less than 60% of the average adult's net worth – in the United States, this would involve providing every citizen with a lump sum of US$231 000 when they reach the age of twenty five.

Piketty's work has been subjected to considerable critique. Is it possible to measure a person's wealth? Why is the rate of capital consistently around 5% throughout recent history? Surely the veracity of the formula $r > g$ is a product of historical and political forces, and is not an immutable fact inherent in capitalism? Is not wealth inequality a proximate determinant of trends in income inequalities, rather than a fundamental progenitor of these inequalities? What political struggles and victories lie beneath patterns of r and g historically? His policy suggestions are by his own admission utopian. But Piketty is right to argue that capitalism produces inequality and to question the inadequacy of only incremental changes to the status quo. And no one can doubt that his indispensable work in building and mining the WID has played a significant role in bringing inequality back to the fore.

A Brief Annotated History of Development Theory and Practice

The Western Tradition of Development Theory and Practice

In the early to mid-twentieth century, a body of thought emerged that held that, left to their own devices, it was unlikely that weak economies would be capable of lifting themselves up by their own bootstraps and becoming globally competitive. Disengagement from the market was not an option, nor indeed wise counsel. Left-behind economic spaces needed to embrace market mechanisms and fight to secure a role in the international division of labor. Catching up meant courting global capitalism and being better integrated into global production networks. But now economic peripheries were to be invested in and supported to enhance their fitness for purpose. "Rollback" neoliberalism (state retreat from the market) needed to be replaced with "rollout" neoliberalism (state intervention to rectify market failure, followed by state retreat from the market). Although construed variously at different scales, state intervention and better management of the space economy were to work to build the capacity of lagging economies, enabling them to catch up with their more affluent neighbors and to reproduce themselves autonomously and sustainably within the market economy.

Challenging Inequalities at the Global Scale

In his 2005 book *The End of Poverty: Economic Possibilities for Our Time*, US economist Jeffrey Sachs argued that with political will, extreme poverty could be eliminated globally by the year 2025. International development aid (IDA) had a critical role to play in helping poor countries to get onto the bottom rung of the ladder; thereafter, they would pull themselves up and integrate into the global economy. But IDA had to be guided by "clinical economics"; generic support would make little impact, only carefully planned bespoke interventions attuned to localized conditions and possibilities would be consequential. In 2019, a total of $136.4 billion was spent on IDA (including humanitarian and military aid). The United Nations asks OECD countries to contribute 0.7% of their GNI to IDA; only Sweden, Luxembourg, Norway, Denmark, and the United Kingdom meet this obligation. The European Union and its member states (especially Germany and also the United Kingdom and France), the United States, and Japan were the largest donors, while India and Turkey, followed by Afghanistan, Syria, Ethiopia, Bangladesh, Morocco, and Vietnam, were the largest recipients. Globally, the rise of political populism and now the domestic economic crisis wrought by the Covid-19 pandemic have resulted in falling IDA budgets.

Challenging Inequalities at the Regional Scale

As it has expanded to include poorer southern and eastern European member states and as globalizing capitalism has left behind poorer "inner peripheries" (most often post-industrial, peripheral rural areas and border regions) in all member states, strengthening social, economic, and territorial cohesion and reducing regional disparities have served as two of the European Union's (EU) chief concerns (Map 7.3). The EU's Multiannual Financial Framework (MFF) 2021–2027 (regular budget) will invest €1.0743 trillion in supporting agriculture and EU regional development, with a focus upon regions most in need. Furthermore, centered on a Recovery and Resilience Fund, a Next Generation EU 2021–2027 fund of €750 billion will help EU countries recover from the Covid-19 pandemic. In addition, the EU will provide three Covid-19 "safety net" funds for workers, businesses, and sovereigns, worth up to €540 billion. Member states contribute circa 0.7% of their GNI to make up the EU budget. The MFF will continue to place at its center European Structural and Investment Funds and in particular the Common Agricultural Policy (CAP), the European Regional Development Fund (ERDF), the Cohesion Fund (CF), and the European Social Fund Plus (ESF+). Investments will focus upon building critical infrastructure (transport, water, energy, and digital), fostering a just decarbonization of the economy, tackling digital poverty and strengthening digital capacity, fortifying migration controls and security, and remediating youth unemployment.

GDP per inhabitant, 2017
(EU-28 = 100, index based on GDP in purchasing power standards (PPS) in relation to the EU-28 average,
by NUTS 2 regions)

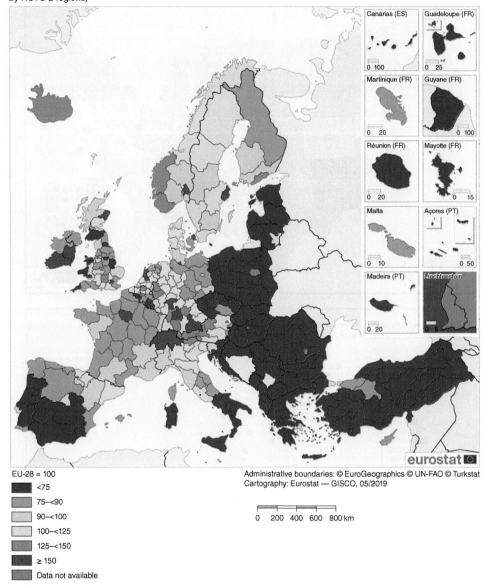

EU-28 = 100

- ■ <75
- ▨ 75–<90
- ▨ 90–<100
- ▨ 100–<125
- ▨ 125–<150
- ■ ≥ 150
- ▨ Data not available

Administrative boundaries: © EuroGeographics © UN-FAO © Turkstat
Cartography: Eurostat — GISCO, 05/2019

0   200   400 600 800 km

Map 7.3   On the importance of Structural Funds in the EU project: GDP per capita expressed as a percentage of the EU average. Source: Eurostat.

Challenging Inequalities at the Intra-Urban Scale

Neither traditional twentieth-century welfare statist nor post-1980 neoliberal market models of urban renewal have proven particularly effective in addressing the complex needs of communities debilitated by low aspiration, poverty, unemployment, family breakdown, poor housing, ill health, and low educational attainment. In particular,

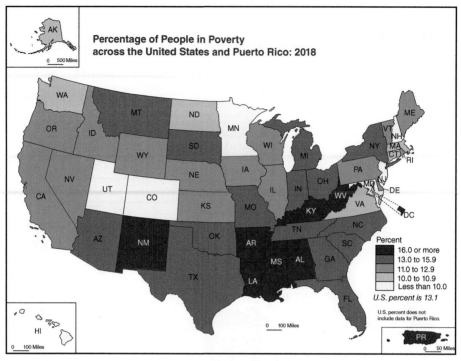

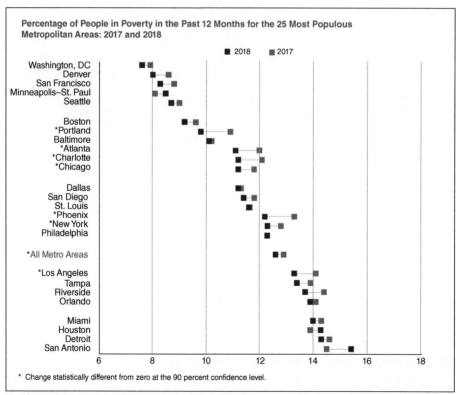

Map 7.4 Regenerating the rustbelt: Geographies of poverty (as measured by the official US poverty line) in the United States, 2018, by state and the 25 most populous cities. Source: US Census Bureau.

forty years of neoliberal urbanism have left in its wake deeply divided cities, in the Global North signaled by gentrified inner-city waterfronts and ghettoized deprived social housing estates (Map 7.4), and in the Global South by informal settlements and slums being encircled by mega property developments and gilded high-rise towers. With its epicenter in the UK, challenging intra-urban inequalities has come to be defined by a "third way" that has sought to chart a novel course between the traditional political landscape of left and right by rethinking the role of the state in market economies. The third way rejects the capacity of neoliberal economics to deliver inclusive urban economic growth and sustainable urban development, but it also rejects state interventions that infantilize communities. The solution is for the state to intervene only to the extent that communities might be rehabilitated so that they can reproduce themselves autonomously and resiliently within the market economy. And so urban regeneration is now dominated by programs that are co-created with communities and that seek to build on local assets, thicken social capital, empower and build capabilities, and increase the resourcefulness of communities.

Challenging Inequalities: An Observation

It is imperative to recognize that spatial planning and development theory and practice have emerged from within Western culture, embody Western values, and evangelize Western recipes for success. Both inside and outside the West, development logics are innately capitalocentric and Eurocentric. Indeed, Swiss development studies scholar Gilbert Rist (2008) has characterized the history of international development using the narrative frame "From Western Origins to Global Faith." And Indian-born US resident urban planner Ananya Roy and American studies scholar Emma Shaw Crane (2015) have called for greater attention to be paid to the Western production of "territories of poverty" at all geographical scales; for Roy and Crane, poverty is less located in space and more constituted spatially by development actors in ways that enable these agents to act on and through poor people and poor places with dedicated ends in mind. Indeed, at its worst, spatial development planning has presented as a thinly veiled moral crusade targeted at disciplining "underperforming" institutions, rogue political leaders, and unruly economic actors. Only by allowing themselves to be open to Western political and economic subjectivities and "normalized," "corrected," and "aligned" will "primitive," "malfunctioning," and "corrupt" Global South institutions, failing regional authorities, and troubled urban communities be made fit for purpose. And so some critics argued that:

- IDA serves as an instrument of Western hard, soft, and smart power, designed to export and propagate Western values, democratic polities, and capitalist economic institutions (see Deep Dive Box 7.4).
- The EU is a harbinger of neoliberal models of development, imposing fiscal austerity on peripheral economies such as Ireland, Portugal, Greece, Spain, and Italy and playing a critical role in transitioning former centrally planned economies in Eastern Europe into the Western fold.
- In its desire to reattach citizens to the "mainstream," third way urban regeneration is creating a new morality of community and judging people and places as "good" and "bad," "right" and "wrong," "just" and "unjust," and "worthy" and "unworthy" based upon their willingness to move away from welfare dependency, embrace market logics, and become competent market actors, active citizens, and consumers.

Deep Dive Box 7.4   International Development Aid (IDA) as Western Hard, Soft, and Smart Power

In the years immediately following independence, many former colonies remained reliant on exports to their European parent. A reflection of its inherited and ongoing dependency on the metropole, the Global South continued to enrich the Global North, exporting low-value-added primary products, which were then manufactured in the core and sold as high-value-added goods, often back in the Global South. In an effort to gain autonomy and move up the value chain, the dominant economic policy followed by many former colonies was one of import substitution. External capital and trade were to be heavily regulated, limited, and policed, and fledgling domestic industries were to be nurtured and protected. Once strong enough, green-shoot indigenous companies would then be in a position to compete globally. Protectionism could cede to liberalization and fuller immersion in the global economy. But over time, it has become clear that import substitution has merely fostered inefficient and non-competitive firms and preserved rather than remediated underdevelopment. Poor governance by an eclectic mix of European-anointed puppet governments, corrupt and rogue kleptocracies, dynastic autocracies, populist and nationalist right-wing military dictatorships, and radical Marxist-inspired states did not help.

From the 1960s in particular, the Global North established a vibrant tradition of international development theory and practice. The West had pioneered a model of development that had worked. Former colonies were failing because their politico-institutional models were flawed. The logical response was to transfer the West's winning formulae (private property, liberal market rule, democratic institutions, and the rule of law) to poorer countries. Liberalized market economies pursuing export-oriented production and fully open to globalizing capitalism were to be the panacea.

Modernization Theory

Walt Whitman Rostow was a US political scientist who served as a national security advisor to US Presidents John F. Kennedy and Lyndon B. Johnson. He was an ardent supporter of US military intervention in Southeast Asia, including in Vietnam and Cambodia. He believed that such intervention was necessary to hold back the tide of socialism and communism. In 1960, Rostow published a book entitled *The Stages of Economic Growth: A Non-Communist Manifesto* (Rostow 1960) (See Figure 7.8.). This book sought to offer a coherent alternative development path to underdeveloped countries who were flirting with non-capitalist models of development. It became a classic text in the modernization tradition. Rostow believed that all countries passed through a number of stages as they developed and that Global South countries could best progress through these stages if they copied the West and embraced a liberal, free-trading, open, and competitive global economy. Follow the formula, and development would automatically occur.

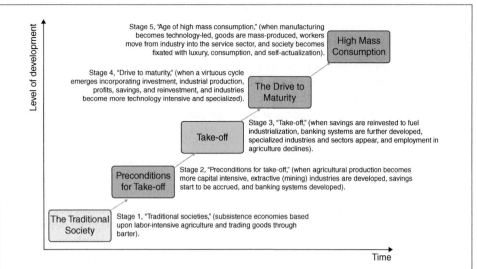

Figure 7.8 Rostow's "noncommunist" model of the stages of economic development. Source: Rostow (1960). Reproduced with permission from Cambridge University Press.

The Washington Consensus

In 1989, Washington-based English economist John Williamson coined the phrase "the Washington Consensus" to refer to 10 principles that ought to define how the international development sector construed development. The phrase has since come to denote pro-market-based approaches to development and, for some, deep market fundamentalism and neoliberalism. Called upon to bail out debt-ridden countries in the Global South from the 1980s, the World Bank, International Monetary Fund (IMF), and World Trade Organization (WTO) (what British-US geographer Richard Peet [2009] pointedly refers to as the "unholy trinity") have used these principles to impose on Global South countries severe structural adjustment (fiscal austerity) programs (SAPs). Amid widespread critique of SAPs, in 1996 US economist Dani Rodrick introduced the idea of an augmented Washington Consensus and identified a further 10 foundational principles for efficacious but also more equitable and sustainable development (see Table 7.2).

Table 7.2 The pillars of the Washington Consensus and the augmented post-Washington Consensus.

Original Washington Consensus	"Augmented" Washington Consensus (the previous 10 items plus)
1. Fiscal discipline	11. Corporate governance
2. Reorientation of public expenditures	12. Anticorruption
3. Tax reform	13. Flexible labor markets
4. Financial liberalization	14. WTO agreements
5. Unified and competitive exchange rates	15. Financial codes and standards
	16. "Prudent" capital-account opening
6. Trade liberalization	17. Non-intermediate exchange rate regimes
7. Openness to DFI	18. Independent central banks/inflation
8. Privatization	targeting
9. Deregulation	19. Social safety nets
10. Secure property rights	20. Targeted poverty reduction

Source: From Rodrik (2006).

Challenging Inequalities Globally in the Twenty-First Century

The efficacy of approaches to territorial development continues to cause concern; too many interventions have toiled (if not failed) to overcome the powerful forces that lie behind uneven geographical development. To often, trillions of US dollars have washed through marginalized spaces at all geographical scales, barely leaving a noticeable legacy.

- US economist Angus Deaton (2013) has argued that although IDA has contributed to improved health in the Global South, it has corrupted already malfunctioning governments, failed to deliver for the poor, and overall done more harm than good.
- European-based regional studies scholars Lewis Dijkstra and his colleagues (Dijkstra et al. 2020) have argued that notwithstanding EU investment in lagging peripheries, local economic and industrial decline in combination with lower employment and a less educated workforce have combined to sow a geography of political populism and EU discontent across the continent manifest in the rise of anti-EU sentiment.
- US urbanists Nijman and Wei (2020) argue that globalizing capitalism has created fresh spatial inequalities in and between cities and among urban populations in North America, Europe, South America, and China. At times, urban regeneration has merely served to "contain" low-income neighborhoods now existing apart from mainstream society.

In response, a new geography of anger and discontent has frothed, and nationalist political populism has mushroomed. In the United States, the phenomenon of Trumpism and the slogans of "America first" and "make America great again" resonated with rustbelt regions who perceive themselves to be left behind by globalizing capitalism. But the rustbelt is in revolt. Populism is sweeping the world and is not an exclusively US development; from Brazil to the United Kingdom and from India to Hungary, elections signal a growing sentiment that representative politics is no longer serving the people. The resulting turn to protectionism seems to signal the death of Smithonian, classical, neoclassical, and neoliberal economic philosophies and two centuries of hyper-globalization. Although anti-globalist isolationism and reduced international aid budgets will do little to reduce international inequalities, populist governments have been elected on platforms committed to leveling up the space economy in their own states. It remains to be seen if any will leave office with reduced within-country inequalities in income and wealth.

In parallel with critiques of the efficacy of territorial development strategies, more attention is now being placed upon "capabilities" and "rights-based approaches" (RBAs) to development – and their interconnection.

In her book *Creating Capabilities: The Human Development Approach* (2011), US philosopher of law and ethics Martha Nussbaum argues that there can be no enduring economic, social, or physical remediation of distressed, failing, or marginalized economic peripheries until there is first a transformation in development theory and practice toward the building of human capabilities. Life itself needs to be rebuilt before new economies can thrive. For economic peripheries to develop, it is imperative that political institutions ensure that a number of critical basic and enhanced human capabilities are permitted to flourish; certainly, they should not be allowed to fall below a given threshold. Meanwhile, RBAs to development propose that achievement of human rights for the world's poor is a prerequisite for development and therefore ought to be the principal objective of spatial development strategies. RBAs make use of the United Nations' 1948

Universal Declaration of Human Rights (UDHR) and the United Nations Commission on Human Rights' two further covenants (the International Covenant on Civil and Political Rights and the International Covenant on Economic, Social and Cultural Rights) to make a moral and legal case for human rights to be placed at the heart of development. Human beings, precisely because they are members of the human family, are endowed with a number of inalienable rights irrespective of their race, age, nationality, gender, class, sexuality, disability, religion, and so on. Underdevelopment arises when these rights are neglected or abused (See Deep Dive Box 7.5).

Deep Dive Box 7.5   The United Nations Development Program (UNDP) MDGs and SDGs

Globally, the UNDP has sought to promote sustainable development, first through its 8 Millennium Development Goals (MDGs) (2000–2015), and most recently through 17 Sustainable Development Goals (SDGs) (2016–2030) (Figure 7.9).

Some commentators suggest that by setting goals, targets, and indicators, the MDGs and SDGs assume and legitimize a globalizing neoliberal development agenda. Rollout neoliberalism and talk of sustainable and resilient development are two sides of the same coin. But this characterization might be unfair. Certainly, with the mantra "leave no one behind," the SDGs appear open to sponsor more holistic people-centered and climate- and ecology-friendly interventions. If there is an underpinning development philosophy to the SDGs it is perhaps likely to be closer to the capabilities and human rights development agenda than that of rollout neoliberalism.

Figure 7.9   United Nations Sustainable Development Goals (SDGs) (2016–2030). Source: Envision2030. United Nations.

(Continued)

Box 7.5   *(Continued)*

According to US geographer Diana Liverman (2018), the MDGs recorded both successes and failures. Although (at least in terms of the metrics set) significant progress was made in remediating poverty, hunger, access to clean water, and debt forgiveness, insufficient progress was made on tackling climate change, (re)emerging infectious disease, infant mortality, maternal deaths, sanitation, and access to primary education. Moreover, progress was uneven, varying from region to region, from country to country, and between social groups (by age, gender, ethnicity, sexuality, disability, and so on). Furthermore, goal setting and target chasing proved to be problematic; targets for poverty and hunger were arbitrary, and success could not always be attributed to the MDGs. There was a lack of attention to human rights, an inadequate focus upon gender inequalities, and only a limited interest in the environment. The governance of the MDGs was to an extent compromised; UN agencies and the nongovernmental organizations (NGOs) they funded had a vested interest in setting a low bar and putting a positive shine on results. The MDGs were also targeted only at the Global South. Finally, they suffered from methodological nationalism or, in other words, a limiting belief that the nation-state scale is the most appropriate entry point for development projects.

The SDGs differ from the MDGs in that they are broader in scope, are built around 17 new goals and a much wider range of targets and indicators, work to better integrate into the UNDP development agenda and the UN Earth Summits, and are intended to apply to both the Global North and the Global South. According to Liverman (2018), the SDGs likewise have a number of strengths and weaknesses. Their sheer number points to intense lobbying of the United Nations by stakeholder groups in the lead-in to their agreement, and at times they present as too all-encompassing to be useful and as bedeviled by complexity and contradiction. The SDGs continue to be goal centered and target oriented, with measurable indicators exerting unproductive discipline on NGOs. They remain limited to working primarily at the scale of the nation–state.

In the 2020 Sustainable Development Report, the UNDP tracked the performance of all UN Member States on the 17 SDGs and measured the extent of the progress that each will need to make if they are to achieve set targets (Sachs et al. 2020).

Advanced capitalist OECD continue to enjoy the highest levels of achievement, with the Nordic countries of Sweden, Denmark, and Finland performing best. But starting from high levels, it is not surprising that OECD countries have displayed only moderate improvement in the past decade. Moreover, collectively they have much work to do on, for example, Goal 10 (responsible Consumption and Production) and Goal 14 (Life Below Water). Encouragingly, although starting further back, all other world regions are making stronger progress toward meeting their SDG goals, and low-HDI regions that had a low SDG score in 2010 would appear to have accelerated toward the 2030 targets fastest. East and South Asian countries in particular have progressed the most. At the country level, Côte d'Ivoire, Burkina Faso, and Cambodia have progressed significantly, but equally countries such as Venezuela, Zimbabwe, and the Democratic Republic of Congo have regressed, mainly due to political difficulties and conflict (Map 7.5).

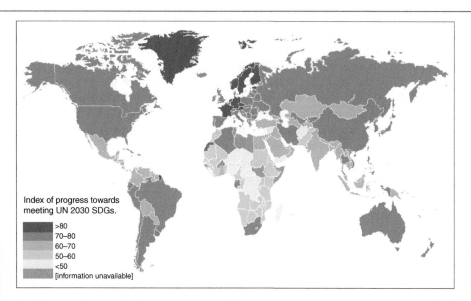

Map 7.5 Overall distance of each country from meeting the SDG 2030 Goals. World map showing countries that are closest to meeting the SDGs (in dark blue) and those with the greatest remaining challenges (in the lightest shade of blue) in 2018. Source: Sustainable Development Report 2019, https://sdgindex.org/reports/sustainable-development-report-2019/

Progress has been particularly rapid with respect to SDG 1 (No Poverty), SDG 9 (Industry, Innovation and Infrastructure), and SDG 11 (Sustainable Cities and Communities). However, there is arising a concerning lack of progress toward achieving SDG 2 (Zero Hunger) and SDG 15 (Life on Land). Moreover, "major performance" gaps have been noted in regard to SDG 12 (Responsible Consumption and Production), SDG 13 (Climate Action), and SDG 14 (Life Below Water), and improvements in SDG 10 (Reduced Inequalities) and SDG 17 (Partnerships for the Goals) in particular appear to be stalling.

In calling for a Decade of Action (2020–2030) to move countries closer to meeting the SDGs, the UNDP have identified six "SDG transformations" – what they call modular building-blocks – to focus the attention of political leaders and galvanize the stakeholder community. These focus upon: (i) education, gender, and inequality; (ii) health, well-being, and demography; (iii) energy decarbonization and sustainable industry; (iv) sustainable food, land, water, and oceans; (v) sustainable cities and communities; and (vi) a digital revolution for sustainable development (Sachs et al. 2020).

Building capabilities for, and RBAs to undergird, development can be regarded as mutually reinforcing. Human rights law provides a juridical framework in which citizens can hold governments to account so that they can acquire, improve, and deploy all the capabilities they have at their disposal. In his book *Development as Freedom*, Indian economist Amartya Sen (1999) argues that the success of spatial development strategies must be measured not in terms of Gross Domestic Product (GDP) or rising standards of living but instead on the extent to which human freedom is enhanced. Poverty should not be construed only in terms of the absence of wealth and income, but should include in addition the extent to which an individual is free to make and realize particular choices – that is, to live the life they have reason to value. Freedom ought to be both the intended outcome of development and the principal means through which development is achieved. Freedom is good in and of itself; it is also good because it leads to human self-actualization, increases social effectiveness, spurs innovation, and creates growth. For Sen, the expansion of five particular freedoms is central to development: political freedoms, economic facilities, social opportunities, transparency guarantees, and protective security. It is the role of human rights law to safeguard these freedoms, enable people to make choices and acquire capabilities, and catalyze development.

Development Alternatives and Alternatives to Development

US geographer Jim Blaut was one of the earliest scholars to claim that development theory and practice was fundamentally a Eurocentric project. In his 1993 book *The Colonizer's Model of the World*, Blaut argued that the West had actively propagated a story about itself, which it then used to justify first colonization and later a very particular development agenda. Europe rose to power by dint of its discovery of the superior model of liberal capitalist modernity; as an act of benevolence, it then gifted this model to the rest of world (where, lacking enlightenment, polities, economies, and cultures remained badly configured and governed, atavistic, uncivilized, stagnant, and backward), first through imperialism and then through neocolonialism and its international development machinery. Blaut challenges this narrative, pointing out that capitalism did not originate in Europe, that Europe emerged by dint of its subjugation of the Global South and is the cause of the latter's underdevelopment, that its postcolonial imposition of market institutions and rule is self-serving, and that the Global South ought to be architect of its own development philosophy and methodology.

For Marxist geographers, a signature characteristic of the capitalist space economy is uneven geographical development at all spatial scales, and there is no reason to assume that the market will distribute economic activity equally over the long run. Indeed, Scottish human geographer Neil Smith (1984) has cogently argued that capitalism, in fact, requires uneven territorial development to prosper. Capitalism actively develops underdevelopment and produces inequalities out of necessity and self-preservation. For Marxists, it is illogical to expect any unfettered market economy to remedy inequalities that it itself has played a significant role in birthing; ongoing blind faith that it can will only make matters worse. Enduring fixes to tenacious spatial inequalities will not be found until the spatial-planning and development industry becomes properly political, doing more than compensating for the status quo. To fix inequalities, it will be necessary to "fix" polities, economies, and societies as much as marginalized economic spaces (Deep Dive Box 7.6).

In the wake of critiques of mainstream development theory and practice, there has arisen debate on the need for both development alternatives and alternatives to development.

Deep Dive Box 7.6 For Spatial Justice: What Makes Justice Spatial? What Makes Space Just?

Human geographers argue that particular political and economic models bequeath inequalities that express themselves spatially in the form of territorial disparities in economic activity, income and wealth, and human welfare. But they go further and claim that uneven geographical development is itself a progenitor of further uneven geographical development. Neoliberal capitalism is constitutive of uneven geographical development at all scales. But equally, uneven geographical development at all geographical scales must itself be understood to be constitutive of neoliberal capitalism. The spaces in which we live and act have significant impacts on what we do and what we can become.

In his 2010 book *Seeking Spatial Justice*, US urban scholar Ed Soja (2013) argues that spatial injustice is not simply an expression of social, economic, and political injustices; it is constitutive of these injustices. It is not enough to see underdevelopment as a product of the capitalist economic model; instead, it is uneven geographical development that enables this model to function. Likewise, in his 2016 book *Limits to Globalization: The Disruptive Geographies of Capitalist Development*, US geographer Eric Sheppard explores both how capitalism organizes the spatial distribution of human activity across the surface of the earth and recursively how the spatial organization of capitalism enables and constrains the possibilities of its development. Spatial (in)justice is both an outcome and a process that results in outcomes.

When uneven geographical development rigidifies into ingrained patterns of privilege and poverty, intervention in the form of territorial planning and development is necessary. Spatial justice – defined as the "equitable distribution in space of socially valued resources and opportunities to use them" – mandates that the space economy be rebalanced and opportunity be leveled up geographically. According to both Soja and Sheppard, it is for this reason that territorial development theory and practice have had to recognize two realities: to remedy our unequal world, it will be necessary to redevise the dominant politico-economic paradigm; and, equally, to redesign the institutional structure of the capitalist economy, it will be necessary to produce space differently.

Spatial justice is a key human geographical concept and one that has the potential to illuminate development theory and practice. But, when applying it, it is necessary to first ask: What makes justice spatial? What makes spaces just?

Development Alternatives

At one level, critics have questioned whether there are better ways to achieve Western standards of living without courting Western development pathways. In her 2020 book *World Making After Empire: The Rise and Fall of Self-Determination*, US political scientist Adom Getachew argues that in the early days of independence, African, African-American, and Caribbean anticolonial movements bubbled with radically new ideas concerning non-European models of development. Alas, the imposition by Europe of Westphalian models of the nation-state repressed these alternative visions, tied development to nation building, and propagated Western development theory and practice. But, of course, many economies have since enjoyed development and prosperity without adopting, in part or in whole, liberal democratic and free market capitalism. In particular, the "Tiger economies" (Hong Kong, South Korea, Taiwan, and Singapore) and "Cub economies" (Thailand, Malaysia, and Philippines) of Southeast Asia and, of course, the Brici countries (Brazil, Russia, India, China, and Indonesia) are now tracking a course of development and in some cases are challenging and even eclipsing their neighbors in the Global North. Moreover, oil revenues have propelled development in many countries in the Middle East, most notably Saudi Arabia, the United Arab Emirates, Qatar, Bahrain, and Oman. But at least to date, while these particular alternatives to Western development have demonstrated an ability to lift human welfare comprehensively and meaningfully, they suffer from human rights limitations and are not especially democratic. As such, for many in the West, they are not alternative models which hold any merit or attraction.

Alternatives to Development

The rise of post-development (PD) is more radical again in that it refuses to contemplate development *per se*, preferring instead to reframe the future in terms of more sustainable ways of living. "Development" is a wholly European idea, mobilized by Western governments in the absence of colonial rule to sustain Western geopolitical advantage. It his 2014 book *Epistemologies of the South: Justice Against Epistemicide*, Portuguese scholar Boaventura de Sousa Santos argues that its influence has stymied much-needed debate over non-European human welfare agendas. But as the West lurches from one economic crisis to the next, watches with alarm as growing inequalities transform into potentially regressive populist movements, and brings the planet to a global climate and ecological crisis, "development" has now lost much of its authority and exists as a "zombie category." In his seminal book on post-development, *The Development Dictionary: A Guide to Knowledge as Power*, German development studies scholar Wolfgang Sachs (1992) famously announced that it was now time to write development's "obituary." According to Colombian and US anthropologist Arturo Escobar, there is no shortage of ideas concerning alternatives to development (within, for example, indigenous knowledge communities and Global South theory that champions "alternative universalisms"), but to unlock the potential of these alternatives there needs to be "alternative thinking on alternatives." In his 2020 book *Pluriversal Politics*, Escobar makes the case for a pluriverse, a world of worlds, each with its own ontological and epistemic logic. The much talked about Latin American "Buen Vivir" philosophy for example provides a vision for good living which construes development as a subordinate (an not especially central) means to and end.

Conclusion

We are now in a position to offer at least a tentative answer to the question we posed at the start of this chapter: can liberalized laissez-faire market economies both (i) increase the wealth of all nations and transform the quality of human life, lifting the world's poor out of poverty; and (ii) correct over time diverging levels of economic prosperity and geographical inequalities in wealth and income at all geographical scales?

Two centuries of classical, neoclassical, and neoliberal economics and economic policy and market rule have spurred human creativity, ingenuity, acuity, and entrepreneurialism and have transformed our species' prospect. Capitalism has made it possible for human beings today to enjoy long, healthy, and prosperous lives that would have been unimaginable to our ancestors. A revolution unlike any other in human history has transformed the quality of human existence, and we are privileged to be living at this historical moment. But this revolution has been unevenly distributed and remains incomplete. Moreover, at times, it has produced grotesque and regressive outcomes. Indian economist Amartya Sen astutely observes that when it comes to people living in precarity and poverty, "a misconceived theory can kill." And in his book *Late Victorian Holocausts*, Mike Davis (2002) has documented the unprecedented human suffering that has been visited on the poor historically by dint of the reticence of some governments to interfere in the machinations of markets.

Development theory and practice, whether it be applied to developing countries in the Global South, underdeveloped "inner peripheries" in developed countries in the Global North, or the territories of concentrated poverty in the "projects" and slums of cities in each, offers tools to remediate territorial inequalities by redistributing economic activity and its fruits more evenly. But reengineering long-established economic geographies has proven to be exceptionally difficult. Market fundamentalists espouse the view that further progress will require wider and deeper penetration of market rule and more aggressive practices of rollout neoliberalism. But while post-imperial and perhaps even post-European globalizing capitalism is in the process of creatively rebuilding the unequal world it has inherited, it is reterritorializing the global economy in ways that are creating new spatial inequalities in human development, poverty, and income inequality at all geographical scales. Far from reflecting the need for more aggressive market rule, perhaps it is now time to acknowledge that although capitalism has achieved immense improvements in human welfare, it is hardwired to produce geographical inequalities and is systemically and structurally constrained to privilege spatial convergence over divergence.

Considered reflection on neoliberal development models and market-centered spatial strategies and territorial planning leads inevitably to the conclusion that to "fix" underdevelopment at all geographical scales, it will be necessary first to fix societies, polities, and the institutional pillars of the economy. And to fix societies, polities and economies it will be necessary to produce anew spatial conditions, patterns and arrangements.

Checklist of Key Ideas

- Classical, neoclassical, and neoliberal economic thought proposes that embracing liberalized free market capitalism is the most effective means of transforming human development, eradicating poverty, and eliminating inequalities.
- The rise from the 1800s of a European-led world capitalist economy created core regions, semi-peripheral regions, and peripheral regions, but globalizing capitalism is reorganizing this world and bequeathing new geographies of development, poverty, and inequality at all spatial scales.
- A critical history of development theory and practice might engage the following observations (see 4 through 8):
- Western development models centered upon "rollout" neoliberal remedies have been judged by some to be inadequate in their efficacy and overly conservative in their politics.

- Emerging capability and human rights approaches to development offer something structurally fresh and potentially transformative, although some critics believe they too are merely another manifestation of "rollout" neoliberalism.
- The United Nations Millennium Development Goals (MDGs) (2000–2015) and Sustainable Development Goals (SDGs) (2015–2030) have ameliorated the extent of poverty, precarity, vulnerability, and inequality that continues to blight the lives of the greater majority of humankind. But they have attracted criticisms, too. A heightened "decade of development" will be required if the SDGs are to be met.
- More radical critics of orthodox development theory and practice argue that meaningful resolution to geographical inequalities at all scales will require a shift in attention to both development alternatives and alternatives to development.
- Nordic social democratic capitalism merits greater attention as we work to challenge inequalities in the twenty-first century.

Chapter Essay Questions

1) "Laissez-faire market capitalism merits greater recognition for the work that it has done in lifting billions of people out of poverty." Discuss.
2) Identify and comment upon the ways in which global poverty has been measured, and describe the geographical patterns of poverty that are discernible on the basis of each measure.
3) Identify and comment upon the ways in which global income inequalities have been measured, and describe the geographical patterns of income inequalities that are discernible at different scales on the basis of each measure.
4) Describe and comment on the contribution of the United Nations Sustainable Development Goals (2015–2039) to the remediation of global poverty and inequality.

References and Guidance for Further Reading

The "Brandt line" was first proposed in:
Brandt, W. (1980). *North–South: A Programme for Survival – The Report of the Independent Commission on International Development Issues under the Chairmanship of Willy Brandt*. London: Pan Books.

Good general introductions to geographies of uneven development, poverty, and inequalities include:
Carmody, P. (2019). *Development Theory and Practice in a Changing World*. New York: Routledge.
Grugel, J. and Hammett, D. (eds.) (2016). *The Palgrave Handbook of International Development*. London: Palgrave Macmillan.
Harvey, D. (2005) *A Brief History of Neoliberalism*. Oxford: Oxford University Press.
Horner, R. and Hulme, D. (2019). From international to global development: new geographies of 21st century development. *Development and Change* 50: 347–378.

Kothari, U. (ed.) (2019). *A Radical History of Development Studies: Individuals, Institutions and Ideologies*. London: Zed Books.

Lawson, V. (2014). *Making Development Geography*. New York: Routledge.

Lawson, V. and Elwood, S. (2018). *Relational Poverty Politics: Forms, Struggles, and Possibilities*. Athens, GA: University of Georgia Press.

MacKinnon, D. and Cumbers, A. (2018). *An Introduction to Economic Geography: Globalisation, Uneven Development and Place*. London: Routledge.

McEwan, C. (2009). *Postcolonialism and Development*. London: Routledge.

Peet, R. (2009). *Unholy Trinity: The IMF, the World Bank and the WTO*. London: Zed Books.

Peet, R. and Hartwick, E. (2015). *Theories of Development: Arguments, Contentions, Alternatives*. New York: Guilford Press.

Potter, R.B., Binns, T., Elliott, J.A. et al. (2019). *Geographies of Development: An Introduction to Development Studies*. London: Routledge.

Power, M. (2019). *Geopolitics and Development*. London: Routledge.

Roy, A. and Crane, E.S. (eds.) (2015). *Territories of Poverty*. Athens, GA: University of Georgia Press.

Roy, A., Negrón-Gonzales, G., Opoku-Agyemang, K., and Talwalker, C. (2016). *Encountering Poverty: Thinking and Acting in an Unequal World*. Berkeley: University of California Press.

Sheppard, E. (2016). *Limits to Globalization: Disruptive Geographies of Capitalist Development*. Oxford: Oxford University Press.

Simon, D. (ed.) (2006). *Fifty Key Thinkers on Development*. London: Routledge.

Smith, N. (1984). *Uneven Development: Nature, Capital and The Production of Space*. London: Verso.

Soja, E.W. (2013). *Seeking Spatial Justice*. Minneapolis: University of Minnesota Press.

Solarz, M.W. (2020). *The Global North-South Atlas*. London: Routledge.

Williams, G., Meth, P., and Willis, K. (2014). *Geographies of Developing Areas: The Global South in a Changing World*. London: Routledge.

Willis, K. (2011). *Theories and Practices of Development*. London: Routledge.

Adam Smith's key text extolling the merits of a market-based economy is:

Smith, A. (1776). *An Inquiry into the Nature and Causes of the Wealth of Nations*. London: W. Strahan and T. Cadell.

Key books on income and wealth inequality written by Thomas Piketty include:

Piketty, T. (2014). *Capital in the Twenty First Century*. Cambridge, MA: Harvard University Press.

Piketty, T. (2020). *Capital and Ideology*. Cambridge, MA: Harvard University Press.

An insightful geographical commentary on the work of Thomas Piketty is offered in:

Sheppard, E. (2015). Piketty and friends: capitalism, inequality, development, territorialism. *The AAG Review of Books* 3 (1): 36–42.

Important scholarly works on inequality and its measurement and implications are:

Alston, P. (2020). *The Parlous State of Poverty Eradication*. New York: United Nations Human Rights Council.

Alvaredo, F., Chancel, L., Piketty, T. et al. (2018). *World Inequality Report*. Cambridge, MA: The Belknap Press of Harvard University Press.

Bourguignon, F. (2015). *The Globalization of Inequality*. Princeton, NJ: Princeton University Press.

Bourguignon, F. and Morrisson, C. (2002). Inequality among world citizens: 1820–1992. *American Economic Review* 92: 727–744.

Casey, A. and Deaton, A. (2020). *Deaths of Despair and the Future of Capitalism*. Princeton, NJ: Princeton University Press.

Deaton, A. (2013). *The Great Escape: Health, Wealth, and the Origins of Inequality*. Princeton, NJ: Princeton University Press.

Dorling, D. (2019). *Inequality and the 1%*. London: Verso.

Kuznets, S. (1955). Economic growth and income inequality. *American Economic Review* 45: 1–28.

Milanovic, B. (2016). *Global Inequality: A New Approach for the Age of Globalization*. Cambridge, MA: The Belknap Press of Harvard University Press.

OECD (2015). *In It Together: Why Less Inequality Benefits All*. Paris: OECD.

Prados de la Escosura, L. (2015). World human development: 1870–2007. *Review of Income and Wealth* 61: 220–247.

Rosling, H., Rosling, O., and Rosling Rönnlund, A. (2019). *Factfulness: Ten Reasons We're Wrong About the World – and Why Things Are Better Than You Think*. New York: Flatiron Books.

Smits, J. and Permanyer, I. (2019). The subnational human development database. *Scientific Data* 6: 190038.

Stiglitz, J.E. (2012). *The Price of Inequality: How Today's Divided Society Endangers Our Future*. New York: W.W. Norton & Company.

UNDESA (2020). *World Social Report*. New York: UNDESA.

United Nations (2019). *Human Development Report: Beyond Income, Beyond Averages, Beyond Today: Inequalities in Human Development in the 21st Century*. New York: United Nations.

United Nations (2020). *Multidimensional Poverty Index 2020: Charting Pathways out of Multidimensional Poverty: Achieving the SDGs*. New York: United Nations.

World Bank (2019). *Annual Report: Ending Poverty, Investing in Opportunity*. Washington, DC: World Bank.

World Bank (PovcalNet). (2020) [Online analysis tool]. Washington, DC: World Bank. http://iresearch.worldbank.org//PovcalNet.

The classical text on modernization theory is:

Rostow, W.W. (1960). *The Stages of Economic Growth: A Non-communist Manifesto*. Cambridge: Cambridge University Press.

The Washington and post-Washington Consensus is discussed in:

Williamson, J. (2003). The Washington consensus and beyond. *Economic and Political Weekly* 38: 1475–1481.

Rodrik, D. (2006). Goodbye Washington consensus, hello Washington confusion? A review of the World Bank's Economic Growth in the 1990s: learning from a decade of reform. *Journal of Economic Literature* 44: 973–987.

Good introductions to international develop aid (IDA) include:

Deaton, A. (2015). *Aid and Politics*. Princeton, NJ: Princeton University Press.

O'Reilly, G. (2019). *Aligning Geopolitics, Humanitarian Action and Geography in Times of Conflict*. Amsterdam: Springer.

Sachs, J.D. (2006). *The End of Poverty: Economic Possibilities for Our Time*. New York: Penguin.

Insights into geographical variations in growing inequalities in the European Union are provided in:

Dijkstra, L., Poelman, H., and Rodríguez-Pose, A. (2020). The geography of EU discontent. *Regional Studies* 54: 737–753.

Iammarino, A., Rodriguez-Pose, A., and Storper, M. (2019). Regional inequality in Europe: evidence, theory and policy implications. *Journal of Economic Geography* 19: 273–298.

Storper, M. (2018). Separate worlds? Explaining the current wave of regional economic polarization. *Journal of Economic Geography* 18: 247–270.

Good introductions to intra-urban inequality and third way urban regeneration can be found in:

Davidson, M. and Ward, K. (2018). *Cities Under Austerity: Restructuring the US Metropolis*. Albany: State University of New York: Press.

Florida, R. (2017). *The New Urban Crisis: How Our Cities Are Increasing Inequality, Deepening Segregation, and Failing the Middle Class – And What We Can Do About It*. New York: Basic Books.

Nijman, J. and Wei, Y.D. (2020). Urban inequalities in the 21st century economy. *Applied Geography* 117: 102188.

Key texts on capabilities perspectives and rights-based approaches include:

Nussbaum, M.C. (2011). *Creating Capabilities*. Cambridge, MA: Harvard University Press.

Sen, A. (1999). *Development as Freedom*. Oxford: Oxford University Press.

Good insights into the United Nations SDGs 2015–2030 are provided in:

Bhattacharya, D. and Ordóñez, L.A. (2016). *Southern Perspectives on the Post-2015 International Development Agenda*. London: Routledge.

Liverman, D. (2018). Geographic perspectives on development goals: constructive engagements and critical perspectives on the MDGs and the SDGs. *Dialogues in Human Geography* 8: 168–185.

Sachs, J., Schmidt-Traub, G., Kroll, C., et al. (2020) *The Sustainable Development Goals and Covid-19. Sustainable Development Report 2020*. Cambridge: Cambridge University Press.

US critical geographer Jim Blaut's influential works include:

Blaut, J.M. (1976). Where was capitalism born? *Antipode* 8: 1–11.

Blaut, J.M. (1993). *The Colonizer's Model of the World*. New York: Guilford Press.

An important work confirming Blaut's claim that the origins of capitalism are earlier and more complex than Western historiographies claim is:

Abu Lughod, J. (1991). *Before European Hegemony: The World System A.D. 1250–1350*. New York: Oxford University Press.

Other critical approaches to development are explored in:

Davis, M. (2002). *Late Victorian Holocausts: El Niño Famines and the Making of the Third World*. London: Verso.

Escobar, A. (1995). *Encountering Development: The Making and Unmaking of the Third World*. Princeton, NJ: Princeton University Press.

Escobar, A. (2018). *Designs for the Pluriverse: Radical Interdependence, Autonomy, and the Making of Worlds*. Durham, NC: Duke University Press.

Getachew, A. (2020). *Worldmaking After Empire: The Rise and Fall of Self-Determination*. Princeton, NJ: Princeton University Press.

Hardt, M. and Negri, A. (2000). *Empire*. Cambridge, MA: Harvard University Press.

Rist, G. (2008). *The History of Development: From Western Origins to Global Faith*. London: Zed Books.

Sachs, W. (ed.) (1992). *The Development Dictionary. A Guide to Knowledge as Power*. London: Zed Books.

Santos, B.D.S. (2014). *Epistemologies of the South. Justice against Epistemicide*. Boulder, CO: Paradigm Publishers.

Santos, B.D.S. (2018). *The End of the Cognitive Empire: The Coming of Age of Epistemologies of the South*. Durham, NC: Duke University Press.

Ziai, A. (ed.) (2017). Post-development 25 years after The Development Dictionary. *Third World Quarterly* 38: 2547–2558.

An important contribution that brought the Latin America Buen Vivir movement to global attention was:

Gudynas, E. (2011). Buen Vivir: todays tomorrow *Development* 54(4): 441–447.

An important body of work, which seeks to unpack capitalism as a variegated and not singular model of political economy and which produces therefore a range of geographies of human development, poverty, and inequality, includes:

Esping-Andersen, G. (1990). *The Three Worlds of Welfare Capitalism*. Cambridge: Polity Press.

Hall, P.A. and Soskice, D. (eds.) (2001). *Varieties of Capitalism: The Institutional Foundations of Comparative Advantage*. Oxford: Oxford University Press.

Milanovic, B. (2019). *Capitalism, Alone: The Future of the System That Rules the World*. Cambridge, MA: The Belknap Press of Harvard University Press.

Website Support Material

A range of useful resources to support your reading of this chapter are available from the Wiley Human Geography: An Essential Introduction Companion Site http://www.wiley.com/go/boyle.

Chapter 8

10 000 000 000: The Modern Rise in World Population from 1750

Chapter Table of Contents

Chapter Learning Objectives

By the end of the chapter you should be able to:
- Describe, illustrate, explain, and comment critically upon the basic tenets of the Demographic Transition Model.
- Produce a typology of the demographic transitions that have occurred in the past, and describe and comment upon the demographic transitions that will unfold in the Global South throughout this century.

- Identify and comment upon the processes and policies that drive changes in mortality as countries pass through demographic transition.
- Identify and comment upon the processes and policies that drive changes in fertility as countries pass through demographic transition.
- Chart and account for China's passage through demographic transition since 1949.

Introduction

For most of its history the population of the human species has remained stationary, and when it has grown its growth has been almost infinitesimal. Around 70 000 YBP no more than an estimated 15 000 human beings lived on earth. As late as 12 000 YBP, on the eve of the Neolithic Revolution, no more than an estimated 1 million people existed. Evidently the Neolithic Revolution created a step change in human population growth and by the time of Christ, the human family had expanded to an estimated 200 million. Growth continued and by 1000 CE an estimated 300 million people inhabited the planet. Whilst the Neolithic Revolution represented a breakthrough in the capacity of the human species to master the environment and swell in numbers, it is clear that something more profound again has occurred with the rise of Western civilization, in particular from around 1750 CE. The planetary population has grown from an estimated 791 million in 1750 to 1.26 billion in 1850, 3 billion in 1960, 4 billion in 1975, 5 billion in 1987, and 6 billion in 1998 (see Figure 8.1). The United Nations estimate that the number of human beings alive in 2020 breached the historic ceiling of 7.7 billion.

A pivotal question therefore presents itself: Why, given that it existed in a state of restraint throughout its long history, did world population suddenly explode in the past 250 years? This chapter places under scrutiny the claim that the *Demographic Transition Model* provides a powerful answer to this question.

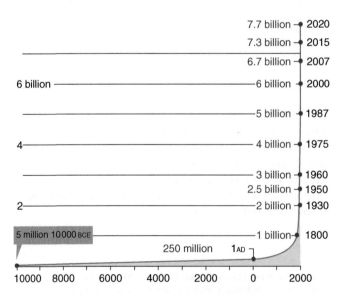

Figure 8.1   World population growth from 10 000BCE. Source: Compiled by the author.

Introduction to the Demographic Transition Model

The Demographic Transition Model was first introduced by US demographer Frank Wallace Notestein in 1945. Notestein's interest in demographic transition stemmed from, and in turn informed, his views on the implications of global population growth for the United States and for the world more generally. His seminal article on demographic transition first appeared in a book edited by US agricultural economist Theodore William Schultz titled *Food for the World* and published in 1945. Notestein feared that advances in medicine and public health might lead to a dramatic decline in mortality levels in the developing world without a corresponding fall in fertility. He promoted the view that the population explosion that would follow would bring dire consequences, not least for the US. Also concerned that world population growth had implications for the US, in 1952 US industrialist and philanthropist John D. Rockefeller 3rd established the Population Council, a nongovernmental organization promoting population control. Notestein was appointed as the first president of the Population Council (1959–1968) and called for measures to be taken to control fertility (controversially, especially among the poor in the Global South) and stem population growth (see Deep Dive Box 8.1).

Deep Dive Box 8.1   Frank W. Notestein's (1964) Address to the Ceylon Association for the Advancement of Science on the Subject of Population Growth and Economic Development

On 22 September 1964, as President of the Rockefeller Population Council, Frank W. Notestein was invited by the Ceylon Association for the Advancement of Science to deliver, in Colombo, Ceylon (present-day Sri Lanka), a lecture on the topic of "Population growth and economic development."

Notestein began his address by noting that life expectancy in Ceylon had increased dramatically from 30 years of age in 1921 to 70 in 1964. Its crude death rate (CDR) had fallen from 29 per 1000 in 1921 to 13.5 per 1000 in 1960. But in 1960, Ceylon's total fertility rate (TFR) stood at 5 and its crude birth rate (CBR) at 37 per 1000. Driven by natural increase alone, the country's population had grown from 5.3 million in 1931 to 10.2 million in 1961.

Notestein argued that science had now resolved all "ideological" and "religious" disputation as to whether population growth was good or bad for a country; it was bad (heaping more misery on the poor) and had to be stopped. Noting developments in modern contraception and the political will to forcibly reduce fertility in countries such as India, China, Korea, Pakistan, Indonesia, and Singapore, Notestein concluded:

> For the first time in human history we appear to have methods that are appropriate to meet the needs of the weakly motivated, illiterate, and impoverished elements of the population. Given suitable organisation for the dissemination of information, service, and supplies, there is no reason why all peoples everywhere cannot restrict their childbearing as they see fit. (Notestein 1983, p. 360)

Ceylon introduced a number of significant fertility control programs from 1965, and by 2000 its TFR stood at 1.9 and its CBR at 18 per 1000. Nevertheless, today, with a CDR of 6.8 per 1000, a TFR of 2.19, and a CBR of 15.5 per 1000, Sri Lanka's population has continued to grow and amounts now to 21.3 million people.

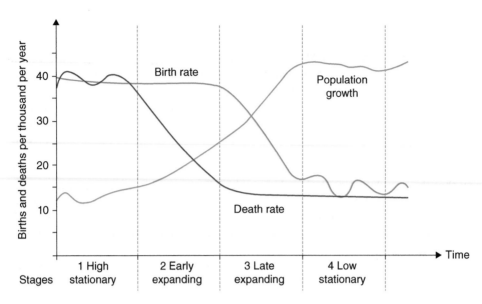

Figure 8.2   The classic Demographic Transition Model. Source: Compiled by the author.

Figure 8.2 provides a summary overview of the Demographic Transition Model as it is normally depicted today. Through time, as countries modernize and/or industrialize and/or develop and/or urbanize, their populations pass through a sequence of transformations normally captured in four separate stages (but see also Deep Dive Box 8.2).

- Stage 1, the "high stationary period," is marked by high CDRs (fluctuating in response to famines and epidemics), high CBRs (fluctuating in response to fluctuating death rates), and, as a consequence, a small and stable population.
- Stage 2, the "early expanding phase," denotes the beginnings of demographic transition; in response to socioeconomic progress, mortality levels fall whilst levels of fertility remain high and population growth begins in earnest.
- In Stage 3, the "late expanding phase," mortality continues to decline, bottoming out at a lower level, but birth rates also now respond to modernization and development and fall, tempering rates of population growth.
- By Stage 4, the "low stationary period," both birth rates and death rates hover around the lower threshold and cancel each other out. Population growth comes to a halt, and populations now stabilize.

Deep Dive Box 8.2   Tim Dyson's Rethinking of Demographic Transition

It has long been assumed that development and/or industrialization and/or modernization and/or urbanization drive countries through demographic transition. But in his book *Population and Development: The Demographic Transition*, British demographer Tim Dyson (2010) has introduced fresh thinking into the theory about what triggers demographic transitions and the relationship between population growth and development.

According to Dyson, because the two tend to unfold concurrently, hitherto too much emphasis has been placed upon economic development as the fuel that fires the demographic transition process. In fact, once it gets started, demographic transition largely powers itself through a series of causally connected stages. Dyson includes in his cause-and-effect chain the following links: mortality decline, rapid population growth, urbanization, reductions in fertility, and an aging population structure. According to Dyson, it is mortality decline that *causes* populations to grow (fewer people are dying), it is growing populations that *cause* fertility to decline (infant mortality declines and there is less of a need for large family sizes to ensure the survival of children) and cities to form (overpopulation of the countryside and the attraction of lower death rates in cities fuel migration to cities), and it is fertility decline that *causes* aging population structures (by creating top-heavy population pyramids).

In other words, once underway, even in the absence of further economic advancement and progress, demographic processes themselves push the transition to completion. Taken to its logical conclusion, it is entirely possible for the world's poorest countries, without further development, to experience medical- and health care–led improvements in life expectancy and reductions in mortality, falling fertility and low birth populations, a slowdown and stabilization of population growth, aging population structures, and rapid urbanization.

Not only does Dyson believe that development is not the key progenitor of passage through demographic transition, but he calls for more attention to be given to the effects of demographic transition on development (i.e. the impact of population on development rather than development on population). What happens when one reverses the causal arrow? Against perhaps the grain of conventional wisdom, and whilst acknowledging the environmental pressures wrought by population growth (especially climate change), Dyson is of the view that by and large the demographic transition has been a positive event for most societies, spurring development and progressive change. For example, he examines the impacts of demographic transition on the rise of democracy. His claim is that the transition from autocracies to democracies was driven in no small way by demographic transition.

Dyson can be accused of placing too much emphasis upon demographic drivers of change; his view of mortality decline as a domino – which, once toppled, in turn topples all other dominoes arranged in the line – ascribes too much importance to "internal" demographic processes. But demographic processes cannot be separated from their social, political, cultural, and economic determinants to this degree. Moreover, Dyson can be accused of failing to acknowledge sufficiently the challenges that demographic transitions have given rise to. Still his work is an antidote to those lazy commentaries that pass off demographic transition as somehow vaguely caused by social and economic progress and neo-Malthusian pessimists who assume that nothing good has ever come from the modern rise in world population.

Whilst these four phases form the backbone of the Demographic Transition Model, it has become common to add a fifth phase when populations actually shrink in response to very low birth rates and when such contraction is accelerated by increasing death rates as diseases of affluence and obesity, a mental health pandemic, and (re)emerging infectious disease claim more lives in affluent societies.

Histories and Geographies of Demographic Transitions

In his 1992 *The Demographic Transition: Stages, Patterns, and Economic Implications*, French demographer Jean-Claude Chesnais presented a longitudinal study of the demographic transitions witnessed by 67 countries, across the period 1720–1984. It is clear that some demographic transitions started much later than others, that the speed of passage through demographic transition varies from place to place, and that peak rates of population growth vary from one case to another.

Chesnais argued that, as of the early 1990s, three types of demographic transition could be found in history.

Developed countries in Europe: These demographic transitions enjoyed peak population growth rates of 2% per year and were very long lasting, taking anything from 75 to 200 years to run their course. Recognizing that demographic regimes change as one moves from the northwest to the southeast of Europe, Chesnais set out further subdivisions within this category:

Nordic Model – A very long transition lasting almost 150 years with maximum population growth occurring in the period 1870–1880; Sweden is the classic example.

Western Model – A long transition of approximately 100 years with population growth peaking around 1900; Germany is the iconic example.

Southern Model – A somewhat long transition running to completion over a period of between 70 and 90 years, with sustained plateaus of maximum population growth occurring after 1900; Italy and the USSR are the best examples.

Interestingly, Chesnais points out that both France (because of its historically low birth rates) and Ireland (because of the great Irish famine of 1848–1852 and mass emigration) are rogue outliers that defy easy incorporation into these three models.

Less developed countries: Countries that remain in the throes of demographic transition. In these countries, transitions started much later, appear to be progressing at an accelerated pace suggesting a passage through transition of between 40 to 80 years, and are dominated by rates of natural increase of between 2 and 4% per annum, which peak across a short 20-year span. Some countries remain in Stage 2 of the transition, whilst others have progressed to Stage 3. Because these transitions remain active, it is difficult to be certain as to how they may turn out and therefore how to differentiate them further. But we might usefully note that there are countries with fairly high rates of population growth (from 2 to 2.5% per annum – for Chesnais, India at that time), high rates of growth (from 2.5 to 3% per annum – for Chesnais, Egypt at that time), and very high rates of growth (3% per annum and above – for Chesnais, Mexico at that time).

Principal countries of immigration: Between the developed countries in Europe and the less developed countries of the world lies an intermediate group of countries that Chesnais calls "principal countries of immigration." These are countries that were primary reception centers for European migrants in the eighteenth, nineteenth, and twentieth centuries. They include the United States, Canada, Australia, New Zealand, Argentina, and Uruguay. Because European emigrants were for the most part young adults, these countries were born with very peculiar population profiles. Their demographic transitions are marked by declines in both death rates and birth rates from the outset; essentially, they began the demographic transition at Stage 3.

Of course, the modern rise in world population has yet to run its course in many regions, and there exist many and variegated demographic transitions still in the making.

Mortality continues to decline. According to the United Nations' *World Population Prospects: 2019 Revision* (United Nations Population Division 2020), globally, life expectancy at birth increased from 47 years of age in 1950 to 65.6 in 2000 to 72.3 today. By 2050 life expectancy is projected to rise to 76.8 years of age and from there to grow to 81.7 by 2100. Africa experienced the greatest increases, where life expectancy rose from 37.5 years of age in 1950 to 62.7 today. Worldwide, life expectancy at birth has increased from 49.5 years of age in 1950 to 75.59 today for females, and from 45.2 years of age to 70.81 years for men. Today life expectancy is 64.1 in Africa, 74.2 in Asia, 79 in Europe, 76 in Latin America and the Caribbean, 79.5 in North America, and 79.2 in Oceania. Countries with a life expectancy of 82 years or older include Australia, Iceland, Italy, Japan, Singapore, Spain, and Switzerland. Countries with a life expectancy of 55 years or younger include the Central African Republic, Chad, Côte d'Ivoire, Lesotho, Nigeria, Sierra Leone, Somalia, and Swaziland.

The Fertility Transition is the point at which TFRs peak (averaging at 7–8 births per woman), after which they endure a sustained decline to replacement levels of 2.2 births per woman, or lower. According to the United Nations' *World Population Prospects: 2019 Revision* (United Nations Population Division 2020), globally TFRs fell from 4.97 births per woman in 1950 to 2.42 in 2019. The TFR in Africa fell from 6.57 births per woman in 1950 to 4.16 today. Over the same period, fertility levels also fell in Asia (from 5.83 to 2.09), Latin America and the Caribbean (from 5.83 to 2.04), North America (from 3.34 to 1.76), Europe (from 2.66 to 1.62), and Oceania (from 3.89 to 2.2). Globally, TFRs are expected to decrease from 2.42 today to 2.18 in 2050 and 1.94 in 2100. In 2021, no less than 83 countries have below-replacement-level fertility (that is, fewer than 2.2 births per woman). TFRs are projected to increase between 2010–2015 and 2045–2050 from 1.60 to 1.78 in Europe and from 1.85 to 1.89 in North America. But in Africa, Asia, Latin America and the Caribbean, and Oceania, fertility is expected to fall between 2010–2015 and 2045–2050, with the largest reductions projected to occur in Africa. In all regions of the world, fertility levels are projected to converge to levels around or below the replacement level by 2095–2100 (see Deep Dive Box 8.3).

The United Nations Population Division provides authoritative population projections to the year 2050, and more speculative forecasts to the year 2100 and even 2300. Because population projections are especially sensitive to changes in fertility levels, the United Nations Population Division paints a variety of scenarios depending upon different fertility forecasts:

- A "constant-variant" scenario, assuming fertility levels continue in the future much as they are today.
- A "medium-variant" estimate, where TFRs behave as expected.
- "High-variant" and "low-variant" projections, where countries exhibit TFRs of 0.5 of a child higher than the medium variant, or 0.5 of a child lower than the medium variant.

In their most recent work, the United Nations have used medium-variant estimates and added confidence intervals of 80 and 95% (that is, how confident they are that population size will land where they forecast).

Deep Dive Box 8.3   Africa's Unique Fertility Transition

The modern rise in world population, which began around 1750 CE in Europe and has spread throughout the world, will run its course across the twenty-first century and will witness a final surge in Africa. But the African experience promises to be exceptional. According to US demographer Jean Bongaarts (2017), Africa's fertility transition is historically unique because it has occurred:

- **Later:** The onset of the transition in Africa occurred on average in the mid-1990s, about two to three decades later than in non-African Global South countries.
- **Earlier:** The level of development at the time of onset of the African fertility transition was lower than at the onset in other countries located in the Global South. In other words, the African transitions occurred earlier than they would have occurred if Africa had followed the non-African relationship between fertility and development.
- **Slower:** The pace of fertility decline at the time of the African transition onset was slower than the comparable pace at the onset of non-African transitions. At the same time, the pace of improvement in development indicators at the time of the African onset was also slower.
- **Higher:** At a given level of development, Africa's fertility is higher, contraceptive use is lower, and desired family size is higher than in non-African countries. Africa's stubborn pro-natalist attitude reflects the ongoing impress of traditional social, economic, and cultural practices on its fertility dynamics.

Still, by 2100 even Africa will experience fertility transition, population growth will come to a halt, and the continent will witness a stable population.

According to the United Nations' *World Population Prospects: 2019 Revision* (United Nations Population Division 2020), world population will continue to grow during the remainder of this century, although the pace of growth may decline after 2050. If birth rates were to remain as they were in the period 2005–2010, remarkably, world population could rise to 10.5 billion in 2050 and 21.6 billion in 2100. The medium-variant projection assumes that the global fertility level will decline from 2.5 births per woman in 2010–2015 to 2.2 in 2045–2050, and will then fall to 2.0 by 2095–2100; and, on this basis, it predicts a growth in world population from 7.7 billion today to 9.7 billion in 2050 and to 10.9 billion by 2100 (Figure 8.3). Global population is expected, with a probability of 80%, to lie between 9.9 and 12.0 billion in 2100. With a certainty of 95%, the size of the global population will stand between 8.5 and 8.6 billion in 2030, between 9.4 and 10.1 billion in 2050, and between 9.4 and 12.7 billion in 2100. In their most recent forecasts, including low-variant and high-variant estimates (2017), the United Nations predicts a high-variant scenario of 10.5 billion in 2050 and 21.6 billion in 2100 and low-variant scenario of 8.8 billion at midcentury, falling to 7.3 billion in 2100.

As the modern rise in world population finally exhausts itself by 2100, the distribution of world population will look different (Map 8.1 and Figure 8.4). Currently, 60% of the world's populace live in Asia (4.6 billion), 17% in Africa (1.3 billion), 10% in Europe (747 million), 9% in Latin America and the Caribbean (653 million), and 6% in Northern America (369 million) and Oceania (42.6 million) combined (United

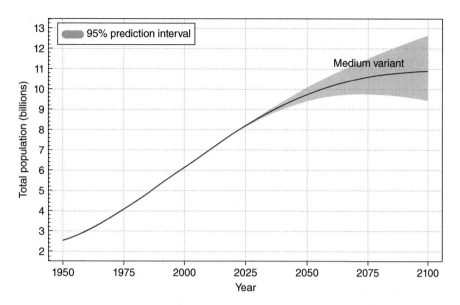

Figure 8.3a World population growth, 1950–2100: medium-variant scenario with 95% confidence interval. Source: United Nations. DESA. Population Division. World Population Prospects 2019. CC BY 3.0 IGO.

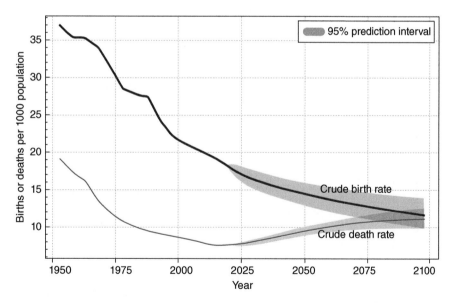

Figure 8.3b Crude death rates (CDRs) and crude birth rates (CBRs), 1950–2100: medium-variant scenario with 95% confidence interval. Source: United Nations. DESA. Population Division. World Population Prospects 2019. CC BY 3.0 IGO.

Nations Population Division 2020). Figure 8.5 shows the world's 10 most populous countries. According to the United Nations (United Nations Population Division 2020), the following developments are likely to occur under the medium-variant scenario:

Asia: Asia in 2100 will remain the most populated continent, but its population will peak around 2055 (at c. 5.3 billion) and gradually decline toward the end of the

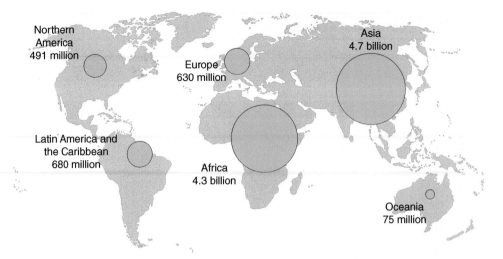

Map 8.1   Distribution of World Population in 2100. Source: United Nations. DESA. Population Division. World Population Prospects 2019. Pew Research Center, https://www.pewresearch.org/fact-tank/2019/06/17/worlds-population-is-projected-to-nearly-stop-growing-by-the-end-of-the-century/.

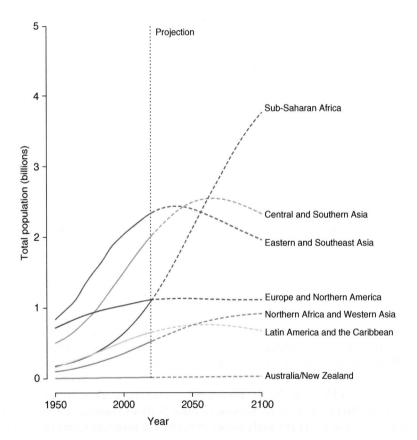

Figure 8.4   Population by United Nations Sustainable Development Group (SDG): estimates, 1950–2020, and medium-variant projection with 95% prediction intervals, 2020–2100. Source: United Nations, Department of Economic and Social Affairs, Population Division (2019). World Population Prospects 2019: Highlights (ST/ESA/SER.A/423). CC BY 3.0 IGO.

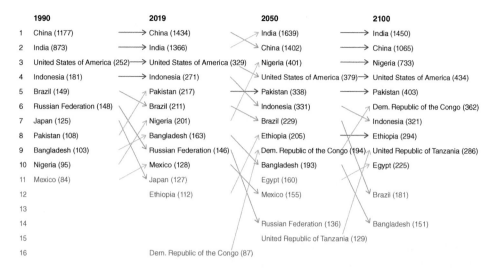

Figure 8.5   Rankings of the world's 10 most populous countries, 1990 and 2019, and medium-variant projections for 2050 and 2100 (numbers in parentheses refer to total population in millions). Source: United Nations, Department of Economic and Social Affairs, Population Division (2019). World Population Prospects 2019: Highlights (ST/ESA/SER.A/423). CC BY 3.0 IGO.

century to 4.7 billion by 2100. China (with 1.44 billion inhabitants) and India (1.38 billion inhabitants) remain the two most populous countries, comprising 19 and 18% of the total global population; by circa 2024, the population of India is expected to exceed that of China. By 2100, India, with a population of 1.45 billion, will be the most populated country in the world (the Chinese population will be 1.06 billion).

Africa: Perhaps the most significant demographic event of this century will be Africa's sustained population expansion, from 1.26 billion today to 4.3 billion by the end of the century. By 2100, from a base today of 206 million, Nigeria will grow rapidly to reach 734 million people and will displace the United States to become the third-largest country in the world.

Europe: The population of Europe has peaked (around 2020) at 747 million and is expected to decline to 630 million by 2100. From its current base of 103 million people, by 2100, the Russian Federation with 145 million people will house Europe's largest population.

Latin America and the Caribbean: The population of Latin America and the Caribbean is projected to reach a maximum around 2065 of 767 million, after which point it will decline to 679 million by 2100. Brazil, currently home to 213 million people, will decline to 181 million by 2100, and Mexico, now at 142 million, will have 129 million people; but both will still be the largest countries in Latin America.

North America: The population of North America will rise from 330 million in 2020 to 425 million by 2050 and finally to 490 million in 2100. By 2100, the United States will be the fourth-largest country in the world, with a population of 434 million.

Oceania: The population of Oceania will rise marginally from 42.6 million in 2020 to 57 million by 2050 and 75 million by 2100. The population of Australia will grow from its current base of 25.5 to 43 million by 2100. But Oceania will remain the least populated of all continents, with only 0.6% of the world's population.

The Demographic Transition Model and Mortality Decline

Explaining Mortality Decline

The World Health Organization's (WHO) International Classification of Disease (ICD) (11th revision) provides a recognized categorization of causes of mortality and morbidity (there are around 55 000 disease codes in ICD-11). Simplifying the ICD, it is possible to aggregate all disease and ill health into one of three groups. Group 1 comprises infectious, contagious, or communicable diseases: these are diseases that involve the transfer of a pathogen between organisms and that can be spread within populations. This group includes airborne disease (for example, influenza, Covid-19 bronchitis, and tuberculosis), water- and foodborne disease (for example, cholera, typhoid, and typhus), and vector-borne disease (for example, malaria transmitted by the female *Anopheles* mosquito). Group 2 incorporates non-communicable or degenerative disease. Here the body suffers an internal malfunction or breakdown. Included in this group would be cancers, heart disease, and strokes. Finally, Group 3 captures all deaths inflicted by people themselves, including death from suicide, warfare, murder, and accidents.

In the 1970s, Egyptian physician Abdel Omran (1971) noticed that as countries developed and as their mortality declined, the structure of the principal causes of death and ill health in those countries changed in a systematic way. Omran proposed that countries progressed through demographic transition only when they learned to master communicable and infectious disease. Rooting his findings within the field of epidemiology (the study of diseases), he introduced to demographic transition theory the parallel idea of epidemiological transition (see Deep Dive Box 8.4).

Deep Dive Box 8.4   Tracking Movement Through Epidemiological Transition Using the DALY Measure

The Disability-Adjusted Life Year (DALY) measure provides one way of tracking the passage of countries through epidemiological transition.

DALYs were first calculated as part of the World Bank's *World Development Report 1993* to assist health planners to prioritize scarce resources by shedding new light on the toll exacted by different diseases and illnesses. Today, the measure is widely used by the World Bank, the World Health Organization (WHO), and the United Nations. Until 1993, the burden of disease suffered by any country was measured largely by subtracting the age of death of each citizen from a defined optimum life expectancy and summing to produce an overall figure. The World Bank sought to add to this standard calculation an additional measure of years of life lost even if lived as a consequence of debilitating ill health and impaired quality of life. DALYs are the sum of years of life lost due to premature death (YLLs) and years lived with disability (YLDs). As such, DALYs can be thought of as years of healthy life lost.

Whilst helpful, the DALY measure continues to excite controversy. When possible, the calculation is made using actual data, but often the necessary data are absent, and in these circumstances estimates from experts are sought. But experts often disagree over how debilitating a particular illness might be to live with. Evidently, such assessments gloss over the fact that particular diseases and illnesses might afflict people with different degrees of severity and be weathered by people in different ways. Moreover, there is disagreement over the optimum life expectancy against which loss of healthy life years should be measured. This ceiling varies between organizations and countries, making comparative analyses difficult. In addition, many applications of the measure inflate the value of a YLL/YLD during the productive period of people's lives (say, between 16 and 64 years of age) and devalue YLL/YLD in infant-hood and in retirement. Whilst this skew might facilitate the targeting of resources so as to save more years of productive life, clearly the judgment that any one year of anyone's life is more valuable than anothers year is morally questionable.

Figure 8.6a–8.6c shows the global burden of disease as measured by the Institute for Health Metrics and Evaluation (IHME) and *The Lancet* in 2016

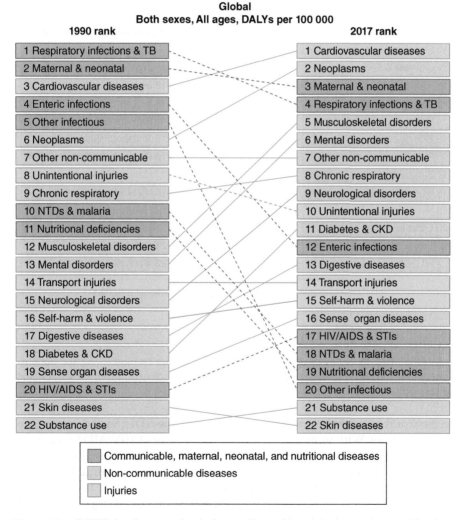

Global
Both sexes, All ages, DALYs per 100 000

1990 rank	2017 rank
1 Respiratory infections & TB	1 Cardiovascular diseases
2 Maternal & neonatal	2 Neoplasms
3 Cardiovascular diseases	3 Maternal & neonatal
4 Enteric infections	4 Respiratory infections & TB
5 Other infectious	5 Musculoskeletal disorders
6 Neoplasms	6 Mental disorders
7 Other non-communicable	7 Other non-communicable
8 Unintentional injuries	8 Chronic respiratory
9 Chronic respiratory	9 Neurological disorders
10 NTDs & malaria	10 Unintentional injuries
11 Nutritional deficiencies	11 Diabetes & CKD
12 Musculoskeletal disorders	12 Enteric infections
13 Mental disorders	13 Digestive diseases
14 Transport injuries	14 Transport injuries
15 Neurological disorders	15 Self-harm & violence
16 Self-harm & violence	16 Sense organ diseases
17 Digestive diseases	17 HIV/AIDS & STIs
18 Diabetes & CKD	18 NTDs & malaria
19 Sense organ diseases	19 Nutritional deficiencies
20 HIV/AIDS & STIs	20 Other infectious
21 Skin diseases	21 Substance use
22 Substance use	22 Skin diseases

Communicable, maternal, neonatal, and nutritional diseases
Non-communicable diseases
Injuries

Figure 8.6a DALYs lost by cause (ranked according to impact 1 = most severe and burden-some), 1990 and 2017: Global. Source: From IHME, Global burden of disease as measured by the IHME. © IHME.

(Continued)

Box 8.4 *(Continued)*

High SDI
Both sexes, All ages, DALYs per 100 000

1990 rank	2017 rank
1 Cardiovascular diseases	1 Neoplasms
2 Neoplasms	2 Cardiovascular diseases
3 Musculoskeletal disorders	3 Musculoskeletal disorders
4 Mental disorders	4 Neurological disorders
5 Neurological disorders	5 Mental disorders
6 Unintentional injuries	6 Chronic respiratory
7 Other non-communicable	7 Diabetes & CKD
8 Chronic respiratory	8 Unintentional injuries
9 Transport injuries	9 Other non-communicable
10 Digestive diseases	10 Digestive diseases
11 Diabetes & CKD	11 Substance use
12 Self-harm & violence	12 Sense organ diseases
13 Maternal & neonatal	13 Self-harm & violence
14 Sense organ diseases	14 Skin diseases
15 Skin diseases	15 Transport injuries
16 Respiratory infections & TB	16 Respiratory infections & TB
17 Substance use	17 Maternal & neonatal
18 HIV/AIDS & STIs	18 Nutritional deficiencies
19 Nutritional deficiencies	19 Enteric infections
20 Other infectious	20 HIV/AIDS & STIs
21 Enteric infections	21 Other infectious
22 NTDs & malaria	22 NTDs & malaria

Communicable, maternal, neonatal, and nutritional diseases
Non-communicable diseases
Injuries

Figure 8.6b   DALYs lost by cause (ranked according to impact 1= most severe and burden-some), 1990 and 2017: World Bank high-income countries. Source: From IHME, Global burden of disease as measured by the IHME. © IHME.

for the world as a whole, and then for high- and low-SDI (Sociodemographic Index) countries. With some exceptions like the scourge of HIV/AIDS (and now, of course, Covid-19, these figures confirm the basic premises of epidemiological transition theory: that poorer countries are burdened by communicable diseases more so than richer countries, and that over time all regions of the world are moving steadily from an age of receding pandemics to an age of degenerative and people-made causes of mortality.

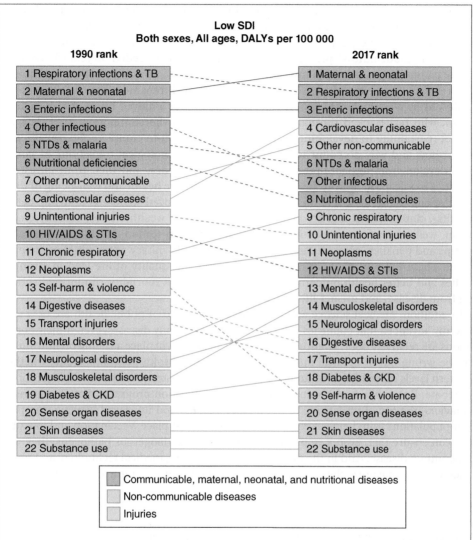

Figure 8.6c DALYs lost by cause (ranked according to impact 1 = most severe and burdensome), 1990 and 2017: World Bank low-income countries. Source: From IHME, Global burden of disease as measured by the IHME. © IHME.

According to Omran, countries yet to experience development or that were in the early stages of development suffered disproportionately from infectious and communicable disease. These countries resided in the "age of pestilence and famine" (where life expectancy was between 20 and 40 years). As development continued apace, steadily countries learned to conquer and cleanse themselves of the principal infectious diseases, and the burden of illness transferred from communicable to noncommunicable disease and to people-made causes of morbidity and mortality. Here, countries passed through an "age of receding" pandemics, and life expectancy improved to 30 to 50 years. During this age, the most profound changes in health and disease patterns are felt by hitherto vulnerable children and young women. Finally, the more developed countries came to purge themselves of infectious disease to the point that degenerative and people-made ill health dominated; now the "age of degenerative and people-made illness" prevailed, and life expectancy exceeded 50 years. Variations in the pattern, pace, determinants, and consequences of mortality decline combine to differentiate three basic models of the epidemiological transition: the classical model (followed by Western countries), the accelerated model (enjoyed by countries such as Japan, who fast tracked throughout the process), and the contemporary or delayed model (pertaining to developing countries who have still to complete the transition).

But how and why do countries move through epidemiological transition? As countries develop from agricultural societies to industrial ones, and then from industrial societies to advanced high-technology ones, inevitably the health and well-being of their populations change too. Initially, prosperity brings improved standards of living: better housing, clothes, diets, and so on. As countries become wealthier, they steadily create better welfare systems: education improves, social protection for more vulnerable groups and pensions are introduced, and sewerage and sanitary water supplies are provided. These improvements in material well-being and quality of life lead to stronger bodies and better health outcomes – lower rates of infant mortality, diminished suffering from disease, and greater life expectancy.

In addition, medical science and knowledge in the field of public health are now sufficiently advanced that countries can rid themselves of many diseases, even if their populations are relatively poor. Eventually, standards of living will place a ceiling on how much can be wrung out of medical interventions and public health programs alone; whether Western levels of health can be attained by all without Western standards of living is open to debate. But public health initiatives, including education programs that change risky behaviors, the use of insecticide to kill disease vectors, and improved hygiene, can prevent populations from being exposed to illness in the first place. Vaccination and inoculation can ensure that, even if populations are exposed to dangerous pathogens, they will not contract disease. And in the event that disease occurs, medical technology and pharmacology can maximize the chances of survival and promote a return to well-being.

Policies for Improved Global Health

As countries continue to move through epidemiological transition, it would appear that two core challenges remain.

Creating an age of delayed degenerative disease: The world will continue to pass through epidemiological transition, and the global burden of disease will increasingly center on degenerative disease and people-made illness. For developed countries, and increasingly also for developing countries, the priority will be to prolong life to its biological maximum by delaying the onset of degenerative disease. Preventative health care, the search for cures for diseases such as cancer and Alzheimer's, the promotion of exercise and well-being, and improved palliative care will become more important. Addressing problems caused by alcoholism, the use of tobacco, and substance abuse will be particular priorities. Improving air quality will be critical. But the growing epidemic that is obesity, and the medical complications associated with this disease, will restrain the capacity of rich countries to strive for an age of delayed degenerative disease. In addition, addressing people-made forms of mortality and morbidity will become a more pressing concern. Reducing accidents, in particular transport-related fatalities and injuries, and tackling suicide will be important areas for policy intervention.

Addressing the problem of (re)emerging infectious disease: For all countries in the world, the problem of (re)emerging infectious disease will become a more pressing concern. Diseases such as malaria, ebola, dengue fever, HIV/AIDS, tuberculosis, influenza, severe acute respiratory syndrome (SARS), Lassa fever, and Escherichia coli (E. coli) will become more prevalent (see Deep Dive Box 8.5). In 2020, the world was stopped

---

Deep Dive Box 8.5   The West Africa Ebola Outbreak, 2013–2016

Ebola is a communicable or infectious disease spread by the exchange of bodily fluids (principally blood, stools, and vomit, but also urine, sweat, tears, saliva, semen, and breast milk). It is a disease that can prove fatal in a significant number of cases; death frequently occurs 7–14 days after symptoms first emerge.

Ebola was first discovered in 1976. Between 1976 and 2013, WHO charted a total of 24 outbreaks of Ebola comprising 1716 cases. Prior outbreaks have tended to occur in Equatorial Africa and have tended to be brought under control quickly with limited damage. In December 2013, an 18-month-old boy in a small village in Guinea West Africa fell ill from an unknown disease and died shortly thereafter. Subsequently, this boy was recognized to be the Index Case (the point of origin) for the latest Ebola outbreak in West Africa. It is believed that the boy contracted the disease by coming into contact with the waste of a fruit bat – contact made more likely because of recent deforestation in the area. By March 2014, the WHO declared that an Ebola outbreak was unfolding in West Africa and, on 8 August 2014, deemed this outbreak to be an international public health emergency and more lethal than all prior outbreaks.

A total of 28 616 Ebola cases were reported (mainly Guinea 3804 cases; Liberia 10 675; and Sierra Leone 14 122) with 11 310 deaths (mainly Guinea 2536; Liberia 4809; and Sierra Leone 3955) – see Map 8.2. Whilst Senegal, Nigeria, and Mali were also exposed to imported cases, through preparation and vigilance they managed to avoid significant outbreaks. Meanwhile an outbreak in the Democratic Republic of Congo was revealed to be dissociated from the West African outbreak and declared as a "normal" or "classic" Equatorial African outbreak – dangerous but not as severe.

Why did Ebola emerge in West Africa with such lethality?

1) At the heart of the West African outbreak was a new, virulent mutant virus of the already-potent Zaire Ebola virus.
2) Because past outbreaks occurred in Equatorial Africa, West Africa had no prior experience of dealing with the disease and was unprepared.
3) Poor health care infrastructure and weak public institutions rendered West Africa unable to cope with the initial rapidly expanding spread of the disease.
4) West Africa is experiencing urbanization and significant population migration across porous borders, which is creating new incubators for the disease and new channels for its speedy diffusion.
5) Traditional funeral and burial rituals mean that many locals carried dead bodies over distances, increasing the chances of infection and the radius of the disease.

(Continued)

Box 8.5 *(Continued)*

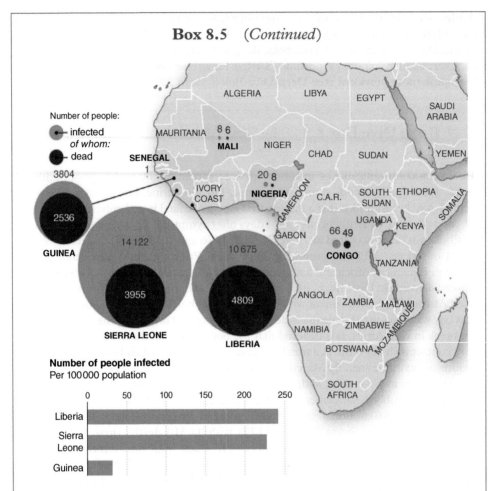

Map 8.2   The West Africa Ebola Outbreak, 2013–2016. Source: Redrawn from data published by the World Health Organisation - http://www.who.int/csr/disease/ebola/en/

6) Suspicion and fear of Western medical science and health care professionals (according to some locals, "invading villages wearing spacesuits!") resulted in a greater reliance on traditional medicine, which proved ineffective.

7) Because Ebola emerged in an area that has suffered from cognate outbreaks of cholera, Lassa fever, and Marburg disease, it took longer to identify that Ebola was responsible (and not these other diseases), and thus it took longer to trigger a response.

8) At the time there existed no vaccine for Ebola (but note that in December 2019, the U.S. Food and Drug Administration approved the world's first vaccine - Ervebo).

9) The virus spread over a wider geographical area than prior outbreaks (which tended to be more localized and clustered geographically), and this made it more difficult to target resources.

10) The disease behaved like "fire in a peat bog" – that is, it proved resistant to complete eradication in specific places, and after a period of subdual it often resurfaced.

No one country or international organization could solve the Ebola crisis alone. To address the outbreak, WHO worked with a number of agencies to implement a phased plan.

Phase 1 (August–December 2014) focused upon contact tracing, early reporting of symptoms, quarantining, adherence to recommended preventative and protective measures, and safe burials. Phase 2 (January–July 2015) embraced new developments in Ebola control, including researching vaccines, improving diagnostics, and fostering survivor counseling and care. Phase 3 (August 2015–midyear 2016) focused upon driving progress to ensure a "resilient zero" number of new cases.

This multi-agency response was critical in bringing the epidemic to an end. By June 2016 the Ebola outbreak was announced over across most of West Africa. Still, fresh outbreaks in 2018 and 2020 (especially in the Democratic Republic of Congo) suggest that the availability of a new vaccine notwithstanding, Ebola continues to be a live threat in the region.

in its tracks by Covid-19, the biggest global pandemic for a century. We will have much more to say about Covid-19 later in this book in our closing coda. Quite why these diseases appear to be (re)emerging is open to debate.

- Perhaps erroneous assumptions that medical science and public health have largely conquered infectious diseases have resulted in a degree of complacency.
- Perhaps pharmaceutical companies are concentrating most attention on degenerative diseases because these burden the more developed countries, where most profit is to be made.
- Perhaps through lack of ongoing investment and/or reluctance to use insecticides (because of the environmental harm they can do), campaigns to eradicate pathogens and their breeding grounds are losing the battle.
- Perhaps excessive usage of antibiotics is creating a new family of more potent "superbugs."
- Perhaps vaccination schemes have failed to inoculate populations at sufficiently frequent intervals.
- Perhaps rapid socioeconomic and socioecological changes in the developing world (forest clearance, mining, rapid urbanization, etc.) have released dormant viruses from the natural habitats they were trapped in and created a more hospitable environment for their incubation and proliferation.
- Perhaps climate change has altered disease regimes and enabled dangerous pathogens to incubate and spread more easily.

Whatever the cause, (re)emerging infectious diseases will make it difficult for poor countries to progress beyond the age of receding pandemics, whilst for rich countries these diseases may reverse progress and, in the worst-case scenario, move countries back into an "age of (re)emerging communicable disease."

The Demographic Transition Model and Fertility Decline

Explaining Fertility Decline

How industrialization and/or development and/or modernization and/or urbanization work to temper fertility is not self-evident; the mechanisms involved are far from obvious and vary from one country to another.

Welsh population geographer Huw Jones offers a useful framework for thinking about the factors that determine levels of fertility in any country (see Figure 8.7). According to Jones, the number of births that occur in any country is a reflection of three factors: the amount of sexual intercourse that occurs, the amount of intercourse that results in conception, and the amount of conceptions that result in actual live births. There exist a number of immediate or direct factors that determine all three; Jones calls these the proximate determinants of fertility. But there also exist a number of deep or ultimate or fundamental determinants of fertility; for Jones, these include socioeconomic, cultural, and environmental factors. These fundamental determinants of fertility work through the proximate determinants to shape fertility.

It is easy to see how changes in the proximate determinants of fertility might lead to lower levels of fertility. For example, if the average age of marriage was to increase, through social coercion or legal sanction, even allowing for sexual intercourse outside of marriage, it is likely that the level of sexual intercourse and the number of conceptions in a society would decrease. Moreover, if various forms of contraception were made afford- able and accessible in a country, it is likely that more people would use them and that fewer children would be born than otherwise would be the case. According to the United Nations, in 2020 over 225 million women in the Global South wished to use contracep- tion such as hormonal interventions (oral contraceptives [the "Pill"], injections [Depo- Provera], and implants), barrier methods (spermicides, male condom, female condom, diaphragm, and cervical cap), intrauterine devices (IUDs, or the Coil), permanent steri- lization (female tubal ligation and male vasectomy), emergency contraception, and abor- tion, and sought information provision (on practices of withdrawal, natural family planning, and abstinence), but lacked the capacity to access these resources.

But what about the fundamental determinants? In his book *Theory of Fertility Decline*, Australian demographer John C. Caldwell (1982) argues that households make rational choices about family size based upon prevailing economic conditions. Caldwell's theory centers upon the idea of intergenerational wealth flows, defined as net transfers of wealth from parents to children and children to parents that occur throughout a parent's

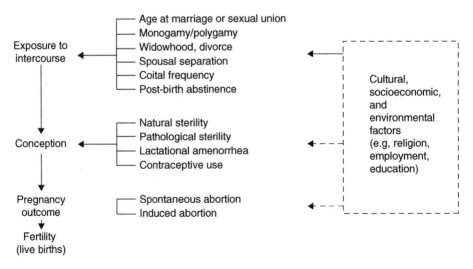

Figure 8.7   Huw Jones on the determinants of fertility. Source: Jones (1990). © 1990 SAGE Publications.

life course. Transfers can comprise money, goods, services, and guarantees. Caldwell argues that it is possible to discern two types of society: primitive or traditional societies where net wealth flows firmly to the advantage of parents, and modern developed societies where net wealth flows predominantly from parents to children. It is not surprising that fertility levels are higher in traditional societies; as they contribute to the wealth of the household, people choose to have as many children as possible, constrained only by biological limits. A culture of large family sizes emerges. Likewise, we should not be surprised that fertility is low in modern societies as people rationally choose to have fewer or no children, constrained now only by their psychological and emotional need to reproduce and raise children. A culture of reduced family sizes emerges. The shift from high to low fertility, then, is driven by the changing decisions made by households in the context of changes in the direction in which wealth flows between generations. But Caldwell's theory has been criticized: (i) for being too "economic" (cultural norms and values concerning family size have more complex origins); (ii) because it gives insufficient attention to the role of wealth transfers from parents to children in primitive societies, and from children to parents in modern societies; and (iii) because of the little empirical evidence Caldwell has presented to substantiate his case.

Other fundamental determinants that lead to lower levels of fertility include:

- **City living**: Rural populations tend to have large families as children can be a vital source of labor on farms. Mass migration to cities reduces the number of people working on the land and therefore the number of people who need to have large families for this reason.
- **Pensions and social insurance**: Increasingly, state organizations take care of people's needs, especially when they are old, unemployed, or sick. People no longer need to have children to help them in times of need.
- **Compulsory schooling**: Schooling reduces the economic exploitation of young children and introduces new costs for parents (school uniforms and books). Both induce smaller family sizes.
- **Education**: Education opens people's eyes to career opportunities and increases people's awareness of the existence and benefits of contraception.
- **Changing status of women in society**: Greater female participation in the labor market leads to changing views on the role of women in society and diminishes antiquated ideas which construe women as principally home makers.
- **Consumption**: As the consumption of luxury goods becomes an increasing priority, having fewer mouths to feed means families can dedicate more of the household budget to servicing wants rather than needs.
- **Secularism**: Some religions promote large family sizes and/or approaches to human sexuality that lead by default to larger family sizes. Modernization often leads to secularism. As religions ebb in their significance, secular attitudes to marriage, child bearing, contraception, and abortion emerge. People are better able to control the size of their families.

Policies for Lowering and Increasing Fertility Levels

As countries continue to move through the fertility transition, it would appear that two core challenges remain:

Fertility control projects: If we were to wait for fertility levels to fall as a natural response to development, we might be waiting a long time; in the interim, population growth might accelerate to unmanageable levels. Fertility levels most often fall only when governments intervene and convince, co-opt, and coerce populations into exercising fertility control. Here attention is paid to state engineering and manipulation of the determinants of fertility. The artificial repression of fertility through fertility control programs means that low birth rates will emerge even in poorer countries with large rural populations and pro-family cultures. But these programs come with significant moral conundrums and social and financial costs. Fast-working fertility control programs are especially difficult to introduce in democratic countries (which more readily embody the popular will of the people) and very poor countries (where socioeconomic and cultural factors often promote "naturally" high fertility levels).

Pro-natalist policies: Some countries that have moved through demographic transition now have below-replacement fertility levels and suffer from both a shrinking of their populations and an aging of their population profiles. Increasingly, a smaller working population is being asked to pay for the pensions of a larger pool of retirees. Pension arrangements made when population profiles were much more youthful no longer seem tenable. Moreover, with more people living longer, demand for health care is growing and pressure on health services is become especially acute. Faced with the problem of low birth rates, governments have generally responded in two ways. Pro-natalist population policies are policies designed to increase birth rates. These policies target both proximate determinants of fertility (for example, by banning abortions, making contraception illegal, tightening state laws on divorce, reducing the legal age of marriage, etc.) and fundamental determinants of fertility (provision of child care facilities, promoting gender equality, flexible working practices, tax allowances, generous child care allowances, promotion of virtues of marriage, housing subsidies, family-friendly service delivery, and placing heavy financial penalties on couples who choose to remain childless).

In addition to, or in place of, pro-natalist policies, some countries have sought to address the problems created by an aging society by: raising the age of retirement; placing a new focus on savings, private insurance, life assurance, and private pensions; promoting preventative health care, ensuring that the aging cohort enjoys a healthier old age; promoting selective immigration to plug skill shortages; encouraging flexible working arrangements that facilitate a return to work for stay-at-home parents; up-skilling the population so that the workforce is capable of generating more wealth per capita; and supporting and leveraging the invaluable work the elderly do in the care sector (for each other and for children and grandchildren).

Demographic Transition: The Case of China from 1949

The Communist Party came to power in China in 1949 and has ruled ever since. In 1949, it was possible to locate China in the second stage of demographic transition. Within only 30–40 years, it was to progress to Stage 4. The rapidity of China's passage through demographic transition – and, by implication, epidemiological and fertility transition – marks it as historically unique (Deep Dive Box 8.6). Why was China able to pass through demographic transition so rapidly?

From 1949 to the late 1970s, attitudes to fertility control in China changed as political power within the Chinese Communist Party oscillated between the radical faction (symbolized by the policies of Mao Tse-tung [1893–1976]) and the more moderate wing (symbolized by the policies of Deng Xiaoping [1904–1997]) (Plate 8.1). However, since

Deep Dive Box 8.6   China's Movement Through Demographic Transition, 1949–Present

Figure 8.8 provides an overview of key changes in CBRs, CDRs, and overall population size in China across the period 1949–2021. Key trends to note include:

The speed and scale of population growth

China's population grew from 547 million in 1949 to 662 million in 1960, 830 million in 1970, 987 million in 1980, 1.14 billion in 1990, 1.27 billion in 2000, 1.34 billion in 2010, and now stands at 1.44 billion today. The annual rate of growth of the Chinese population peaked in the 1960s – in 1966 it reached 2.8% – but this rate gradually tapered and in 2017 the population grew by only 0.6%.

The rapidity of falling mortality rates and improvements in life expectancy

The CDR in 1949 was 22.6 per 1000; by 1963 this figure had fallen to 12.7 per 1000, and by 1990 it had fallen again to 6.7 per 1000, the level at which it has effectively stabilized (in 2021, the figure is 7.1 per 1000). Infant mortality declined from 129 per 1000 births in 1950 to 35 per 1000 in 2000 and 10 per 1000 in 2020. Between 1949 and the early 1980s, the increase in life expectancy achieved by China was two to three times higher than the norm across the Global South. Life expectancy was 42.5 years in 1959; by 2000 this had increased to 70.6, and in 2020 it stood at 76.6. A child born today can expect to live more than three decades longer than forebears 50 years ago, a gain in life expectancy that current Global North countries took on average twice as long to achieve.

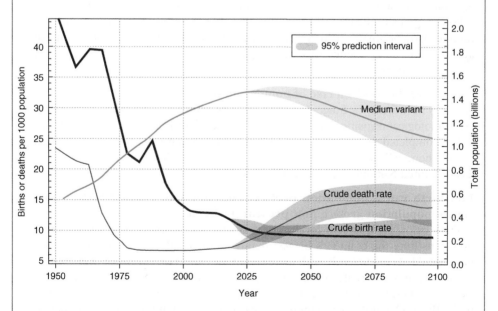

Figure 8.8   China's passage through demographic transition, 1949–2100. Source: Data from CSP, Chinese Statistical Yearbook.

(Continued)

Box 8.6   (Continued)

The changing structure of causes of mortality
Immediately prior to 1949, the principal killers in China included typhoid fever, bacillary dysentery, cholera, hookworm disease, childhood measles, smallpox, diphtheria, tuberculosis, malaria, schistosomiasis, tetanus, venereal diseases, and natural disasters such as floods and droughts. Whilst infectious disease accounted for the vast majority of deaths in 1949, by the mid-1970s, China had decisively moved through epidemiological transition and communicable disease then accounted for only 27.8% of deaths. Today, contagious diseases account for only 5.2% of deaths, whilst chronic diseases account for 77.1%. Strokes (cerebrovascular disease), coronary heart disease, lung disease, lung cancer, liver cancer, stomach cancer, road traffic accidents, hypertension, and diabetes now constitute the principal causes of mortality.

The rapidity of fertility decline
Measured in 1950 at 42.5 per 1000, as late as 1970 China's CBR still stood at 39.5 per 1000; by 1979, however, it had fallen to 17.8 per 1000, and today it is 11.9 per 1000. China is reaching the end of Stage 3 of demographic transition, and population growth is easing. The TFR, which stood at 6.11 in 1950, fell to 2.73 by 1990. Today it stands at 1.6 and whilst steadily recovering will rise to only 1.77 across this century to the year 2100.

The demographic disaster of 1958–1962 and its long shadow
Interrupting China's passage through epidemiological and fertility transitions is the disaster that unfolded in the period from 1958 to 1962. Here a national economic disaster and famine led to a significant spike in mortality, which was followed by a significant fall in fertility and then a rebound in fertility (a post-famine baby boom) to a historical all-time high. This demographic event cut across China's experience of demographic transition and helped to give it a unique character.

The end of the modern rise of population in China?
China is now firmly in Stage 4 of demographic transition. It has passed through epidemiological transition and is now in the age of delayed degenerative disease. It has completed fertility transition. China's population will stabilize in 2030 at 1.46 billion, will decline to 1.2 billion by 2075, and will fall to 1.064 billion by 2100 (United Nations Population Division 2020).

Low birth rates, population decline, and population aging rather than rate of population growth now present the biggest challenges
China's below-replacement-level fertility regime and aging population structure now constitute the country's biggest demographic challenge. Falling birth rates have unfolded concurrently with increases in life expectancy from 68.9 in 1990 to 76.6 today; moreover, life expectancy is projected to rise to 78.3 by 2050 and 87.6 by 2100. Whilst the median age in China in 1980 was 23, by 2030 it will have risen to 43. By 2050 half the population will be over 45, and a quarter of the population will be 65 or older. The absolute number of children and young workers will decline markedly. The total working-age population peaked in 2015 at about 1 billion (nearly 75% of the total population) and will fall to 870 million by 2050 (62%) and 610 million (56%) by 2100.

Plate 8.1   Mao Tse-tung and Deng Xiaoping. Source: World History Archive / Alamy Stock Photo.

the early 1980s, moderate rule has largely prevailed. Mao Tse-tung was an ardent critic of those who frette over the implications of China's large and growing population. Mao viewed people more as producers than as consumers; more people meant faster economic growth and development. In his view, those who feared that China's growing population would constrain economic prosperity were guilty of propagating false Western ideas. When Mao and the radicals were in power, fertility control programs were rejected. In contrast, Deng Xiaoping feared China's surging population greatly; people consumed more than they produced, and population growth would dilute the effects of economic development and undermine standards of living. When Deng and the moderates were in power, fertility control programs were promoted with increasing vigor (Figure 8.9).

Having brought the Communist Party to power in China in 1949, it is unsurprising that Mao Tse-tung and the radicals held the balance of power initially. Mao immediately improved the health of the nation by introducing law and order, increasing food supplies, and building health and medical facilities. In 1953, the first census under the Communist Party was collected. The results, published throughout 1954 and 1955, announced that, with a population of over 580 million, China was the most populous country on earth. Initially, this result was met with jubilation. Further reflection on China's CBR of 37 per 1000 and the challenges that a rapidly expanding population might bring, however, brought a more considered response. In the years 1956 and 1957, Deng Xiaoping and the moderates captured power, and under their rule initiatives were taken to curb fertility. Badly conceived, poorly financed, and weakly implemented, this first fertility control scheme was ineffectual. By late 1957, China's CBR still stood at 34 per 1000.

Communist Party Radical Wing	Communist Party Moderate Wing	Population Policies: Key Developments
• Mao Tse-tung **Paramount Leader 1949–1976** • Lin Biao **Vice** • **Chairman of the Communist Party of China 1966–1971**	• Zhou Enlai **Premier PRC 1949–1976** • Liu Shaoqi **Chairman PRC 1959–1968**	• 1949–1955 Radicals in power, Mao rejects Malthus • First National Population Census of PRC 1953 • 1956–1957 Moderates secure influence, first fertility control project attempted • 1958–1962 Mao's Great Leap Forward/Great Leap Backward • 1962–1965 Moderates control policy, second fertility control project attempted • Second National Population Census of PRC 1964
Hua Guofeng **Paramount Leader 1976–1978**		• 1966–1969 Mao's Cultural Revolution • 1970–1976 Mao steadily loses authority in the PRC and dies in 1976.
	• Deng Xiaoping **Paramount Leader 1978–1992** • Hu Yaobang **Chairman and General Secretary Communist Party 1982–1987** • Jiang Zemin **Paramount Leader 1992–2002** • Hu Jintao **Paramount Leader 2002–2012** • Xi Jinping **Paramount Leader 2012–Present**	• 1970–1979 Moderates seize power, third fertility control project attempted *Wan-Xi-Shao* • 1979–2015 Moderates (especially Deng) in control. Sweeping economic reforms introduced. Fourth fertility control project attempted: *One Child Policy* • Third National Population Census of PRC 1982 • Fourth National Population Census of PRC 1990 • Fifth National Population Census of PRC 2000 • 2015–2018 Fifth fertility control project attempted. Steady relaxation of the One-Child Policy and move to Two Child Policy • Sixth National Population Census of PRC 2010 • Seventh National Population Census of PRC 2020 • From 2018 - End of the One-Child Policy. End of fertility control in China and transition towards a pro-natalist future?

Figure 8.9   Politics, leaders, and population policies in China from 1949. Source: Compiled by the author.

The years from 1958 to 1962 mark a dark period in China's history. By 1958, Mao Tse-tung had regained supreme influence and, impatient with the progress the country was making, devised an ambitious plan to transition China from an underdeveloped society to a global superpower. According to the radicals, people were China's greatest asset and a dedicated focusing of the country's labor force would thrust China to the forefront. Mao's Great Leap Forward was to be accomplished by taking people off the land (reducing the number of farmers) and putting them to work in small rural industries whilst at the same time scaling up the productivity of those who continued to work in agriculture. The results were disastrous. Rural factories were primitive, and the products they churned out were scarcely fit for purpose. Meanwhile, with fewer workers and less dedicated land, agricultural output collapsed, and famine and starvation ensued. Mao's Great Leap Forward became China's Great Leap Backward.

For a long time concealed from the world by the Chinese Communist Party, the famine of 1958–1962 represented a calamitous moment in China's demographic history. Between 1958 and 1962, foodgrain production in China fell from 200 million tons (MT) to 143.5 MT. Given that 90% of the Chinese diet derived from food grain at that time and that an estimated 10 MT of foodgrain was required to feed 30 million people, the loss of 60 MT of food grain meant that, by definition, 180 million Chinese citizens were left in a state of chronic food insecurity. The famine resulted in an estimated 30 million excess deaths. Official figures suggest that CDRs rose to over 25 per 1000 during this period; unofficial estimates suggest death rates soared to as high as 43 per 1000. Meanwhile, as a result of its effects on marriage rates, male fertility, and female reproductive capacity, the Chinese famine proved itself to be a major contraceptive, and an irony of Mao's project was that CBRs fell to an artificial low in 1961 of 18 per 1000.

Quite why Mao thought it possible to redirect a significant percentage of the Chinese population and land surface from agriculture to industry whilst at the same time expanding agricultural output and exports remains a mystery. That he failed to avert the crisis once it commenced continues to be one of the great conundrums of history. Perhaps Mao placed unwarranted confidence and faith in the optimistic forecasts emanating from pioneering and experimental test farms he had established; perhaps his attention was distracted by the international conflicts he was waging with India and Taiwan; and, certainly, in 1958–1960 he was ill informed about the scale of the impending disaster by fearful and eager-to-please junior cadres. Mao himself was later to blame the Great Leap Backward on an unhappy coincidence of freak natural hazard events (flooding, drought, and earthquakes) that beset China at this time. Whilst undoubtedly an aggravating factor, natural hazards in fact are normally local in their effects. That famine was geographically distributed across the entire country would tend to support the view that it was politics and not nature that derailed Mao's great political experiment.

By 1962, Deng and the moderates had regained control of the Chinese Communist Party. Health further improved through better nutrition, hygiene, medical care, housing, education, vaccination, water supply, and sanitation. China's growing population was once again viewed as a problem, and fertility control was a policy priority. A second fertility control program was launched. This program was presented to the Chinese population as a health promotion campaign designed to improve maternal and infant health. The mortality of women and newborn children improved accordingly. On this occasion, the campaign was better conceived, furnished with more funds, and better rolled out across the country. Its central strategy was to make contraception more widely available and to increase its usage. Notwithstanding the better execution of the strategy, CBRs rose from 18 per 1000 in 1961 to 43 per 1000 in 1963. Unwittingly, Deng and the moderates had sought to reduce fertility in the midst of a post-famine baby boom; after the famine the physical reproductive capacity of both males and females improved, and marriages that had been postponed during the famine now took place.

In any event, by 1966 Mao Tse-tung and the radicals had regained supreme influence, and fertility control was no longer a political objective. Suspecting that China's commitment to communist ideals was faltering, Mao called for an intensive (re)education of the masses in the merits of Chinese communism. Mao gave license to a revolutionary group of younger communist activists – the Red Guard – to aggressively promote communist principles, and produced a *Little Red Book* outlining the basis of communist ideals. People suspected of Western leanings or liberal beliefs – including the educated classes – were to be "corrected"; in practice, this often meant victimization, exile, torture, and even murder. During this period preeminent attention was given to politics rather than population, but throughout, Mao's belief in the virtues of a large Chinese population was unwavering. Mao officially declared his Cultural Revolution ended in 1969, but the damage had been done; China was in chaos. Steadily, throughout the 1970s, the moderates sought to regain control of the country. With the death of Mao Tse-tung in 1976, the period of radical control over the Chinese Communist Party was effectively over, and from 1978 Deng and the moderates prevailed.

With the moderates on the ascendancy, throughout the 1970s China pursued a vigorous fertility control campaign. For the moderates, China's progress was being hindered by its rapid population growth; any advances that were being achieved were not translating into better standards of living when measured in per capita terms. Given the title *Wan, Xi, Shao* or "Later, Longer, Fewer," this campaign sought to encourage later marriage, longer spacing between births, and fewer births overall. Its central premise was increasing access to and usage of contraception (especial IUDs) and abortion. During this period, the power of the Chinese Communist Party to effect change at the grassroots level became evident. Initially, target family sizes were set at two for urban households and three for rural households; by 1977, however, a single target of two for all households was set. First through coercion and then through compulsion, fertility levels were dramatically reduced; CBRs came down from 36 per 1000 in 1968 to 17.8 per 1000 by 1979.

Notwithstanding the success of the *Wan, Xi, Shao* campaign, by the late 1970s and early 1980s, Deng and the moderates concluded that population growth remained a crippling problem for China and that it was essential that even more drastic measures be taken. Why did the moderates decide to embark upon an even stricter fertility control policy? Two reasons present themselves. First, China recognized that, in spite of the development of its economy from 1949 to 1979, the standard of living of the average person in the country had not improved. Food consumption per capita, literacy levels, and overcrowded housing all remained significant problems. Secondly, China discovered the power of computer forecasting and now, glimpsing the potential of its population to grow exponentially, became even more alarmed. Computer programmers and software engineers emerged as key lobbyists for more draconian fertility control policies.

China's response was the *One-Child Policy*. China calculated its optimum or ideal population size to be 750 million, and set out to restrict fertility so as to reach this target by 2080. China's solution was to introduce a one-child (per family) policy from 1980 to 2000, to gradually relax restrictions to reach a replacement-level 2.2-children (per family) policy from 2000 to 2020, and to maintain a 2.2-children policy from 2020 to 2080. To achieve this outcome, the Communist Party introduced a series of economic rewards and penalties. Those who stuck to the policy could expect an increased salary; better access to food, health care, and housing; and other types of prioritizations for their children. In support of the campaign, contraception was made available widely. Its usage was mandated. Once again, the Communist Party mobilized its nationwide administrative machinery to effect change at the grassroots level. Strong moral pressure and coercion were brought to bear on those who resisted or flouted the policy, and through "enforced submission," "correction," and "re-education" many opponents were forced to conform.

China evidently struggled to implement its One-Child Policy effectively. Throughout the 1980s, birth rates rose to as high as 23 per 1000 as a consequence of the so-called echo effect of the post-famine (after the Great Leap Backward) baby boom of the early 1960s – as that generation entered adulthood, they too became parents. A surge in fertility resulted. Moreover, China was unable to implement the policy nationally and presided over a series of regional policies that varied between and within provinces. One-child policies did exist in some (mainly urban and coastal) regions in China, but two-child policies, three-child policies, one-son policies, and no policies at all prevailed in other (more rural interior) regions. Notwithstanding its variable application, however, it is clear that China's coercive fertility control programs have exacted an enormous toll on Chinese society. For cultural and economic reasons, many families preferred a son to a daughter, and because of selective abortion and the murder of infant girls (hideous female infanticide), significant gender imbalances now present themselves. China will suffer into the future from old-age dependency with a small cohort of workers looking after an increasingly bulging retired population (Figure 8.10). China will also have fewer children going through its schooling system, meaning the redundancy of a proportion of trained teachers. Meanwhile, the long-term sociological implications of rearing a generation of single children remain to be understood.

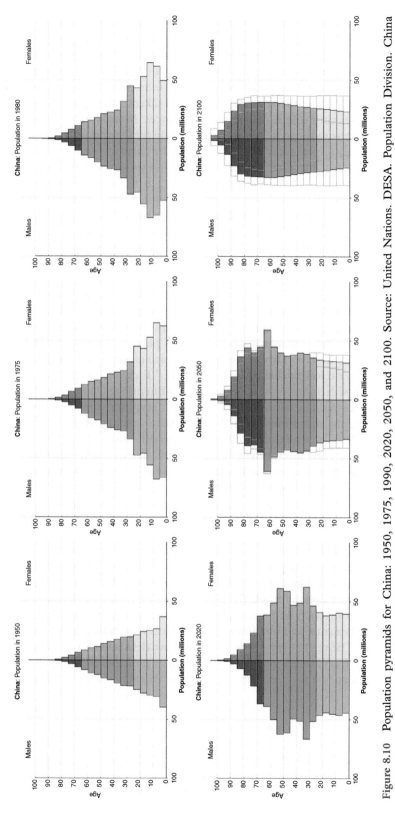

Figure 8.10 Population pyramids for China: 1950, 1975, 1990, 2020, 2050, and 2100. Source: United Nations. DESA. Population Division. China Population Pyramids 2019. CC BY 3.0 IGO.

The One-Child Policy also came under sharp attack from human rights activists and pro-life lobby groups in the West. China has been accused of breaching human rights by dictating people's family size, often in a coercive or compulsory way including mass state-sponsored abortion and enforced contraception. Other commentators note that whilst human rights abuses have occurred, perhaps some Western criticisms fail to do justice to the complexity of the approach China has taken. China's approach has been flexible, and policies vary within and between provinces. Moreover, whilst the Chinese Communist Party has in the past attempted to restrain fertility through top-down party rule and the brutal enforcement of family planning (the "hard Leninist method"), across the past decade a softer approach has emerged that has sought to encourage citizens to restrict family sizes out of volition so that they might benefit from China's emergence as a global economic powerhouse (a "soft neoliberal approach"). Today the emphasis is upon quality as much as quantity. The idea of self-restraint for personal gain is being promoted alongside the need to make mandatory sacrifices for the good of the collective or for the generations to come. Carrots, it is assumed, work better than sticks.

Meanwhile, economic reforms introduced by Deng deepened, and China pioneered a hybrid capitalist/communist and hyperglobalized economy that proved capable of exceptional growth rates. Health improvements were further consolidated because of the country's impressive economic growth, income and wealth accrual, and welfare improvements, especially in urbanized southern and eastern coastal regions. More wealth, better diets, and stronger immune systems played a key role in the ongoing demise of infectious disease. Diseases of poverty have given way to diseases of affluence, and China is no longer the "sick man of Asia." Today, China faces health challenges associated with an aging and increasingly affluent population, combined with problems caused by rapid urbanization (not least air pollution and accidents), emerging and re-emerging infectious diseases (especially sexually transmitted diseases [STDs]), and widening inequalities in health and health care (with poorer communities and those in the rural interior lagging in terms of health and health care).

Having pursued severe fertility control initiatives since 1979, in 2015 China decided to revisit its approach to family size. Motivating this return to first principles was a recognition that the country was now facing the problem of low birth rates, a falling proportion of the population in the working-age bracket, and old-age dependency. China now seems more concerned with the negative consequences of a contracting population. In response to these problems, in late 2013 China finally decided to relax its One-Child Policy. Children whose parents adhered to the policy were now to be permitted to have two children. And in 2015 China finally disbanded the policy completely and moved to a two-child policy for all. By 2025 or sooner, China could remove fertility control measures completely and might even shift to become a pro-natalist country. But the country's fundamental and proximate determinants of fertility are not supportive of large family sizes. The rising Chinese middle class have developed favorable attitudes to small family sizes. It will take a generation for any new policy to address concerns such as growing dependency ratios, pressures in the labor market, and population aging.

Conclusion

The rise of the West from 1500 began a process of world population growth without parallel in human history. From its epicenter in Europe, this process has steadily fanned out across the globe, and there exists no region untouched by it. In 1750, the global population stood at 741 million; today it exceeds 7.7 billion. The modern rise in world population will end in Africa around 2100. By then the size and geography of the human population will be dramatically different from that which prevailed in 1750. The Demographic Transition Model – and attendant ideas such as epidemiological transition and fertility transition – are immensely useful for making sense of this world historical burst in population growth. Whether the world will be able to live indefinitely with a population of circa 10 billion by 2100 remains a key question for the future.

Checklist of Key Ideas

- The Demographic Transition Model has its origins in the work of US demographer Frank W. Notestein and his fear of the impact of global population growth, not least its ramifications for the United States.
- The Demographic Transition Model suggests that as countries industrialize and develop, so too their populations pass through a number of stages; some models recognize only four stages, others five. Recently, British demographer Tim Dyson has queried the need for countries to industrialize and develop if they are to pass through demographic transition.
- French demographer Jean-Claude Chesnais has provided a typology of different passages through demographic transition, noting variations in start dates, durations from start to finish, and peak rates of population growth. Beginning in Europe in the eighteenth century, the modern rise in world population continues and will abate only around the year 2100. As this century wears on, Africa will replace Asia as the epicenter of world population growth.
- Egyptian physician Abdel Omran's concept of epidemiological transition provides a useful lens through which to examine declining mortality rates. A combination of economic development, improvements in standards of living and diet, medical advances, and public health campaigns have pushed countries through epidemiological transition.
- Welsh population geographer Huw Jones offers a useful framework for thinking about the determinants of fertility. Fertility levels fall as a lag effect of socioeconomic development and cultural change, as these fundamental determinants drive down fertility by recalibrating the workings of a number of proximate determinants of fertility.
- China's progression through demographic transition from 1949 provides a good example of the ways in which demographic processes are entwined with social, economic, cultural, political, and environmental processes.

Chapter Essay Questions

1) Describe, explain, and comment upon the utility of the Demographic Transition Model when making sense of the rise in world population from 1750.
2) Describe and explain the processes and policies that have driven global trends in *either* mortality *or* fertility since 1750.
3) Outline the key challenges the world will face as the demographic transition runs its course to its final conclusion across the remainder of this century.
4) Describe and explain China's passage through demographic transition from 1949.

References and Guidance for Further Reading

Important work on and by some foundational thinkers in population geography are:
Glenn T. Trewartha
Trewartha, G.T. (1953). A case for population geography. *Annals of the Association of American Geographers* 43: 71–97.
Trewartha, G.T. (ed.) (1972). *The Less Developed Realm: A Geography of Its Population*. New York: Wiley.
Trewartha, G.T. (1978). *The More Developed Realm: A Geography of Its Population*. Oxford: Pergamon.
Graham, E. (2004). The past, present and future of population geography: reflections on Glenn Trewartha's address fifty years on. *Population, Space and Place* 10: 289–294.

John Innes Clarke
Clarke, J. (1965). *Population Geography*. Oxford: Pergamon.
Findlay, A.M. and Evans, I. (2018). Population geography, scale, environment, and sex-ratio imbalances: an appreciation of the contribution to population geography of Professor John Clarke (1929–2018). *Population, Space and Place* 24: e22207.

Frank W. Notestein
Notestein, F.W. (1945). Population: the long view. In: *Food for the World* (ed. T.W. Schultz), 36–57. Chicago: University of Chicago Press.
Notestein, F.W. (1964/1983). Frank Notestein on population growth and economic development. *Population and Development Review* 9: 345–360.
Coale, A. (1983). Frank W. Notestein, 1902–1983. *Population Index* 49: 3–12.

Good general introductions to population geography include:
Bailey, A. (2005). *Making Population Geography*. London: Routledge.
Barcus, H.R. and Halfacre, K. (2017). *An Introduction to Population Geographies*. London: Routledge.
Dorling, D. (2013). *Population 10 Billion*. London: Constable.
Findlay, A. and Findlay, A. (2012). *Population and Development in the Third World*. London: Routledge.
Gould, W.T. (2015). *Population and Development*. London: Routledge.
Hassan, M.I. (2020). *Population Geography: A systematic exposition*. Delhi: Routledge.
Holdsworth, C., Finney, N., Marshall, A., and Norman, P. (2013). *Population and Society*. London: Sage.
Jones, H. (1990). *Population Geography*, 2e. London: Sage.
Livi-Bacci, M. (2017). *A Concise History of World Population*, 6e. Oxford: Wiley Blackwell.
Newbold, K.B. (2021). *Population Geography: Tools and Issues*, 4e. Toronto: Rowman & Littlefield Publishers.
Poston, D.L. Jr. (ed.) (2019). *Handbook of Population*. Amsterdam: Springer.
Weeks, J.R. (2020). *Population: An Introduction to Concepts and Issues*. London: Cengage Learning.
Woods, R. (1982). *Theoretical Population Geography*. London: Longman.

For an overview of demographic transition theory, see:
Bongaarts, J. (2009). Human population growth and the demographic transition. *Philosophical Transactions of the Royal Society B: Biological Sciences* 364 (1532): 2985–2990.
Bongaarts, J. (1978). A framework for analyzing the proximate determinants of fertility. *Population and Development Review* 6: 105–132.
Caldwell, J.C. (1976). Toward a restatement of demographic transition theory. *Population and Development Review* 4: 321–366.
Chesnais, J.-C. (1992). *The Demographic Transition: Patterns, Stages and Economic Implications: A Longitudinal Study of Sixty-Seven Countries Covering the Period 1720–1984*. London: Clarendon Press.

Coale, A.J. and Hoover, E.M. (2015). *Population Growth and Economic Development*. Princeton, NJ: Princeton University Press.

Guilmoto, C.Z. and Jones, G.W. (eds.) (2016). *Contemporary Demographic Transformations in China, India and Indonesia*. Dordrecht, the Netherlands: Springer.

Kirk, D. (1996). Demographic transition theory. *Population Studies* 50: 361–387.

The essence of Tim Dyson's work on demographic transition can be found in:

Dyson, T. (2010). *Population and Development: The Demographic Transition*. London: Zed Books.

Dyson, T. (2013). On demographic and democratic transitions. *Population and Development Review* 38: 83–102.

Dyson, T. (2018). *A Population History of India: From the First Modern People to the Present Day*. Oxford: Oxford University Press.

Up-to-date information on population trends across the world can be found in:

United Nations Population Division (2020). *World Populations Prospects: 2019*. New York: UN Population Division.

For an overview of epidemiological transition theory (including foundational works), see:

Gwatkin, D.R. (1980). Indications of change in developing country mortality trends: the end of an era? *Population and Development Review* 8: 615–644.

Gwatkin, D.R. and Guillot, M. (2000). *The Burden of Disease Among the Global Poor: Current Situation, Future Trends, and Implications for Strategy*. Washington, DC: World Bank.

Gribble, J.N. and Preston, S.H. (eds.) (1993). *The Epidemiological Transition: Policy and Planning Implications for Developing Countries*. Washington, DC: National Academy Press.

Gatrell, A.C. and Elliott, S.J. (2015). *Geographies of Health: An Introduction*. Oxford: Wiley.

Olshanki, S.F. and Ault, A.B. (1986). The fourth age of the epidemiological transition: the age of delayed degenerative transition. *The Millbank Quarterly* 64: 509–538.

Omran, A.R. (1971). The epidemiological transition: a theory of the epidemiology of population change. *The Milbank Quarterly* 49: 509–538.

Mercer, A. (2014). *Infections, Chronic Disease, and the Epidemiological Transition: A New Perspective*. Rochester, NY: University of Rochester Press.

McKeown, T. (1976). *The Modern Rise in Population*. London: Edward Arnold.

Preston, S.H. (1975). The changing relation between mortality and level of economic development. *Population Studies* 29: 231–248.

Santosa, A., Wall, A., Fottrell, E. et al. (2014). The development and experience of epidemiological transition theory over four decades: a systematic review. *Global Health Action* 7 (1): 23574.

The Disability-Adjusted Life Years (DALYs) measure was first outlined in:

The World Bank (1993). *World Development Report 1993: Investing in Health*. Washington, DC: World Bank.

Up-to-date data on DALYs can be obtained in:

GBD (2018). Disease and Injury Incidence and Prevalence Collaborators. Global, regional, and national incidence, prevalence, and years lived with disability for 354 diseases and injuries for 195 countries and territories, 1990–2017: a systematic analysis for the Global Burden of Disease Study. *Lancet* 392 (10159): 1789–1858.

Institute for Health Metrics and Evaluation (IHME) (2020). *The Global Burden of Disease: Generating Evidence, Guiding Policy*. Seattle: IHME.

A good critical introduction to DALYs is:

Laurie, E.W. (2014). Who lives, who dies, who cares? Valuing life through the disability-adjusted life year measurement. *Transactions of the Institute of British Geographers* 40: 75–87.

Scholarship on the 2013–2016 West African Ebola outbreak includes:
Crawford, D.H. (2016). *Ebola: Profile of a Killer Virus*. Oxford: Oxford University Press.
Kaner, J. and Schaack, S. (2016). Understanding Ebola: the 2014 epidemic. *Global Health* 12: 53.
Richards, P. (2016). *Ebola: How a People's Science Helped End an Epidemic*. London: Zed Books.
Wilson, R. (2020). *Ebola and the Global Scramble to Prevent the Next Killer Outbreak*. Washington, DC: Brookings Institute.

For an overview of the fertility transition, see:
Caldwell, J.C. (1982). *Theory of Fertility Decline*. London: Academic Press.
Day, L. (1995). *The Future of Low Birth-Rate Populations*. London: Routledge.
Demeny, P. (1986). Pronatalist policies in low-fertility countries: patterns, performance, and prospects. *Population and Development Review* 12: 335–358.
Demeny, P. (2015). Sub-replacement fertility in national populations: can it be raised? *Population Studies* 69: S77–S85.
Rindfuss, R.R., Choe, M.K., Kravdal, O. et al. (2016). *Low Fertility, Institutions, and Their Policies*. Cham, Switzerland: Springer.
Timæus, I.M. and Moultrie, T.A. (2020). Pathways to low fertility: 50 years of limitation, curtailment and postponement of childbearing. *Demography* 57: 267–296.
UN Population Division (2020). *World Contraceptive Use 2020*. New York: United Nations.

Africa's fertility transition is explored in:
Bongaarts, J. (2017). Africa's unique fertility transition. *Population and Development Review* 43: 39–58.
Canning, D., Raja, S., and Yazbeck, A.S. (eds.) (2015). *Africa's Demographic Transition: Dividend or Disaster?* Washington, DC: The World Bank.
Casterline, J.B. (2017). Prospects for fertility decline in Africa. *Population and Development Review* 43: 3–18.

China's population trends have been the subject of investigation in:
Cai, F. (2020). *Demographic Perspective of China's Economic Development*. London: Taylor & Francis.
Greenhalgh, S. (2010). *Cultivating Global Citizens: Population in the Rise of China*. Cambridge, MA: Harvard University Press.
Greenhalgh, S. and Winckler, E.A. (2005). *Governing China's Population: From Leninist to Neoliberal Biopolitics*. Stanford, CA: Stanford University Press.
Riley, N.E. (2016). *Population in China*. New York: Wiley.
Zhigang, G., Feng, W., and Yong, C. (2017). *China's Low Birth Rate and the Development of Population*. London: Routledge.
Zhou, M., Wang, H., Zeng, X. et al. (2017). Mortality, morbidity, and risk factors in China and its provinces, 1990–2017: a systematic analysis for the Global Burden of Disease Study 2019. *Lancet* 394 (10204): 1145–1158.

Up-to-date data on China's population can be sourced from:
Chinese Statistical Press (2020). *Chinese Statistical Yearbook*. Beijing: Chinese Statistical Press.

Website Support Material

A range of useful resources to support your reading of this chapter are available from the Wiley Human Geography: An Essential Introduction Companion Site http://www.wiley.com/go/boyle.

Chapter 9

A Planet in Distress: Humanity's War on the Earth

Chapter Table of Contents

Chapter Learning Objectives

By the end of the chapter you should be able to:
- Outline and comment upon the future that neo-Malthusians (population pessimists) forecast if population and economic growth and environmental degradation are not arrested and reversed. Define and comment upon the concept of the Anthropocene.

Human Geography: An Essential Introduction, Second Edition. Mark Boyle.
© 2021 John Wiley & Sons Ltd. Published 2021 by John Wiley & Sons Ltd.
Companion website: www.wiley.com/go/boyle

- Describe and comment upon cornucopian perspectives (those of population optimists), which place confidence in the ability of human beings to continually extend the carrying capacity of the earth.
- Describe and comment upon the views of political ecologists, including Marxist thinkers, who believe that institutions, politics, and poverty, and not population or economic growth *per se*, lie at the heart of the world's principal environmental problems. Define and comment upon the concept of the Capitalocene.
- With reference to the problems presented by global warming, biodiversity loss, poor air quality, growing waste, and water quality and insecurity, document why many believe that we are on the brink of a global climate and ecological emergency.
- Describe and comment upon the ways in which cities are responding to the global climate and ecology emergency.
- Explain how optimists, pessimists, and political ecologists interpret the causes, implications, and solutions to the global climate and ecological emergency. Comment on what a Green New Deal might look like.

Introduction

In time, 2019 may well come to be known as the year when the world – or at least (with some notable exceptions) its political leaders – woke up to the full extent of the risks posed by the global climate and ecological emergency. According to British environmental and political activist George Monbiot, for too long the fossil fuel industry has enforced a "pollutocrats charter" and legalized the crime of "ecocide": the conscious destruction of the earth's natural environment. In fact, since the Industrial Revolution, human interference in the natural environment has grown to the extent that human beings have become "geological agents," etching onto the earth a stratigraphic record many times more impactful than any other species. This is the age of the Anthropocene – or better still, given the political-economic model that has brought us to this point, the Capitalocene. It is now difficult to identify any remaining pristine or first nature; there exist only human-modified natures that are volatile, unstable, and unpredictable. We have breached some, and risk breaching further, crucial life-sustaining planetary boundaries. A historical and for some (especially the world's poor) existential crisis beckons.

The purpose of this chapter is to examine humanity's growing war on planet earth: how we have reached this point, what the challenges are, what we are doing to confront them, and what needs to happen next if we are to accelerate and scale, as we must, our efforts to conserve a planet that is increasingly in distress.

Perspectives on Humanity's War on the Earth

The Pessimists: Rediscovering Malthus in the Age of the Anthropocene

Given that the modern rise in world population first began in England and Wales in 1750, it is perhaps not altogether surprising that concerns over the sustainability of population growth first arose there too. In 1798, the English Anglican curate, demographer, and economist Thomas R. Malthus published a book entitled *An Essay on the Principle of Population as It Affects the Future Improvement of Society*, in which he developed the view that although food production is capable of expanding at an arithmetic rate, population growth tends to occur at a geometric rate. The result is that, if left unchecked, in time populations will rise to levels that the planet is unable to support. Malthus's view was that, unless populations could exercise sexual restraint, population growth would inevitably lead to vice (which for Malthus meant "immoral" forms of birth control) and/or misery (war, famine, and disease).

When Malthus wrote his essay, the population of England and Wales was about 8.5 million and the population of the planet was around 900 million; the present population of each is 56.1 million and over 7.7 billion, respectively. Although the earth is not without its problems, it has enjoyed 250 years of population growth without mass starvation, economic collapse, or a deterioration of human welfare. There might be "vice" and "misery," but it is not on the scale imagined by Malthus, and it has not stopped population from surging. Evidently, even though he was writing at the start of the Industrial Revolution in England and Wales, Malthus failed to appreciate that through technology humankind might be able to avert crises, innovate itself out of trouble, and keep one step ahead of scarcity and environmental blight.

But will Malthus be proved correct in the long run? Has technology merely delayed crisis and not eradicated it? Are there limits to technological fixes?

From the 1950s onward, there have emerged a number of population pessimists – or neo-Malthusians – who believe that it is indeed time to retrieve Malthus from the dustbin of history and consider his warnings anew. Neo-Malthusians believe that people consume more resources than they produce and therefore that, if left unchecked, over time population growth will eventually lead to a depletion of the earth's resources, will breach the earth's carrying capacity, and will lead to a collapse of civilization as we know it today (see also Deep Dive Box 9.1). The result will be the breakdown of society, social and political anarchy, and perhaps even mass famine and warfare waged to secure the few scarce resources that continue to exist.

In his 1956 address to the American Petroleum Institute in Texas, entitled "Nuclear Energy and Fossil Fuels," petroleum geologist and geophysicist Marion King Hubbert advanced the theory of "peak oil," forecasting a threshold tipping point at which the rate of extraction of oil would decline. US marine biologist and conservationist Rachel Carson's book *Silent Spring*, published in 1962, marked another turning point in the rehabilitation of Malthus. Accusing the chemical industry of spreading disinformation, Carson contended that synthetic insecticides such as dichlorodiphenyltrichloroethane (DDT) were poisoning the environment, leading to the indiscriminate death of animals, the creation of a new breed of hardier super-pests, and an increased risk of ecosystem colonization by hostile invasive species. In 1968, US biologist Paul R. Ehrlich and his wife Anne H. Ehrlich (who was never credited) published a book entitled *The Population Bomb* (Ehrlich 1968). Later, in 1990, they followed up with a further book, *The Population Explosion* (Ehrlich and Ehrlich 1990). *The Population Bomb* prophesied that due to rapid population growth, the world stood on the brink of a great economic, environmental, and humanitarian disaster. Population growth was leading to poverty, and imminently the "death rate solution": mass starvation, the mining of scarce environmental resources to exhaustion, and a dying planet.

Ideas such as "peak oil," "silent spring," and the "population bomb" proved to be midwife to an influential think tank that has championed the cause of Malthus across the past forty years and prophesied doom if human consumption of the earth's resources is not restrained. In 1968, Italian industrialist Aurelio Peccei and Scottish scientist Alexander King convened a unique brainstorming workshop in Rome to consider the question: is it possible that relentless population and economic growth and consumption will eventually

Deep Dive Box 9.1   In Search of the Earth's Carrying Capacity: The Thoughts of US Population Biologist Joel E. Cohen

Is it possible to reach scientific consensus on the maximum number of people the earth can sustain?

In his 1995 book *How Many People Can the Earth Support?* US population biologist Joel E. Cohen collected and analyzed over 65 estimates of earth's carrying capacity that have been advanced since Dutchman Antoni van Leeuwenhoek first calculated the figure at 13.4 billion people in 1769. Estimates made in the past 50 years alone range from less than a billion to over 1000 billion, with a median low estimate of 7.7 billion and a median high estimate of 12 billion. Confusion has not reduced through time; as late as 1994, five estimates were produced within the space of a year that ranged from less than 3 billion to over 44 billion people.

Why so much variation and uncertainty? According to Cohen, all estimates of the earth's carrying capacity explicitly or implicitly are calculated on the basis of:

- An assumed level of material welfare (standard of living)
- An assumed distribution of material welfare (preparedness to accept the coexistence of degrees of poverty and affluence)
- An assumed technology (capacity to identify and harness resources)
- An assumed set of domestic and international political institutions
- An assumed set of domestic and international economic arrangements
- An assumed set of domestic and international demographic arrangements (population distribution and migration)
- An assumed biogeochemical context (resource base)
- An assumed degree of variability of material welfare needs (degree of tolerance of seasonal variations)
- An assumed set of risks from natural hazards and catastrophic events
- An assumed time span over which a population is to be sustained
- An assumed set of tastes, values, and fashions

Because these assumptions require judgments of a political nature to be made, it is not surprising that different forecasters calibrate them in different ways. Cohen's work leads to the conclusion that the search for scientific consensus on the earth's carrying capacity is a futile one; ultimately the question is a political one that is open only for debate, not resolution.

Cohen himself refuses to put an exact number on the earth's carrying capacity. Nevertheless, he does note that if the United Nations' median high and low estimates of peak world population (in 1995, 7 billion and 12.5 billion) set boundaries that are in any way meaningful, humankind might indeed be venturing into dangerous territory.

Cohen proposes three courses of action: the "bigger pie" solution argues that technology should be used to maximize the production of those resources required to support the human race, the "fewer forks" solution advocates fertility control and population control, and the "better manners" solution advocates the creation of societies that preside over resources and their distribution with greater reverence.

Cohen argues that universal provision of both primary and secondary education is the key to all three; better minds create better technology, educated women better understand the gains to be accrued by regulating their fertility and controlling the size of their family, and educated citizens are more able to call their leaders to account and lobby for better governance. More recently, Cohen has placed attention on the plight of newborn infants whose intellectual development is impaired from the very start of their lives by starvation and malnourishment. For better schooling to be effective, learners first need to present themselves as fit and able.

deplete the earth's finite natural resources to levels that threaten the future of humankind? Following the workshop, an independent think tank was born, the Club of Rome, comprising concerned world citizens with professional expertise and influence from all corners of the earth, dedicated toward building long-term solutions to the global challenges that confront humankind.

In 1972, at the behest of the Club of Rome, scientists from the Massachusetts Institute of Technology (MIT) published a book entitled *The Limits to Growth* (Meadows et al. 1972). Selling 12 million copies and published in 37 languages, this book both caught the attention of the world and defined the environmental agenda for a generation.

The Limits to Growth placed under scrutiny interactions that occur between five key variables: population growth and economic growth, which place pressure on the earth's ecosystems; food and natural resources, which are the key "inputs" humankind needs to survive; and environmental pollution, which threatens the sustainability of ecosystems in the long run. *The Limits to Growth* devised a computer model (World3) to create scenarios of possible futures for humankind, dependent upon the behavior of these five variables. A total of 12 scenarios were fashioned (charting trends between 1900 and 2100). Limits to growth arise when the rate at which nonrenewable resources are extracted from the earth leads to an exhaustion of those resources, and when the earth's tolerance of and capacity to absorb the pollutants deposited when these resources are used are exceeded. The earth can be said to be in "overshoot" when critical thresholds have been crossed. If not handled properly, overshoot can lead to societal collapse.

The Limits to Growth proposed that although further uncontrolled growth would result in critical environmental thresholds being breached in the twenty-first century and a derailment of growth, with some restraint and within certain limits further growth was possible. But two decades later, in 1992, an updated book entitled *Beyond the Limits* was published in which it was argued that, due to inaction, growth had overshot in a number of critical areas, leading to environmental stresses (Meadows et al. 1992). Remedial action was now required to reverse the damage that had been done and to put society back on a sustainable footing. Further pessimism pervaded the 2004 *Limits to Growth: The 30-Year Update*, which argued that society was now in a dangerous state of overshoot (Meadows et al. 2004). Many of the predictions of 1972 were now coming true. The opportunity to correct problems had been squandered, and nothing less than a revolutionary transformation to sustainable growth was required.

In 2012, at the request of the Club of Rome, Norwegian management scientist Jørgen Randers provided a forecast of what the world might look like in 2052. Randers's book, *2052: A Global Forecast for the Next Forty Years* (Randers 2012), argued that although food shortages and resource constraints will continue to be a problem, especially for the world's poor, it will be pollution, and in particular climate change and global warming, that will lead to a painful collapse of the entire global system in the second half of this century. Randers argued that in spite of hopes to the contrary, there will be no reduction in the usage of fossil fuels and carbon emissions in the foreseeable future, and climate change and global warming will emerge as significant burdens. The global temperature will rise by 2 °C by 2050 (from the pre-industrial mean), peaking at 2.8 °C in 2080. This peak will be sufficient to create "runaway global warming," which in turn will adversely affect global human society through sea-level rise, desertification, wildfires, water shortage, crop failure, extreme weather, disease, climate refugees, and increased risk of wars and conflicts.

Believing that the impact of the human species on the planetary system has reached a tipping point, in the 1980s, Dutch Nobel Prize–winning atmospheric chemist Paul Crutzen decreed that we are now entering a new geological time period, the Anthropocene. The Anthropocene is the "age of man" [*sic*]. According to Crutzen, from the 1800s onward, and in particular with the invention, in Scotland by James Watt, of steam power and the European Industrial Revolution that followed thereafter, human transformation of planetary ecosystems has been so consequential that we are now witnessing the birth of a new geological epoch. Over the last 500 000 years, the earth's natural system – what British environmentalist James Lovelock once vividly referred to as Gaia – has functioned within key ecological parameters. US chemist Will Steffen and colleagues argue that in this age of the Anthropocene, four out of nine of these "planetary boundaries" (climate change, biosphere integrity, land-system change, and biochemical flows) have now been pushed significantly beyond their range of natural variability, driving the earth system into a new "non-analogue state" (Steffen et al. 2015) (See also Deep Dive Box 9.2.) Today it is impossible to find any remaining pristine or first nature; there exists only human-modified or "cyborg" or "Frankenstein" natures that are without historical precedence.

Deep Dive Box 9.2   Doughnut Economics and Planetary Boundaries

In *Doughnut Economics: Seven Ways to Think Like a 21st Century Economist*, British economist Kate Raworth (2017) argues that the prevailing political-economic model is ill-placed to tackle the scale of the challenge that now presents. This model overlooks the ecological damage it is doing, fails to reward parenting and unpaid work, and produces inequality. For Raworth, Gross Domestic Product (GDP) growth is a flawed ambition; there is a need to measure human flourishing – or what Hannah Arendt once referred to as human "natality" – using alternative measures of well-being and prosperity. To move toward a more sustainable and inclusive world, Raworth proposes a doughnut model designed to protect key social foundations without breaching the planet's ecological ceiling (Figure 9.1). Humanity requires a basic minimum quantity of resources to meet its social foundations, and provided it conserves those resources, it can thrive. Around the "doughnut" are nine planetary boundaries, which delimit ecological ceilings: too much resource extraction and pollution will diminish the very ecosystems we need to thrive. At that point, the earth may not be able to sustain essential social foundations. For Raworth, the boundary limits for climate change, biodiversity loss, land conversion, and nitrogen and phosphorus loading have already been breached.

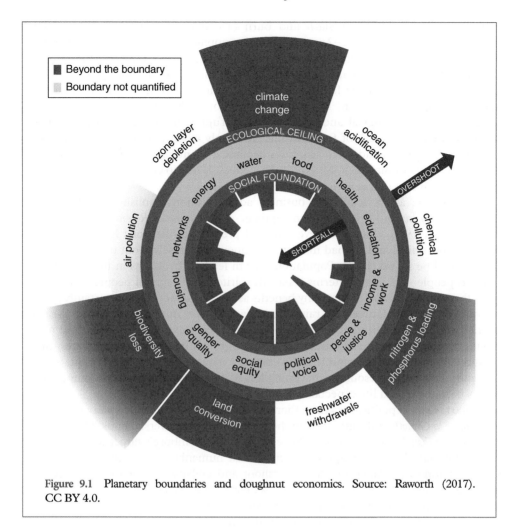

Figure 9.1 Planetary boundaries and doughnut economics. Source: Raworth (2017). CC BY 4.0.

The Optimists: Cornucopians and the Age of Green Technology and Clean Growth

Cornucopians, or population optimists, have taken issue with the claim that humanity is approaching the limits of the earth's carrying capacity and is about to overshoot and collapse. Cornucopians believe that human beings produce more than they consume. To the extent that population growth increases global consumption and exerts pressures on natural resources, human beings can always innovate themselves out of trouble by discovering new reserves of nonrenewable resources and harnessing renewable resources to a greater extent than they have hitherto. Moreover, insofar as population growth and increases in global consumption leave in their wake more waste and pollution, human beings can create new technologies to ameliorate the damage being done.

In 1965, Danish economist Ester Boserup published a short book entitled *The Conditions of Agricultural Growth: The Economics of Agrarian Change Under Population Pressure* (Boserup 1965). Boserup's focus was the development of agriculture, particularly in Southeast Asia. Boserup noted the steady intensification of agriculture that has occurred throughout human history. Five phases in this intensification process were recognized according to the length of fallow periods between periods of cultivation: forest fallow, or slash and burn (15–20 years of fallow); bush fallow (6–10 years); short fallow (1–2 years); annual cropping (a few months); and multi-cropping (no fallow). As countries moved from one phase to another, they had to invent and apply new technologies and reform their agricultural systems. More importantly, because of the law of diminishing returns, passage from one phase to another required societies to apply more effort for each unit of additional output secured. It is both illogical and prohibitively disruptive for societies to make these changes in circumstances of food surplus. It is only when population growth takes societies to the brink, and when scarcities become evident and risk life and limb, that they act to revolutionize their agricultural practices. Boserup concluded that it is population growth that drives agricultural revolution.

Julian Simon, a US scholar of business administration, published his book *The Ultimate Resource* in 1981 and an update, *The Ultimate Resource 2*, in 1996 (Simon 1981, 1996). Simon believed that all previous civilizations had fretted over resource constraints and yet had managed to innovate themselves out of crisis. Society today would be no exception. Population growth leads to problems in the short term, but it is people – in Simon's terms, "skilled, spirited, and hopeful people" – that will solve these problems and create many benefits aside. Population growth certainly places pressure on scarce resources and raises their market price. But price rises spur a search for new supplies, substitute resources, better ways to use existing resources, and techniques for recycling these resources. Population growth in the long run creates more resources than it consumes and leads to cheaper resources. It is for this reason, Simon contends, that natural resources are getting less scarce, food supply is expanding, pollution is decreasing, and life expectancy is expanding.

In 2001, Danish political scientist Bjørn Lomborg brought cornucopian critiques of *The Limits to Growth* to the attention of the world in his controversial book *The Skeptical Environmentalist: Measuring the Real State of the Earth*. Lomborg's target was the "environmental litany" – the emerging consensus that humankind is super-exploiting and super-polluting the planet and that overshoot and collapse are imminent. Skeptical about the accuracy of this litany, Lomborg set himself the goal of weighing the evidence and discovering the truth about the current state of the world.

The Skeptical Environmentalist develops four central arguments:

- Forecasts such as those offered in *The Limits to Growth* were simply wrong; technology has ensured that the world has not run out of resources, per capita food availability has increased, life expectancy has expanded, and human welfare has prospered.
- There remains scope to dramatically increase agricultural productivity and per capita food consumption, there will be more forests in the future than ever before, the discovery of unconventional sources of fossil fuels will boost supplies, alternative energy sources will be developed and mainstreamed, mineral resources will be available in abundance, and water shortages will not deteriorate.
- Trends in pollution and pollution control suggest that, if anything, the planet will become less polluted and less toxic through time; there is no reason to expect that air pollution, acid rain, water pollution, and waste will poison planetary ecosystems to the extent that these ecosystems are terminally compromised.
- Apocalyptic warnings of environmental problems yet to come are overblown; problems with the use of insecticides are exaggerated, biodiversity is not declining, and although human-induced climate change will occur, global warming does not constitute as big an existential threat as say, world poverty does.

The Skeptical Environmentalist continues to provoke the ire of environmental scientists, who accuse Lomborg of harboring ideological motives, practicing poor science, deliberately courting notoriety, and/or sponsoring complacency and inaction. Certainly, a substantial section of the scientific community believe that the book offers a partial, inaccurate, and at times erroneous reading of contemporary environmental trends, threats, and impacts. Lomborg has continued to defend the work nevertheless and has accused critics of cultivating and propagating the environmental litany out of self-interest. Lomborg insists that population growth and economic growth have not destroyed the earth's resources nor polluted planetary ecosystems to the point of toxification, that environmental pressures and threats are not as severe as is commonly believed, that what threats do exist can be dealt with in a cost-effective way using emerging technologies, and that humanity faces more pressing challenges than those posed by the environment. "*Cool it!*" warns Lomborg (2007); global warming does not rank in the list of the world's top problems or priorities, and decarbonization is an unfair burden to load onto newly industrializing and peripheral economies.

The Political Ecologists: Marx in the Age of the Capitalocene

Political ecologists believe that population and economic growth is in and of itself neither inherently good nor bad. Relationships between people and the natural environments they occupy are mediated not simply by population and economic pressures but also by social, economic, technological, political, and cultural institutions. According to US political and environmental geographers Paul Robbins, Richard Peet, and Michael Watts, political ecologists concern themselves with the relationships that exist between society – not population – and nature (Robbins 2004; Peet et al. 2011). They place under scrutiny those economic systems that plunder nature and pollute ecosystems most, the role of social inequalities in determining access to resources and rendering vulnerable people more exposed to pollution and its effects, and the politics of environmental blight (that is, who ought to be held accountable for environmental destruction and for finding solutions).

Political ecologists use the idea of the "metabolic" needs of society to better understand why some societies place more pressure on the earth's resources than others.

In his 1984 book *Uneven Development: Nature, Capital, and the Production of Space*, Scottish-born and then US resident Marxist geographer Neil Smith argued that how human beings think about their relationship with the natural environment shifted abruptly following the emergence of Western civilization, and in particular following the European Enlightenment and the rise of the European capitalist economy from the fifteenth century. Prior to the rise of the West, it was not uncommon for human beings to think of themselves as just one, albeit privileged, species eking out an existence from nature much the same as every other animal. Even as the Neolithic Revolution unfolded and great civilizations brought humankind to new heights, human beings understood that they had to live in harmony with nature as they were part of nature. With the rise of the West has come an era of unprecedented anthropocentricism; some human beings now even consider themselves to be a species apart, capable of dominating over nature and sheltering themselves from worst of nature's extremes.

Political ecologists believe that there has existed no society in human history as exploitative of nature as Western society and no social group in human history as burdensome on the earth's resources as the capitalist class. European capitalism has incessantly raided, plundered, reworked, and transformed the earth's ecosystems for its own ends. Capitalism has an endless need for raw materials and has striven to tame and conquer nature. Capitalism has made it profitable for human beings to realign river basins, genetically modify crops and animals, reclaim polders from the sea, desalinate sea water, dam great rivers, flood irrigation channels in deserts, defend coasts from erosion, mine quarries, clear and plant vast tracts of forest, and so on.

According to Belgian-born and British resident Marxist geographer Erik Swyngedouw, capitalism has tampered with nature to the extent that nature is not what it was; it is difficult to see what remains authentic and what is artificial. Perhaps it is no longer appropriate to even speak of nature *per se*; capitalism has reworked planetary ecosystems today to the extent that all "nature" exists only in a socially modified form. For Swyngedouw (1999), today it might be more accurate to speak only in terms of "socionature." This idea represents a political ecologist reworking of the notion of the Anthropocene, and is much more attentive to the role of social structures and power relations in the transformation of first or pristine nature than mainstream scientists would be.

For US environmental historian Jason Moore, the idea of the Anthropocene is indeed misleading. In his 2015 book *Capitalism in the Web of Life: Ecology and the Accumulation of Capital*, Moore argues that the climate and ecological crisis is at root a product of political economy and its metabolic relation to nature. For Moore, the idea of the Anthropocene is overly benign and misses the point; it is not humanity generally but the capitalist economic system specifically that is sending our natural world into crisis. There needs to be a political ecology that grasps the importance of the insatiable metabolic needs of the capitalist mode of production and that understands that any remediation will require deep system change in prevailing political and economic structures. The nature we now know has its origins in the rise of European capitalism; the incorporation of the rest of Asia, Latin America, and Africa into a world economic system dominated by Europe; and the transformation of nature in the New World into a source of raw materials (sugar, tobacco, cotton, and silver). Our very ideas of nature are inescapably and dialectically rooted in this historical and materialist reality. It is most accurate, then, to speak in terms of the Capitalocene.

In using nature's bounty, capitalism has also created and released into the environment pollutants with chemical compositions that are largely foreign to nature. At root, global warming, depletion of the ozone layer, acid rain, and the leaking of poisons when disposing of waste are all threats that exist because of capitalism's relentless plundering of the natural environment. Political ecologists contend that environmental pollution most often affects those who are already most vulnerable. Pollution, of course, emanates most from Western societies, but these societies are often more able to address and ameliorate the effects of environmental contamination, especially when its consequences come back to haunt them. As less developed countries embrace capitalist forms of development and begin to industrialize, they too are becoming important progenitors of pollutants. But given their weak institutional capacities and their especially impoverished populations, they are more vulnerable to the environmental degradation that pollution brings. Meanwhile, even within countries, it is often the weakest and most vulnerable peoples who are left to live with the pollution created by the consumption practices of, and industries owned by, business elites. Given that it is the West that has the greatest ecological footprint and it is the West that has grown affluent by abusing the environment, Western countries should shoulder most of the responsibility for solving the world's environmental problems.

A Planet in Distress: The Global Climate and Ecology Crisis

In May 2019, 16-year-old Swedish environmental activist Greta Thunberg featured on the front cover of *Time* magazine. Less than a year earlier, Thunberg had risen to international prominence by dint of her school strike for climate protest held on the doorstep of the Swedish Parliament (Plate 9.1). "No one," she insists, "is too small to make a difference" (Thunberg 2019). Following Thunberg's address to the United Nations Climate Change Conference (COP24) in Katowice in December 2018, school strike for climate quickly captured the attention of the world's youth, and similar strikes were called in many countries. Addressing the United Nations Climate Action Summit in September 2019, Thunberg lamented world leaders who she claimed were "not mature enough to tell it like it is." Bearing a direct, blunt, and at times angry message, throughout 2019 Thunberg and likeminded green activists (in the United Kingdom and elsewhere – for example, the Extinction Rebellion movement) have mobilized popular opinion in a spectacular way. Most (but of course not all) world leaders have come to grasp appreciate the perilous state of the natural environment and a rush to declare: a rush by national, regional, and local governments to declare a climate and ecological emergency has followed.

Plate 9.1   Greta Thunberg's skolstrejk för klimatet (school strike for climate) at the Swedish Parliament. Source: Anders Hellberg, Wikimedia Commons, https://commons.wikimedia.org/wiki/File:Greta_Thunberg_4.jpg. CC BY-SA 4.0.

Global Warming: Decarbonizing Our Overheating Planet

According to the United Nations (UN) Intergovernmental Panel on Climate Change (IPCC) (Masson-Delmotte et al. 2018), the mean surface temperature of the earth is now 1 °C higher than in the pre-industrial era. Although no specific limit constitutes a critical threshold, the IPCC concludes that rises above 1.5 °C from preindustrial temperatures and especially rises above 2 °C constitute "dangerous human interference" in the global climate system. Driven by a still dominant carbon-fueled economy (oil, natural gas, and coal), the world is on track to exceed the 1.5 °C threshold by the year 2030. Figure 9.2 shows trends in energy consumption from 1965 to (projected) 2035.

In his 2013 book *Overheated: The Human Cost of Climate Change*, US expert in international law Andrew T. Guzman argues that even with a rise in global temperature of 2° C, this century will witness a climate-induced collapse of society and planetary ecosystems. Rises in sea level will drown some low-lying islands and will flood low-elevation delta regions, leading to mass population displacement. Extreme weather events (violent storms, hurricanes, floods, landslides, land loss, blizzards, heat waves, droughts, crop failure, wildfires, desertification, and tornadoes) will create localized hardships. The result will be increased domestic and international tension over food and resources, and mass migration to "climate-refugee camps" and slums in already overcrowded developing world cities. By 2050 there could be nearly 150 million climate change refugees driven from Latin America, sub-Saharan Africa, and Southeast Asia. New concentrations of impoverished refugees will in turn create a fertile breeding ground for a new generation of more virulent infectious disease, making global epidemics more likely. Time is short. Urgent action is needed to decarbonize the economy and reduce emissions (see Deep Dive Box 9.3). It is necessary to shift toward renewable energy sources such as tidal, wind, hydro, wave, and solar power (Plate 9.2); waste-to-energy, which is energy captured by burning waste in incinerators; and biomass, geothermal, and hydrogen energy. For some, nuclear energy should also be added to this list.

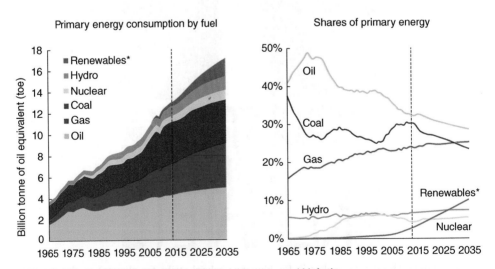

* Renewables includes wind, solar, geothermal, biomass, and biofuels

Figure 9.2 Trends in worldwide energy consumption between 1965 to 2035. Source: Redrawn from BP, Energy Outlook 2017 edition.

Deep Dive Box 9.3   Decarbonizing Our Fossil Fuel–Addicted Economy

Currently, nonrenewable energy resources or fossil fuels provide for 81.6% of the world's energy needs (with oil meeting 34% of total demand, coal 27%, and natural gas 24%). Meanwhile, energy procured from renewable supplies (such as from hydroelectric schemes, biomass used as biofuel [burning crops], wind farms, tidal and wave sources, solar power, hydrogen power, geothermal springs, and waste-to-energy initiatives) meets only 11% of global demand. A third category, in principle infinite in supply but differentiated due to the environmental risks it poses, is nuclear energy (from uranium and thorium). Currently nuclear power meets 4% of the world's energy needs.

Fossil fuel consumption is the biggest source of carbon/greenhouse gas emissions and the primary cause of human-induced global warming. Figure 9.3 shows global CO_2 emissions from 1900 through the present, and Figure 9.4 breaks down the emissions from 1970 to 2017 by country.

For this reason, unhooking the world from its reliance on fossil fuels presents as perhaps the greatest environmental challenge of the twenty-first century. This task will be far from easy. Although the West will continue to consume the greatest amount of energy, growth in demand for power will come from the developing world, where industries remain founded on fossil fuel energy sources (the so-

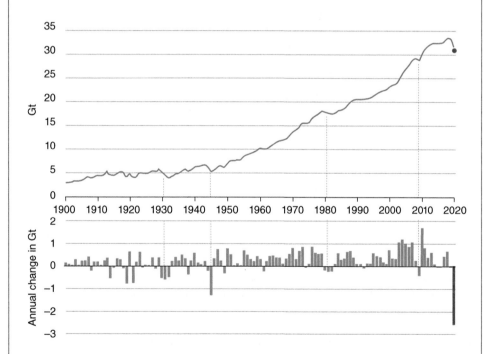

Figure 9.3   Global energy-related CO_2 emissions (gigatons, Gts)   1900–2020. Source: Redrawn from International Energy Agency, Global energy-related CO2 emissions, 1900-2020.

(Continued)

Box 9.3   (*Continued*)

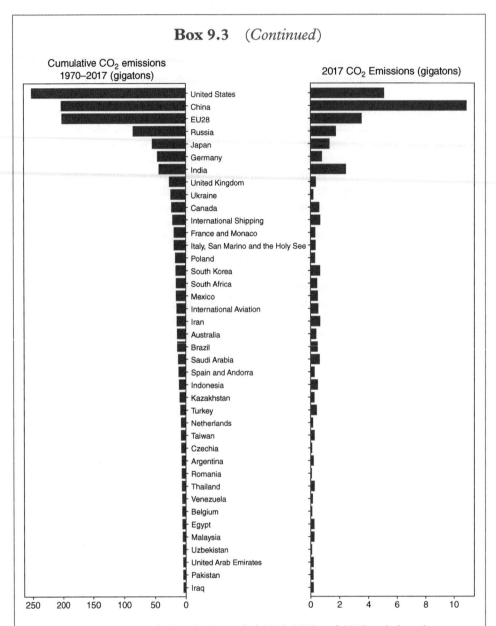

Figure 9.4   Total CO$_2$ emissions for the period 1970–2017 and 2017 emissions, by country/region. Source: Redrawn frm Mgcontr, Total CO2 emissions for the period 1970 to 2017 and 2017 emissions.

called "coal-fired" economies). Meanwhile, the West will embrace greener technologies and clean growth, but only when costs and competitive pressures allow. While technologies to exploit renewable energy will improve, it is questionable whether they will harness enough energy to replace fossil fuels in the short term. Moreover, people will continue to fear the potential hazards presented by nuclear energy.

The United Nations Framework Convention on Climate Change (UNFCCC) convenes an annual meeting of the Conference of Parties (COP); the first (COP1) was held in Berlin in 1995, and the 2021 meeting will be in Glasgow (COP26). Important agreements on the governance of climate have been signed at COP meetings, most recently being the signing of the Paris Agreement at COP21 in 2016, committing countries to reducing carbon emissions and checking further temperature rises. The IPCC produced its *Fifth Assessment Report* (AR5) in 2013–2014. Its next report, AR6, will be published in 2022.

In 2018, the IPCC, the UN body for scrutinizing the science related to climate change, published *The Special Report on Global Warming of 1.5 °C* (Figure 9.5). Based on an assessment of over 6000 scientific papers, the report warned that the window for avoiding the worst climate change impacts could close by 2030 unless urgent action is taken to keep global warming within the 1.5 °C ceiling (Masson-Delmotte et al. 2018).

Adopted in 2015, the Paris Agreement brings together 195 countries within the UNFCCC. It commits governments to four binding objectives:

- Limit increases in global average temperatures to less than 2 °C above pre-industrial levels.
- If possible, limit the increase to 1.5 °C, as this would reduce the risk and impact of climate change.
- Ensure that global emissions peak and then decline as soon as possible, recognizing that this will take longer for developing countries.
- Undertake rapid reductions in emissions thereafter in accordance with the best science.

Although the Covid-19 global pandemic resulted in drastic reductions in energy consumption in 2020 (International Energy Agency 2020), the consensus is that the world is far from being on course to meet its Paris ambitions.

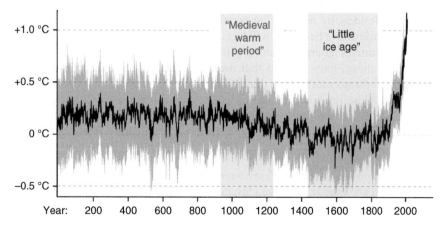

Figure 9.5 Global average temperature change during the past 2000 years. Source: Ed Hawkins, Global average temperature change across the past 2000 years. https://www.climate-lab-book.ac.uk/2020/2019-years/. CC BY 4.0.

Plate 9.2   Solar farm, Región de Antofagasta, Chile. Source: Antonio Garcia on Unsplash.

The impact of climate change on global GDP remains under dispute. The *Stern Report* famously argued that failure to remediate climate change would be equivalent to losing at least 5% of global GDP each year, now and forever. With wider associated risks and impacts, this could rise to 20% of global GDP (Stern 2007). According to Burke (2015), if future adaptation mimics past adaptation, global warming will reduce average global incomes by 23% by 2100 and will widen global income inequality, relative to scenarios without climate change. Diffenbaugh and Burke (2019) likewise argue that, although there is uncertainty as to whether historical global warming has actually benefitted some temperate Global North countries, for most poor countries there is a >90% likelihood that per capita GDP is lower today than if global warming had not occurred. Moreover, they argue that although some Global North countries might in fact benefit from global warming in the next 80 years – the United Kingdom, Norway, and Canada, for instance – the GDP of Global South countries could be reduced by between 17 and 31% by 2100. Besides, by not sharing equally in the direct benefits of fossil fuel use, many poor countries have been and will be significantly harmed by the warming arising from wealthy countries' energy consumption. Critics, however, including Danish statistician Bjørn Lomborg, have insisted that these estimates are exaggerated and that climate change will do no more harm than one large global economic recession between now and 2100. Although a significant problem, global warming will not prove to be fatal. Even if GDP is lower by 2100 than it might otherwise have been, this must be understood against the register that the world will be significantly richer by then.

Defining questions for the twenty-first century: How can we decarbonize the economy? How do we enable effective climate adaptation and build resilience? Who might low-carbon transitions and adaptation leave behind?

Biodiversity: Avoiding a Sixth Mass Extinction

According to British-American biologist and theoretical ecologist Stuart Pimm (Pimm et al. 2014), the pre-human rate of extinctions on earth was around 0.1 species per year for every million species. Today, this rate has increased to between 100 and 1000 species per year for every million species in existence. Reduced biodiversity presents a threat to humanity because our survival ultimately depends upon healthy ecosystems, not least for food, carbon capture, medicines, and healthy lives. According to some in the scientific community, we are now on the brink of a sixth mass extinction event. The trigger will not be, as in the past, natural changes in climate or showers of meteorites, but instead human recklessness, deforestation, population growth, economic development, urbanization, global warming, increased movement of invasive species, and overfishing and overharvesting from the oceans. It is imperative that all species – and in particular those on the International Union for Conservation of Nature's (IUCN) "Red List of Threatened Species," especially the "Priority Species" – are saved from extinction through the conservation and management of ecosystems and habitats, rewilding projects, and the re-naturing of cities (Deep Dive Box 9.4). The legally binding Convention on Biological Diversity (CBD) provided the first-ever framework for collective international action on biodiversity when it was signed in 1992, with the current Strategic Plan for Biodiversity (incorporating the "Aichi Targets") covering the period of 2011–2020. In its Global Assessment Report on Biodiversity and Ecosystem Services, the Intergovernmental Science-Policy Platform on Biodiversity Ecosystem Services (2019) concluded that most of the Aichi biodiversity targets would not be met by 2020. Given the collective failure of governments so far even to slow down the pace of ecological collapse, there is an expectation that the post-2020 Global Biodiversity Framework (to be agreed in summer 2021) may need to be significantly more ambitious.

Defining questions for the twenty-first century: How can we arrest species decline? How can we conserve and rewild habitats and restore urban nature?

Deep Dive Box 9.4   The Degradation and Loss of the Amazonian Rainforest: Implications for IUCN Red List Species

The IUCN's Red List of Threatened Species (Figure 9.6) is the world's most authoritative data source on global trends in biodiversity. In 2020, this list recorded the status of more than 116 000 species; the IUCN wishes to raise this number to 160 000 by 2021, producing what they call a "barometer of life." The IUCN classifies species into nine categories: Not Evaluated, Data Deficient, Least Concern, Near Threatened, Vulnerable, Endangered, Critically Endangered, Extinct in the Wild, and Extinct. The Red List identifies no less than 31 000 species threatened with extinction, including 41% of amphibians, 25% of all mammals, 34% of conifers, 33% of reef-building corals, 14% of birds, 30% of all sharks and rays, and 27% of all selected crustaceans (that is, all crustacean species that have been assessed so far).

Globally, the degradation and loss of forests constitute perhaps the gravest threat to planetary biodiversity. Over 80% of the world's terrestrial biodiversity can be found in forests, and forests produce 40% of the world's oxygen. They protect watersheds, which release fresh water into rivers. Forests serve as carbon sinks and are an essential ally in the battle against soil erosion, and 1.6 billion people rely on forests for an income. They provide raw material for medicines (the Amazon River basin alone is home to more than 1300 plant species used by the pharmaceutical industry).

(Continued)

Box 9.4   *(Continued)*

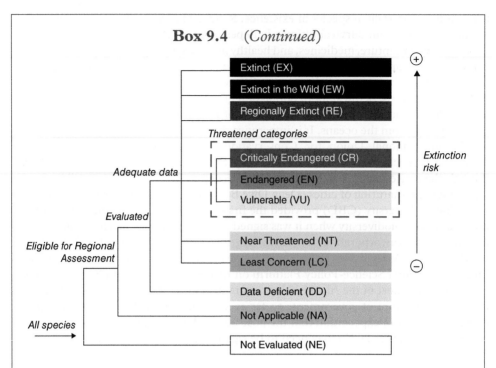

Figure 9.6   The IUCN Red List of Threatened Species Taxonomy, Version 2020-1. Source: Redrawn from International Union for Conservation of Nature and Natural Resources methodology, https://www.iucnredlist.org.

According to Global Forest Watch, in 2019 alone, 11.9 million hectares of tree cover across the world were lost, 3.8 million hectares (a third) from tropical rainforests. For context, this amounts to losing the equivalent of a football pitch of forest every six seconds across the full calendar year. Amazonia is perhaps the single most important world region where deforestation has occurred at scale (in 2019, Brazil lost 1.36 million hectares, Bolivia 29 000 ha, and Peru 162 000 ha), followed by the Democratic Republic of Congo (DRC; 465 000 ha) and Indonesia (324 000 ha). In the case of the Amazon, deforestation is being driven by agriculture, mining, logging, and power generation projects.

Despite covering only around 1% of the planet's surface, the Amazon forest is home to 10% of all known wildlife species. The Brazilian Ministry of Environment's Chico Mendes Institute for Nature Conservation (ICMBio) has worked with the IUCN to identify Red List species within the country. From 2010 to 2014, a total of 12 256 species were assessed, including 8924 vertebrates, 732 mammals, 1980 birds, 732 reptiles, 973 amphibians, 4507 fishes (3131 freshwater, 1363 marine species, and 5 myxines), and 3332 invertebrates (including crustaceans, mollusks, and insects). The assessment found that 1182 species (9.6% of all species assessed) were threatened with extinction, including 110 mammals (15% of all mammals assessed), 234 birds (12%), 80 reptiles (11%), 41 amphibians (4%), 353 bony fishes (8%), 55 elasmobranchs (32%), one myxine (20%), and 299 invertebrates (9%). Among these species, one is Extinct in the Wild, 318 are Critically Endangered, 406 are Endangered, and 448 are Vulnerable. Furthermore, 10 species were already Extinct. Finally, 314 (2.5%) species were assessed as Near Threatened, and 1669 (13.6%) were regarded as Data Deficient.

What can be done to arrest forest loss and degradation in the Brazilian Amazon? A plethora of conservation initiatives are afoot. Protected areas with rich biodiversity have been established, the Central Amazon Conservation Complex (a United Nations World Heritage Site) being the largest protected area in the Amazon River basin (over six million hectares). More recently, indigenous reserves and community forests have enjoyed enhanced legal protection. Communities have been paid to provide ecosystem services. Companies (for example, logging firms) have been more formally regulated and monitored. And reforestation has been instituted, most notably in the Atlantic forest region.

But with the election of climate skeptic Jair Messias Bolsonaro as President of Brazil in early 2019, the country has witnessed more rather than less forest loss and degradation. Bolsonaro has stated his commitment to extending agribusinesses, ranching, commercial mining, and oil and gas extraction, including within indigenous territories. His policies have been accused of precipitating extensive (and deliberate) forest fires across the Amazon basin; Bolsonaro in turn has accused environmental nongovernmental organizations (NGOs) of starting fires as a political ruse to strengthen their case for government funding. Because the Amazon River basin is dubbed the "lungs of the world" (Plate 9.3), it is little surprise that politicians around the world have sought to pressure Brazil into scaling forest conservation. There have been calls to show solidarity with conservationists by reducing global meat consumption (and hence reducing land clearance for animal husbandry). But there is little evidence that Bolsonaro is intending to change his policy agenda any time soon.

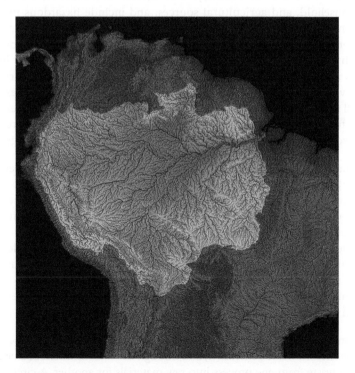

Plate 9.3 The "lungs of the earth": the Amazon River basin rainforest. Source: NASA image created by Jesse Allen, using Shuttle Radar Topography Mission (SRTM) data provided courtesy of the University of Maryland's Global Land Cover Facility, and river data provided courtesy of the World Wildlife Fund HydroSHEDS Project.

Air Quality: Detoxing the Air We Breathe

Poor air quality derives from the release of pollutants such as particulate matter (PM), ozone (O_3), nitrogen dioxide (NO_2), and sulfur dioxide (SO_2) from sectors such as agriculture, energy, manufacturing, construction, and transport. Poor air quality is recognized as one of the largest environmental risks to public health. Globally, the World Health Organization (WHO 2016) estimates that ambient air pollution causes in excess of three million deaths per year. In the short term, air pollution exacerbates chronic respiratory conditions such as asthma. In the longer term, it contributes to the prevalence of lung cancer and cardiovascular diseases, including strokes and heart attacks, with emerging evidence linking poor air quality to the onset of dementia. Air pollution has significant adverse impacts on the environment and biodiversity, and is a major contributor to global climate change. Policies to reduce smog and clean the air – particularly within cities – need to be scaled.

Defining questions for the twenty-first century: How can we tackle urban smog and clean the air we breathe? How can we respond to the health impacts and health inequalities that arise from poor air quality?

Growing Waste: From "Cradle to Cradle"

Waste, or materials are that are residual to societal needs at a given moment in time and require disposal, derive from industry, commercial, construction and demolition, municipal, household, and agricultural sources, and include hazardous materials and end-of-life vehicles. Owing to the ongoing reliance on landfill and incineration, waste creates environmental, health, and economic problems. The relationship between economic growth and waste generation varies according to the waste stream in question. Nevertheless, according to the World Bank (Kaza et al. 2018), without urgent action, global waste will increase by 70% from current levels by 2050. Plastic wastes (constituting 12% of all solid waste) are especially damaging; if not managed effectively, they have the potential to contaminate oceans, waterways, and ecosystems for hundreds of years. Around the world, a number of principles are being applied to help manage waste:

- A waste management hierarchy (prevention is preferred over reuse, then recycling, then energy recovery, with finally disposal being a last resort)
- The polluter-pays principle and extended producer responsibility (those who create waste take ownership of and pay for its redemption, even if that occurs further down the production chain)
- Self-sufficiency and proximity (waste produced locally should be handled locally)
- The precautionary principle (in governing waste, human health and environmental protection should always take precedence)
- Waste management planning (competent authorities should be mandated to develop area-based strategic waste management plans).

As waste continues to grow, a key challenge facing the world economy will be to decouple economic growth from waste generation, and through the establishment of a circular economy to convert waste from one process into raw materials for another. A circular economy is an alternative to a traditional linear economy (which is predicated upon the formula make, use, and dispose) in which resources are kept in use for as long as possible by recovering, and regenerating materials at the end of each product life-cycle and making them available as raw materials for other industries and production processes. (Deep Dive Box 9.5).

Deep Dive Box 9.5   Waste Pickers and the Very Particular Circular Economy in Delhi

What to do with the garbage that society creates has become a key environmental concern in Indian cities today; the capital city of Delhi is no exception. The Delhi metropolitan region has grown rapidly from 6.2 million in 1981 to 25 million in 2020. Growing consumption has resulted in significant increases in municipal solid waste (MSW). In 2007, Delhi produced 5000 t of MSW per day; in 2020, this figure stood at 10 000 t per day.

Currently, only an estimated 70–80% of Delhi's MSW is collected, the rest being discarded in illegal sites in every corner of the city. Of the MSW that is collected, over 90% is disposed of in largely uncontrolled landfill sites at the outskirts of the city; the rest is composted. Because these landfill sites lack leachate and landfill gas collection systems, contamination of the water table, air pollution, and soil pollution are normal, producing adverse health outcomes for groups forced to live in close proximity.

The Ghazipur landfill opened in 1984 (Plate 9.4). According to Indian law, waste can only be piled to a height of 65 ft before a landfill facility must be shut down. Ghazipur reached this milestone in 2002, and yet waste has continued to arrive. Now referred to as the Mount Everest of garbage, Ghazipur sprawls as wide as 40 soccer fields and is higher than the towers on London's Tower Bridge. It is rising by 32 ft every year and in 2020 became as tall as the Taj Mahal (240 ft). In 2017, a 50-t "avalanche" of waste killed two local residents.

There exist approximately 200 000 waste pickers in Delhi (0.8% of the population) and as many as four million waste pickers in India (0.29% of the country's population). These waste pickers come from the lowest caste groupings and are often referred to as the "untouchables". The sight of entire families of waste pickers, knee-deep in garbage, trawling their way through trash mountains in landfill sites in Delhi – in

Plate 9.4   Waste pickers at the Ghazipur landfill site, Delhi, India. Source: © Source: Hindustan Times/Getty Images.

(Continued)

Box 9.5   *(Continued)*

search of items that might be recovered and used, eaten, or sold, from dawn to dusk, for little more than US$2 or 3 per day – is harrowing. But through scavenging, these waste pickers are by default recycling between 15 and 20% of India's waste.

Action is needed, but care must be taken not to introduce technological solutions without attending to the poverty of India's underclass. Plans to privatize waste management in Delhi and to build an incinerator near the Ghazipur landfill site threaten the livelihoods of the waste pickers and in the short term may reduce the amount of recycling that does occur.

Ultimately, the solution is not to defend the right of waste pickers to live a life of squalor and ill health but instead to work to lift waste pickers out of abject poverty. It would seem sensible that Indian cities work to formalize the informal recycling that already takes place in the country, and to harness the skills of those waste pickers who have eked out a living doing precisely this for many years, by providing a livable wage and improved health and safety training and protection. It is here that India's middle classes need to take responsibility for the role they play in tarnishing the Indian environment by funding waste management schemes that both clean up the city and address the social and economic precarity of the untouchables.

Defining questions for the twenty-first century: How can we better manage waste? How can we move to zero waste and a circular economy? How can we maximize the environmental, economic, and social benefits of a circular economy?

Water Insecurity: Water, the New Gold?

Currently, over one billion people do not have access to clean drinking water, more than two billion people do not have access to adequate sanitation, and as many as five million people die every year from preventable, waterborne infectious disease (United Nations 2020). Tensions over the equitable sharing of water resources are aggravating international conflicts between, among others, Sudan, Ethiopia, and Egypt; India and Pakistan; Turkey and Syria; China and Tibet; and Israel and Palestine. In 2010, US water resources scientists Peter H. Gleick and Meena Palaniappan argued that population growth, economic development, and global warming have conspired to deplete and/or pollute the world's stock of freshwater resources to the point that it is meaningful to speak in terms of "peak water." Inspired by Hubbert's idea of "peak oil," "peak water" is the point at which the volume of available freshwater begins to decline through overuse, contamination, and climate change. According to Gleick, we might be approaching three kinds of peak water:

- Peak renewable water (where water is drawn from hydrological systems faster than it is replaced)
- Peak nonrenewable water (where water is pumped from underground fossil aquifers faster than it is being replaced)
- Peak ecological water (where the ecological and economic costs of transporting water from areas of surplus to areas of deficit are too prohibitive to countenance)

> *Defining questions for the twenty-first century*: How can we achieve secure access to safe and affordable water for all? How can we clean our rivers, lakes, and oceans? How can we ensure universal access to sanitation?

Case Study: Tackling the Global Climate and Ecological Crisis in the Liverpool City Region

In May 2019, the Liverpool City Region Combined Authority's metro-mayor Steve Rotheram declared a "climate emergency" and affirmed Liverpool City Region Combined Authority's commitment to undertaking proportionate remediating actions. The metro-mayor has set his sights on the Liverpool City Region becoming net zero-carbon by 2040; local authorities and anchor institutions from the public, private, and third sectors have likewise set net zero-carbon targets by or before 2040. This 2040 target will undoubtedly prove difficult to meet.

Liverpool (Map 9.1) grew as a port city serving both British imperial expansion and the UK Industrial Revolution. Like many rustbelt port cities, the collapse of empire and deindustrialization led to a spiral of decline. Throughout the twentieth century, the city struggled to reinvent itself and has been described as the classic "left-behind place." But today, the

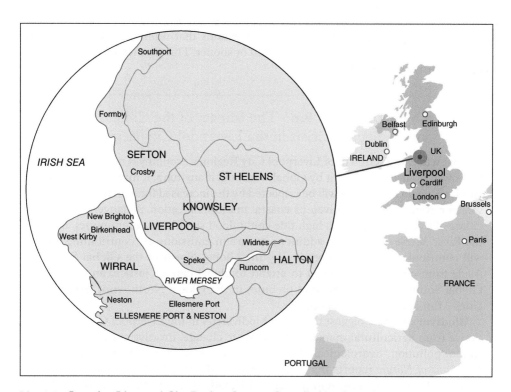

Map 9.1   Locating Liverpool City Region. Source: Compiled by the author.

Table 9.1 Headline historic and projected growth rates, Liverpool City Region and the United Kingdom.

	Liverpool City Region			United Kingdom		
Indicator	*2018*	*Growth (2009–2018)*	*Growth (2018–2040)*	*2018*	*Growth (2003–2018)*	*Growth (2018–2040)*
Population	1 552 000	4%	1%	66 436 000	11%	7%
Employment	713 000	10%	5%	35 081 000	14%	7%
GVA	£32 bn	14%	30%	£1803 bn	28%	37%

bn: Billion; GVA: gross value added.
Source: Data provided by the Liverpool City Region Combined Authority.

Liverpool City Region has turned the corner and is once again on an upswing. Whilst there is still much to do and the economy remains fragile, from the mid-1990s it has enjoyed a degree of regeneration. Liverpool will continue to grow as a city between now and 2040 – to a limited extent in terms of population, to some extent in terms of employment, and to a large extent in terms of gross value added (GVA) (Table 9.1). At the same time, it is aspiring to become the greenest city in the United Kingdom (See Deep Dive Box 9.6.)

How might the Liverpool City Region grow the local economy while reducing its ecological footprint?

The Liverpool City Region, Combined Authority has placed "clean growth" at the center of its new Local Industrial Strategy (LIS) and has identified "clean technology" as one of the city's critical "sector accelerators." In November 2019, the Authority established a Climate Partnership to coordinate the City Region's response to the climate emergency and bring together all organizations that want to play their part in achieving the goal of net zero-carbon by 2040 or sooner. This partnership is developing a comprehensive Climate Action Plan.

Deep Dive Box 9.6   The Impact of the Climate and Ecological Crisis in the Liverpool City Region

Global warming: Growth of Liverpool City Region's economy is not expected to be slowed to any great extent by global warming; if anything, it might be boosted. But Liverpool City Region will be impacted by the increased frequency of extreme weather events, especially given its coastal maritime location. A dangerous coinciding of a storm surge and a high tide could flood significant communities along the River Mersey estuary. In addition, its carbon emissions are contributing and will contribute to the immiserating by the Global North of the Global South. Aside from ethics and the need to attend to global climate justice, an unstable Global South is liable to rebound on cities in the Global North – not least port cities – through climate refugees.

Biodiversity: The biggest threats to biodiversity in the Liverpool City Region derive from agricultural management, climate change, invasive species, urbanization, pollution, hydrological change, and woodland management. The 2019 UK *State of Nature* reports that with respect to the IUCN Red List categories, of the 8431 species assessed in the United Kingdom, 1188 (15%) can be classified as threatened and at risk of extinction. In addition, 2% of species are known (133) or considered likely (29) to have gone extinct since 1500. Meanwhile, the abundance and distribution of species have declined over recent decades. There is no reason to suppose that this national picture is not replicated in Liverpool. The local Biodiversity Action Plan identifies six birds, four mammals, four amphibians and reptiles, six invertebrates, and eight plant species that are under particular threat within the region.

Air quality: Liverpool City Region has some of the highest levels of air pollution in the country, with Public Health England estimating that it contributes to around 700 deaths a year locally. Pollutants such as particulate matter (PM), ozone (O_3), nitrogen dioxide (NO_2), and sulfur dioxide (SO_2) are most culpable. The impacts of poor air quality are unequally distributed across the region, with poorer communities disproportionately affected. Low-income communities are not the primary generators of high air pollution levels (car ownership is significantly higher in more affluent areas) yet they suffer excessively due to their disproportionate proximity to busy inner-city roads, through which heavy commuting traffic passes and along which housing is in consequence more affordable.

Growing waste: As the local economy has regenerated, waste streams have grown. In 2018, the Liverpool City Region area was estimated to have generated almost 4.45 m t of waste, comprising: Local Authority Collected Waste (MSW) 860 000 t, Commercial Waste 750 000 t, Industrial Waste 360 000 t, Construction and Demolition Waste 2 300 000 t, Hazardous Waste 160 000 t, and Agricultural Waste 20 000 t. The City Region continues to rely on landfill, incineration, and recycling. There has yet to emerge a substantial circular economy; only 45% of municipal waste is recycled. Moreover, the Port of Liverpool exports significant quantities of waste to other parts of the world, who, notwithstanding the self sufficiency principle, are paid to accept the United Kingdom's garbage.

Water quality: The River Mersey was severely polluted with a deadly cocktail of raw sewage and toxic chemicals during the Industrial Revolution and was known as the dirtiest river in Europe. In 1985, the Mersey Basin Campaign was established to improve water quality and encourage waterside regeneration. By 2009 it was announced that the river was now one of the cleanest in the United Kingdom, with aquatic life including dolphins, humpback whales, octopus, salmon, gray seals, and large cod. Water quality, including drinking water, across the City Region is now very high.

To date, key actions taken to decarbonise the City-Region economy and remediate threats to the natural environment include:

- The River Mersey running through the Liverpool City Region has the second highest tidal range in the United Kingdom, and Liverpool City Region Combined Authority plans to build Europe's largest tidal barrage project by 2030, supplying enough power to heat 1 million homes.
- The Liverpool City Region has also been designated as one of six UK Centres for Offshore Renewable Engineering (COREs) and plans to triple the volume of energy generated by offshore wind in Liverpool Bay by 2032.
- Liverpool City Region has unique research and technology expertise in hydrogen power. It has an ambition to replace all methane with hydrogen from the City Region's gas grid by 2035, and in 2020 it introduced to the public transport system 17 new hydrogen buses emitting nothing more than water vapor.

- The city's pioneering new Knowledge Quarter and local universities have joined forces to birth the Liverpool City Region Low Carbon Eco-Innovatory (LCEI) and the Centre for Global Eco-Innovation (CGE) to help local small and medium-sized enterprises (SMEs) shift toward low-carbon energy consumption.
- A Mersey Forest Plan has recently been expanded to include an ambitious proposal to create a "Northern Forest" joining Liverpool, Chester, Manchester, Leeds, Sheffield, and Hull by planting 50 million new trees – among other "nature-based benefits," the Northern Forest will constitute a national carbon-offsetting resource. Recognizing its global responsibilities, the Liverpool City Council has also partnered with the Poseidon Foundation to fund carbon offsetting in lieu of its carbon footprint in the Global South.
- The Liverpool City Region Combined Authority has invested in the first phase of a £16m 600 km Greenways cycling and walking network.
- A "brownfield first" approach to development has witnessed a re-greening of industrial wasteland.
- The Liverpool City Region Combined Authority has established a £5 million fund to support the local social economy (third sector) to deliver for poorer communities, including enabling these communities to adapt to climate change.
- The Royal Town Planning Institute (RTPI) is working with the Liverpool City Region to develop a climate resilience policy that will be incorporated into the City Region's Spatial Development Strategy to push up building standards and safeguard against flooding and extreme weather events alongside other climate threats.
- Renovation and retrofitting of the city's housing stock remains a work in progress, and the Liverpool City Region has among the least energy-efficient housing stock in the United Kingdom. The city has proposed a £230m Green City Deal to national government focusing on improving household energy consumption.

Plate 9.5 Protesters from the climate change group Extinction Rebellion hold a "die-in" protest at Liverpool's Anglican Cathedral underneath Gaia 2019. Source: Christopher Middleton / Alamy Stock Photo.

A New Model of Political Economy for a Cleaner and Greener Planet Earth?

To tackle the climate and ecological crisis effectively, it will be necessary to reach consensus on its causes. We return therefore to the three perspectives introduced at the beginning of this chapter: pessimist, optimist, and political ecologist.

- Neo-Malthusians might contend that the present crisis is the greatest example of the profound effects of humankind on planet earth, and stress the extent to which the future of humanity is in doubt. Slowing population and economic growth – and, indeed, even moving to a new period of de-growth – may be the only viable response.
- Cornucopians, in contrast, might question the extent to which scientific discussion of climate change is scaremongering. New clean technologies and a greening of capitalism will remediate the crisis.
- Political ecologists, meanwhile, might question the extent to which the capitalist economic system has changed the global climate system. It is no accident that humanity's war on the earth has coincided with the rise of the West and the emergence of the world capitalist economy. Solutions to climate change require political action. Even a root-and-branch overhaul of the capitalist system will not suffice. Humanity needs to discover an economic system that is more environmentally sustainable.

Arguably, cornucopian perspectives currently hold sway. Many governments continue to view the market as the primary driver of a green transition toward clean growth. In its September 19 2019 editorial preface to its special issue entitled "A Warming World: The Climate Issue," *The Economist* warns starkly, "if capitalism is to hold its place, it must up its game," but proceeds to argue that "to infer climate change should mean shackling capitalism would be wrong-headed and damaging. There is an immense value in the vigour, innovation and adaptability that free markets bring to economies." For those who base solutions on market reform, carbon pricing (taxes, caps and trades, feebates, and regulations), subsidies, and offsetting provide the main policy tools. In a recent report, the International Monetary Fund (IMF 2019) argued that a carbon tax of $75 per ton by the year 2030 – a quantum leap from the present $2 per ton – could limit global warming to 2 °C. Revenue raised might be rerouted to subsidize green projects, especially to help poor communities adapt. The state, in other words, needs to use fiscal levers to create conditions to catalyze green entrepreneurs to innovate and deliver cleaner growth.

But the cornucopian agenda invites debate on whether a transformed and reregulated market alone will be able to remediate environmental damage for which it itself carries significant culpability, or whether any mission to "green" capitalism runs the risk of "greenwashing" capitalism. Will the market alone (or even principally) be sufficiently self-starting and socially responsible to generate the scale of renewable energy we need, achieve net zero-carbon, fortify (especially vulnerable) communities by promoting climate adaptation and mitigation, clean our air, protect and enhance biodiversity and reverse species extinction, deliver zero-waste and establish a circular economy, and purify our water? These are complex and large-scale challenges. They are also challenges dogged by persistent market failure and social injustices. Is it prudent to expect or assume that the market as currently ordered is up to the job?

Other commentators argue that the "status quo" will no longer do, and to suppose that the present emergency will be solved through technical adjustments to present policy agendas is to fundamentally misconstrue the enormity, urgency, and intractability of the problem. A new paradigm is needed; deeper structural reform and systemic change will be required. It is against this backdrop that much discussion has recently arisen concerning the concept of a Green New Deal – a new social contract in the spirit of Franklin D. Roosevelt's 1933 New Deal – to transition politics, economy, and society in favor of models of sustainable development. In her new book *On Fire: The Burning Case for a Green New Deal*, Canadian activist Naomi Klein (2019) argues at length that it will only be possible to confront the climate and ecological emergency effectively if we are willing to transform the systems that produced this crisis. A Green New Deal is necessary to reform political and economic institutions and create a fairer and more sustainable economic model (Deep Dive Box 9.7). Governments, not markets, need to lead the transition; social justice needs to work in tandem with environmental justice; and the market needs to be accompanied by alternative economic models and logics and disciplined so that it serves the public good.

Deep Dive Box 9.7   Beyond the Status Quo: A Green New Deal?

Calls for a Green New Deal argue that there needs to be a deeper systemic reform to the prevailing capitalist political-economic model – in the form of a new social contract for sustainability and a just transition. At the heart of this contract might be:

- Government-led solutions and a green public works program
- A government that leads by example by green-proofing its own practices and buildings
- Stronger institutions and enhanced environmental governance capacity
- Fostering community and citizen participation and developing localized eco-community responses
- Promoting a mixed economy and harnessing social enterprise and the third sector
- Government action to discipline and incentivize the market to deliver clean growth
- Government support for research and innovation into green technologies
- Ensuring affordable finance for soft and hard green infrastructure
- Skills strategies to build a green workforce
- Strengthening the ability of vulnerable groups to cope with the impacts of climate change
- Improving carbon literacy and heightening awareness of the carbon footprint of organizations, businesses, and people
- Spatial planning for eco-friendly cities, for example by promoting biodiversity and encouraging public transit and cycling/walking
- New performance metrics of well-being that prioritize welfare outcomes and social justice, not simply economic growth
- Engaging the data revolution and smart technology to solve pressing environmental problems

Conclusion

Across the past 60 years, population pessimists have argued that humanity is depleting nonrenewable resources to the point of extinction and is polluting and poisoning the earth's ecosystems. Overshoot and collapse at some point in the twenty-first century are inevitable. While it is true that pressures on the earth's resources have never been greater, it is also the case that human beings have survived and prospered thus far by using technology to find more resources, more efficient ways of using resources, and usable substitutes, and to clean up their worst environmental disasters. For cornucopians, this proves that the earth can support a far larger population and economy than pessimists suppose. Political ecologists argue that it is more appropriate to focus upon relationships between politics, poverty, and resources than upon population growth and the earth's carrying capacity *per se*. Today, planet earth is on the brink of a climate and ecology crisis, marked by global warming, biodiversity loss, poor air quality, growing volumes of waste, and water insecurity. Neo-Malthusians, cornucopians, and political ecologists offer competing interpretations of the global climate and ecology emergency and its causes, extent, consequences, and solutions. For many political ecologists, remediation will require nothing less than a new social contract for sustainability and a just transition or a Green New Deal.

Checklist of Key Ideas

- Across the past 60 years, neo-Malthusian population pessimists have repeatedly warned that population growth and economic development have reached such a level that the earth's carrying capacity has now been passed and that overshoot and collapse will bring chaos to the world at some point in the twenty-first century.
- Cornucopians, in contrast, argue that there is no evidence that the human species is plundering and polluting the earth to the degree that the future of humankind is at risk, and that human ingenuity allied with the virtually unlimited resources that exist on planet earth will ensure that the earth's carrying capacity will not be reached anytime soon.
- Political ecologists meanwhile argue that it is politics, society, and economics, not population growth *per se*, that create environmental pressures, threats, vulnerabilities, and responsibilities.
- Humanity's war on planet earth is leading to a climate and ecology crisis, marked by human-induced climate change, species decline, air pollution, expanding mountains of waste, and water insecurity. Governments around the world are now acting to decarbonize their economies, develop climate adaptation and resilience strategies, arrest biodiversity decline, clean the air we breathe, embrace a circular economy, and improve access to clean water.
- Debate continues between population pessimists, optimists, and political ecologists as to what has caused the global climate and ecology emergency and what changes will be needed to remedy it. For some, nothing short of a Green New Deal will do.

Chapter Essay Questions

1) The Club of Rome's *The Limits to Growth* conscripted and energized a whole generation of environmentalists, but its conclusions are largely redundant today. Discuss.

2) It is politics, society, and economics, not population growth *per se*, that create environmental pressures, threats, vulnerabilities, and responsibilities. Discuss.

3) The climate and ecological crisis is the biggest challenge facing humankind today, and yet to date government responses have been wholly inadequate. Discuss.

4) How far do you agree that there is a "burning case for a Green New Deal," and what would this deal ideally comprise?

References and Guidance for Further Reading

Thomas R. Malthus' seminal essay was:
Malthus, T.R. (1798). *An Essay on the Principle of Population*. London: J Johnson.

The relevance of Malthus today is considered in:
Bashford, A. and Chaplin, J.E. (2016). *The New Worlds of Thomas Robert Malthus: Rereading the Principle of Population*. Princeton, NJ: Princeton University Press.

For early and particularly alarmist and controversial views on the dangers of population growth, see:
Carson, R. (1962). *Silent Spring*. Boston: Houghton Mifflin.
Ehrlich, P.R. (1968). *The Population Bomb*. London: Ballantine Books.
Ehrlich, P.R. and Ehrlich, A.H. (1990). *The Population Explosion*. London: Hutchinson.
Hubbert, M.K. (1956) *Nuclear Energy and the Fossil Fuels*. Publication No. 95, June. Houston, TX: Shell Development Company, Exploration Production and Research Division.

One can trace the evolution of the Club of Rome's thinking in the period from 1972 to 2012 by reading:
Meadows, D.H., Meadows, D.L., Randers, J., and Behrens, W.M. III (1972). *The Limits to Growth*. Chicago: Universe Books.
Meadows, D.H., Meadows, D.L., and Randers, J. (1992). *Beyond the Limits*. White River Junction, VT: Chelsea Green.
Meadows, D.H., Randers, J., and Meadows, D.L. (2004). *Limits to Growth: The 30-Year Update*. White River Junction, VT: Chelsea Green.
Randers, J. (2012). *A Count-Up to 2052: An Overarching Framework for Action*. Munich: Chelsea Green.

The most careful and definitive study of the earth's carrying capacity to have been conducted is:
Cohen, J.E. (1995). *How Many People Can the Earth Support?* New York: Norton.

An important neo-Malthusian account linking population growth and climate change is:
Bongaarts, J. and O'Neill, B.C. (2018). Global warming policy: is population left out in the cold? *Science* 361 (6403): 650–652.

A seminal work on the Anthropocene is:
Crutzen, P. (2002). Geology of mankind. *Nature:* 415. https://doi.org/10.1038/415023aermont.

Works by James Lovelock include:
Lovelock, J. (1979). *Gaia: A New Look at Life on Earth.* Oxford: Oxford University Press.
Lovelock, J. (2014). *A Rough Ride to the Future.* London: Allen Lane.
Lovelock, J. (2019). *Novacene: The Coming Age of Hyperintelligence.* London: Penguin.

Books that offer cornucopian perspectives on population growth and its impacts include:
Boserup, E. (1965). *The Conditions of Agricultural Growth: The Economics of Agrarian Change Under Population Pressure.* Chicago: Aldine.
Lomborg, B. (2001). *The Skeptical Environmentalist: Measuring the Real State of the World.* Cambridge: Cambridge University Press.
Lomborg, B. (2007). *Cool It: The Skeptical Environmentalist's Guide to Global Warming.* London: Cyan and Marshall Cavendish.
Lomborg, B. (2010). *Smart Solutions to Climate Change: Comparing Costs and Benefits.* Cambridge: Cambridge University Press.
Simon, J. (1981). *The Ultimate Resource.* Princeton, NJ: Princeton University Press.
Simon, J. (1996). *The Ultimate Resource 2.* Princeton, NJ: Princeton University Press.

Good introductions to political ecology and Marxist perspectives can be found in:
Krueger, R. and Gibbs, D. (eds.) (2007). *The Sustainable Development Paradox: Urban Political Economy in the United States and Europe.* New York: Guilford Press.
Moore, J.W. (2015). *Capitalism in the Web of Life: Ecology and the Accumulation of Capital.* London: Verso.
Moore, J.W. (ed.) (2016). *Anthropocene or Capitalocene? Nature, History, and the Crisis of Capitalism.* Oakland, CA: PM Press/Kairos.
Moore, J.W. (ed.) (2017). *Capitalism's Ecologies: Culture, Power, and Crisis in the 21st Century.* New York: PM Press.
Patel, R. and Moore, J.W. (2017). *A History of the World in Seven Cheap Things: A Guide to Capitalism, Nature, and the Future of the Planet.* Berkeley: University of California Press.
Peet, R., Robbins, P., and Watts, M. (2011). *Global Political Ecology.* London: Routledge.
Peet, R. and Watts, M.J. (eds.) (1996). *Liberation Ecologies: Environment, Development, Social Movements.* London: Routledge.
Raworth, K. (2017). *Doughnut Economics: Seven Ways to Think Like a 21st-Century Economist.* London: RH Business Books.
Robbins, P. (2004). *Political Ecology.* Oxford: Wiley-Blackwell.
Smith, N. (1984). *Uneven Development: Nature, Capital, and the Production of Space.* Oxford: Blackwell.
Swyngedouw, E. (1999). Modernity and hybridity: nature, *Regeneracionismo,* and the production of the Spanish waterscape 1890–1930. *Annals of the Association of American Geographers* 89: 443–465.

Excellent introductions to various perspectives on both the Anthropocene and planetary boundaries can be found in:
Castree, N. (2014a). The Anthropocene and geography I: the back story. *Geography Compass* 8: 436–449.
Castree, N. (2014b). The Anthropocene and geography III: future directions. *Geography Compass* 8: 464–476.
Castree, N. (2014c). Geography and the Anthropocene II: current contributions. *Geography Compass* 8: 450–463.

Castree, N. (2015). The Anthropocene: a primer for geographers. *Geography* 100: 66–75.

Steffen, W., Richardson, K., Rockström, J. et al. (2015). Planetary boundaries: guiding human development on a changing planet. *Science* 347 (6223): 472–475.

Whitehead, M. (2014). *Environmental Transformations: A Geography of the Anthropocene*. London: Routledge.

An introduction to the thinking of Greta Thunberg is provided in:

Thunberg, G. (2019). *No One Is Too Small to Make a Difference*. London: Penguin.

Important insights into various dimensions of the current global climate and ecological emergency can be found in:

Diffenbaugh, N.S. and Burke, M. (2019). Global warming has increased global economic inequality. *Proceedings of the National Academy of Sciences* 116: 9808–9813.

Ellen MacArthur Foundation (2018). *What Is a Circular Economy?* Cowes: Ellen MacArthur Foundation.

Gleick, P.H. and Palaniappan, M. (2010). Peak water: conceptual and practical limits to freshwater withdrawal and use. *Proceedings of the National Academy of Sciences* 107: 11155–11162.

Guzman, A.T. (2013). *Overheated: The Human Cost of Climate Change*. Oxford: Oxford University Press.

Pimm, S.L., Jenkins, C.N., Abell, R. et al. (2014). The biodiversity of species and their rates of extinction, distribution, and protection. *Science* 344 (6187): 1246752.

A popular book on the global climate and ecological crisis written for a general audience is:

Monbiot, G. (2017). *Out of the Wreckage: Finding Hope in the Age of Crisis*. London: Verso.

Important publications on the global climate and ecological crisis and public policy include:

Burke, M., Hsiang, S.M., and Miguel, E. (2015). Global non-linear effect of temperature on economic production. *Nature* 527 (7577): 235–239.

Kaza, S., Yao, L.C., Bhada-Tata, P. et al. (2018). *What a Waste 2.0: A Global Snapshot of Solid Waste Management to 2050*. Washington, DC: World Bank.

International Union for Conservation of Nature (IUCN) (2020). *The IUCN Red List of Threatened Species. Version 2020-1*. Gland: IUCN.

International Energy Association (2020). *World Energy Outlook 2020*. Paris: International Energy Association.

International Monetary Fund (2019). *Fiscal Monitor: How to Mitigate Climate Change*. Washington, DC: International Monetary Fund.

IPCC (2020). *AR6 Synthesis Report: Climate Change 2022*. Geneva: IPCC.

Masson-Delmotte, V., Zhai, P., Pörtner, H.-O. et al. (eds.) (2018). *Global Warming of 1.5°C: An IPCC Special Report*. Geneva: IPCC.

Stern, N. (2007). *Stern Review on the Economics of Climate Change*. London: HM Treasury.

World Health Organization (2016). *Ambient Air Pollution: A Global Assessment of Exposure and Burden of Disease*. Geneva: WHO Press.

An overview of Liverpool's economy and response to the global climate and ecological emergency is offered by:

Boyle, M., Crone, S., Endfield, G. et al. (2020). Cities and the climate and ecological emergency: the Liverpool City Region response. *Journal of Urban Regeneration and Renewal* 13: 365–379.

Parkinson, M. (2019). *Liverpool Beyond the Brink: The Remaking of a Post Imperial City*. Liverpool, Liverpool university Press.

The case for a Green New Deal is set out in:

Klein, N. (2019). *On Fire: The Burning Case for a Green New Deal*. London: Allen Lane.

Rifkin, J. (2019). *The Green New Deal: Why the Fossil Fuel Civilization Will Collapse by 2028, and the Bold Economic Plan to Save Life on Earth*. New York: St. Martin's Press.

Website Support Material

A range of useful resources to support your reading of this chapter are available from the Wiley Human Geography: An Essential Introduction Companion Site http://www.wiley.com/go/boyle.

Chapter 10

Homo urbanus: Urbanization and Urban Form from 1800

Chapter Table of Contents

Chapter Learning Objectives

By the end of this chapter you should be able to:
- Describe the origins of contemporary urbanization in nineteenth- and early twentieth-century Europe and in European settler colonies, and explain the

---

Human Geography: An Essential Introduction, Second Edition. Mark Boyle.
© 2021 John Wiley & Sons Ltd. Published 2021 by John Wiley & Sons Ltd.
Companion website: www.wiley.com/go/boyle

ways in which the modern rise of the city has been embroiled, entangled, and intertwined in the Industrial Revolution and the development of the capitalist economy.

- With reference to the Chicago School of Urban Sociology, reflect critically upon models that purport to describe and explain the spatial organization of the twentieth-century industrial city.
- With reference to the work of British-born, US-based, Marxist geographer David Harvey and the case of the city of Glasgow in Scotland, discuss the meaning and implications of thinking about the industrial city as a "spatial fix" which was destined to be created only then to be destroyed by capitalism.
- Describe rates and levels of urbanization in world regions from the mid-twentieth century to the present, comment on forecasts of likely trends in urbanization to 2050, and explain the significance of the claim that we now live in an "urban age."
- Outline and comment on the monstrous agglomerations and complex urban forms that are being etched onto the surface of the earth today. Explain what is meant by "megalopolis." Describe the chief characteristics of the "postmetropolis." Reflect critically upon why slums exist in cities in the Global South. Discuss the role of the Chinese state in directing urbanization in that country.
- With reference to the idea of "planetary urbanization," explain the claim that the city is a problematic category of analysis. Reflect critically upon the status and future of urban studies as a field of enquiry if we are to take seriously the claim that planetary urbanisation has given rise to an age "after and beyond" the very idea of the city.

Introduction

The purpose of this chapter is to describe and explain the urbanization of the earth's surface that has occurred from 1800, and to reflect critically upon the urban forms that have been imprinted onto the face of the earth as a consequence.

Although cities have featured in many prior civilizations, it has only been with the ascendance of Western society, and in particular with the rise of the capitalist economic system and the Industrial Revolution, that the modern city has become home for the majority of humankind. Urbanization, the movement of people from the countryside to the city, began in earnest in the late eighteenth and early nineteenth centuries in European societies and in countries birthed by European emigration (principally the settler colonies of the United States, Australia, Canada, and New Zealand). The Industrial Revolution gave birth to the industrial city (variously labeled the "smokestack," "blue-collar," "rustbelt," "nineteenth-century," "Victorian," or "modern" city), an agglomeration with a particular form and function. Subsequently, cities grew to be, and still remain, the dominant habitat for human beings in the Global North and now the Global South. But new relationships between capitalism and urbanization are depositing, in their wake, built landscapes – glistening carpets of urban sprawl – which bear little relation to late nineteenth- and early twentieth-century spatial conditions. The traditional industrial city increasingly looks to be an obsolete spatial form. And so we need to reconsider how to think about the relationship between capitalism and urbanization, and the spatiality of agglomeration and urban form after the age of the industrial city.

In this chapter, we will examine the function, form, and fate of the industrial city before considering the form and function and future of the various novel, complex, and expansive agglomerations that are being etched onto the surface of the earth today.

The Modern Rise of the City from 1800

Europe, Capitalism, Industrialization, Urbanization, and the Industrial City

This rise of *homo urbanus* began in earnest in Europe around 1800 and spread first to European settler colonies.

Swiss-based Belgian economic historian Paul Bairoch and US political scientist Gary Goertz provide an authoritative overview of historical trends in urbanization (Bairoch and Goertz 1986). According to Bairoch and Goertz, the percentage of the population in Europe and its offshoot countries living in cities with more than 5000 inhabitants was only 7 to 9% in 1300 CE. By 1800, this figure remained stagnant at around 10.7%. By 1880, however, 23.6% of people had become urbanized, and this figure grew to 35.7% by 1914, 47.1% by 1950, and 66.4% by 1980. Table 10.1 highlights variations in levels of urbanization across Europe and the rest of the developed world from 1800 to 1910. Urbanization began first in the countries of western Europe and was pioneered most vigorously by the United Kingdom. Meanwhile, the countries of Scandinavia and

Table 10.1   Urbanization levels: percentage of population living in settlements >5000 inhabitants, by country (1800–1910).

Country	1800	1830	1850	1880	1900	1910
Europe	10.9	12.6	16.4	23.5	30.4	32.8
United Kingdom	19.2	27.5	39.6	56.2	67.4	69.2
Belgium	20.5	25	33.5	43.1	52.3	56.6
Netherlands	37.4	35.8	35.6	44.5	47.8	50.5
Germany	8.9	9.1	15	29.1	42	48.8
Italy	18	19	23	28	35.5	40
France	12.2	15.7	19.5	27.6	35.4	38.5
Spain	17.5	17.5	18	26	34	38
Switzerland	7	7.5	11.9	20.4	30.6	37.1
Denmark	15.6	14.1	14.6	23	33.5	35.9
Austria–Hungary	6.5	7.1	9.7	16	25.6	28.5
Norway	7	7.2	9	16	24.3	25.1
Bulgaria	5.5	5.5	6	11	15	22.1
Sweden	6.6	6.6	6.8	12.5	19.3	22.6
Greece	11.5	12	14	16	21	22
Romania	7.5	7.5	11	14	17.3	16
Portugal	15.5	15	15	15	15.7	15.6
Russia	5.9	6	7.2	10.6	13.2	14.3
Finland	3.5	3.5	3.7	6.1	10.4	12.6
Serbia	10	10	10	10	9.8	10
Other DCs	5.5	7.9	13.9	24.4	35.6	41.6
TOTAL	10.7	12.3	16.2	23.6	31.3	34.4

DCs: developing countries. Source: From Bairoch, P., & Goertz, G. (1986). © 1986 SAGE Publications.

Eastern Europe were left somewhat behind, and urbanized only later and at a slower pace. Starting as overwhelmingly rural countries in 1800, the rest of the developed world at that time experienced rapid and sustained urbanization across the nineteenth century. Urbanization rates (defined by Bairoch and Goertz as the rate of growth per year of populations living in a settlement with greater than 5000 inhabitants) peaked in Europe at the end of the nineteenth century at around 2.2%. But urbanization rates were much higher in the United States, Canada, Australia, and New Zealand, persisting at above 4% for most of the nineteenth century, and peaking at 7.2% in 1850.

How might one account for this historically unprecedented transition to urban living? Was it an accident that as Europe transitioned from a feudal and agricultural society to a capitalist and industrial one, it was first to bear witness to the rise of the modern city? In their efforts to make sense of ongoing processes of urbanization, human geographers have sought to situate urban agglomerations with the wider economic and political contexts of which they are part and to which they contribute. Specifically, they have explored the ways in which the capitalist economic system has become entangled, embroiled, and intertwined with urbanization processes in different regions and at different times (see Deep Dive Box 10.1).

The rise of capitalism changed the structure of Europe's economy, from one based around agricultural production to one predicated upon industrial production. Capitalism brought to agriculture new farming methods and technologies, and the production of food expanded dramatically. Having figured out a way of producing secure food surpluses, it became possible for society to "release" significant numbers of people from the binds of the land. Landless laborers could now (and were now forced to) migrate to cities and become part of the urban industrial workforce. Capitalism brought with it the factory and smoke-filled landscapes of the Industrial Revolution and created conditions in which cities became not only possible but necessary. The esthetics, essence, and spirit of life in mid-twentieth-century industrial cities were famously captured by English artist Laurence Stephen Lowry (1887–1976). Lowry's paintings were set in the industrial cities of northwest England, but they serve well to depict what life was like in industrial cities more generally at this time (Plate 10.1).

The Industrial Revolution both birthed and, in turn, depended upon cities to provide:

- Employment for landless laborers forced from the countryside by mechanization, and thereby a safety valve to protect society from threats posed by mass unemployment, poverty, social chaos, and revolution.
- A large pool of cheap labor that factory owners could exploit and that could more easily be replenished if unruly.
- A means through which factory owners could collaborate with other industries (suppliers, competitors, and customers) in support of their production processes.
- A key node in transportation networks that factory owners could use to get raw materials and workers to the factory and goods to the market.
- An infrastructure through which factory owners could source utilities (energy, water, and communications) at viable costs.
- An effective channel through which factory owners could retail their goods to large quantities of people.
- A cost-effective way for industrial and political leaders to promote the welfare of factory workers, providing them with public services such as schools, housing, hospitals, and sewerage and sanitation, thereby sustaining their labor power.

Deep Dive Box 10.1 David Harvey on Capitalism and Urbanization

Within the discipline of geography, undoubtedly the first and most significant author to relate the emergence, growth, and mutation of cities to the rise and development of the wider capitalist economic system, and to broader processes of industrialization and economic development, was British-born and US resident Marxist geographer David Harvey.

A graduate of Cambridge University, David Harvey has worked at Bristol University, Oxford University, Johns Hopkins University in Baltimore, and most recently the City University of New York. When Harvey moved to Baltimore in 1969, it was a blue-collar city subject to industrial collapse and burdened with significant poverty and racial tension. This move proved to be a pivotal moment in Harvey's career. Baltimore led him to question the utility of existing perspectives within urban geography, many of which he had previously advocated, but which now seemed to be incapable of making sense of the city's predicament or informing policy makers what ought to be done in response. Harvey believed that Baltimore's fate was inextricably linked to its changing role in the US capitalist economy and the new international division of labor. He believed that geographers needed to develop a much deeper understanding of how capitalism produces spaces, including cities. He argued that the problems faced by many urban communities had their roots in contradictions, inequalities, and malfunctions inherent within the capitalist system. He concluded that the only deep-seated and durable solution to urban disadvantage was to overturn this system.

Increasingly, Harvey became drawn to the ideas of Karl Marx and sought to develop a Marxist theory of urbanization. Harvey's Marxist approach was first introduced in his 1973 book *Social Justice and the City*. In his 1982 book *The Limits to Capital*, Harvey sought to bring the discipline of geography into a conversation with Marxism and ventured a sophisticated Marxist theory of how space is both produced by and in turn shapes the course of capitalism. He applied this Marxist theory of the role of space in capitalist societies to the study of the city in his books *The Urbanization of Capital* (1985a) and *Consciousness and the Urban Experience* (1985b), and sought to illustrate his approach in his magisterial book *Paris, Capital of Modernity* (2003). In his book *Rebel Cities: From the Right to the City to the Urban Revolution* (2012), Harvey sketched out the ways in which fundamental changes in society might lead to fairer societies and more just cities.

According to Harvey, every societal formation imprints its own spatial form onto the earth's surface, and in the capitalist era, the city constitutes the principal spatial expression. At any moment in time, capitalism needs to create particular agglomerations of people, offices, industries, transport facilities, shops, streets, energy facilities, sewerage and sanitation schemes, waste management facilities, open spaces, and so on to function. Harvey refers to these temporary and transitory geographical arrangements as capitalism's "spatial fix." Spatial fixes crystallize onto the landscape when they are needed, but these built environments quickly dissolve when they no longer help the economic system to work. For Harvey, old spatial fixes end up frustrating capitalism's ability to do new things. Eventually change becomes necessary, and through the "creative destruction" of the old built environment, a new spatial fix is created that serves capitalism's needs, at least for another while. For Harvey, capitalism is a restless economic system, and spatial fixes have a short lifespan.

Plate 10.1   Laurence Stephen ("LS") Lowry, 1887–1976, *Industrial Landscape* (1958) in the north of England. Source: https://www.wikiart.org/en/l-s-lowry/industrial-landscape-1958.

We might say, then, that urbanization began when it did and where it did, in response to the rise of the capitalist economic system, and in particular to the Industrial Revolution's need for urban agglomeration.

If the European industrial city was the first to rise, then it was also the first to fall. As a spatial fix that enabled the Industrial Revolution to take place, the industrial city is increasingly becoming obsolete, if not an impediment to future growth of the capitalist economy. It is experiencing a period of "creative dismantling" as capital works to create new urban agglomerations better suited to its emergent needs.

Since the early 1950s in particular, public discussion has increasingly centered upon the claim that deindustrialization has effectively sounded the death knell for the industrial city. Cities whose economies have traditionally relied upon shipbuilding, heavy engineering, the production of locomotives, and chemical industries have found themselves unable to compete with rival industries based in low-wage regions. Cities such as Glasgow, Liverpool, Newcastle, Pittsburgh, Philadelphia, Baltimore, Cleveland, and Detroit have for some time now been forced to contend with deindustrialization and its attendant problems of factory closure, unsightly and dangerous derelict and vacant land, high unemployment and poverty, and outmigration and depopulation. Moreover, this hemorrhaging of industrial jobs continues to this day. With the ongoing collapse and restructuring of the US automobile industry, it is even possible, for instance, that

Detroit – Motor City – may well be in the throes of an irreversible decline. There now exist "left-behind places," ruins in the landscape, communities rendered redundant by global capital, and abandoned it seems to managed decline and terminal marginality.

For the foreseeable future, all declining industrial cities are liable to be rendered especially vulnerable by the heightened mobility of capital. The capacity of transnational corporations to switch their operations from location to location around the world has created a truly global economy and one that has engendered a competition between places to secure investment. This inter-locality competition has grown to include competition for government facilities, tourists, the consumer dollar, and, perhaps most importantly, skilled and talented workers. Increasingly, the task facing city leaders is the creation of urban conditions sufficiently attractive to lure prospective firms, civil servants, tourists, shoppers, and creative workers. For industrial cities in particular, the image of the city has become of paramount importance. Rustbelt cities are trying to alter their appeal by manipulating their soft infrastructure (cultural and leisure amenities, for instance) and by refashioning their economic attractiveness (through provision of grants, property, transport facilities, or tax abatements, for instance).

According to David Harvey (1989), city leaders are now behaving more as urban entrepreneurs than as urban managers. They are diverting ever more of their budgets away from the provision of welfare services to the poor and toward glossy, hyped-up city marketing campaigns. They are turning cities into commodities to be sold to the highest bidder. In so doing, both the façade and the vibe of cities are being given a make-over. By (re)building "prestige," "flagship," or iconic landmarks; expanding cultural, heritage, and museum amenities; and hosting major leisure, cultural, sporting, or political events (such as the Olympic Games, meetings of heads of government, comedy and jazz festivals, cultural events, and so on), cities are trying to put themselves on the map at the expense of others. Urban entrepreneurs may well salvage some cities ravaged by deindustrialization. But according to Harvey, not every city can win. For less fortunate cities, city marketing projects will erect a public face that will conceal the reality of the rot that lies underneath. Many city marketing projects, then, amount to little more than applying lipstick on the gorilla!

The Form of the Industrial City: The Chicago School of Urban Sociology and Beyond

What form did the industrial city take?

The Chicago School of urban sociology (peaking from the 1920s through the 1960s) was a pivotal moment in the rise of urban geography as a systematic branch of enquiry within human geography. While recognizing that the spatial organization of the city varied from place to place, Chicago School scholars believed that it was possible to discern a number of universal features and set forth a series of generic models of urban form.

According to Howard Becker, who completed his PhD in sociology at the University of Chicago in the late 1940s and early 1950s, an "origin myth" has grown around the Chicago School that is not entirely accurate. This myth comprised a number of assumptions. The school was founded by Albion Small, William Isaac Thomas, and George Herbert Mead, who jointly created a unified schema for studying society.

Propelled by the energy and vision of these founders, in the 1920s and 1930s, Robert E. Park and Ernest W. Burgess then developed a specific sociology of the city and trained a new generation of disciples such as Everett C. Hughes and Herbert Blumer. After World War II, the University of Chicago experienced an influx of students whose education was funded by the G.I. Bill. They were veterans of war, well-traveled, talented, and energetic; many became students of Hughes and Blumer, and created a "second Chicago School" comprising an extensive series of studies of the city. In this later work, echoes of Mead, Park, and Blumer resonated strongly (Figure 10.1).

According to Becker, this origin myth conveys a false sense of coherence, unity, and consistency. There was no "school of thought." But there was a "school of activity" – a cluster of people all vigorously interacting as they sought to make sense of the urbanization of Illinois and what this meant for the city of Chicago and its residents.

Perhaps the most famous attempt to describe and explain the spatial organization of the nineteenth- and early twentieth-century city was that provided by Chicago School scholars Robert E. Park, Ernest W. Burgess, and Roderick D. McKenzie in their book entitled *The City*, published in 1925. *The City* was written at a time when Chicago was rapidly expanding. This context shaped the thinking of the book's authors. In 1850, Chicago had a population of circa 29 000 and was the 24th-largest city in the United States. By 1900, its population had increased to 1.7 million, making it the second-largest city. Further growth ensured that Chicago retained this ranking, and by 1930 the city's population amounted to circa 3.4 million. Chicago grew as a result of significant immigration, incorporating the arrival of Irish migrants in the 1840s; late nineteenth-century

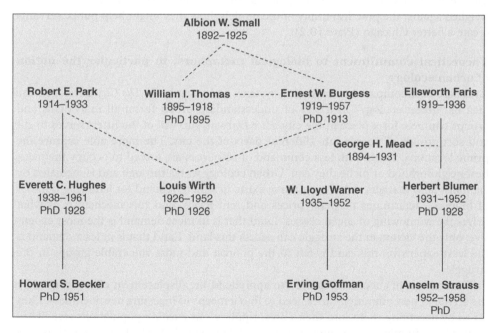

Figure 10.1 The Chicago School urban sociologists and their relationships (advisor-student relations are denoted by a solid line, and other influential relations by a dashed line). Source: Redraw from Wayne G. Lutters, An Introduction to the Chicago School of Sociology, Interval Research Proprietary.

immigrants from Europe (principally Germany, the United Kingdom, Scandinavia, and the Netherlands); turn-of-the-twentieth-century arrivals from Poland, Lithuania, Ukraine, Hungary, the former Czechoslovakia, Greece, and Italy; early twentieth-century Jewish immigrants from the Russian-held territories; and "internal" African-American migrants from southern states.

Reflective of the approach adopted by the wider Chicago School, *The City* was underpinned by five key intellectual foundations: a commitment to the philosophy of classical pragmatism, a theoretical interest in urban ecology (incorporating biological metaphors such as "survival of the fittest," "urban metabolism," and "succession and invasion"), an investment in cartographic innovations that attempt to visualize urban forms and layouts, a methodological proclivity toward ethnographic research methods and the case study, and an (unconscious) weddedness to a type of progressive liberalism that some now construe as reflecting a degree of political naïveté.

Adherence to the philosophy of Pragmatism

Pragmatism was a philosophical tradition that emerged in the United States and particularly in Chicago, which held that scholars needed to be modest in their efforts to make sense of complex puzzles such as why cities exist and why they develop the spatial forms they do. Instead of searching for a grand truth (a big story that renders the city legible), more modest "working theories" were needed. These theories could be held to be true only for as long as they worked in practice. Instead of searching for big and final explanations, urban scholars should search only for humble ideas that work in practice. Theories are always provisional and tentative. The measure of a theory is always its practical usefulness. And so the value of the land-use models the School generated was always weighed against the practical utility of the models: how they might help public servants create a better Chicago (Plate 10.2).

Theoretical commitment to biological metaphors, in particular the notion of urban ecology

Cities are best compared to biological systems and organisms. *The City* developed the idea of "urban ecology" as a way of understanding urban form: all individuals and groups compete for a place in the city. *Pace* Darwin, survival of the fittest serves to sift and sort groups of people into different parts of the city. The more able capture the prime locations. Those with less command of resources are forced to occupy the next-best neighborhood or niche they can. Urban ecology works through and is mediated by capitalist land markets. A bid–rent curve exists in which demand for land (the number of bidders) determines property prices and rent levels, and this mechanism is what drives the winnowing of social classes. Land that is in most demand is the most expensive; only the victors in the struggle can access this land. Land that is in least demand is the least expensive; this land is left to the poorest and most vulnerable groups in the struggle.

The concept of succession is used to apprehend the displacement effects caused by the arrival of new migrants. Cities need to find a means of ingesting new incomers. They do so by accepting them into their cores, rippling out existing populations steadily to their edges. Immigration begins a process of "invasion" in which new social groups colonize the poorest neighborhoods in the Zone of Transition, or "inner city." Through time, "invasion" leads to "succession," and the prior occupants are forced to take flight.

Plate 10.2   An aerial view of Chicago. Source: Siqbal, https://commons.wikimedia.org/wiki/File:
Chicago_Downtown_Aerial_View.jpg. Public Domain.

These occupants, when resourceful, contribute to this process by actively searching out
a different and better niche. The city then is driven to expand by immigrants who create
a wave of invasion, succession, and displacement – which radiates from a downtown or
Zone of Transition outward.

Cartographic innovations and mappings of morphology of cities

Perhaps the most famous chapter in *The City* was that written by Chicago-based sociologist Ernest W. Burgess. Burgess argued that, because of the workings of urban ecology, cities are organized on the basis of a series of concentric zones.

Central Business District (CBD). At the heart of the city is the CBD, where commerce and retailing activities, as well as recreational and political land uses, congregate.
These actors are among the most resourceful and can successfully bid for the most
accessible, and therefore most expensive, locations in the conurbation. Beyond impoverished homeless people, few inhabitants reside in the CBD.

Zone of Transition. Here, light industry, warehouses, and industrial factories
cluster; these businesses need a central location to maximize their access to pools of
labor but, given that they occupy significant quantities of land, they prefer not to pay
the premium rates for land and property that mark the CBD. The Zone of Transition
is also home to the newest and poorest immigrant populations (living in overcrowded
residences at high densities) and bohemian populations courting alternative lifestyles.

Zone of Workingmen's Homes. Second-generation immigrants who have advanced beyond the Zone of Transition dwell in the Zone of Workingmen's Homes. Close enough to the city center to walk to work, this population lives at relatively high densities, making rent affordable. This zone is continually being invaded by incomers from the Zone of Transition but depleted by yet more established populations who search for a better life in neighborhoods farther afield.

The Residential Zone. This zone is home to the aspiring middle classes bent on upward social mobility. Normally educated and working in clerical and even professional and managerial roles, these residents live in lower-density accommodation and enjoy superior residential amenities, gardens and open spaces.

The Commuter Zone. The upper classes live in the Commuter Zone at a distance from the city center. Enjoying the sense of living in smaller communities with a village atmosphere, commuters require salaries commensurate with purchasing a detached home with a spacious garden and investing in an automobile for work and leisure purposes. The Commuter Zone is reserved normally for only the most established and longest-existing immigrant groups and for those capable of the greatest upward social mobility.

Methodological focus upon ethnography and the case study

The Chicago School tested its ideas in practice using Chicago as a laboratory. Although they did employ a range of quantitative methods, they were among the first proponents of the virtues of exploring the city using qualitative methods. In particular, ethnographic immersion in field sites exploring the lives of the people they were studying from their perspective (from the bottom up) was a hallmark of the School. And so it became famous for its series of anthropological, rich, immersive, field-based mappings of social and spatial life in Chicago.

Political commitment to Liberalism

Possibly the biggest feature – and arguably problem – with the Chicago School's urban ecology approach is its lack of attention to politics. It is assumed that cities are organized on the basis of a Darwinian-like struggle for survival in which only the fittest win. But this is not an especially "natural" starting point. That competition for survival exists at all reflects the fact that societies are choosing to live according to this principle. The free market is a socially engineered institution – not a neutral point of departure. Burgess and Park fail to take seriously the importance of urban politics. They offer themselves as, at best, liberal reformers – trying to improve the lot of the most marginal and vulnerable within the existing system. Clearly, deeper structural change within this system is an option too; it is entirely possible to socially engineer new organizing rules for cities and therefore new geographies of the city (See Deep Dive Box 10.2). This is never contemplated in a serious way in *The City*.

The Creative Destruction of the Industrial City: Insights from Glasgow

In 1717, Glasgow was a small village with a population of circa 15 000 perched on the River Clyde in west-central Scotland. By 1800, the population of the city had grown to 80 000. Behind this growth were the Glasgow "tobacco lords" – entrepreneurs who

Deep Dive Box 10.2   Land-Use Change and the Declining Relevance of the Burgess Concentric Zones Model

Many attempts have been made to apply the Burgess Concentric Zones model (see Figure 10.2) and to test its validity with respect to Chicago and other cities. Although it offers a useful starting point, it is clear that land-use patterns in the city have changed. Later in this chapter you will see the vastly more complex urban agglomerations that are emerging as the twenty-first century unfolds. Here we simply point to at least five distinctive processes that have radically transformed the traditional industrial city and shaped its present layout. In combination, these five processes have rendered the Burgess Concentric Zones model steadily less relevant.

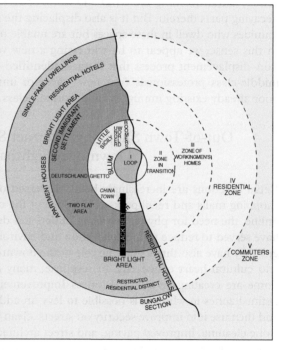

Central Business District (CBD). At the heart of the city is the Central Business District (CBD) where commerce, finance, and retailing activities, as well as recreational and political land uses congregate. These actors are among the most resourceful and can successfully bid for the most accessible, and therefore most expensive, locations in the conurbation. Beyond impoverished homeless people, few inhabitants reside in the CBD.

Zone of Transition. Here, light industry, warehouses, and industrial factories cluster; these businesses need a central location to maximize their access to pools of labor but, given that they occupy significant quantities of land, they prefer not to pay the premium rates for land and property that mark the CBD. The zone of transition is also home to the newest and poorest immigrant populations (living in overcrowded residences at high densities) and bohemian populations courting alternative lifestyles.

Zone of Workingmen's Homes. Second-generation immigrants who have advanced beyond the Zone of Transition dwell in the Zone of Workingmen's Homes. Close enough to the city center to walk to work, this population lives at relatively high densities, making rent affordable. This zone is continually being invaded by incomers from the Zone of Transition but depleted by yet more established populations who search for a better life in neighborhoods further afield.

The Residential Zone. This zone is home to the aspiring middle classes bent on upward social mobility. Normally educated and working in clerical and even professional and managerial roles, this group of people live in lower-density accommodation and enjoy superior residential amenities, including more gardens and open spaces.

The Commuters Zone. The upper classes live in the commuter zone at a distance from the city center. Enjoying the sense of living in smaller communities with a village atmosphere, commuters require salaries commensurate with purchasing a detached home with a spacious garden and investing in an automobile for work and leisure purposes. The commuter zone is reserved normally for only the most established and longest existing immigrant groups and for those capable of the greatest upward social mobility.

Figure 10.2   Chicago-based sociologist Ernest W. Burgess' 1925 Concentric Zones model of urban form. Source: Based on Park, R.E., Burgess, E.W., and McKenzie, R.D. (1925). *The City.* Chicago: The University of Chicago Press.

Urban Planning, Slum Clearance, Social "Projects," New Towns, and the High-Rise Solution

Urban planners have played a significant role in transforming the layout of cities. In particular, in a bid to deal with the late nineteenth- and early twentieth-century drift of inner-city neighborhoods (in the Zone of Transition and the Zone of Workingmen's Homes) into slums, urban planners have cleared slum housing from inner-city areas and relocated populations to high-rise tower blocks, large public housing estates (for example, "the projects" in the United States and public

(Continued)

Box 10.2 (*Continued*)

housing estates in British cities), and so-called "satellite towns" or "new towns" built on greenfield sites at a distance from the city. Many public housing projects/estates and high-rise towers have themselves become problems, though, and some have now been demolished.

Gentrification and the Middle-Class Invasion of the Waterfront

Young, white-collar, middle-class professionals are returning to dilapidated inner-city areas (and, in particular, waterfront locations along the banks of rivers) to live near social and cultural amenities in the city center. This process, called gentrification, is transforming the CBD and Zone of Transition, breathing new life into decaying parts therein. But it is also displacing the remaining working-class communities who dwell in these spaces but are unable to resist development pressures. In this sense, we appear to be witnessing a new variant of the invasion–succession–displacement process that Burgess identified – on this occasion, it is young middle-class professionals (not new waves of immigrants) who are displacing poor already existing immigrant and working-class communities.

Out-of-Town Shopping, Internet Shopping, and the Downtown Fightback

Retail functions are increasingly being dispersed out of the city center and into shopping malls and retail parks at the edge of the city. Also, as more people shop online, the need for physical retailing outlets has declined. Although these trends have served to reduce traffic congestion and environmental pollution in city centers, they have also threatened the role of the downtown as the commercial, social, and cultural heart of the city. In response, many city centers are fighting back. Some are creating so-called Business Improvement Districts (BIDs). BIDs are defined zones in which it is possible to levy an additional tax on local businesses and then use it to improve security on streets, clean streets, and remove graffiti; for stone cleaning, improved paving, and street architecture; and to enhance the general atmosphere of the city center. The idea is to make the city center a more attractive prospect for shoppers.

Counter-Urbanization, Metropolitan Villages, and Exurbia

Some middle-class inhabitants, especially those with families and those nearing retirement, are deciding to leave cities and are choosing to locate in small satellite commuter towns and villages (called variously metropolitan villages and exurbias), sometimes at considerable distances from the city. Driven by lifestyle factors and lower property prices, such "counter-urbanization" is also being lubricated by higher rates of multi-car households, growing work-at-home opportunities, and improved transportation options. This flight to the countryside brings both advantages (helps sustain local shops and employment) and disadvantages (places pressures on local services, increases traffic congestion, raises house prices, and changes the character of local places) to surrounding towns and villages.

Extensive Suburbanization and Edge Cities

Particularly in the case of Australian cities, but also in sprawling cities in North America and Europe, extensive suburbanization is stretching the edge of cities so far from the traditional city center that it is changing the character of suburban communities. As cities sprawl toward the rural hinterland (some cities stretch over 100 miles from one edge to another), some suburban areas are developing mini city centers themselves – often called Edge Cities – which are becoming increasingly self-contained and enjoying their own character, culture, and sense of community.

exploited Britain's colonization of North America, imported tobacco from, among other places, Virginia and North and South Carolina, manufactured tobacco products in Glasgow, and exported these products to the rest of Europe. But the American War of Independence halted these tobacco imports. In the ensuing period of 1776 to 1850, the riches that had been accumulated by the tobacco lords were reinvested in new textile factories. Using raw materials imported from lands that were falling under British colonial rule, Glasgow became a city specializing in the manufacturing of cotton, wool, flax (linen), and silk, and thereafter a magnet for migrant workers. By 1850, its population had grown to 350 000.

Recognizing the virtues of Glasgow's location on the River Clyde and the abundance of local resources such as coal and iron ore, from 1850 onward Glasgow then turned its attention to shipbuilding, heavy engineering, machine making, and the production of locomotives (Plate 10.3). Family dynasties that emerged at the time of the tobacco trade, and that consolidated their power during the period when textiles dominated, now funded the expansion of a rapid industrialization of the city. Migrants flooded to the city from the Scottish Highlands and Lowlands and from Ireland, and between 1860–1910 Glasgow's population expanded rapidly and peaked at nearly 1.2 million in 1911. By 1920, Glasgow was crowned the second city of the British Empire and was held in esteem as a world-leading industrial city.

Alas, for most of the past century, Glasgow has been dealing with the legacy of this past. First, the city's rapid expansion created overcrowding and inner-city slums; resolving these occupied much of the period from World War I to 1977. Planning solutions have fundamentally transformed the form of the city. Second, de-industrialization eroded the economic base of the city, and from 1977 to the present economic regeneration has been the primary priority. Although the city has steadily recovered, some areas, neighborhoods, and communities have benefited from regeneration more than others. Much remains to be done.

Plate 10.3   The *RMS Empress of Britain*, built at John Brown's Shipyard, Glasgow, between 1928 and 1931 and owned by the Canadian Pacific Steamship Company. Alas, the *Empress* was sunk by a Nazi U-boat in 1940.

Glasgow is a city that has been profoundly wounded by the failures of the postwar planning system. Throughout the twentieth century, but especially from 1945 to 1977, the city's priority has been to remedy the problem of significant overcrowding in the city's downtown core. Inner-city slum clearance paved the way for a massive expansion of new build within the social rented sector. This was a hybrid outcome of both the Abercrombie and Matthew Clyde Valley Regional Plan (1946), which proposed clearing the inner city and displacing people beyond the city's boundaries, and the Bruce Plan (1949), which proposed demolishing the city center but retaining people within the city's boundary. In the end, demolition of 29 Comprehensive Development Areas (CDAs, such as the Hutcheson Gorbals, Possil Park, and Anderston) in the inner city was coupled with a large-scale construction of peripheral housing estates ("projects," such as Easterhouse, Castlemilk, and Drumchapel), new towns (East Kilbride, Cumbernauld, Livingston, and Irvine), and high-rise towers (including the infamous Red Road Flats) in both the city and its hinterland.

Planning solutions transformed the geography of the city. First, the redistribution of population to peripheral housing estates and the policy of overspill to the various new towns surrounding the city were socially differentiated. Those who relocated tended to be among the city's younger, more able, and more skilled population. This selective migration process was to leave behind in the inner city a disproportionate concentration of elderly, unemployed, and people with limiting long-term illnesses. Second, overspill was to reduce the city's population from its maximum of 1.2 million in 1920 to the new low of 680 000 by 1980. In itself mourned as a sign of the decline of the city and an indication of its loss of prestige, population decline was also to have important implications for the local tax base. Third, at first a utopian landscape of hope and ambition, peripheral housing estates in particular quickly became problems in and of themselves. Occupying monotonous and bleak landscapes, removed from the amenities of the town center, and lacking the basic services required for everyday life, communities failed to germinate and prosper.

In 1976, British-Canadian economic historian Sydney Checkland famously conjured the idea of "The Upas Tree" (a tree which poisons everything under its canopy) to apprehend the impact of shipbuilding and heavy engineering in Glasgow; the decline of both had served to pull the city more generally into economic crisis. Underinvestment, disinvestment, capital flight, and unemployment came to blight the local economy. No matter which cocktail of measures was used, Glasgow topped virtually every national survey of deprivation and poverty. Glasgow's leaders visited similar rustbelt and smokestack cities in the United States, including Minneapolis, Baltimore, and Pittsburgh. Impressed by the post-industrial renaissance these cities were experiencing, the City Council recognized the importance of transforming the image of the city and putting it back on the investment map. From 1977, local counselors began to divert more local resources from welfare, or urban managerialism, to local economic development and city marketing, or urban entrepreneurialism. Glasgow has launched and invented every conceivable form of place marketing from the construction of cultural attractions to the hosting of hallmark events, the renovation of public space, street lighting, slogans, formal advertising campaigns in the UK press and beyond, and so on.

While Glasgow has enjoyed some success in transitioning to a post-industrial city, it is important to be conscious of the limitations of the city's recovery. Today, Glasgow is "Miles Better," but only for some (Plate 10.4). There remains much to do if the city's recovery is to be enjoyed by all its residents. Glasgow today is home to 600 000 people.

Plate 10.4   Glasgow's Miles Better campaign: regeneration, civic boosterism, and place marketing in Glasgow in the early 1980s. Source: https://en.wikipedia.org/wiki/Glasgow%27s_miles_better#/media/File:Glasgows_Miles_Better.png.

Mapping the Urban Age

Globally, an "urban age" has been boldly declared. We are now, it seems, an urban species; more people across the entire world live in cities than in rural areas. And so the plight of humanity has come to be seen as inextricably tied to the welfare of cities. Making cities work efficiently, sustainably, and equitably is the primary task of our times. Now more than ever, our future cannot be discussed without reflection upon the fate of the city. Foregrounding the primacy of the city to human flourishing and prosperity, at Habitat III in Quito in 2016, 193 UN member states signed the Quito Declaration on Sustainable Cities and Human Settlements for All and set what has been termed a "New Urban Agenda" for the world.

Every four years since 1988, the United Nations Population Division (UNPD) releases statistical updates on global urbanization trends. In its *2018 Revision* of its biannual *World Urbanization Prospects*, the UNPD charts world urbanization from 1950 to the present and offers estimates and projections of likely trends in world urbanization to 2050 (United Nations Population Division 2018).

The difficulty of compiling global statistics on urban populations is that national censuses define what is and what is not an "urban" area in widely varying ways. Because the United Nations does not collect data directly, it has been forced to follow the definitions that are used in each country. Although unavoidable, this of course raises questions as to whether comparisons between countries are meaningful.

- The world's urban population has grown rapidly from 751 million in 1950 to 4.2 billion in 2018. Globally, more people live in urban areas than in rural areas, with 55% of the world's population residing in urban areas in 2018 (Figure 10.3). In 1950, 30% of the world's population was urban, and by 2050, 68% of the world's population is projected to be urban.

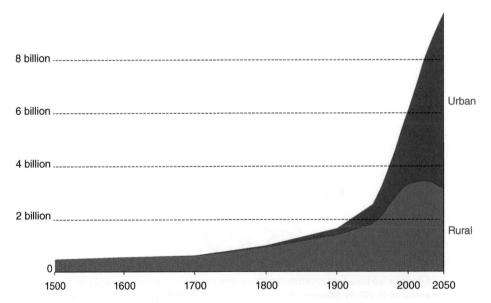

Figure 10.3 World urban and rural populations, 1950–2050. Source: Redraw from Hannah Ritchie and Max Roser (2018) - "Urbanization" OurWorldInData.org. CC BY 4.0.

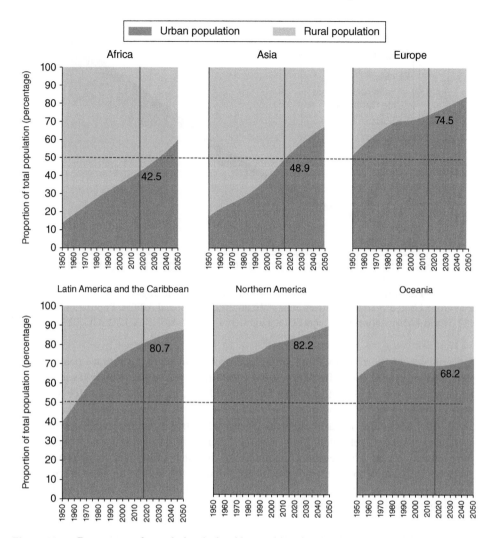

Figure 10.4 Percentage of population 'urban' by world region 1950–2050. Source: Redraw from United Nations, Department of Economic and Social Affairs, Population Division (2019). World Urbanization Prospects 2018: Highlights (ST/ESA/SER.A/421). CC BY 3.0 IGO.

- Today, the most urbanized regions include Northern America (with 83% of its population living in urban areas in 2018), Latin America and the Caribbean (81%), Europe (74%), and Oceania (68%) (Figures 10.4 and 10.5). The level of urbanization in Asia is now approximately 51%. In contrast, Africa remains mostly rural, with 44% of its population living in urban areas. Asia, despite its comparatively lower level of urbanization, is home to 54% of the world's urban population, followed by Europe and Africa with 13% each.
- North America, already the most urbanized region in the world in 1950 (at 64% urban), continued to urbanize and today it is 83% urban. It will further urbanize in the future, although at much slower rates (between 0.5 and 1% per annum), and it will remain the most urbanized region by 2050 (over 89% urban). In 1950, Europe was 53% urban; today it is 74% urban. Europe, too, will continue to urbanize, again at a slower pace (around 0.25% per annum), and by 2050 will be nearly 84% urban.

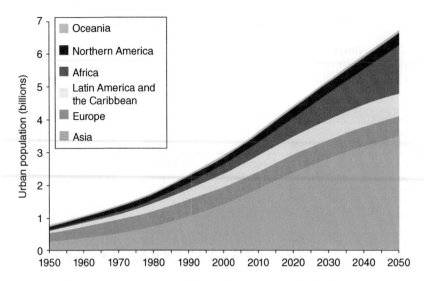

Figure 10.5 The 'urban' population of the world by geographic region, 1950–2050. Source: Redraw from United Nations, Department of Economic and Social Affairs, Population Division (2019). World Urbanization Prospects 2018: Highlights (ST/ESA/SER.A/421). CC BY 3.0 IGO.

- Oceania was 68% urban by 1964, and urbanization rates appear to have peaked, settled, and stagnated thereafter. It remains 68% urban today. Oceania will see further urbanization by 2050 (when it will be 72% urban), but it will not catch up with North America and Europe.
- By 1960 Latin America and the Caribbean was already more than 50% urbanized, and from this relatively high base its urban centers have grown consistently, by between 2 and 4% per annum, meaning that it is now over 81% urbanized. Although urbanization rates will fall to less than 1%, by 2050 Latin America and the Caribbean's population will be 88% urban, making it the second most urbanized region in the world after North America.
- Both Asia and Africa began the period (1950) with the lowest levels of urbanization, at 17 and 15%, respectively. Significant rates of urbanization (between 3 and 5% per annum) mean, however, that today Asia is approximately 51% urban, while Africa is 44% urban. By 2050, Asia's population will be approximately 66% urban, but urbanization rates will be reduced to less than 1%. The percentage of the population of Africa that is urban will increase sharply to around 59%, and urbanization rates of between 2 and 3% will persist until 2050.
- In 1950, the largest city in the world was New York–Newark, with a population of 12.34 million. Only two cities had more than 10 million inhabitants. For the most part, the largest cities in the world (18 of the top 30) were to be found in the United States, Europe, Japan, and the Soviet Union. Today Tokyo is the world's largest city with an agglomeration of 37 million inhabitants (Plate 10.5), followed by Delhi with 29 million, Shanghai with 26 million, and Mexico City, Mumbai, and São Paulo with around 20 million inhabitants each. By 2035, the world is projected to have 41 megacities, which are cities with more than 10 million inhabitants. 28 of the 30 largest cities in the world will be in Asia, Africa, and Latin America. Delhi will be the most populous city with 43 million inhabitants, Tokyo will house 36 million, Shanghai 33 million, Dhaka 31 million, Cairo 29 million, Mumbai 27 million, and Kinshasa 27 million (Figure 10.6).

Plate 10.5   Tokyo, with a population of 37 million, is the largest city on earth. Source: Image by Pierre Blaché from Pixabay.

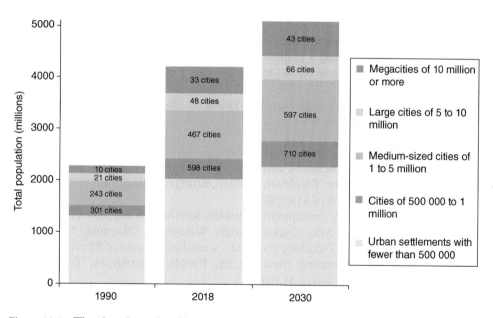

Figure 10.6   The changing urban hierarchy, 1990–2030. Source: Redraw from United Nations, Department of Economic and Social Affairs, Population Division (2019). World Urbanization Prospects 2018: Highlights (ST/ESA/SER.A/421). CC BY 3.0 IGO.

Urban Form After the Age of the Industrial City: The Shape of Things to Come?

Whilst it might be true that more people are living in urban areas than at any period in human history, arguably the city – or at least the industrial city and the city as envisaged by the Chicago School – is being eclipsed as a significant urban form. If capitalism created the traditional city, capitalism is also destroying the traditional city. Specifically, the neat idea of a circular city, radiating out from a CBD, passing through a Zone of Transition, with residential districts giving way to leafy suburbs, looks oddly anachronistic when one thinks of the monstrous and complex urban forms that are crystallizing on the earth's surface today. It is clear that although capitalism still needs urban agglomeration to prosper, the kinds of urban forms that it requires are much more complex than those that characterized the industrial city. Contemporary urbanization is creating historically novel agglomerations – sprawling carpets of glistening buildings and vast urban galaxies – which bear little relation to the city as conventionally understood. And so, we might ask, what might urban form look like after the age of the industrial city?

Here, we consider four developments of consequence.

Megalopolis: From Cities to Networks and Urban Galaxies?

In 1961, French-born and US resident geographer Jean Gottmann published *Megalopolis: The Urbanized Northeastern Seaboard of the United States*. In this book Gottmann argued that cities were now expanding in some regions to the extent that they were coagulating into a single overall agglomeration and becoming physically integrated and, to a degree, functionally integrated. Gottmann referred to this new type of urban space as a "megalopolis."

Gottmann's focus was upon the megalopolis forming on the northeastern seaboard of the United States incorporating Boston, New York City, Philadelphia, Baltimore, and Washington, DC, and boasting a population in 1960 of 27 million (today 52 million). The term "Boswash" was later coined to capture the sprawl that Gottmann had first discovered.

Since Gottmann's seminal insight, numerous claims have been made regarding the existence of other megalopoli. Examples include:

- The *Blue Banana* in Europe (incorporating Liverpool, Manchester, Leeds, Birmingham, London, Brussels, Antwerp, Amsterdam, Rotterdam, The Hague, Luxembourg, Rhine-Ruhr, Frankfurt, Munich, Stuttgart, Basel, Zürich, Turin, and Milan – a total population of 110 million)
- The *Taiheiyō Belt* in Japan (incorporating Ibaraki, Saitama, Chiba, Tokyo, Kanagawa, Shizuoka, Aichi, Gifu, Mie, Osaka, Hyōgo, Wakayama, Okayama, Hiroshima, Yamaguchi, Kitakyūshū, Fukuoka, and Ōita – a total population of 80 million)
- *Central Mexico* (incorporating Mexico City, Puebla, Cuernavaca, Toluca, and Pachuca – a total population of 35 million)

Los Angeles: The 100-Mile City and Our Postmetropolis Future?

In the 1980s and 1990s, there emerged a Los Angeles School, which sought to track emerging trends and to map and account for the layout of the cities of today. Just as the Chicago School had used Chicago as its laboratory in the 1920s, 1930s, and 1940s, the Los Angeles School has searched the city of Los Angeles for clues as to how cities might be changing today (Plate 10.6).

Referring to Los Angeles as a postmetropolis, Angeleno urban geographer and planner Edward Soja identified six geographies of urban restructuring in the city. In *Postmetropolis: Critical Studies of Cities and Regions*, published in 2000, Soja labels these

Plate 10.6   Los Angeles' sprawling urban carpet: view from Mulholland Drive, Santa Monica Mountains of southern California. Source: Image by Schwingi from Pixabay.

"Six Discourses on the Postmetropolis" (Soja 2000). Of the six, one discourse takes priority over the others:

1) **Exopolis:** We might say that the cities of the past were marked by the dominance of an urban core – both functionally (in orchestrating how the wider conurbation worked) and symbolically (through its impressive skyscrapers, important buildings, and public plazas) – over the rest of the conurbation. Today it is sprawling urban hinterlands and their various edges and suburbs that organize metropolitan regions, including former downtown cores. For Soja, the center of Los Angeles is as much in Orange County, the corridor from Malibu to Long Beach, the San Fernando Valley, and San Bernardino and Riverside Counties as it is in the traditional downtown.

   Soja observes how a raft of concepts have been used to capture this fact: "post-suburbia," "metroplex," "technopoles," "technoburbs," "urban villages," "county-cities," "regional cities," and "the 100-mile city." He himself prefers the idea of **exopolis** – literally meaning a city turned inside out.

   Soja's remaining five discourses explain, derive from, and/or supplement the key idea that Los Angeles is an exopolis:

2) **Flexicity**: According to Soja, changes in the economic geography of Los Angeles have transformed the geography of the city. In particular, as clusters of high-technology firms have emerged, industry has moved from the center to the suburbs – forming distinctive "technoburbs."

3) **Cosmopolis**: Remarkably, however, global movements of capital, people, and trade have saved the downtown from extinction. The global film industry in Hollywood continues to prosper. Los Angeles also serves as a headquarters for some of the most powerful transnational corporations in the world and has well-developed

financial services and producer services sectors. Meanwhile, the former city center has become a magnet for immigrants (Mexicans and, to a lesser extent, migrants from El Salvador, China, the Philippines, Guatemala, Korea, Iran, Vietnam, Taiwan, and Armenia). Driven by these groups, a successful small light-industry sector (producing clothing, jewelry, and furniture) has arisen. Although no longer the center of the agglomeration, the Los Angeles downtown has become a part of the new polynucleated city and served as an incubator of this emerging immigrant community.

4) **Metropolarities**: Los Angeles is also a city that exemplifies the growing social divisions and polarities that exist between society's haves and have-nots. The neighborhoods of Bel Air and Beverly Hills somehow coexist alongside almost developing-world neighborhoods such as those found in South-Central Los Angeles.

5) **Carceral Archipelagos**: The 1992 Los Angeles riots revealed the potential for social inequalities and racial tensions to boil over into social chaos, violence, and crime. In response, Los Angeles has become something of a fortified city, with a premium placed upon surveillance and safety. Gated communities, private policing firms, and high-technology security devices work to preserve an uneasy peace, but at times they themselves can be experienced as menacing.

6) **Simcity**: One does not need to enter Universal Studios in North Hollywood or Disneyland in Anaheim to view almost imaginary landscapes recreated and presented as reality. The suburban streets and shopping malls of Orange County and Irvine, for instance, look almost like forged or fake replicas of the suburbs envisaged in popular fantasy-lore. Soja refers to Los Angeles, therefore, as a hyper-real city, built in part around irreal "dreamscapes."

Slums in the Global South: Urbanization Without Industrialization?

Although emerging megacities in the Global South boast complex urban forms, to date much attention has been focused upon the haphazard and unplanned mushrooming of urban slums or shantytowns at the edges of, and sometimes at the heart of, cities. This phenomenon points to at least one critical distinction between Global North and Global South cities: while industrialization was a progenitor of urbanization in nineteenth-century Europe, today in parts of Africa, Asia, and Latin America, urbanization is occurring in the absence of industrialization.

In the early 1970s, the most famous of these were the *favelas* that were developing in Rio de Janeiro, Brazil, providing super-impoverished migrants with barely livable conditions. As urbanization spread to other regions in the Global South, the problem of urban slums became more universal and urgent. An estimated one billion people now dwell in shantytowns in such megacities as Jakarta, Lagos, Karachi, Mumbai, Delhi, and Dhaka; this figure is predicted to increase to two billion by 2030 and three billion by 2040 (Plates 10.7 and 10.8).

In his 2006 *Planet of Slums*, US urban scholar Mike Davis argues that urbanization in the Global South is distinctive because it appears to be decoupled from industrialization. Urbanization, Davis argues, has arisen as a consequence of prior colonial exploitation and failed postcolonial transitions; transformations in agriculture and the rise of a new class of super-poor landless laborers; and the inability of governments to provide social protection, a consequence of austerity forced on the Global South by the Global North and also a result of corruption and kleptocratic government. Urban slums are the direct result of urbanization without growth or development, overseen by poor governance. Davis warns that leaving vast numbers of people in such poverty, insecurity, and

Plate 10.7 Dharavi Mumbai, India, considered to be one of Asia's largest slums. Source: Nishantd85 https://de.wikipedia.org/wiki/Datei:Salaam_Bombay!.jpg. CC BY-SA 3.0.

Plate 10.8 Dharavi Mumbai, India: everyday life. Source: https://upload.wikimedia.org/wikipedia/commons/3/3f/Dharavi_DSC3254_(5842973175).jpg.

hopelessness is not only wrong, and not only threatens the peace and stability of megacities, but also presents a threat to the security of the entire planet. Dwelling on the brink of survival, slum populations can turn to and are often victims of violence – gang wars, drug racketeers, and human traffickers preside over domestic abuse, sexual assault, violent assault, theft, kidnapping, and murder. Moreover, slums are breeding grounds for militant and fundamentalist groups, including those drawn to sectarian hatred.

Dealing with urban slums remains the most pressing challenge facing municipal authorities in the Global South today. For some, only by changing the position of the Global South in the global economy and promoting development will authorities be able to effect meaningful change. Slums are of course a source of embarrassment to aspiring Global South cities with ambitions to go global. They are also a barrier to elite capital investment urban development projects. Clearing slums for new infrastructure has been promoted as a panacea. Others however warn that the grand urban visions which are now being espoused by some municipal authorities are further impoverishing slum populations and that new and more progressive development plans are needed. Either way, political activism by social groups from shantytowns will be needed if the voice of the urban poor is to be heard (See Deep Dive Box 10.3).

In the meantime, Davis argues, slums in the Global South will continue to attract missionaries from all religious hues who view the urban poor as a significant audience and one receptive to evangelization. Paraphrasing Karl Marx, Davis argues that religion continues to be the opium of the people.

Deep Dive Box 10.3   Decolonizing Urban Studies:
Urban Studies in, by, and for the Global South

In his 2011 book *African Cities: Alternative Visions of Urban Theory and Practice*, US-based scholar Gareth Myers counters Ed Soja's claim that Los Angeles is the prototype of the twenty-first-century city and wryly asks, "What if the post-metropolis was Lusaka?"

Whether it be the Chicago School, the Los Angeles School, or any other school of thought with an interest in the city, it would appear to be the case that Anglo-American urban geographers call the shots: it is they who decide how we are to make sense of the city. And yet historically novel types of urbanization are creating complex urban forms throughout the Global South. Surely new theories – and not the same old Western theories that get recycled – are needed to make sense of urbanization and urban forms in the Global South? By calling attention to the importance of the geography of urban theory itself (from where the theory origi-nates), and in particular the Western- and metro-centricity of many urban theo-ries, concepts, and ideas, a number of British and US-based postcolonial scholars, such as Jenny Robinson, Ayona Datta, Eric Sheppard, Simone Abdoumaliq, Helga Leitner, Gareth Myers, and Ananya Roy have foregrounded the need to engage critically with the politics of universalism within urban studies and have called for a provincialization, decolonization, and historicization of urban studies.

What would a decolonized urban studies look like?

Whilst recognizing the crucial importance of global economic and political pro-cesses and the connections between capitalism, industrialization, and urban agglomeration, an urban studies produced in, by, and for the Global South might embrace theories that:

- Are more attentive to context, recognizing that cities in Africa, Asia, and Latin America have experienced colonialism, decolonization, independence, neoco-lonialism, and incorporation into the global economy in different ways.

- Construe connections between political economy, urbanization, and spatial conditions as locally nuanced and mediated by cultural conditions and unique institutional environments.
- Approach cities as messy amalgams of land uses, perhaps even anarchic, not easily understood, and certainly not open to interpretation through simplistic urban models that claim to be universal.
- Resist recycling urban concepts from the Global North that bear a questionable relationship with cities in the Global South (such as CBDs, Zones of Transition, suburbs, new towns, and so on), and possibly work to create a new lexicon more suited to context and the heterogeneity of the urban processes at play.
- Refuse dominating and imperial gazes on the city – aerial and bird's-eye views from above that seek to control urban space – preferring ways of knowing the city that derive from deep immersion in everyday life.
- Reject teleologies that locate cities on a continuum from underdeveloped to developed, preferring instead to view cities as following their own unique historical trajectories.

China's Instant Megacities: State-Orchestrated Urbanization?

China's encounters with urbanization differ from both the classic Western experience and that which marks the Global South today. Chinese cities contributed 85% of the country's total Gross Domestic Product (GDP) in 2018, and by 2025 this will rise to 95% of total GDP, with the eight largest cities generating almost 25% of the nation's wealth. Against this backdrop, to a far greater extent than has been the case in Western cities, the Chinese state has actively sought to trigger and steer urbanization, to control its speed in different regions, and to shape and plan the spatial organization of the city (as a reflection of urban planning, some Chinese cities boast world-class, futuristic, and innovative iconic buildings, landscapes, and public spaces). It does so through central planning, urban planning, the selective allocation of resources, and the strategic deployment of monetary and fiscal levers. China's urbanization, then, has been shaped by the muscular interventions of the Chinese government and its provincial tiers; its result is the speed of creation of its "instant" megacities.

China's experiment with rapid mass urbanization is one fraught with challenges as well as opportunities (See Deep Dive Box 10.4). The Chinese government continues to wrestle with the competing options of encouraging the development of a small number of megacities or fostering a spreading of growth in a more balanced way across the urban system. In addition, there is a fear that Chinese cities will not be able to provide sufficient housing, transportation, health, and educational services to meet the needs of their growing populations. Moreover, given that rates of urbanization are so closely tied to economic performance, there is a concern that overly rapid urbanization might contribute to an overheating of the Chinese economy. All the while, the plight of hundreds of millions of impoverished migrants who flock from the countryside to urban locations in search of factory employment remains troubling. China continues to serve as a low-wage economy only because these migrants accept wages and living conditions that are

Deep Dive Box 10.4   Chinese Encounters with Urbanization

It is especially important to be careful when considering data on Chinese urbanization. Through time, the proportion of the Chinese population officially designated as "urban" has grown – and certainly migration and to a lesser extent natural increase (more births than deaths) have driven this expansion. But increases have also arisen simply as a product of repeated changes in official definitions of "urban," the widening of city boundaries to include hinterlands, and the official registration of settlements hitherto overlooked. Moreover, some migration to cities, while viewed as permanent, is in fact seasonal and circular.

These qualifications notwithstanding, China has undoubtedly witnessed a momentous redistribution in its population over the past 30 years.

- In 1949, only 11% of China's population lived in an urban area. By 1978, that figure had increased only marginally, to 18%, and even as late as 1992 China's urban population amounted to only 23% of the country's total population. By 2005, nonetheless, a rapid surge in urbanization had raised this figure to 44%, and forecasts suggest that by 2025 the proportion of the Chinese population living in cities will be 64%.

- From a base level of 254 million in 1990, China's urban population more than doubled to 572 million by 2005. By 2025, China will have added a further 350 million to its urban stock (equivalent to the population of the United States), and by 2030 China's urban population will reach 1 billion.

- Between 2005 and 2025, China is projected to increase its number of megacities from two to eight (Beijing, Shanghai, Tianjin, Shenzhen, Wuhan, Chongqing, Chengdu, and Guangzhou), its big cities (cities with 5 to 10 million inhabitants) from 12 to 15, and its midsized cities (cities with 1.5 to 5 million inhabitants) from 69 to 115. Today, China has 113 urban areas that surpass the 1 million population threshold. North America and the EU combined have 114 million-plus cities.

- China is witnessing the rise of its own peculiar megalopolis – the *Pearl River Delta Megalopolis* in the southern province of Guangdong (incorporating Hong Kong, Shenzhen, Dongguan, Guangzhou, Foshan, Jiangmen, Zhongshan, Zhuhai, Macau, and Huizhou), which in 2020 was estimated to be home to over 70 million inhabitants (Map 10.1). Although Guangzhou sits at the epicenter of this megalopolis, Shenzhen, dubbed by some "the instant city," provides perhaps the most dramatic example of China's rapid urbanization. From a population base of 94 000 in 1980, Shenzhen grew to house 8.3 million people by 2005, reflecting an annual growth rate of 19.6% a year. From 2005, the city has continued to expand at around 5% per year. By 2010, the city was home to 10.3 million people, and forecasts suggest Shenzhen's population could rise to 12–15.5 million by 2025.

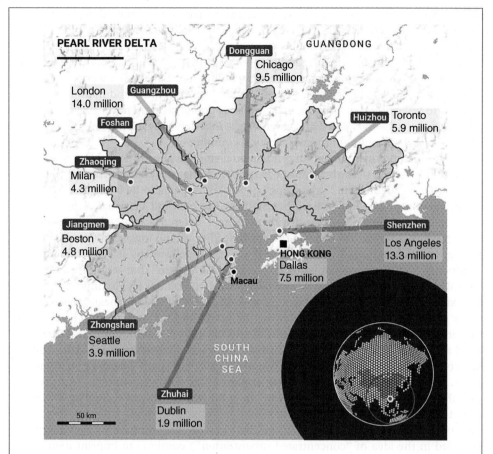

Map 10.1 The Pearl River Delta Megalopolis in southern China. Source: *Chinese Statistical Yearbook 2020*, https://www.chinayearbooks.com.

barely tolerable. Meanwhile, the Chinese state has misjudged its capacity to enforce urbanization in some provinces, leading to the construction of large, eerie ghost cities – lying empty and waiting for migrants who have yet to arrive.

Planetary Urbanization: Urban Studies After the Age of the City?

Is it sufficient to document and explain the variegated and novel urban forms that are being etched onto the face of the earth today? Is there not a need to probe deeper into the very status of the "city" as a category of analysis and object of study (see Rickards et al. 2016)?

At the heart of the idea of the "urban age" is the claim that more people now live in cities than live in rural areas. This claim is in turn predicated upon the assumption that it is possible to define what counts as "urban" and what is properly "non-urban" or countryside. Of course, this concern has vexed the field of urban studies since its inception. Much ink has been spilled trying to establish agreed statistical parameters; should population size and density be the critical determinants or not, and if so, what metrics should be used? But in a new twist, recent scholarship has cast suspicion on "urban age" thinking. These dissenters begin with the claim that all attempts to define the urban and to count the number or estimate the proportion of people who live in urban areas are doomed to failure from the outset; in fact, these attempts simply further confuse and muddle thinking. We need to recognize that contemporary urbanization is creating novel agglomerations, configurations, networks, flows, and assemblages that defy easy capture, evade simplistic nomenclature, and confound simple statistical counting.

What lies behind this moment of fundamental stock taking? French Marxist sociologist Henri Lefebvre's declaration in the early 1970s, that the entire surface of the earth is now *de facto* caught up in "urbanizing" processes, appears to be being given new life by, among others, US and Swiss urbanists Neil Brenner and Christian Schmid. They are currently wrestling with the idea that we have entered a contemporary epoch of "planetary urbanization" in which city-centric studies will no longer do (Brenner 2014; Brenner and Schmid 2014). The idea of the "urban age" sweeps aside terminological, definitional, and measurement issues concerning the "urban" that are, in fact, not only unresolved in the literature but also unresolvable. For Brenner and Schmid, the pervasive extent of urbanization processes at work in the world today means that it is no longer appropriate to attempt to delineate the urban and the non-urban. There is now "nothing outside the urban." For them, the implication is stark: questions need to be raised about the future of the field of urban studies as a consequence. Urban studies, they conclude, is best construed as a twentieth-century "spatial ideology" without an object. There is a need to jettison urban studies as currently constituted, identify and cast off the "city lens," and try to develop an entirely new "way of seeing" more suited to contemporary spatial conditions.

According to Brenner and Schmid, mechanisms of concentration or agglomeration (captured in the idea of "concentrated urbanization") continue to imprint monstrous built environments onto the surface of the earth. But these agglomerations are drawing into their mesh regions and populations formerly considered to be non-urban and in turn are extending and radiating back into these hinterlands. Entangled and imbricated in urbanizing processes, hinterlands cannot now be said to lie beyond the urban; they bear the stamp of the agglomerations of which they are part and become constitutive of them. There is no part of nature, the countryside, or wilderness, no matter how remote or isolated from urban agglomerations or how untouched in appearance (including even, say, Alpine mountains, the Sahara Desert, Arctic ice sheets, Brazilian forests, or the Atlantic marine environment), that has escaped the consequences of urbanization. Somehow, all these ecosystems have become part of the operational landscapes of the world's cities (captured by the phrase "extended urbanization"). Mechanisms of globalization would appear to be further undoing traditional city–hinterland relationships, creating new deployments, withdrawals, and penetrations of global capital into hinterlands, and carving out new multi-scalar and multi-city–hinterland relations.

Although there might be no need for a field of urban studies, Brenner and Schmid continue to see a future for the "field formerly known as urban studies." Perhaps better entitled "the field of urbanization studies," this branch of knowledge would seek to apprehend the geographical variegated, complex, hierarchical, multi-scalar, and always processual sociospatial assemblages and territorial configurations that are crystallizing and dissolving today. To this end, they call for a new focus upon relationships between concentrated urbanization and extended urbanization, and a new canon of work exploring what they term the "hinterland *problematique*." There is a need for an exploration of cartographic techniques, visualization tools, and interpretive lexicons that might yield better insights into the contemporary urbanization of the entire world. The mission of hinterland studies will be to map the recursive relationships that exist between agglomerations and the operational landscapes from which they draw nourishment but upon which they scar with pollutants.

Brenner and Schmid's call for revolutionary change and a clean slate for urban studies is certainly provocative. For some, the idea that cities no longer exist in any meaningful sense is a step too far. Urban agglomeration does matter and mediates social, political and economic processes in definitive ways. But this call does raise a number of important questions. Is "the city" or more broadly the idea of "agglomeration" still a meaningful category of process and object of analysis? What exactly is overlooked or overemphasized when we assume and use "the city" as an analytical starting point rather than questioning its existence? What exactly does lie beyond the reach and influence of the city?

Conclusion

This chapter has recounted the story of the globalization of urbanization from 1800, beginning with its modern-day birth in Europe and initial diffusion to countries formed through European emigration (principally the United States, Australia, Canada, and New Zealand), and incorporating its extension to Latin America, Asia, and Africa. It has placed under scrutiny the different relationships that have existed between the capitalist economic system, industrialization and development, urbanization, and spatial agglomeration at different moments in time and in different regions of the world. It has reflected upon the claim that we are now living in an "urban age" in which the future of humanity and the future of cities will become one and the same thing. But it has also noted that urbanization is creating vastly more complex agglomerations of human activity and that as sprawling urban carpets emerge today, we need to think anew about the spatial organization of urban space. Traditional ideas of the city, inherited from nineteenth-century spatial conditions and twentieth-century models of urban form, need to be questioned, and the morphology of urban spaces considered afresh. More radically, if there is any merit in the idea of planetary urbanization, it might be that it is even necessary to consider again the entire mission of the field of urban studies.

Checklist of Key Ideas

- Human geographers have come to understand that the fortunes of urban agglomerations are inextricably embroiled, entangled, and intertwined with wider economic and political processes, specifically rooted in the capitalist economic system. Although cities have featured in many past civilizations, mass urbanization began only around 1800 in (West) European countries and in countries established through European emigration. It is impossible to understand nineteenth- and early twentieth-century urbanization apart from the rise of capitalism and the Industrial Revolution. The mechanization of agriculture and industrialization of the economy made cities both possible and necessary.

- The Chicago School (illustrated most clearly in the Burgess Concentric Zones model) sought to describe the spatial organization of the nineteenth- and twentieth-century industrial city using the central idea of urban ecology. Land-use change has rendered the work of the Chicago School somewhat obsolete.
- The city of Glasgow in Scotland provides keen insights into the creative destruction of the industrial city and the impact of changing land-use patterns in the city.
- Western countries remain highly urbanized and (with the exception of Oceania) urbanizing countries today; Latin America was among the first of the remaining continents to urbanize and is highly urbanized today; Asia and Africa were among the last continents to urbanize but are rapidly urbanizing today and will soon be majority urban. As more people live in cities, we are entering a new "urban age" in which the plight of humanity and the fate of our cities will become irrevocably connected.
- The twenty-first century will witness ever larger and more complex urban agglomerations. Urban studies will need to invent a whole new slew of concepts if it is to apprehend and render intelligible these emerging urban forms. The ideas of megalopolis, postmetropolis, urbanization without industrialization and state orchestrated instant cities provide a useful start. There is a need in particular to be aware of the fact that many urban theories and concepts originate from scholarship undertaken in the Global North, and that urban studies will need to be decolonized and Global South theories and concepts developed if we are to make sense of contemporary urbanization and urban form in Asia, Africa, and Latin America.
- The concept of "planetary urbanization" raises questions about the status of the idea of the city as a category of analysis and object of study, and the future of urban studies as a field of enquiry.

Chapter Essay Questions

1) Describe the historical geography of the urbanization of the earth's surface that has occurred since 1800.
2) Capitalism, industrialization, and urbanization are all inextricably interlinked. Discuss.
3) In the "urban age", the fate of the city and the fate of humankind are one and the same thing. Discuss.
4) Outline, illustrate, and comment upon the idea of "planetary urbanization."

References and Guidance for Further Reading

Good general introductions to urban geography can be found in:

Hall, T. and Barret, H. (2018). *Urban Geography*, 5e. New York: Routledge.

Jonas, A.E., McCann, E., and Thomas, M. (2015). *Urban Geography; A Critical Introduction*. Oxford: Wiley.

Kaplan, D.H. and Holloway, S. (2014). *Urban Geography*. New York: Pearson.

Knox, P.L. and McCarthy, L. (2013). *Urbanisation: An Introduction to Urban Geography*, 3e. London: Prentice Hall.

Latham, A., McCormack, D., McNamara, K., and McNeil, D. (2009). *Key Concepts in Urban Geography*. London: Sage.

Pacione, M. (2009). *Urban Geography: A Global Perspective*, 3e. New York: Routledge.

Peake, L. and Bain, A.L. (eds.) (2017). *Urbanization in a Global Context*. Oxford: Oxford University Press.

Short, J.R. (2017). *An Introduction to Urban Geography*. London: Routledge.

Ward, K., Jonas, A.E.G., Miller, B., and Wilson, D. (eds.) (2018). *The Routledge Handbook on Spaces of Urban Politics*. London: Routledge.

A valuable guide for students undertaking urban research is:
Ward, K. (ed.) (2020). *Researching the City: A Guide for Students*. London: Sage.

A useful overview of nineteenth- and early twentieth-century urbanization in Europe and countries established through European migration can be found in:
Bairoch, P. and Goertz, G. (1986). Factors of urbanisation in the nineteenth century developed countries: a descriptive and econometric analysis. *Urban Studies* 23: 285–305.

Important works by David Harvey on the city include:
Harvey, D. (1973). *Social Justice and the City*. London: Edward Arnold.

Harvey, D. (1982). *The Limits to Capital*. Oxford: Blackwell.

Harvey, D. (1985a). *The Urbanization of Capital*. Oxford: Blackwell.

Harvey, D. (1985b). *Consciousness and the Urban Experience*. Baltimore: Johns Hopkins University Press.

Harvey, D. (2012). *Rebel Cities: From the Right to the City to the Urban Revolution*. London: Verso Books.

Other recent works that see to understand the nexus between capitalism and city include:
Rossi, U. (2018). *Cities in Global Capitalism*. London: Polity.

Scott, A.J. (2017). *The Constitution of the City: Economy, Society, and Urbanization in the Capitalist Era*. London: Palgrave MacMillan.

The fate of the industrial city is examined in classic urban scholarship including:
Bluestone, B. and Harrison, B. (1982). *The Deindustrialization of America*. New York: Basic Books.

Cooke, P. (ed.) (1989/2016). *Routledge Revivals: Localities: The Changing Face of Urban Britain*. London: Routledge.

Cox, K.R. and Mair, A. (1988). Locality and community in the politics of local economic development. *Annals of the Association of American Geographers* 78: 307–325.

Hackworth, J. (2007). *The Neoliberal City: Governance, Ideology, and Development in American Urbanism*. Ithaca, NY: Cornell University Press.

Hall, T. and Hubbard, P. (1998). *The Entrepreneurial City: Geographies of Politics, Regime, and Representation*. Chichester: Wiley.

Harvey, D. (1989). From managerialism to entrepreneurialism: the transformation in urban governance in late capitalism. *Geografiska Annaler* B 71: 3–17.

Hudson, R. (1989). *Wrecking a Region: State Policies, Political Parties and Regional Change*. London: Pion.

Lever, W.F. (1988). *Industrial Change in the UK*. London: Longman.

Martin, R. and Rowthorn, B. (eds.) (1986). *The Geography of Deindustrialization*. Basingstoke: Macmillan.

Power, A., Plogger, J., and Winkler, A. (2010). *Phoenix Cities: The Fall and Rise of Great Industrial Cities*. Bristol: Policy Press.

Robson, B. (1987). *Managing the City: The Aims and Impacts of Urban Policy*. London: Routledge.

Key readings on Glasgow's experience of deindustrialization include:
Keating, M.J. (1988). *The City That Refused to Die. Glasgow: The Politics of Urban Regeneration.* Aberdeen: Aberdeen University Press.
Pacione, M. (1995). *Glasgow: The Socio-Spatial Development of the City.* London: Routledge.

Other case study cities are covered in:
Beynon, H., Hudson, R., and Sadler, D. (1994). *A Place Called Teesside: A Locality in a Global Economy.* Edinburgh: Edinburgh University Press.
LeDuff, C. (2014). *Detroit: An American Autopsy.* London: Penguin Books.
Parkinson, M. (2019). *Liverpool Beyond the Brink: The Remaking of a Post Imperial City.* Oxford: Oxford University Press.

The ongoing marginalization of former industrial cities is charted in:
McQuarri, M. (2017). The revolt of the Rust Belt: place and politics in the age of anger. *The British Journal of Sociology* 68: S120–S152.

The key text in the Chicago School tradition is:
Park, R.E., Burgess, E.W., and McKenzie, R.D. (1925). *The City.* Chicago: The University of Chicago Press.

Other important works in this tradition include:
Harvey, L. (1987). *Myths of the Chicago School of Sociology.* Brookfield, VT: Gower Publishing Company.
Fine, G.A. (1995). *A Second Chicago School? The Development of a Postwar American Sociology.* Chicago: The University of Chicago Press.
Raushenbush, W. (1979). *Robert E. Park: Biography of a Sociologist.* Durham, NC: Duke University Press.
Short, J.F. (1971). *The Social Fabric of the Metropolis: Contributions of the Chicago School of Urban Sociology.* Chicago: The University of Chicago Press.

The topics of slum clearance, social housing estates, changing geographies of retail, gentrification, and suburbanization are examined in:
Fullilove, M.T. (2016). *Root Shock: How Tearing Up City Neighborhoods Hurts America, and What We Can Do About It.* New York: New Village Press.
Gleeson, B. (2006). *Australian Heartlands: Making Space for Hope in the Suburbs.* London: Allen and Unwin.
Hubbard, P. (2017). *The Battle for the High Street: Retail Gentrification, Class and Disgust.* London: Springer.
Lees, L., Shin, H.B., and López-Morales, E. (2016). *Planetary Gentrification.* Chichester: Wiley.
Keil, R. (2017). *Suburban Planet: Making the World Urban from the Outside In.* Cambridge: Polity.
Wylie, E.K., Lees, L., and Slater, T. (2008). *Gentrification.* London: Routledge.
Power, A. (1993). *Hovels to High Rise State Housing in Europe Since 1950.* London: Psychology Press.
Power, A. (1999). *Estates on the Edge.* London: Palgrave MacMillan.

The most recent global statistics on urbanization can be found in:
United Nations Population Division (UNPD) (2019). *World Urbanization Prospects: 2018 Revision.* New York: UNPD.

A productive way to start the process of engaging with Edward Soja's work on Los Angeles is to read Part 2 of his *Postmetropolis: Critical Studies of Cities and Regions*, entitled "Six Discourses on the Postmetropolis":
Soja, E. (2000). *Postmetropolis: Critical Studies of Cities and Regions.* Oxford: Wiley Blackwell.

See also:
Scott, A.J. and Soja, E.W. (eds.) (1998). *The City: Los Angeles and Urban Theory at the End of the Twentieth Century.* Berkeley, CA: University of California Press.

The seminal text introducing the idea of the megalopolis is:

Gottmann, J. (1961). *Megalopolis: The Urbanized Northeastern Seaboard of the United States*. New York: The Twentieth Century Fund.

Critical accounts of the problem of slums in the megacities of the Global South can be found in:

Davis, M. (2006). *Planet of Slums*. New York: Verso.

Datta, A. (2016). *The Illegal City: Space, Law and Gender in a Delhi Squatter Settlement*. London: Routledge.

A good overview of China's experience of urbanization is provided in:

Shiqiao, L. (2014). *Understanding the Chinese City*. London: Sage.

Kirkby, R.J. (2018). *Urbanization in China: Town and Country in a Developing Economy 1949–2000 AD*. London: Routledge.

Routley, N. (2020) Meet China's 113 cities with more than one million people. *Virtual Capitalist*, February 6.

Wong, D.W., Wong, K.K., Chung, H., and Wang, J.J. (2018). *China: A Geographical Perspective*. London: Guilford.

Good introductions to both decolonizing urban studies and postcolonial urban studies are:

Leitner, H. and Sheppard, E. (2016). Provincializing critical urban theory: extending the ecosystem of possibilities. *International Journal of Urban and Regional Research* 40 (1): 228–235.

Robinson, J. (2006). *Ordinary Cities: Between Modernity and Development*. London: Routledge.

Roy, A. and Ong, A. (2011). *Worlding Cities: Asian Experiments and the Art of Being Global*. Oxford: Blackwell.

Roy, A. (2016). Who's afraid of postcolonial theory? *International Journal of Urban and Regional Research* 40 (1): 200–209.

Introductions to the concept of "planetary urbanization" can be found in:

Brenner, N. and Schmid, C. (2014). The urban age in question. *International Journal of Urban and Regional Research* 38: 731–755.

Brenner, N. (ed.) (2014). *Implosions/Explosions: Towards a Study of Planetary Urbanisation*. Berlin: Jovis.

Brenner, N. (2019). *New Urban Spaces: Urban Theory and the Scale Question*. Oxford: Oxford University Press.

Good reviews of debates on the status of the city and the future of urban studies can be found at:

Rickards, L., Gleeson, B., Boyle, M., and O'Callaghan, C. (2016). Urban studies after the age of the city. *Urban Studies* 53 (8): 1523–1541.

Storper, M. and Scott, A.J. (2016). Current debates in urban theory: a critical assessment. *Urban Studies* 53 (6): 1114–1136.

Up-to-date data on China's population can be sourced from:

Chinese Statistical Press (2020). *Chinese Statistical Yearbook*. Beijing: Chinese Statistical Press.

Website Support Material

A range of useful resources to support your reading of this chapter are available from the Wiley Human Geography: An Essential Introduction Companion Site http://www.wiley.com/go/boyle.

Chapter 11

The Walling of the West: Migration, Hospitality, and Settling

Chapter Table of Contents

Chapter Learning Objectives

By the end of this chapter you should be able to:
- Define the terms "diaspora," "migrant," "refugee," "asylum seeker," "internally displaced person," "human trafficking," and "stateless person."
- Provide a typology of diasporas; identify the great diasporas in world history; and discuss the causes, scale, geographies, experiences, and legacies of the African diaspora in Americas.

Human Geography: An Essential Introduction, Second Edition. Mark Boyle.
© 2021 John Wiley & Sons Ltd. Published 2021 by John Wiley & Sons Ltd.
Companion website: www.wiley.com/go/boyle

- Map global migration stocks and flows, and discuss recent patterns and trends; compare and contrast the types of migrations that occur within and between countries in the Global South and Global North.
- Distinguish between integration policies and outcomes, and comment on recent histories and geographies of each.
- Appreciate that there are multiple ways to understand host country–migrant relationships (including through ideas such as assimilation, multiculturalism, diaspora space, securitization, and settling), and reflect upon the different policy implications implied by each understanding.
- Document and discuss the causes, geographies, and implications of the Syrian refugee crisis.

Introduction

> Those who build walls will become prisoners of the walls they put up. This is history.
>
> Pope Francis, July 2019

Although migration has been omnipresent in human history, its scale, pace and extent has grown steadily over the past 500 years, and perhaps today we live in the most migratory of all times.

Human geographical work on migration began in earnest with German-British geographer Ernst Georg Ravenstein's 1885 book *The Laws of Migration*. Ravenstein sought to identify the scientific determinants of the scale, direction, and distance of population movements within a given urban hierarchy and hinterland. Nearly a century later (1971), US geographer Wilbur Zelinsky introduced a temporal dimension and proposed that internal migration within any given system varies as countries modernize and develop. Zelinsky introduced a five-stage Mobility Transition Model to augment the more established Demographic Transition Model and claimed that as countries progress through demographic transition, they also bear witness to changing migration regimes: specifically, as countries develop, (increasingly longer distance) urban-to-urban migration comes to replace rural-to-urban migration. This shift is implicated in, and in turn impacted by, rates of fertility, mortality, and natural increase. From these seminal works, there has arisen within human geography a surging literature on (i) internal migration (within a country), comprising all inter-regional, rural–urban, urban–rural, and urban–urban movements and even intra-city residential mobility; and (ii) drivers, patterns, and trends in international migration stocks and flows.

The focus of this chapter is upon global migration, although we will also note the ways in which internal migration regimes nest within international migration regimes, and we will argue that both are best studied together under a common framework – an approach Singaporean geographer Elaine Ho (2019) calls "contemporaneous migration."

The core supposition of this chapter is that it is indeed impossible to understand the scale and geography of twentieth-century and early twenty-first-century global migration flows (and therein associated internal, or intra-country, mobility) without recognizing the long shadow that the rise of the world capitalist system and the history of European empire continue to cast. One of the chief legacies of the uneven world that the rise, reign, and faltering of the West has bequeathed has been a global migration system structured around flows of people within and between the Global South and the Global North. Notwithstanding the ongoing legacies sown by this history, and the fact that a significant quota of Global South migrants move only within the Global South itself, within the Global North there has emerged a heightened concern over the scale of immigration and the impacts of immigration on national economies, welfare systems, and cultures. A primary beneficiary of 500 years of globalization and a longtime advocate of liberal global trade regimes, with some exceptions, the West, it seems, is now pulling up the drawbridge. Once supportive of liberal tolerance and multicultural diversity, the West is increasingly witnessing a rising tide of right-wing populist nationalist movements, a tightening of borders, and demands for tougher immigration policies. This is the age of fortress building and taller walls. The immigrant is now once again the stranger. A new politics of hospitality beckons.

The Great Human Diasporas

Diaspora (Διασποράς)

Of Greek origin and commonly thought of as popularized by Jewish religious history, the idea of "diaspora" eventually worked its way into the social scientific literature in the 1950s, first through African studies and then Armenian studies. Taken to refer in principle to the scattering or scrambling of a particular population, in the past 20 years the concept has diffused widely throughout the social sciences and humanities, and in so doing has lost much of its original meaning.

US political scientist William Safran (1991) conceives of diaspora in terms of only one form of mass migration, that involving forced exile, a fraught and lengthy period of resettlement, and a failure to establish new roots speedily in regions of destination. Classically, diasporas form when people flee from persecution and natural disasters. It is unsurprising, then, that diasporic communities continue to identify as co-ethnics with kin they leave behind; there are no greater patriots, it is said, than disaffected exiles. For Safran, a hallmark of diaspora is a shared interest in preserving a common national, civic, or ethnic identity. Even when they plant firm roots in destination countries and even after many generations, many global diasporic peoples and their descendants continue to display a strong affinity to their ancestral homelands; some lead transnational lives, hold dual citizenship, and even co-locate. At their purest, diasporas are characterized by migrations where:

- Original communities have spread from the homeland to two or more countries.
- These communities are bound to their original geographical locations by a common vision, memory, or myth about their homeland.
- These communities harbor a belief that they will never be accepted by their hosts and therefore develop autonomous cultural and social lives.
- They believe that they or their descendants will return to the homeland should conditions prove favorable.
- They are strongly motivated to maintain support for their homeland, and continue to take an interest in the affairs of their homeland.

South African–born migration scholar Robin Cohen (1997), now a British resident, has meanwhile developed a wider concept, identifying five types of diaspora:

- Victim diasporas (for example, classic diasporas forced into exile by a traumatic historical event or series of events, such as the Jewish, African, and Armenian diasporas)

- Labor diasporas (for example, mass migration in search of work, such as the Indian and Turkish diasporas)
- Trade diasporas (for example, migrations seeking to open trade routes and links, such as the Chinese and Lebanese diasporas)
- Imperial diasporas (for example, migration among those keen to serve and maintain empires, such as the British and French diasporas)
- Cultural diasporas (for example, those who move through a process of chain migration, such as the Caribbean diaspora).

Case Study: The Atlantic Slave Trade and African Diaspora in the Americas

The "discovery" by Christopher Columbus of the Americas in 1492 marked a watershed in human history. Exploration was quickly followed by exploitation and colonization. The trafficking of African slaves formed part of an Atlantic trading system that connected Europe, Africa, and the Americas (Map 11.1). As pioneers of seafaring, it was perhaps unsurprising that it was Portugal that was first to instigate the trading of slaves. But the Dutch, French, Spanish, and British soon joined the endeavor. These powers created a trade triangle. Slave ships began in Europe and followed a familiar pattern. European-made goods, such as guns and textiles, were traded with African merchants in return for African slaves. Slaves were then forcibly shipped for sale in south, central, and north America. The Americas in turn provided raw materials to Europe for manufacturing.

Slaves were kidnapped not only from coastal regions that European traders had contacts with, but through time also from the interior of the continent. European merchants rarely captured slaves themselves, relying instead on stoking friction between tribal groupings and paying a tribute to tribal chiefs who undertook raids on rival settlements on their behalf. Captured slaves were then marched (often over great distances) to the nearest port. It is estimated that more than 10 million Africans were ensnared into bondage and forced to migrate across the Atlantic to be traded as commodities. Many (perhaps up to one-fifth or more) lost their lives during the slave raids and during transit in the horrific "coffin ships" that took those captured to the Americas. Although slaves were traded in central and south America from the early 1500s (Table 11.1), lore holds that the arrival in 1619 at the colony of Virginia (Jamestown) of the *San Juan Bautista* carrying "twenty or so slaves" (350 had boarded the ship in Africa) taken captive in Ndongo (present-day Angola) marked the beginning of the slave trade as an industrial system (Plate 11.1).

The psychological and physical torture visited upon enslaved Africans through capture and transit and on arrival in the Americas was unconscionable. Separated from family and community, slaves were sold through public auction to the highest bidder, branded, and given a new name. Most were put to work in sugar and coffee plantations; others toiled to produce cotton, ginger, and aloe. Later, plantations also introduced tobacco, rice, and indigo. Slaves were treated as subhuman and were expected to work for up to 18 hours a day during peak season for 7 days a week. A slave coveting retribution against a plantation owner or who had damaged the property of a landowner was put to death. For less serious offenses such as sloth or insubordination, beatings, whippings, neck collars, and leg irons were deployed. Sexual violence was commonplace. Attempts to escape were rarely successful, and failure meant death by hanging, burning, or violent beating. Children born of slave

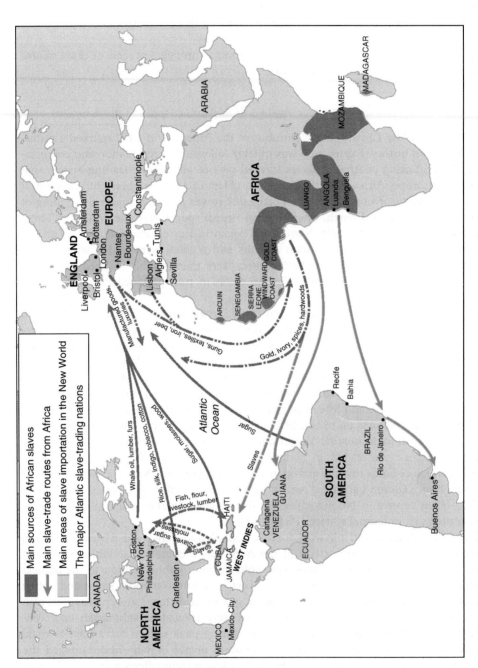

Map 11.1 The Atlantic Slave Trade, 1501–1867. Source: Compiled by the author with the support of the International Slavery Museum, National Museums Liverpool.

Plate 11.1   Slaves cutting sugar cane on a plantation on the Island of Antigua, 1823. Source: Photo by British Library on Unsplash.

Table 11.1   Principal phases in the Atlantic slave trade.

1501–1601	
Trade to Africa	From Portugal, Spain, and the UK to West Africa and West Central Africa
Slave trade from Africa	**Enslaved Africans – 270 000 – from West Africa and West Central Africa to Cartagena in Spanish America and Brazil**
Trade from the Americas	From Cartagena in Spanish America and Brazil to Portugal, Spain, and the UK
1601–1701	
Trade to Africa	From Spain, Portugal, UK, South America, and Brazil to Senegambia, Gold Coast, Bight of Benin, and Bight of Biafra
Slave trade from Africa	**Enslaved Africans – 1.25 million – from Senegambia, Gold Coast, Bight of Benin, Bight of Biafra to Brazil, Spanish America, Dutch Caribbean, Jamaica**
Trade from the Americas	From Brazil, Spanish America, Dutch Caribbean, and Jamaica to UK, Denmark, Netherlands, France, Portugal
1701–1807	
Trade to Africa	UK, Portugal, Denmark, France, Netherlands and Brazil trade goods to West Central Africa and Bight of Biafra
Slave trade from Africa	**Enslaved Africans – 6.1 million – traded from West Central Africa and Bight of Biafra to British Caribbean, Guianas, North America, Jamaica, Santa Domingue**
Trade from the Americas	From British Caribbean, Guianas, North America, Jamaica, Santa Domingue to UK, Portugal, Denmark, France, and Netherlands

(Continued)

Table 11.1   (Continued)

1807–1867	
Trade to Africa	From France, Portugal, Spain, and South East Brazil, Cuba, USA, Bahia to the Gold Coast, Bight of Benin, Bight of Biafra, West Central Africa, and South East Africa
Slave trade from Africa	**Enslaved Africans – 3.5 million – traded from West Central Africa, Sierra Leone, South East Africa to Cuba, Jamaica, South East Brazil, Bahia, Guianas**
Trade from the Americas	From Cuba, Jamaica, Northeast Brazil, Bahia, Guianas to UK, Portugal, Denmark, France, and Netherlands

Source: Compiled by the author with the support of the International Slavery Museum, National Museums Liverpool.

parents themselves entered a life of slavery and endless toil. Not surprisingly, life expectancy was short: on many plantations, slaves died on average only seven to nine years after arrival.

In 1783, Quakers, Evangelical Christians, and Enlightenment scholars joined forces, first in Europe and then in the Americas, to found the abolitionist movement. Their mission was to abolish slavery in European colonies, and in particular to outlaw the Atlantic slave trade. Olaudah Equiano (renamed "Gustavus Vassa" by his British "master") was born in the Kingdom of Benin (present-day Nigeria) in 1745, kidnapped at the age of 11, sold to British traders, shipped in a grueling five-week voyage with 244 other slaves to Barbados and then onward to the Colony of Virginia, and traded as a slave several times. He bought his freedom (an exceptional event), moved to London, and with the support of the abolitionist movement, in 1789 wrote a book about his life – *The Interesting Narrative of the Life of Olaudah Equiano, or Gustavus Vassa, the African*. The world was awakened. Meanwhile, between 1680 and 1776, approximately 800 000 Africans were sold into bondage in the colony of Haiti. The Haitian revolution (1791 to 1804) witnessed the rise of slaves against French, British, and Spanish colonists; led by Toussaint Louverture, slaves not only set themselves free but also secured independence for Haiti. A transformational moment in the Atlantic slave trade, the Haitian revolution provided a warning that the indefinite bondage of slaves could prove perilous for their captors. Many European and American industrialists who had supported the slave trade in the eighteenth century came to regard it as ethically and economically detrimental to their interests.

Denmark (1792) and Britain (1807) were among the first European countries to ban the slave trade. The United States (1808), France (1811), and the Netherlands (1814) followed thereafter. While banning the trade of slaves in 1811 and 1819 respectively, Spain and Portugal continued to see African slaves sold to landowners in South America (especially in Cuba and Brazil) until near the end of the nineteenth century. Of course, although banned, an illegal slave trade continued and British, French, and Dutch navy ships were required to tackle piracy until as late as the 1890s. Moreover, the banning of slavery itself (beyond the trade in slaves) came much later. Spain abolished slavery in 1811, but the ban took decades to become effective. Britain (1833), Denmark (1846), France (1848), Portugal (1858), the Netherlands (1861), Cuba (1886), and Brazil (1888) followed suit. The legality of enslavement, of course, was a progenitor of the American Civil War fought between the "northern" United States of

America and the "southern" Confederate States of America from 1861 to 1865. Although northern states outlawed the practice in the early 1800s, the country as a whole abolished the slave trade only after victory over the Confederate states: in 1863 after President Abraham Lincoln issued his Emancipation Proclamation, and in particular in 1865 with the Thirteenth Amendment to the US Constitution. For many slaves, emancipation was heavily conditioned; freedom was earned but at the cost of transitioning to indentured labor.

The abolition of slavery did not end the hardships faced by the African diaspora in its new American homelands. In many countries, racist and discriminatory laws continued to hamper the progress of indentured laborers and free slaves and their descendants.

In the United States, the post–Civil War Reconstruction period (1865–1877) sparked a counter-reaction from the former Confederate states; a new series of "black codes" worked to limit the rights and opportunities of African Americans. 1877 witnessed the rise of the Southern Ku Klux Klan (KKK). In the same year, the Jim Crow laws were enacted when the US Supreme Court rendered actions by southern states to segregate black people from white people lawful. From 1877 to as late as the passing of the Voting Rights Act in 1965, African Americans were treated as subordinate and inferior citizens. During this period (1916–1975), a population movement known as the Great Migration occurred; as many as four million (indeed, some estimate six million) African Americans sought to flee racism in the south by migrating to cities in the west and north (Map 11.2). In these cities, however, they also confronted varying degrees of racism. In his 2015 book *Spatializing Blackness: Architectures of Confinement and Black Masculinity in Chicago*, US geographer Rashad Shabazz has unveiled the ways in which, even in more "progressive" northern cities, black existence was overwhelmed by discrimination, oppression, and confinement.

As elsewhere, although presented as a liberal humanitarian gesture of support, state programs that did aspire to the noble goal of integration and upward social mobility were often underpinned by deeply racist assumptions about defects in the character of the black community and the importance of fostering a culture of self-improvement. They often did more harm than good. According to US geographer Mona Domosh (2019), in the late nineteenth century many of the mentalities guiding European colonizers in Africa were recycled to inculcate white WASP values in African-American communities in the United States itself. This amounted to an *internal* "mission civilisatrice." Established in 1881, the Tuskegee Institute (now Tuskegee University, Alabama) for instance was founded explicitly to provide practical and technical higher education for African Americans. Education should meet African Americans where they were most comfortable; farming should be taught on farms.

Institutional racism and discrimination led to the birth of the civil rights movement, which, throughout the 1950s and 1960s in particular, agitated for equality of opportunity for all irrespective of race, color, or creed. Inspired by campaigners such as Frederick Douglass (1818–1895), a freed slave who became a leading figurehead in the anti-slavery movement, and Harriet Tubman (1822–1913), a former slave who organized the escape of many more slaves through the famous "Underground Railroad" to the north, black activists triggered a search for equality and justice. Luminaries included W.E.B. Du Bois (1868–1963), one of the founders of the National Association for the Advancement of Colored People (NAACP); Baptist minister Martin Luther King (1929–1968), a powerful orator championing nonviolent civil rights protests; Malcolm X (1925–1965), a more controversial member of the Nation of Islam, which

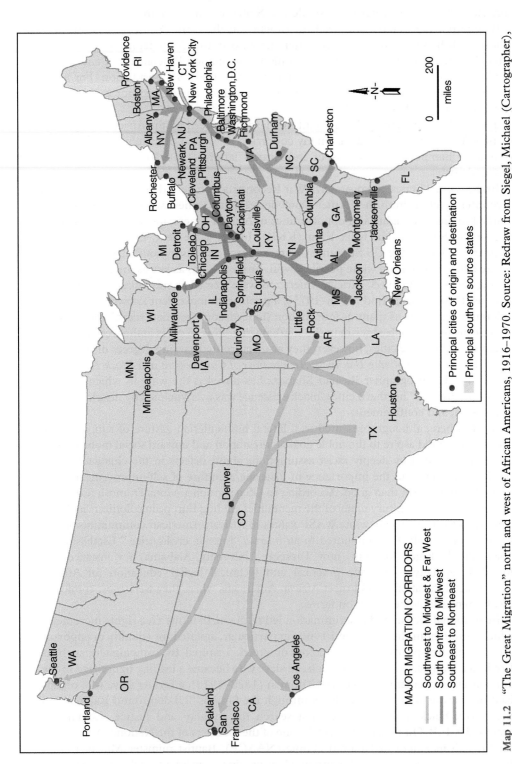

MAJOR MIGRATION CORRIDORS
— Southwest to Midwest & Far West
— South Central to Midwest
— Southeast to Northeast

• Principal cities of origin and destination

  Principal southern source states

Map 11.2 "The Great Migration" north and west of African Americans, 1916–1970. Source: Redraw from Siegel, Michael (Cartographer), "The second great migration, 1930–1980".

advocated black self-determination and the separation of black and white people; and Rosa Parks (1913–2005), who mounted a protest against segregated buses in 1955. The label "African-American" came to be viewed by some as problematic; to underscore the truth that every citizen was fully American, some activists expressed a preference for the identity "black."

Today, the African-American diaspora in the United States is estimated to amount to 40 million people – mainly, although not exclusively, descendants of slaves - with cities such as Detroit, Jackson, Miami, Baltimore, Birmingham, Memphis, New Orleans, Montgomery, Flint, Atlanta, Philadelphia, and Washington, DC, being home to significant black communities. Progress has been made, but the surging tide of populism has served to sustain significant residual racist attitudes and behaviors. In these cities, black people continue to encounter institutional racism and less than full citizenship rights and legal protections. The Black Lives Matters movement has sought to draw attention to police brutality, the role of the prison industrial complex in incarcerating black people, and the injustice of killings of black people and other ethnic minorities in police custody. Some consider police violence to be best understood as a public health crisis (Cooper and Fullilove 2020).

While recognizing that the identity of the black community is a fluid and ever-changing one, British cultural theorist Paul Gilroy argues that histories of slavery remain ever present. When used in context, the label African-American diaspora still denotes a historical truth. In Europe and the United States, black identity represents an amalgam of African, American, Caribbean, and British identities. But it would be wrong to think that over time, the African heritage of the African-American diaspora has become overlooked in favor of a series of new and disconnected localized political struggles, civil

Plate 11.2   Signs carried by marchers during the 'March on Washington' for equal rights August 28, 1963. Later Martin Luther King Jr. delivered his famous "I Have a Dream" speech at the Lincoln Memorial. Source: Library of Congress Prints and Photographs Division.

rights movements, and single-issue protests and campaigns. In his 1995 book *The Black Atlantic*, Gilroy argues that, although creolized, black identity always needs to be understood against the backdrop of the Atlantic slave trade and its continuing impression in the culture and memory of the black community.

In 2019, the world paused to take stock of and to recall "400 years of slavery" (or at least 400 years of black oppression following the arrival of the *San Juan Bautista*). In May 2020, the murder by asphyxiation in Minneapolis of African-American George Floyd during a police arrest created a global outcry, riots, and renewed support for civil rights. Floyd's last words, "I can't breathe," have come to signify the black experience of racism and has fueled further calls for direct action for equality; "no peace without justice" has become the rally cry (Plate 11.3). Connections between memories of slavery and the murder of George Floyd testify to the salience of Gilroy's thesis. There has followed a raft of destructions of colonial, slavery-related, and Confederate statues across the world. Cultural vandalism to some, decolonizing public space to others – it is clear from these acts that histories of slavery provide an important lens for those resisting racism in US cities.

Meanwhile, former metropoles and settler colonies are being forced to re-examine their colonial histories. Slavery was a source of great wealth for European and American colonists. The beneficiaries of slavery and their forebears largely failed to recognize the extent of the barbarism perpetrated in their name, at their command, or from which they continue to secure advantage. For some, notwithstanding some misdeeds, the prowess of Europe's empires is to be marveled at and merits ongoing celebration. For others, however, ongoing pride (or even ambivalence) toward past imperial wrongs signals a pathological tone deafness and is morally reprehensible. Certainly the project of decolonizing the West (changing street names, altering language, removing statues,

Plate 11.3 Sidewalk in Minneapolis where George Perry Floyd Jr. (October 14, 1973–May 25, 2020) was killed by a police officer during an arrest in May 2020. Source: Photo by munshots on Unsplash.

erasing symbols, and so on) remains at a nascent stage. And although plantation owners were "compensated" as emancipation acts were passed and slaves were freed, reparation to slaves and their ancestors has yet to be seriously countenanced.

Global Migration Stocks and Flows: Definitions, Patterns, and Trends

Data on migration patterns and trends need to be approached with caution: differences in definitions and data collection methodologies between countries mitigate against straightforward comparisons. The International Organization for Migration's (IOM, now formally part of the United Nations) *World Migration Report 2020* (IOM 2020) provides an authoritative guide to key migration-related definitions and statistics and will be used hereinafter (see also United Nations Population Division 2019).

Migrant
The IOM defines a "migrant" as any person who moves across an international border (international migrant) or within a state (domestic migrant) away from her or his habitual place of residence, regardless of the person's legal status and whether the movement is voluntary or involuntary. Moves of less than three months are more frequently construed as tourism or business travel than migration. The "stock" of migrants in any country is defined as the total number of international migrants (most often foreign born, sometimes those who are legally citizens of another country) present at a particular point in time, and migrant "flow" is defined as the number of people entering or leaving a given country during a specific period of time. While data on migrant stock are widely available, data on global migration flows are more difficult to gather and are limited in availability (but see Deep Dive Box 11.1). Calculating flow by comparing changes in migrant stock over a period is problematic, complicated by birth and death rates within migrant communities and intercensal circular moves.

According to the IOM, in 2020, globally there were 272 million international migrants, or 3.5% of the world's population; this is up from 84 million (2.3%) in 1970, 174 million (2.8%) in 2000, and 221 million (3.2%) in 2010 (Table 11.2). India was the leading country of origin of international migrants, with 17.5 million persons living abroad, followed by Mexico (11.8 million), China (10.7 million), the Russian Federation (10.5 million), and Syria (8.2 million). In 2019, 66% of all international migrants lived in just 20 countries. The largest number of international migrants lived in the United States (51 million, or 19% of the world's total), followed by Germany and Saudi Arabia (each home to circa 13 million migrants), the Russian Federation (11 million), and the United Kingdom (10 million). In 2019, Europe and Asia hosted circa 82 million and 84 million international migrants respectively (61% of the total global international

Table 11.2   Summary of international migration stocks, 2000 and 2020.

	2000 report	*2020 report*
Estimated number of international migrants	150 million	272 million
Estimated proportion of world population who are migrants	2.8%	3.5%
Estimated proportion of female international migrants	47.5%	47.9%
Estimated proportion of international migrants who are children	16.0%	13.9%
Number of refugees	14 million	25.9 million
Number of internally displaced persons	21 million	41.3 million
Number of stateless persons	n/a	3.9 million

Source: From IOM World Migration Reports.

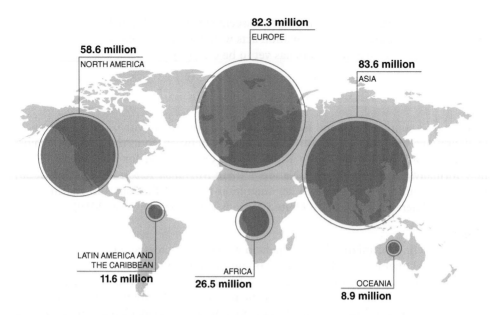

Map 11.3   International migration stocks by geographic area in 2019. Source: From International Organisation for Migration World Migration Report.

migrant stock). These regions were followed by North America, with almost 59 million international migrants in 2019 (or 22% of the global migrant stock), Africa at 10%, Latin America and the Caribbean at 4%, and Oceania at 3% (Map 11.3).

Refugee
The UN Refugee Convention of 1951 offered a now well-known definition of a refugee:

> A person who owing to a well-founded fear of being persecuted for reasons of race, religion, nationality, membership of a particular social group or political opinion, is outside the country of his nationality and is unable or, owing to such fear, is unwilling to avail himself of the protection of that country; or who, not having a nationality and being outside the country of his former habitual residence as a result of such events, is unable or, owing to such fear, is unwilling to return to it.

At December 2018, a total of 25.9 million migrants were classified as refugees, with 20.4 million under UNHCR's mandate and 5.5 million refugees registered by the United Nations Relief and Works Agency for Palestine Refugees (UNRWA). Of the refugees under UNHCR's mandate. the principal countries of origin were Syria, Afghanistan, South Sudan, Myanmar, Somalia, Sudan, the Democratic Republic of Congo, the Central African Republic, Eritrea. and Burundi (these 10 countries accounted for circa 16.6 million, or 82%, of the total refugee population). Three countries Syria, Afghanistan, and South Sudan generated 6.7 million, 2.7 million, and 2.3 million refugees, respectively. Refugees mostly move to immediately adjacent countries. Reflecting the significance of Syrian refugees in the global refugee population, it is little surprise that Turkey with 3.7 million refugees remains the world's principal host country, with Jordan, Lebanon, Pakistan, Iran, Uganda, Sudan, Germany, Bangladesh, and Ethiopia making up the remaining top ten host countries. The least developed countries in the

Deep Dive Box 11.1   Tracking Global Migration Flows Using Circular Plot Visualization

Guy Abel and Nikola Sander are statistical and migration researchers formerly based at the Wittgenstein Centre at the University of Vienna in Austria. They pioneered "circular plot visualization," a new method for presenting global migration flows. Circular migration plots undoubtedly help convey massive amounts of data in a clear and digestible way, and represent a significant breakthrough in the visualization of migrant flow data. Trends revealed by a circular plot visualization devised by Abel and Sander in 2014 showing global migration flows between 1990 and 2015 include:

- Asia and Africa are source regions for the majority of global migrants, followed by Europe, Latin America, and the Caribbean; comparatively few migrants exit the United States and Oceania.
- Migrants from sub-Saharan Africa (the majority of African migrants) move predominantly within the continent of Africa. In particular, moves between member states of the West Africa Economic and Monetary Union drive African migration patterns. When African migrants do leave Africa, they move predominantly to Western Europe.
- Migrants from Asia also tend to stay in Asia, and in particular in the region of Asia from where they come. Flows from South Asia and East Asia tend to have as destinations West Asia, North America, and to a lesser degree Europe.
- Migration flows in Latin America tend to leave the continent and be directed toward North America and Southern Europe.
- Migrants from European countries tend to move to other European countries; when they leave Europe, they scatter widely to all the principal world regions.
- Very few migrants leave North America; when they do, West Asia, Europe, and Latin America and the Caribbean are preferred destinations.
- Very few migrants leave Oceania; when they do, Europe is the preferred destination.
- Migrants who move between rather than within continents are in a minority but tend to move from Asia, Africa, and Latin America and the Caribbean to North America, Europe, and Oceania, making the latter continents more culturally diverse than the former continents.
- Migration flows that cross vast distances and exchange peoples between regions far away at the other side of the world tend to be structured by levels of development (the majority of such very long-distance flows being from poorer countries to richer countries).

(Continued)

Box 11.1 (*Continued*)

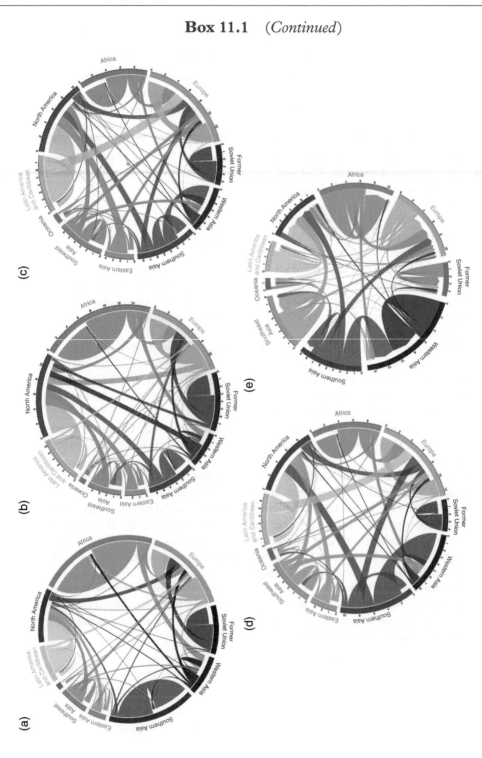

Figure 11.1 (a) The global flow of people, 1990–1995. (b) The global flow of people, 2000–2005. (c) The global flow of people, 1995–2000. (d) The global flow of people, 2005–2010. (e) The global flow of people, 2010–2015. Source: From Abel, G. J., & Sander, N. (2014). © 2014 AAAS.

Table 11.3 Types of migration within and between the Global South and Global North.

Direction of Migration	Nature of Migration	Progenitors of Migration
Global South to Global North movements	High-skilled technical workers (e.g. doctors, nurses, engineers, ICT specialists)	Differential labor market opportunities. Migrants' desire to access education, better employment opportunities and improved standards of living. Skill shortages in the Global North.
	Low-skilled service sector workers (e.g. cleaning, child minding, collecting garbage, fruit picking).	Differential labor market opportunities. Poverty in the Global South. Hard-to-fill service sector jobs in the Global North (with low pay and limited social protections).
	Refugees and asylum seekers	Flight to evade (often civil) war, famine and starvation, political oppression and victimization, and natural hazards (tsunamis, hurricanes, earthquakes, etc.).
	Modern-day slaves	Organized criminal gangs, human trafficking for sexual exploitation and forced labor.
Global South to Global South movements	Refugees and asylum seekers	Flight to evade (often civil) war, famine and starvation, political oppression and victimization, and natural hazards (tsunamis, hurricanes, earthquakes, etc.).
	Low-skilled manual workers (e.g. in agriculture, mining, construction)	Differential labor market opportunities. Migrants from ultra-poor countries moving to countries that, while still low to middle income, are comparatively more prosperous.
Global North to Global North movements	Highly skilled managerial and professional workers	Differential labor market opportunities and internal labor markets of TNCs. Global mobility essential for career progression.
	Retiree snowbirds	Seasonal migration of retirees from Global North countries with harsh winters to Global North countries enjoying warmer climes.
Global North to Global South movements	(Senior) administrators in International Organizations (IOs)	Administrators working for postcolonial "commonwealth" political bodies and charitable, religious, and aid organizations.
	High-skilled technical workers	Company expansions into the Global South. TNCs establishing branch plants. Workers with technical skills required by the primary and extractive industries (oil and gas companies, and agribusinesses).
	Gap-year students and workers on sabbatical leave	Sojourners seeking personal self-development and enrichment by exploring other cultures.

Source: Compiled by the author.

world, including Bangladesh, Chad, the Democratic Republic of the Congo, Ethiopia, Rwanda, South Sudan, Sudan, the United Republic of Tanzania, Uganda, and Yemen, host no less than 6.7 million refugees, or 33% of the global total.

Resettlement to a third country is often used to help recipient countries cope with refugee movements. In 2018, 92 400 refugees were accepted for resettlement, a 10% decrease from 2017 (102 800). The traditional resettlement countries of Canada, the United States, and Australia continued to receive the majority of the world's refugee resettlements. Syrian, Congolese, and Eritrean refugees were the key beneficiaries.

Asylum Seeker

An asylum seeker is a person who seeks to be recognized as a refugee under the terms of the 1951 UN Refugee Convention but whose claim has yet to be adjudicated on. In 2018, there were approximately 3.5 million people seeking asylum and therefore refugee status. The United States (with 254 300 new asylum applications) was the principal country of application, followed by Peru (192 500). Numbers vary greatly from year to year. For example, in 2018 Germany received the third largest number of applicants, with 161 900 applicants, a significant number but much lower than the 722 400 applicants who applied in 2016.

Internally Displaced Persons (IDPs)

In December 2018, the number of internally displaced persons (IDPs) – people forced to flee their homes but relocating to another location within the same country – was 41.3 million, up from 38.2 million people in 2014 and 20.5 million in 2000. Over 30 million (nearly 75%) of the global IDP population live in just 10 countries; Syria had the highest number of IDPs (6.1 million), followed by Colombia (5.8 million), the Democratic Republic of Congo (3.1 million), Somalia (2.6 million), and Afghanistan (2.6 million). Many more people are displaced by natural disasters in any given year, compared with those displaced by political conflict and violence.

Human Trafficking

Human trafficking is the "the bondage of people and enforcement of migration through the use of violence, deception, coercion and for the purposes of sexual exploitation, indentured labor, begging, crime (drugs), domestic servitude, marriage or organ removal," and is becoming a growing global problem. Estimates of the number of people labeled "modern-day slaves" range from around 21 million to 46 million. Between 600 000 and 800 000 people are trafficked across an international border each year: 80% are women, and 20% are children. Countries judged by the US State Department to be the worst offenders (Tier 3 countries with insufficient policing of trafficking) include Russia, China, Belarus, Venezuela, Iran, North Korea, the Central African Republic, Sudan, Algeria, and Eritrea.

Stateless People

Statelessness refers to the condition of an individual who is not considered a national by any state. It is difficult to calculate the number of people who are stateless currently, but the United Nations estimates figure to be circa 3.9 million (included in figure are the Rohingya refugees in Bangladesh and IDPs in Rakhine State), with Bangladesh leading with 906 000 stateless people, followed by Côte d'Ivoire (692 000), Myanmar (620 000), Thailand (490 000), Latvia (225 000), the Syrian Arab Republic (160 000), Kuwait (92 000), Uzbekistan (80 000), Estonia (78 000), and the Russian Federation (76 000).

Climate Change–Related Migration

According to some, climate change has already rendered homeless between 20 and 50 million, and will lead to as many as 200 million people being displaced from their homes by 2050 (some estimates even forecast as many as two billion climate refugees and IDPs by 2100). Currently, climate refugees come mainly from sub-Saharan Africa, the Middle East, and small Pacific islands, but in time these regions will be joined by countries such as Bangladesh, Indonesia, India, and China, where for example shrinking coastlines and delta flooding are significant issues.

Rethinking Integration: On the Politics of Hospitality

Migrant settlement in host countries is a deeply political and contested process. Historically, the concept of migrant assimilation was the dominant policy frame; steadily this was replaced by the idea of multiculturalism (and in scholarship the associated radical idea of reframing host–migrant relationships in terms of the idea of "diaspora space"); today securitization is the preeminent policy agenda. Within the Global North, the rise of anti-immigrant sentiment and emergence of what some have called a "new politics of hospitality" are creating a historically novel operational environment for policy makers at all levels (supra-national nation, state, region, city, and community) with responsibility for promoting migrant integration, intercultural dialogue, social inclusion, and community cohesion. In the context of the growing politicization of immigration across the Global North and renewed calls for tougher immigration controls and diminished hospitality toward immigrants, there is an urgent need to articulate a new vision for immigrant integration and integration policy. This section will conclude that there is conceptual, political, and policy value in reframing migrant integration in terms of "settlement" or "settling."

Host Country Integration: Policies and Outcomes

It is to be expected that how open, cosmopolitan, and welcoming (as opposed to closed, hostile, and xenophobic) a place is will play a role in shaping how quickly migrants are able to build a fresh life in that place. A distinction is often made between *integration policies*, policies pursued by host governments to lubricate the integration of migrants, and *integration outcomes*, the experiences of immigrants and the extent to which they achieve parity with the host populations.

Sponsored by the European Union, published in 2004, and updated in 2007, 2011, 2015 and 2020 the EU MIPEX project evaluates the performance of the integration policies of 52 countries: all EU Member States (including the UK), other European countries (Albania, Iceland, North Macedonia, Moldova, Norway, Serbia, Switzerland, Russia, Turkey and Ukraine), Asian countries (China, India, Indonesia, Israel, Japan, and South Korea), North American countries (Canada, Mexico and US), South American countries (Argentina, Brazil, Chile), and Australia and New Zealand in Oceania. MIPEX2020 uses 58 indicators covering eight policy areas.

- **Access to Nationality**: How easily can immigrants become citizens?
- **Anti-discrimination**: Is everyone effectively protected from discrimination on the basis of race, ethnicity, nationality, and religion in all areas of life?
- **Education**: Is the education system responsive to the needs of the children of immigrants?
- **Family Reunion**: How easily can immigrants reunite with family?
- **Health**: Is the health system responsive to immigrants' needs?
- **Labor Market Mobility**: Do immigrants have equal rights and opportunities to access jobs and improve their skills?
- **Permanent Residence**: How easily can immigrants become permanent residents?
- **Political Participation**: Do immigrants have comparable rights and opportunities to participate in political life?

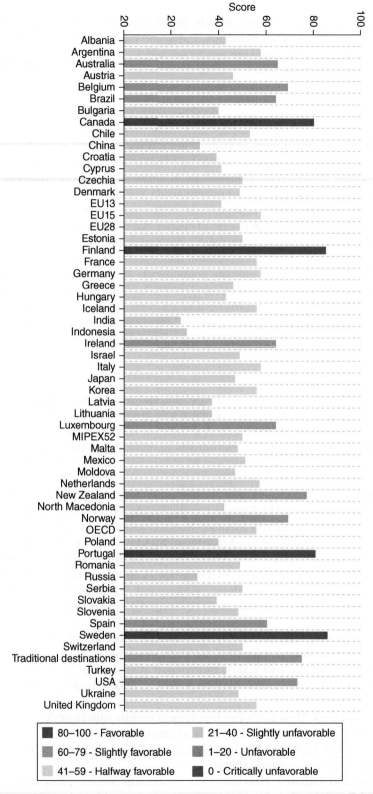

Figure 11.2 MIPEX2020 Migration Integration Policies: Status of policies (highest score is best performer) Source: From MIPEX2020 Migration Integration Policies. © 2020 MIPEX.

Table 11.4   Migration Integration Outcomes by OECD country type.

Typology of countries (OECD)	Migration integration outcomes
Group 1: Settlement countries (Australia, Canada, Israel, New Zealand)	Strong integration
Group 2: Long-standing destinations with many recent and highly educated migrants (Luxembourg, Switzerland, the United Kingdom, the United States)	Moderate integration
Group 3: Long-standing destinations with many poorly educated migrants (Austria, Belgium, France, Germany, the Netherlands)	Poor to moderate integration
Group 4: Destination countries with significant recent and humanitarian migration (Denmark, Finland, Norway, Sweden)	Moderate integration
Group 5: New destination countries with many recent, low educated migrants (Cyprus, Greece, Italy, Portugal, Spain)	Poor to moderate integration
Group 6: New destination countries with many recent highly educated immigrants (Iceland, Ireland, Malta)	Moderate integration
Group 7: Countries with an immigrant population shaped by border changes and/or by national minorities (Croatia, the Czech Republic, Estonia, Hungary, Latvia, Lithuania, Poland, the Slovak Republic, Slovenia)	Moderate integration
Group 8: Emerging destination countries with small immigrant populations (Bulgaria, Chile, Japan, Korea, Mexico, Romania, Turkey)	Mixed integration fortunes

Source: Data from OECD (2018 p.25).

Meanwhile, efforts have been made to measure migrant integration outcomes using indicators pertaining to the labour market (for example, employment rates, wages or income, occupation, activity rate and over qualification rate); education (for example, highest level of education attainment, dropout rate, grades and skills); health (for example, healthy life years and life expectancy); social inclusion (for example, property ownership, housing cost overburden, child poverty and social exclusion); civic inclusion/engagement (for example, voting rights, representation in the political arena, public employment, naturalization rate, share of long-term residence and volunteering); cultural inclusion (for example, customs, traditions, language and religion); financial inclusion (for example, banking, savings, credit, insurance and advice); spatial inclusion (for example, residential segregation by socio-economic status); public opinion (for example, ability to integrate highly heterogeneous and culturally diverse groups of people); and the role of media (for example, inclusion and diversity in public service media) (OECD 2018).

Bringing these two data sets together, it is clear that the relationship between integration policies and integration outcomes is far from simple (see table 11.4). There is a need to better understand the mechanisms that affect integration. Evidently migrants settle into host societies in complex ways as a function of the factors that led them to migrate in the first instance: ongoing political events and happenings in the country of origin, host country policies, and conditions in places of destination. Integration is rarely a linear process. More deeply again, perhaps there is a need to rethink the very idea of integration itself.

Thinking Integration: Assimilation, Multiculturalism, Diaspora Space, and Securitization

It is possible to recognize four ways in which scholars and policy makers have sought to make sense of host–migrant relationships: assimilation, multiculturalism, diaspora space, and securitization.

Traditionally, the concept of assimilation has been used to think about migrant encounters with host societies. Assimilation here refers to the absorption into a majority population of a foreign minority population as the latter acculturates, acclimatizes, and adapts to resemble the host. Assimilation frames migrants as initially strangers in a strange new land, then planters of roots and builders of a new life, and finally full members of the host country and indistinguishable from the domestic population. Four types of assimilation are recognized.

- Social assimilation occurs when migrants forge new social networks in the host society and secure equal access to housing, education, health care, and recreational facilities.
- Economic assimilation occurs when migrants plant new roots in the host society and secure employment. Through upward socioeconomic mobility, they begin to achieve parity with the socioeconomic profile of the host population.
- Political assimilation occurs when migrants gain full citizenship rights in destination countries and participate as equal members in the political life of the nation (vote in elections, stand for elections, participate in public debate, and so on).
- Cultural assimilation occurs when migrants are willing and able to embrace local ways of life and take on the cultural habits of the host (values, traditions, clothing, dance, religion, food, national belonging, etc.), identifying with and becoming patriotic nationals of host countries.

In the second half of the twentieth century, the doctrine of multiculturalism came to dominate debate on how best to accelerate migrant integration. Social, economic, and political integration is best encouraged not by demanding that migrants adopt the cultural practices of the local population but instead by fostering cultural diversity, tolerance, and pluralism. Accordingly, many host countries have sought to protect and cherish several distinct cultural and ethnic communities, believing that from tolerance of difference comes unity.

Canada presents itself as a tolerant and multicultural society with a progressive and open attitude to immigrants. It points to its two principal Citizenship Acts, of 1947 and 1977, and its 1988 Multiculturalism Act as having enabled immigrants to enjoy access to Canadian citizenship while promoting their right to celebrate their own culture, religion, language, and customs. The 1988 Act in particular recognizes that: Canada has a multicultural heritage, and this heritage should be protected; the rights of Aboriginal peoples should be promoted; although English and French remain the only official languages, other languages may be used; every group is equal under the law, regardless of origins, race, or creed; and ethnic minorities have the right to enjoy their own cultures. After a period in which the Harper Conservative Party government (2006–2015) sought to tighten access to citizenship, support for immigration and multiculturalism has been reinforced by Prime Minister and leader of the Liberal Party Justin Trudeau (from 2015), whose mantra has been "a Canadian is a Canadian is a Canadian." In 2017 Trudeau introduced Bill C-6, an act to reverse restrictions on citizenship claims and make it easier for immigrants to apply for citizenship recognition. Canada continues to buck the trend and to adopt pro-immigration policies and support multiculturalism even when it is under attack elsewhere. It has committed to fast tracking applications for citizenship from migrants who have helped Canadian society navigate through the Covid-19 crisis. But if it is to sustain this commitment, it will need to find compelling responses to critics of multiculturalism.

Elsewhere in the world, multiculturalism is now a policy in crisis. From the early 2000s onward, mainstream politics has increasingly turned its back on it in preference for nationalist populism and majoritarian rule. Only in a select number of diverse countries and diverse global cities is multiculturalism continuing to be championed. Critics from the Right argue that multiculturalism has diluted nationalist values and devalued citizenship, enabling it to be acquired too cheaply. Critics from the Left argue that multiculturalism is simply a cover for free movement associated with neoliberal globalization and is gendered in its consequences. Critics of all political hues argue that the micro-geography of migrants in cities in destination regions influences processes of assimilation. Many migrant groups, out of both choice and necessity, cluster into ethnic neighborhoods, enclaves, and ghettoes. According to some, the clustering of immigrants in their own communities, speaking their own languages and practicing their own cultures, is a fundamental problem. Ethnic enclaves permit migrants to live in a goldfish bowl, looking out at the wider community but having no contact with that community. Integration it is supposed cannot occur when migrants chose to live apart from mainstream society and fail to respect and embrace the cultural traditions of the host. But is this really true? (see Deep Dive Box 11.2).

In her book *Cartographies of Diaspora*, British-based sociologist Avtar Brah (1996) argues that to think in terms of *diaspora spaces* is to radically transform the conversation concerning host–migration relationships. Both migrants and host communities inevitably experience cultural hybridity as they become entangled and there is no centered mainstream into which migrants must dissolve. As they plant roots in new host countries, migrants neither retain their existing identities nor assume the identity of the new host, but instead develop a series of hybrid or in-between identities. Crucially, being in diaspora is also a state that the host country population can reside in, even when staying put. As they enter into a Manichean tango with migrants, it is possible for indigenes to be at home but also in diaspora. As migrants and hosts intermingle, these very categories dissolve, and both groups are inescapably transformed. Essentialized identities are mythical. To think of strangers integrating into indigenous host communities is to misconstrue the cultural dynamic at work. However, intersections of migrants and hosts are structured by race, gender, class, sexuality, ethnicity, generation, and nationalism. Powerful structural forces often conspire to limit the extent to which identity experimentation is possible. The luxury of being in diaspora is not equally shared by all social groups. Policies need to work, then, to help everyone dwelling in diaspora space to recognize the creolized journeys they are part of and the ways in which they are existentially co-dependent.

In his 2018 book *Whiteshift: Populism, Immigration, and the Future of White Majorities*, Hong Kong/Canadian but British-based political scientist Eric Kaufmann examines the question of racial and ethnic change in the West wrought by immigration. The term Whiteshift is used to refer to the process through which white majorities absorb an admixture of different peoples through immigration and become a "majority-minority" and in time a mixed race. Immigration, multiculturalism, and left-wing postmodernism

Deep Dive Box 11.2   Chinatowns: Beachheads Lubricating Migrant Assimilation, or Buffers Slowing Down Integration?

Perhaps the most iconic cultural freight that Chinese migrants have introduced into host cities is the phenomenon of the Chinatown. Many Chinatowns in North America were built in the nineteenth century. Those in Europe tended to form later, from the 1950s onward. Because Chinese migrants in Southeast Asia often enjoy better economic and political conditions and feel less culturally estranged from the host population, Chinatowns there (although they do exist) tend to be less common and less pronounced. The largest Chinatown in the United States is in San Francisco (Plate 11.4); in Canada, it is in Vancouver; in Australia, Melbourne; and in Europe, Paris and London vie for the top spot. In Asia, Singapore has a particularly significant Chinatown.

Chinatowns can be thought of as a barrier limiting integration into the host city. Chinatowns permit migrants to live with only minimum contact with local people; they can dine in Chinese restaurants, shop in Chinese stores, surf Chinese websites, and gamble in Chinese casinos. Moreover, Chinatowns can support the work of human traffickers like the infamous Snakeheads gangs from the Fujian region who route illegal migrants via Chinatowns and exploit such migrants by forcing them to work for low wages or to serve as prostitutes to pay off their "debt." But Chinatowns also provide Chinese migrants with a beachhead into host societies. Initially, restaurant and catering services (and also barbers and laundries) sustained many Chinatowns. These services did not demand start-up capital and afforded unskilled Chinese migrants with limited language skills a means of earning a living. They also provided familiar food to the overseas Chinese community, which eased assimilation.

Plate 11.4   Chinatown in San Francisco. Source: Gavin Hellier/Jon Arnold Images Ltd/ Alamy Stock Photo.

Subsequently, other services, including travel agencies, health clinics, legal consultancies, Chinese schools, and Chinese grocery stores, have flourished.

In the United States, Western Europe, and Australia, Chinese migrants (especially following the abolition of discriminatory laws in the 1940s and 1950s) have enjoyed upward social mobility and have joined the flight of the wealthy to the suburbs. This process has rendered Chinatowns vulnerable. Some have been turned from functioning ethnic enclaves into more superficial tourist attractions. Moreover, today migrants leave China to pursue economic opportunities, to study abroad, to improve their language skills, and to acquire a better quality of life. As private investors and Chinese state-owned enterprises undertake projects overseas (not least in Africa), Chinese workers are also being relocated to oversee operations. These migrants have less need for support. In addition, some recent migrants have left to escape political persecution (democratic and human rights activists) and religious discrimination (like members of the dissident Falun Gong movement). These migrants might benefit from support, but Chinatowns can be inappropriate places for them to dwell.

and political correctness have led to a "sublimation of white identity." Unable to defend their white racial and ethnic identities and to express their worries about immigration for fear of being labeled "racist," white communities in the West are resorting to extreme right-wing populist movements who are sensing an "Overton window" shift for their message.

Although stirring controversy and criticized by both the Left and the Right, Kaufmann contends that "we need to talk about white identity" if we are to help politicians to address the anxieties created by immigration and multicultural policy. White majorities in the West have a right to defend their ideas of place and valorize their identities, and vilifying them as racists and imperialists will only embolden right-wing populism. He argues that four possible strategic responses now present:

- **Fight**: A white majority resists further Whiteshift by closing borders, rejecting multiculturalism, and reasserting white privilege.
- **Repress**: Governments reject Right populism as racism, defend liberal open borders and cultural diversity, and repress white opposition.
- **Flee**: Whiteshift leads to "white flight," as white people establish regions, cities, and neighborhoods where they remain a supermajority.
- **Join**: Where the white majority accepts Whiteshift and celebrates white identity as an important strand in a mixed-race hybrid and creolized West.

Kaufmann presents his arguments objectively and in great detail using a wide range of data sets, but his conclusion is normative. His goal appears to be to "draw the sting" of right-wing populism by presenting resistance to Whiteshift as futile. But he also recognizes that repression is unlikely to succeed and could be counterproductive. Conservative whites need to have their concerns dignified and to be part of the public conversation. Otherwise white flight, ghettoization, and segregation will intensify, leading to greater racial polarization and tension. "Join" then appears to be the logical conclusion, but the white majority needs to be reassured that any mixed-race future will protect its interests and identity too.

In this age of revived nationalism and political populism, a chorus of voices now anxiously debate the existence of an "immigration crisis" in which migrants stand indicted of taking local jobs, abusing welfare systems, increasing terrorist attacks and crime, building parallel societies, and compromising existing ways of life. Critics argue that Global North states have lost control over immigration and have failed to manage the costs, tensions, and controversies associated with free movement. In a particularly forthright book entitled *The Strange Death of Europe: Immigration, Identity, Islam,* British neoconservative critic Douglas Murray (2017) argues that the "liberal conceit" that whites must be post-ethnic cosmopolitans has outlived its usefulness. "Europe is committing suicide," Murray proclaims, by welcoming migrants from other cultures who "hold less liberal views than the majority of people in the countries they have come into" and promoting multiculturalism; paradoxically, fostering diversity could lead to a less tolerant society. Already guilty about its past, jaded, and withering, European culture is no match for the strong and assertive cultures that migrants bring with them. If cherished European traditions and values are to be upheld, stronger controls over migration will be necessary.

For Algerian-born French philosopher Jacques Derrida, there is arising in many Western host countries a new politics of hospitality that is leading to a closing and securing of borders; elevated suspicion of migrants' motivations, loyalties, and commitments; new tests of citizenship; and increased surveillance and policing of migrant communities. In his book *Of Hospitality,* Derrida (2000) questions the growth of "conditional" hospitality – which has slid too easily, it seems, into the "crime of hospitality" – and in contrast to Murray, argues for the merits of "unconditional" hospitality and an ethics of care for the stranger.

Rethinking Integration: Migrant Settling and Settling Services

It is clear, then, that there are multiple ways of thinking about host country–migrant relationships. So what are we to conclude? Reflecting new thinking, a number of pioneering host states (Canada being the prime example) now construe integration as a state of becoming and not simply an outcome, and think and act in terms of migrant "integration and settlement policy" or simply "settlement services." Migrant settlement/resettlement refers to the restless and contingent processes of planting and replanting roots that migrants undertake (marked by oscillations between, and varieties of, belonging and estrangement) in their country of destination, as they move to a third country or return to a country of previous residence. This shift from "integration" to "settlement" thinking constitutes a potentially powerful response to the securitization of migration policy and furnishes the policy community with a new vocabulary with which to speak about the necessity and virtues of enacting policy interventions at times and in locations where immigration is highly politicized. It helps policy makers to suspend, void, escape, and work around at least some of the thorny controversies that emerge when integration is understood as "outcome driven." Speaking in terms of migrant settling helps us to better understand why integration is nonlinear and why some second-, third-, and fourth-generation diasporeans feel more drawn to their ancestral homeland and estranged from their county of birth. It helps to draw attention to the needs of migrant communities routinely neglected in existing integration policy approaches, including circular migrants, bi-nationals, footloose or transient migrants, and diasporic communities. Thinking in terms of settling calls attention to the need to extend services to migrants who experience forms of racism and sectarianism in everyday life that fall beyond the antennae of conventional measures of integration. It also helps us to understand why policies designed to mainstream migrants often fail.

In a speech in 1965, in the context of significant immigration to Britain from the British Commonwealth, British Labour MP Roy Hattersley (in)famously proclaimed: "without integration limitation is inexcusable; without limitation, integration is impossible." Hattersley was convening the ideas of assimilation and securitization as pivotal to British immigration policy. A number of demands were then placed on policy makers to accelerate the integration of migrants and on migrants to acculturate. But given that assimilation is not a meaningful category of analysis or process, it appears to make no sense to treat it as a category of policy. The dangers of giving this idea life was seen just three years later in 1968, when British MP Enoch Powell delivered his "Rivers of Blood" speech at the Conservative Political Centre in Birmingham, United Kingdom, strongly rejecting a race relations bill. In the wrong hands, analytical categories can cause harm. The ideas of multiculturalism and diaspora space offer more compelling points of entry, but they only gesture to what is required. If migrant settling is a meaningful category of analysis, referring to a contingent and nonlinear process of settling, then policies to support migrant integration need to attend to that reality. Migration policies will only gain traction if they properly understand their object of concern.

Case Study: The Syrian Refugee and IDP Crisis

Since 2011 there has been a civil war in Syria, a product of the fallout of the wider Arab Spring. This war has been waged by opponents of sitting President Bashar al-Assad and his Ba'ath government – held by opponents to be a corrupt, sectarian, and authoritarian regime. However, it has been complicated by the competing interests of a wide variety of factions who have exploited and/or been caught up in the conflict. The Syrian problem is indeed a wicked problem.

While rebels gained control of parts of the country (with the help of, among others, the West and especially the United States), the Ba'ath government has remained in power (with the help of, among others, Russia, China, and Iran) and has used the Syrian army to crush insurrection. However, amid the turmoil there emerged in northern Syria and Iraq a radical Islamic fundamentalist group who attempted to create a new caliphate – the Islamic State of Iraq and Syria, or ISIS. Both the government and rebel forces contend that war atrocities have been committed against them. Peace talks between both sides have proved difficult, and divisions seem intractable. And as it gained ground, ISIS committed further atrocities and abhorrent human rights abuses. From 2015 Russia has helped President al-Assad militarily and driven ISIS and other rebel forces significantly back, such that they now have a presence in only small patches of the national territory. Meanwhile in 2018, the United States, which had been protecting rebel forces and the Kurdish population living in the north of the country, withdrew largely from Syria. In the vacuum that has resulted, Turkish and Kurdish forces have fought along the northern border, while the Russian-backed Ba'ath army have sought to recover land still held by rebel groups, most recently in late 2019/early 2020 in a major assault on rebel-held positions in the northwestern Province of Idlib. With a pre-war population of 22 million, estimates

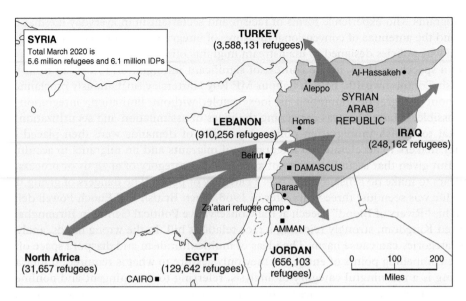

SYRIA
Total March 2020 is
5.6 million refugees and 6.1 million IDPs

TURKEY
(3,588,131 refugees)

Al-Hassakeh ●

Aleppo ●

**SYRIAN
ARAB
REPUBLIC**

LEBANON
(910,256 refugees)

Homs ●

IRAQ
(248,162 refugees)

Beirut ■

■ DAMASCUS

Daraa
●

Za'atari refugee camp ●

■
AMMAN

North Africa
(31,657 refugees)

EGYPT
(129,642 refugees)

CAIRO ■

JORDAN
(656,103
refugees)

0   100   200

Miles

Map 11.4   The Syrian refugee crisis. Source: Redraw from UNCHR, The Syrian refugee crisis.

suggest that the Syrian civil war has claimed the lives of between 384 000 and 586 100 people (across the period 2011–2020).

The turmoil of the conflict has triggered in its wake the greatest humanitarian disaster of the early twenty-first century. As of March 2020, 5 563 951 million refugees have crossed the border from Syria. Turkey has received the greatest number at 3 588 131 (64.5% of all migrants), followed by Lebanon with 910 256 (17.4%), Jordan with 656 103 (11.8%), Iraq with 248 162 (4.5%), and Egypt with 129 642 (2.3%). A total of 288 267 Syrian refugees live in UNHCR camps. As of January 2020, only 234 817 refugees had returned to Syria. Meanwhile, within Syria, 11.7 million people are in need of humanitarian assistance, including 6.1 million IDPs, the biggest internally displaced population in the world. Some 5 million people are in acute need due to a convergence of vulnerabilities resulting from displacement, exposure to hostilities, and limited access to basic goods and services. There are also 1.1 million people in need in hard-to-reach locations. The December 2019 government assault in Idlib has led to another 900 000 IDPs, reinforcing the claim that the problem is far from over (Map 11.4).

In an attempt to bring planning and order to efforts to deal with the growing refugee crisis, the international community has launched a series of "Syria Regional Response Plans" (updated periodically, and from 2015 labeled 3RP). Led by the UNHCR, these plans attempt to bring together all partners working in the region (including more than 270 national and international development and humanitarian actors) into a "regionally coherent planning process."

The ambition and scope of the plans have changed as the scale of the problem has grown. The first Syria Regional Response Plan was published in March 2012 and sought US$84.1 million from the international community to introduce measures to deal with only 96 500 refugees and only over a period of six months. The Regional Refugee and Resilience Plan, 3RP 2020, sought US$5.5 billion to assist 4.4 million refugees.

3RPs are innovative in that they are built around not only a refugee protection and humanitarian component but also a resilient/stabilization component.

- The 3RP refugee protection and humanitarian component works to protect and meet the needs of refugees living in camps, in settlements, and in local communities, as well as the most vulnerable members of impacted communities.
- The 3RP resilience and stabilization-based development component works to build the capacities of national and subnational service delivery systems; strengthen the ability of host governments to formulate effective crisis responses; and provide strategic, technical. and policy support and advice.

To date, most of the Syrian Regional Response Plans have failed to generate the resources they require for full implementation. Donations and commitments have fallen well short (only circa 60% of what is needed has been raised thus far), and lack of funds remains a key issue. Still, circa US$14 billion has been spent through 3RPs since 2015. The objective has been to ease the pressures on host countries, enhance refugee self-reliance, support conditions for a safe return to Syria for return in safety and expand third-country resettlement. Resources have been used to safeguard security, safety and protection, food security, education, health and nutrition, shelter, and livelihoods.

The UNHCR has worked, for example, with international aid agencies and local charities to build and service refugee camps. The Za'atari refugee camp in northern Jordan (12 km from the Syrian border) is a good example. The camp is so vast that it takes an estimated 20 minutes to drive around its 8-km radius. It opened in July 2012 to house refugees mainly from Dara'a Governorate in Syria. The Za'atari Camp is under the joint administration of the Syrian Refugee Affairs Directorate and UNHCR. Built to house around 60 000 residents, the camp is (as of March 2021) home to 78 812 refugees, nearly 20% of whom are under five years of age (at its peak, it housed 172 000 people), living in over 27 000 shelters (tents). In addition to providing displaced Syrians with makeshift accommodation, the Za'atari Camp provides food (over 0.5 million loaves of bread per day), access to water and sanitation (the camp needs 4.2 million liters of water per day), and hospitals (there are three on the site). 19 243 children are enrolled in 32 schools. In addition, there has emerged a vibrant retail sector (there exist 3000 shops and 850 food outlets – many along a main road that has been retitled the Champs-Élysées!). All of this is coordinated by over 1700 administration points and costs US$0.5 million each day to run.

From March 2012, in an effort to bring planning and order to the varied responses to the growing IDP crisis within Syria, the international community has also launched a series of Syrian Humanitarian Assistance Response Plans (SHARPs) – updated periodically and now called the Syria Humanitarian Response Plans (HRPs). Led by the UN Office for the Coordination of Humanitarian Affairs (UNOCHA), the Government of Syria, the Syrian Arab Red Crescent (SARC), and the IOM, SHARPs have coordinated the work of over 170 international nongovernmental organizations (INGOs) active in Syria, and have undertaken projects in all 14 of the country's governorates. In 2019, HRPs supported 11.7 million Syrians by providing food security, health, education, washing facilities, shelter, protection, nutrition, livelihoods, camp coordination and camp management, logistics coordination, and emergency telecommunications.

Like 3RPs, HRPs have failed to generate and gather the scale of resource needed. In 2020, of the US$3.48 billion requested, only US$2.31 billion was received, leaving an unmet requirement of US$1.17 billion, or 32.8% of the planned spend. Furthermore, given ongoing conflict, the dangers faced by aid workers, and the fact that new IDPs are being created daily by the conflict, HRPs have faced enormous challenges.

Given the intractability of the conflict, it will be necessary to find permanent homes for those unable or unwilling to return to Syria. If left to absorb the majority of refugees, it is likely that Syria's neighbors will be overwhelmed, leaving refugees to deal with long-term social, economic, and political marginalization. The UN General Assembly's 2018 Global Compact on Refugees is a framework for more equitable responsibility sharing, recognizing that international solidarity in support of host countries on the front lines is essential. Here, the countries of the Global North need to play a more active role. Resettlement involves the selection and transfer of refugees from a state in which they have sought protection to a third state that has agreed to admit them as refugees with permanent residence status. Thus far, resettlement and humanitarian offers and schemes have been woefully inadequate. UNHCR estimate that 10% of Syrian refugees (480 000 people) are in need of resettlement and meet resettlement criteria, but only 0.5% of Syrian refugees will be submitted for resettlement. Humanitarian admission (or "complementary routes") is a similar, but expedited, process providing protection in a third country for refugees in greatest need in the region. Although high-income countries such as Germany, Sweden, Canada, and until recently the United States have pledged a significant quota of places, countries such as Russia, Singapore, South Korea, Qatar, United Arab Emirates, Saudi Arabia, Kuwait, and Bahrain have failed to step up.

Many Syrians have left the region by traveling across the Mediterranean and through Europe in search of opportunities to claim asylum. Between April 2011 and December 2019, 1.2 million Syrians lodged claims for asylum in EU countries. Germany, Sweden, Norway, France, the United Kingdom, Italy, and Hungary have attracted the greatest number of applicants (Map 11.5).

For a while, EU countries seemed to grasp the enormity of the problem and their need to shoulder responsibility. In part, awareness was raised with the death of refugees making fatal journeys to Europe across the Mediterranean Sea. According to the UNHCR, nearly 2 million refugees in total traveled to Europe via the Mediterranean between 2014 and 2020, and nearly 20 000 drowned or died *en route*. In his books *Lights in the Distance: Exile and Refuge at the Borders of Europe*, British journalist Daniel Trilling (2018) brings to life the human face of the trauma that has resulted. The image of three-year-old Aylan (Alan) Kurdi from the Syrian town of Kobani washed up on a beach in Bodrum, Turkey (September 2015), drowned whilst traveling with his family to the Greek island of Kos in their efforts to reach the European Union proved to be a potent mediator of public opinion – at least for a time (Plate 11.5). But racist-fueled attacks have increased again in countries such as Denmark, Germany, Sweden, Luxembourg, Finland, Hungary, Italy, Malta, Austria, and Ireland.

Arguably, the Global North generally and Europe specifically (with some laudable exceptions) have failed Syria's neighboring countries, leaving them to take responsibility for managing the Syrian refugee crisis. They have sought to put distance between themselves and the crisis.

- In December 2015, the European Commission announced that Frontex (an EU border and coastguard agency with responsibility for policing the borders of the EU Schengen area) would be strengthened and given more powers. Frontex supports member states to securitize Schengen's frontiers, supervise asylum claims, and detain and deport failed asylum seekers. With a staff of 45 and a budget of €5.5 million in 2005, in 2020 Frontex had 750 employees and a budget of €333 million.

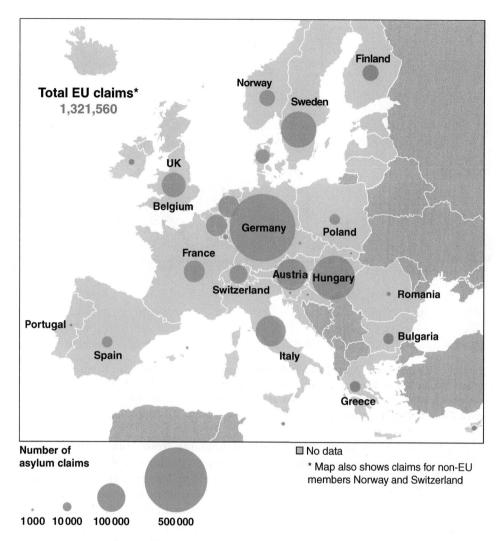

Map 11.5   Asylum claims in Europe, 2015 (Syrian asylum seekers are the largest group of claimants). Source: Eurostat.

- Operation Mare Nostrum was an Italian government–led year-long naval and air search-and-rescue operation that commenced on October 2013 to assist refugees making the perilous trip across the Mediterranean. It was replaced in November 2014 with Operation Triton led by Frontex. Although this shift was presented as necessary to help Italy cope, in fact it has turned a search-and-rescue operation into a security and surveillance project – leading, it is asserted by some, to more deaths at sea. In further support of Italy, in 2018 Operation Triton was replaced by Operation Themis. Frontex-rescued migrants must now be taken to the nearest EU port rather than to only Italian ports, and added efforts will now be made to stop terrorists from entering the EU.
- In November 2015, the EU agreed to pay Turkey €6 billion (and provided other concessions) to support the 3.58 million Syrian refugees residing in Turkey and to help Turkey stop the onward movement of these refugees to the EU. The EU also

Plate 11.5 Mural of Aylan (Alan) Kurdi in Frankfurt am Main, Germany. Aylan Kurdi was a three-year-old Syrian boy of Kurdish ethnic background who drowned while fleeing the conflict on September 2, 2015, in the Mediterranean Sea. His body was washed up in a beach in Bodrum, Turkey. Source: Plenz, https://en.wikipedia.org/wiki/File:Alan_Kurdi_Graffiti_(cropped).jpg. CC BY-SA 3.0.

agreed to resettle a proportion of Syrian migrants from Turkey. But Turkish President Erdogan has repeatedly accused Europe of failing to uphold their side of the deal and in March 2020 declared that he would no longer stop Syrian refugees from fleeing to Europe. Of the €6bn, €4.7bn has been earmarked for initiatives and €2.3bn has been paid out since 2016. With the agreement strained, a fresh wave of Syrian refugees are crossing into Greece in particular, placing added pressure on host communities, including at the Moria Camp in Lesbos. Greece in turn has secured the support of Frontex to heighten security at the Turkey–Greece border.

- Increasingly, there have been calls for an effective suspension of the Schengen Agreement on open borders across Europe, and demands that identity checks be carried out at internal borders (borders within the EU and Schengen) that were previously open to free movement. This proposal would be difficult to operationalize but is gathering support from some EU countries.

The Syrian refugee crisis looks set to be a humanitarian agony that will persist for many years to come. A significant number of refugees from, and IDPs in, Syria remain in exile. The conflict shows no sign of abating, the burden continues to fall on Syria's immediate neighbors to deal with refugees, and plans to address the problems faced by these host countries and IDPs within Syria remain underfunded. The response of the rest of the international community to assist immediately adjacent countries seems to be wholly inadequate and, in the case of the EU, defensive. Only when a political solution to the crisis is found and when countries of the Global North take more responsibility (directly and by supporting Syria's neighbors more fully) will it be possible to even think about a future for Syria and Syrians.

Conclusion

Mass migrations and the formation of great diasporas that have fanned out across the world have been a feature of big human history. But these are especially migratory times. Human geographers consider the global migration system that exists today to be a product of uneven geographical development across the face of the earth and as such a legacy of history. It is impossible to understand global migrant stocks and flows both within and between the Global South and Global North without an understanding of the rise, reign, and faltering of the West from the fifteenth century. The paradigmatic example is the African diaspora in the Americas, a product of the Atlantic slave trade and an historical event which has profoundly affected the history of both Africa and the Americas (in particular the United States). Human geographers are also interested in the experiences of migrants as they settle into their new host societies. Whilst migrant integration policy and integration outcomes share a complex relationship, host country attitudes to immigration play a key role in how migrant 'assimilation' is understood and planned for. Multiculturalism, hitherto the dominant policy approach in host countries, is increasingly giving way to nationalist populism, a new politics of hospitality, and a walling of the West. In countries which continue to embrace liberal attitudes to immigration, the concept of assimilation is steadily being replaced by that of "settling." But these countries are now in the minority. The Syrian refugee crisis constitutes the largest refugee emergency of the twenty-first century and yields detailed insights into the securitization of migration and the walling of Europe specifically.

Checklist of Key Ideas

- In studying global migration stocks and flows, human geographers use the terms "diaspora," "migrant," "refugee," "asylum seeker," "internally displaced person," "human trafficking," and "stateless person."
- Classically, diasporas form when populations are scattered through forced migration. The emergence of the Atlantic slave trade and Africa-American diaspora has been of world historical import.
- According to the IOM, globally today there are circa 272 million international migrants, 25.9 million refugees, 3.5 million people seeking asylum, between 600 000 and 800 000 people trafficked across an international border each year, and 3.9 million stateless people.
- Migrant integration policies and outcomes do not correlate particularly strongly, suggesting that integration is a nonlinear and contingent process. Existing migrant integration policy frames based upon ideas of assimilation, multiculturalism, and securitization fail to understand processes of settling. A new generation of scholarship, policy, and politics is needed that grasps what is actually happening when migrants plant roots down in host countries.
- The Syrian IDP and refugee crisis stands as the greatest refugee and mass displacement emergency of the early twenty-first century. Immediately adjacent countries have absorbed the vast majority of refugees; Global North countries have increasingly sought to securitize their borders and have failed to take their share of resettlement quotas.

Chapter Essay Questions

1) The African-American diaspora is an example of a classic diaspora. Discuss.
2) Describe and explain patterns and trends in global migrant stocks and flows.
3) Write an essay on the merits and demerits of multiculturalism.
4) The Global North has failed Syria's neighboring countries, leaving them to take responsibility for managing the Syrian refugee crisis. Discuss.

References and Guidance on Further Reading

Seminal human geographical writings on migration are:

Ravenstein, E.G. (1889). The laws of migration. *Journal of the Royal Statistical Society* 52: 241–305.

Zelinsky, W. (1971). The hypothesis of the mobility transition. *Geographical Review* 61: 219–249.

Good general (human geographical) introductions to migration include:

Bastia, T. and Skeldon, R. (eds.) (2020). *Routledge Handbook of Migration and Development*. London: Routledge.

Bauder, H. (2016). *Migration Borders Freedom*. London: Routledge.

Boyle, P. and Halfacree, K.H. (2014). *Exploring Contemporary Migration*. London: Routledge.

de Haas, H., Castles, A., and Miller, M.J. (2019). *The Age of Migration: International Population Movements in the Modern World*. Surrey: Guilford Press.

Gatrell, P. (2020). *The Unsettling of Europe: The Great Migration, 1945 to the Present*. London: Penguin.

Ho, E.L.E. (2018). *Citizens in Motion: Emigration, Immigration, and Re-migration Across China's Borders*. Stanford, CA: Stanford University Press.

Inglis, C., Li, W., and Khandria, B. (2020). *The SAGE Handbook of International Migration*. London: Sage.

Koser, K. (2016). *International Migration: A Very Short Introduction*. Oxford: Oxford University Press.

Liu-Farrer, G. and Yeoh, B.S. (eds.) (2018). *Routledge Handbook of Asian Migrations*. New York: Routledge.

Manning, P. and Trimmer, T. (2020). *Migration in World History*. London: Routledge.

Mavroudi, E. and Nagel, C. (2016). *Global Migration: Patterns, Processes, and Politics*. London: Routledge.

Mitchell, K., Jones, R., and Fluri, J.L. (2019). *Handbook on Critical Geographies of Migration*. London: Edward Elgar.

Portes, A. and Rumbaut, R.G. (2014). *Immigrant America: A Portrait*. Berkeley: University of California Press.

Samers, M. and Collyer, M. (2016). *Migration*. London: Routledge.

Teixeira, C., Lee, W., and Kobayashi, A. (eds.) (2012). *Immigrant Geographies of North American Cities*. Oxford: Oxford University Press.

Tumbe, C. (2018). *India Moving: A History of Migration*. New Delhi: Penguin Viking.

Important contributions to scholarship on global refugees, IDPs, and human trafficking include:

Betts, A. and Collier, P. (2017). *Refuge: Rethinking Refugee Policy in a Changing World*. New York: Oxford University Press.

Chowdhory, N. and Mohanty, B. (eds.) (2020). *Citizenship, Nationalism and Refugeehood of Rohingyas in Southern Asia*. New Delhi: Springer.

Cohen, R. and Van Hear, N. (2019). *Refugia: Radical Solutions to Mass Displacement*. London: Routledge.

DeJesus, K. (2018). Special issue: forced migration and displacement in Africa: contexts, causes and consequences. *African Geographical Review* 37 (2).

Mountz, A. (2010). *Seeking Asylum*. Minneapolis: University of Minnesota Press.

Wennersten, J.R. and Robbins, D. (2017). *Rising Tides: Climate Refugees in the Twenty-First Century*. Bloomington: Indiana University Press.

Yea, S. (2019). *Paved with Good Intentions? Human Trafficking and the Anti-Trafficking Movement in Singapore*. Singapore: Palgrave MacMillan.

Seminal works on "diaspora" include:

Brah, A. (1996). *Cartographies of Diaspora: Contesting Identities*. London: Psychology Press.

Cohen, R. (1997). *Global Diasporas: An Introduction*. London: Routledge.

Cohen, R. and Fischer, C. (eds.) (2018). *Routledge Handbook of Diaspora Studies*. London: Routledge.

Fisher, M.A. (2013). *Migration: A World History*. Oxford: Oxford University Press.

Safran, W. (1991). Diasporas in modern societies: myths of homeland and return diaspora. *A Journal of Transnational Studies* 1: 83–99.

Important human geographical works on empire, slavery, and the African-American diaspora include:

Domosh, M. (2018). Race, biopolitics and liberal development from the Jim Crow South to post-war Africa. *Transactions of the Institute of British Geographers* 43: 312–324.

Dwyer, O.J. and Alderman, D.H. (2008). *Civil Rights Memorials and the Geography of Memory*. Atlanta: University of Georgia Press.

Lambert, D. (2013). *Mastering the Niger: James MacQueen's African Geography and the Struggle over Atlantic Slavery*. Chicago: University of Chicago Press.

Nelson, C. (2017). *Slavery, Geography and Empire in Nineteenth-Century Marine Landscapes of Montreal and Jamaica*. London: Routledge.

Ogborn, M. (2019). *The Freedom of Speech: Talk and Slavery in the Anglo-Caribbean World*. Chicaco: University of Chicago Press.

Shabazz, R. (2015). *Spatializing Blackness: Architectures of Confinement and Black Masculinity in Chicago*. Urbana: University of Illinois Press.

Other important works on empire, slavery, and the African-American diaspora are:

Beckert, S. (2015). *Empire of Cotton: A Global History*. New York: Penguin.

Blackmon, D.A. (2008). *Slavery by Another Name: The Re-enslavement of Black Americans from the Civil War to World War II*. New York: Knopf Doubleday Publishing Group.

Cooper, H.L.F. and Fullilove, M.T. (2020). *Ending Police Violence: A Public Health Primer*. Baltimore: Johns Hopkins University Press.

Gilroy, P. (1995). *The Black Atlantic: Modernity and Double-Consciousness*. Cambridge, MA: Harvard University Press.

Hahn, S. (2003). *A Nation Under Our Feet: Black Political Struggles in the Rural South from Slavery to the Great Migration*. Cambridge, MA: Harvard University Press.

Litwack, L.F. (2009). *How Free Is Free? The Long Death of Jim Crow*. Cambridge, MA: Harvard University Press.

Morgan, K. (2007). *Slavery and the British Empire: from Africa to America*. Oxford: Oxford University Press.

Tolnay, S.E. (2003). The African American "great migration" and beyond. *Annual Review of Sociology* 29: 209–232.

See also the Special Issue in *American Journal of Public Health* of 400 years of slavery:

Brown, T.M. (ed.) (2019) Special issue: four hundred years since Jamestown. *American Journal of Public Health* 109(10).

Canadian geographer Audrey Kobayashi's thinking on human geography and race is outlined in:

Kobayashi, A. (2014). The dialectic of race and the discipline of geography. *Annals of the Association of American Geographers* 104: 1101–1115.

Statistics on migration, migration policies, and migrant integration are provided in:

Internal Displacement Monitoring Centre (IDMC) (2020). *Global Report on Internal Displacement 2020*. Oslo: IDMC.

International Organisation for Migration (IOM) (2020). *World Migration Report 2020*. Geneva: IOM.

OECD/European Union (2018). *Indicators of Immigrant Integration*. Brussels: OECD Paris/ European Union.

The latest data from MIPEX, including its 2020 update, can be found at http://www.mipex.eu.
Circular plot visualization charting global migration flows (1960–2015) can be found in:

Abel, G.J. and Sander, N. (2014). Quantifying global international migration flows. *Science* 343: 1520–1522.

Abel, G.J. (2018). Estimates of global bilateral migration flows by gender between 1960 and 2015. *International Migration Review* 52: 809–852.

An informative commentary on recent international migration flows can be obtained from:

de Haas, H., Czaika, M., Flahaux, M.L. et al. (2019). International migration: Trends, determinants, and policy effects. *Population and Development Review* 45: 885–922.

The relationship between migration, integration, populism, migration policy, and strong borders is examined in:

Alba, R. and Foner, N. (2015). *Strangers No More: Immigration and the Challenges of Integration in North America and Western Europe*. Princeton, NJ: Princeton University Press.

Anderson, R. (2019). *No Go World How Fear Is Redrawing Our Maps and Infecting Our Politics*. Berkeley: University of California Press.

Blanca, G. and Penninx, R. (2016). *Integration Processes and Policies in Europe*. Delhi: Springer.

Conlon, D. and Hiemstra, N. (eds.) (2016). *Intimate Economies of Immigration Detention: Critical Perspectives*. London: Routledge.

Derrida, J. and Dufourmantelle, A. (2000). *Of Hospitality*. Stanford, CA: Stanford University Press.

Farer, T. (2019). *Migration and Integration: The Case for Liberalism with Borders*. Cambridge: Cambridge University Press.

Gilmartin, M. and Wood, P. (2018). *Borders, Mobility and Belonging*. Bristol: Policy Press.

Kaufmann, E. (2018). *Whiteshift: Populism, Immigration and the Future of White Majorities*. London: Penguin UK.

Murray, D. (2017). *The Strange Death of Europe: Immigration, Identity, Islam*. London: Bloomsbury Publishing.

Vallet, E. (2014). *Borders, Fences, and Walls: States of Insecurity?* London: Routledge.

Multiculturalism is placed under the microscope in   books such as:

Amin, A. (2012). *Land of Strangers*. Cambridge: Polity Press.

Brownlee, P. (2020). *The Political Economy of Migration and Post-industrialising Australia: Valuing Diversity in Globalised Production*. New Delhi: Routledge.

Chin, R. (2019). *The Crisis of Multiculturalism in Europe: A history*. Princeton, NJ: Princeton University Press.

Forbes, H.D. (2019). *Multiculturalism in Canada: Constructing a Model Multiculture with Multicultural Values*. Amsterdam: Springer.

Gilroy, P. (2004). *After Empire: Melancholia or Convivial Culture?* London: Routledge.

Keith, M. (2005). *After the Cosmopolitan?: Multicultural Cities and the Future of Racism*. London: Routledge.

Kymlicka, W. (1995). *Multicultural Citizenship: A Liberal Theory of Minority Rights*. Oxford: Clarendon Press.

Kymlicka, W. (2012). *Multiculturalism: Success, Failure, and the Future*. Washington, DC: MPI.

Vertovec, S. (2016). *Super-Diversity*. London: Routledge.

The role of ethnic segregation and the ethnic enclave as buffer or beachhead of migrant integration is examined in classic works such as:

Dikeç, M. (2011). *Badlands of the Republic*. London: Wiley.

Portes, A. and Manning, R.D. (1986). The immigrant enclave: theory and empirical examples. *Social Stratification* 576: 47–68.

Marcuse, P. (1997). The enclave, the citadel, and the ghetto: what has changed in the post-Fordist US city. *Urban Affairs Review* 33: 228–264.

Wirth, L. (1956). *The Ghetto*. Chicago: University of Chicago Press.

Wacquant, L. (2013). *Urban Outcasts: A Comparative Sociology of Advanced Marginality*. London: Wiley.

Scholarship on "Chinatowns" includes:

Anderson, K.J., Ang, I., Del Bono, A. et al. (2019). *Chinatown Unbound: Trans-Asian Urbanism in the Age of China*. Singapore: Rowman & Littlefield International.

Ang, I. (2020). Chinatowns and the Rise of China. *Modern Asian Studies* 54: 1367–1393.

Anderson, K.J. (1991). *Vancouver's Chinatown: Racial Discourse in Canada, 1875–1980*. Montreal: McGill-Queen's Press-MQUP.

Yeoh, B.S.A. and Kong, L. (2012). Singapore's Chinatown: nation building and heritage tourism in a multiracial city. *Localities* 2: 117–159.

Wong, B.P. and Chee-Bing, T. (eds.) (2013). *Chinatowns Around the World: Gilded Ghetto, Ethnopolis, and Cultural Diaspora*. Leiden: Brill.

An important overview of the Syrian refugee crisis is provided in:

Trilling, D. (2018). *Lights in the Distance: Exile and Refuge at the Borders of Europe*. London: Picador.

A good introduction to life in the Za'atari refugee camp in Jordan is:

Pasha, S. (2020). Developmental Humanitarianism, Resilience and (Dis) empowerment in a Syrian Refugee Camp. *Journal of International Development* 32: 244–259.

The geopolitics of Syrian migration in Turkey is examined in:

Loyd, J.M., Ehrkamp, P., and Secor, A. (2018). A geopolitics of trauma: refugee administration and protracted uncertainty in Turkey. *Transactions of the Institute of British Geographers* 43: 377–389.

Europe's response to immigration, including the Syrian refugee crisis, is examined in:

Buonanno, L. (2017). *The European Migration Crisis*. London: Palgrave Macmillan.

Crawley, H. and Duvell, F. (2017). *Unravelling Europe's "Migration Crisis": Journeys Over Land and Sea*. Bristol: Policy Press.

Duszczyk, M., Pachocka, M., and Pszczółkowska, D. (eds.) (2020). *Relations Between Immigration and Integration Policies in Europe: Challenges, Opportunities and Perspectives in Selected EU Member States*. London: Routledge.

Weinar, A., Bonjour, S., and Zhyznomirska, L. (eds.) (2018). *The Routledge Handbook of the Politics of Migration in Europe*. London: Routledge.

The latest UNHCR response plans in support of Syrian refugees and IDP are:

UNHCR (2020a). *UNHCR Regional Refugee and Resilience Plan (3RP) for 2019–2020*. New York: UNHCR.

UNHCR (2020b). *UNHCR Syrian Humanitarian Response Plan*. New York: UNHCR.

The UNHCR Global Compact on Refugees:

UNHCR (2018). *UNHCR Global Compact on Refugees*. New York: UNHCR.

Website Support Material

A range of useful resources to support your reading of this chapter are available from the Wiley Human Geography: An Essential Introduction Companion Site http://www.wiley.com/go/boyle.

Chapter 12

At Risk: Hazards, Society, and Resilience

Chapter Table of Contents

Chapter Learning Objectives

By the end of this chapter you should be able to:
- Produce a typology of hazards, and identify the most consequential global hazards by likelihood and impact.
- Describe and explain the significance of Gilbert White's contributions to human geographical studies of hazards.
- Explain what is meant by Risk = Hazard × Vulnerability ($R = H \times V$) and the idea of the *social production of vulnerability* to hazards, and document the social,

Human Geography: An Essential Introduction, Second Edition. Mark Boyle.
© 2021 John Wiley & Sons Ltd. Published 2021 by John Wiley & Sons Ltd.
Companion website: www.wiley.com/go/boyle

economic, cultural, and political processes that put some social groups at greater risk than others.

- Referring to the United Nations University's (UNU) ongoing project of "Mapping the World at Risk," define the concepts of "risk," "exposure," and "vulnerability," and identify the countries that are most and least at risk from natural hazards.
- With reference to the history of disaster risk reduction, describe and comment upon the efficacy of the Sendai Framework for Disaster Risk Reduction 2015–2030.
- Present a case study of disaster risk reduction in Peru, and comment on the efficacy of Peru's approach.
- Define what is meant by resilience politics. Unbundle the variety of ways in which the concept of resilience has been used in disaster risk reduction, and discuss the rival merits of the ideas of robustness, recovery, reform, and redesign.

Introduction

This is the age of risk. With the rise of the West came a new level of optimism and confidence; human beings believed that they no longer needed to cower in the wake of nature's extremes and human-induced societal and technological hazards, for technology itself could afford humankind unprecedented levels of protection. Of course, such arrogance has proven to be ill founded. Hazards continue to threaten humanity, often leaving in their wake death and injury, wreckage of vital infrastructure, social disruption, and economic carnage. Perhaps more significantly, although the emergence of the world capitalist system brought with it new technologies and competencies to shield humanity against nature's extremes and societal- and technological-induced risks, paradoxically it also created as its legacy historically unprecedented levels of vulnerability to hazard events, at least for some people. As US geographer and hazard expert Susan L. Cutter has long argued, it is through the social production of vulnerability that hazards become disasters and disasters become catastrophes.

By why is the world a riskier place for poor people, women, children, ethnic minorities, and those with a disability? What kinds of social relations are at work such that survival is heavily loaded in favor of some, while others dwell in positions of precarity and fragility?

Taking Cutter's work as its cue, the purpose of this chapter is to examine the ways in which the ascendance of the West and the rise of the capitalist political economy, along with the workings of patriarchy, racism, ageism, ableism, and so on, have bequeathed inequalities in vulnerabilities to hazards within and between the Global North and Global South.

Global Risks: Hazards by Likelihood and Impact

It is usual to distinguish two kinds of hazards:

Natural hazards: Earthquakes, volcanoes, tsunamis, hurricanes, floods, droughts, mudslides, wildfires, pandemics, tornadoes, blizzards, heat waves, and lightning strikes.

Societal and technological hazards: Nuclear power and nuclear weapons, (bio) terrorism, cybersecurity, complex financial systems, artificial intelligence, and carbon pollution.

When placed into historical context, it is the case that natural hazards are occurring with increased frequency and severity, and are affecting more people and leading to greater economic losses than ever before, but they are also decreasing in their lethality. At least in this regard, we might say that societies are successfully working to protect their citizens and to produce a safer world. In the period 2010–2020, the economic damage wrought by natural disasters amounted to $2.98 trillion, $1.19 trillion more than the preceding decade. While earthquakes have proven most lethal since 2000, increasingly hurricanes and wildfires have resurged in frequency and impact (Plate 12.1). In the same period of 2010–2020, natural disasters killed on average 60 000 people per year (0.1% of all deaths); in 2019 11 000 people died. By comparison, in the early to mid-twentieth century, the annual death toll from disasters often reached one million per year, and averaged across particular decades averaged at circa 500 000 (Figure 12.1). This decline in natural hazard death rates is even more significant given world population growth since 1900. Whether Covid-19 (which looks set to claim up to 3 million lives) constitutes a new chapter in this story remains to be seen.

But as noted, natural hazards are not the only type of hazards that exist. The World Economic Forum's 15th *Global Risks Report 2020* disaggregates hazards in a novel way and estimates the likelihood and impacts of each of the economic, geopolitical, environmental, societal, and technological categories of risk, covering a total of 40 hazards (Figure 12.2). Based upon a perception study (1047 total responses to the Global Risks Perception Study [GRPS]) capturing the views of a multi-stakeholder community (business, academia, civil society, and the public) and the views of a smaller group of "global shapers," the World Economic Forum concludes that risk is intensifying over time. Environmental risks, particularly those associated with extreme weather events, accounted for three of the top five risks by likelihood and four by impact (Figure 12.3). Technology and in particular data fraud and cyber-attacks were also perceived to be

Plate 12.1   2019: The year of the forest fires. Big Fall Creek Road, Lowell, Oregon, United States. Source: Photo by Marcus Kauffman on Unsplash.

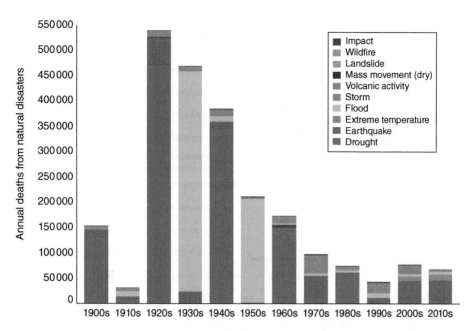

Figure 12.1   Global annual deaths from natural disasters, by decade. Source: EMDAT (2017): OFDA/CRED International Disaster Database. Universite catholique de Louvain- Brussels-Belgium. The data visualization is available at OurWorldinData.Org. Licensed under CC-BY-SA.

Economic	Environmental	Geopolitical
Asset bubble	Biodiversity loss	Global governance failure
Critical infrastructure failure	Climate action failure	Interstate conflict
Deflation	Extreme weather	National governance failure
Energy price shock	Human-made environmental	State collapse
Financial failure	disaster	Terrorist attacks
Fiscal crises	Natural disasters	Weapons of mass destruction
Illicit trade		
Unemployment		
Unmanageable inflation		

Societal	Technological	
Failure of urban planning	Adverse technological advances	
Food crises	Cyberattacks	
Infectious diseases	Data fraud or theft	
Involuntary migration	Information infrastructure breakdown	
Social instability		
Water crises		

Figure 12.2   A typology of risks. Source: From World Economic Forum World Risk Report 2020. © 2020 World Economic Forum.

growing risks. Furthermore, respondents believed that across the next decade, a number of key hazards are highly likely to become riskier. The Covid-19 pandemic affirms that we are living in an increasingly risky world. It also reveals that, like everyone else, the World Economic Forum failed to recognize its likelihood and impact. 2020/1 will witness a dramatic rise in death from hazards and extraordinary economic loss.

Multi-stakeholders		Global Shapers	
Likelihood	**Impact**	**Likelihood**	**Impact**
◼ Extreme weather	◼ Climate action failure	◼ Extreme weather	◼ Biodiversity loss
◼ Climate action failure	◼ Weapons of mass destruction	◼ Biodiversity loss	◼ Climate action failure
◼ Natural disaster	◼ Biodiversity loss	◼ Climate action failure	◼ Water crises
◼ Biodiversity loss	◼ Extreme weather	◼ Natural disasters	◼ Human-made environmental disasters
◼ Human-made environmental disasters	◼ Water crises	◼ Human-made environmental disasters	◼ Extreme weather
◼ Data fraud or theft	◼ Information infrastructure breakdown	◼ Water crises	◼ Weapons of mass destruction
◼ Cyberattacks	◼ Natural disasters	◼ Data fraud or theft	◼ Natural disasters
◼ Water crises	◼ Cyberattacks	◼ Involuntary migration	◼ Food crises
◼ Global governance failure	◼ Human-made environmental disasters	◼ Social instability	◼ Infectious diseases
◼ Asset bubble	◼ Infectious diseases	◼ Cyberattacks	◼ Cyberattacks

◼ Economic ◼ Environmental ◼ Geopolitical ◼ Societal ◼ Technological

Figure 12.3 Top 10 risks by likelihood and impact over the next decade, as judged by multi-stakeholders and global shapers. Source: From World Economic Forum World Risk Report 2020. © 2020 World Economic Forum.

Understanding Risk: What Causes Hazards to Become Disasters and Disasters to Become Catastrophes?

Gilbert White: Pioneering Human Geographical Interest in Natural Hazards

Earlier than most, US geographer Gilbert White came to the realization that natural hazards and extreme weather events ought to be of interest to human geographers and not just physical geographers. White's thinking on hazards was rooted in and informed by the risks posed by flooding and his passionate advocacy of the need for comprehensive floodplain management. But his approaches and insights were pertinent to all hazards, and as a consequence he is widely considered to be the founder of hazard studies as a formal branch of knowledge (Hinshaw 2006).

In an obituary published following his death in 2006, the Association of American Geographers identified five particular thematic areas that White made seminal contributions to:

- Domestic water supply, and in particular the bringing of safe potable water to all people as a matter of human right, and not simply as another good or commodity.
- Natural hazards and disasters, and in particular the reduction of the toll of deaths, damage, and destruction.
- Peace, globally and regionally, through the cooperation of diverse peoples and the development and management of river basins and water resources.
- The affirmation and deployment of science in general (and geography in particular) beyond the bounds of the academy into practical service to humanity.
- The reconciliation of human wants, needs, and greed with the conservation of the environment and global resources, to achieve a sustainable development.

White's interest in flood events was ignited when, as a young civil servant in Washington in the 1930s, he was appointed by US President Franklin D. Roosevelt to serve on the National Resources Planning Board (NRPB). Roosevelt's New Deal response to the Great Depression that then gripped the United States included putting the nation's unemployed to work in grand public works schemes. Roosevelt decreed flood defense and floodplain management a priority and set out to build new protective dams and levees throughout the country. During this period, the Hoover Dam was built on the Colorado River, and the Tennessee Valley Authority (TVA) was created and the Wilson Dam built on the Tennessee River.

White's specific responsibility was to work with officials to improve floodplain management in the Missouri River basin. During this time, he became convinced that "technological fixes" alone were doomed to failure. Building better flood defenses was a necessary but insufficient solution to the problems caused by flooding. Although such defenses might protect floodplains in the short term, eventually they would succumb to extreme flood events. White was convinced that a whole variety of supplementary strategies were needed; alongside protection there needed to be prevention, prediction, mitigation, response, and adaptation.

In his 1942 University of Chicago PhD thesis entitled *Human Adjustment to Floods*, White dismissed the claim that natural hazards are best remediated by engineering solutions. Instead, floods and indeed any other natural disaster could be better mitigated by modifying human behavior. "Floods are 'acts of god', but flood losses are largely acts of man," he famously stated.

Later, working as a geographer at the University of Chicago in 1966, White was appointed by the national Budget Task Force on Federal Flood Control Policy to lead a review of the merits and demerits of establishing a national flood insurance program. White's proposals supporting such a scheme were widely praised and accepted. Although the complexities entailed by introducing a National Insurance Scheme proved overwhelming and slowed progress, a modified system of insurance against losses incurred due to flooding was eventually introduced. In 1979, this scheme was integrated into the roles and responsibilities of the Federal Emergency Management Agency (FEMA), the country's principal planning agency for hazards.

All the while, White continued to argue that flood defense systems, though important, could only ever be considered a partial response to the threats posed by extreme flood events. Of equal importance were stronger spatial planning to limit land-use development in floodplains, better forecasting, improved emergency services, expanded social insurance, and superior building regulations and standards (see Deep Dive Box 12.1).

White recognized that his comprehensive approach to minimizing risks posed by extreme flood events was capable of wider application to other hazard events. In 1976, he founded the Natural Hazards Research and Applications Information Center (today simply the Natural Hazards Center [NHC]) at the University of Colorado and served as director of the NHC from 1976 to 1984 and 1992 to 1994. This center continues to serve as a world leading authority on public policy as it relates to natural disasters, and has undertaken research on flooding, drought, earthquakes, mudslides, hurricanes, wildfires, tsunamis, tornadoes, and volcanoes. Inspired by White's insistence on a comprehensive approach to the mitigation of hazards, the NCH places equal importance on disaster avoidance, preparedness, response, and recovery. White was awarded the Public Welfare Medal from the US National Academy of Sciences for his applied geographical research on hazards and flood insurance and was presented with a National Science Foundation "National Medal of Science" by President Bill Clinton. The AAG subsequently introduced a Gilbert F. White Public Service Honors award.

Deep Dive Box 12.1 Westcoat and White's (2003) *Water for Life: Water Management and Environmental Policy*

Gilbert White's final book, *Water for Life* (coauthored with US environmental geographer James Westcoat), provides a summation of his views on the management of floodplains. Good governance, it was argued, required actions in seven particular areas:

- Mapping incidences of flooding in history and estimating the frequency of different kinds of flood events by scale or magnitude: for example, a 2-m rise in the water level may be a 1-in-100-year event (that is, it may have a probability of occurrence in any given year of 1%) in this floodplain but a 1-in-25-year event in another floodplain.
- Applying suitable planning and regulation in particularly vulnerable floodplains (to prevent risky building and development in the first place) and in areas that affect drainage (to ensure development does not create even more dangerous drainage regimes).
- Establishing, with government support, an effective insurance scheme to ensure that those who are affected by flooding can rebuild their lives.
- Improving short-term forecasting and warning systems and upskilling all stakeholders (the general population, community groups, the emergency services, businesses, local governments, and so on) so that they are competent to respond effectively to early warnings.
- Fortifying the ability of architects, builders, construction companies, and property developers to build sturdier structures that are at least better able to withstand flood events, and at best are flood-proof.
- Scaling up support and relief offered by federal government to victims of flood events, to enable these victims to abandon particularly damaged property and move into homes and workplaces that lie at a safe distance from floodplains.
- Undertaking cost–benefit analyses of the impacts of both flooding and flood defense systems on natural ecosystems, both those created by and those destroyed by flood events.

When Hazards Become Disasters: Risk = Hazard × Vulnerability (R = H × V)

A common misconception about natural disasters is that populations most at risk are simply those unlucky enough to have been born in parts of the world where nature's extremes are most manifest; danger is simply a function of the uneven distribution of the magnitude and frequency of hazards across the face of the earth. Increasingly, it is being recognized that although exposure to natural hazards is important, ultimately it is society that puts people at increased risk and, therefore, that solutions to natural hazards need to tackle the root causes of the social production of vulnerability to hazard events.

And so the formula *Risk = Hazard × Vulnerability (R = H × V)* has become of central importance in hazards research.

For nearly 40 years, US geographer Susan L. Cutter has been at the forefront of human geographical explorations of vulnerability to hazards. Her 2003 article "Social vulnerability to environmental hazards" in the journal *Social Science Quarterly* remains one of the most influential papers in the history of hazards research. While recognizing the importance of underlying geological, meteorological, and hydrological processes, Cutter argues that there is nothing particularly natural about natural disasters. She places under scrutiny the ways in which social, political, cultural, and economic processes play a role in increasing the risk that natural hazards more easily become natural disasters for specific social groups, and catastrophes for others. This rings true also for societal and technological hazards. Hazards are inherently riskier for impoverished, powerless, and disadvantaged populations – in reality, the poor, women, the less educated, those socially isolated, ethnic minorities, people with disabilities, and children, especially, although not exclusively, in the developing world. Social, political, and economic precarity renders some populations especially vulnerable, leaving them without the capacity to limit danger through forward planning, and making them more susceptible to the harmful effects of a shock and less able to cope when perturbation strikes.

The pioneering work of Canadian geographer Kenneth Hewitt also encouraged human geographers to begin to take seriously the role of society in increasing and aggravating the lethality of natural hazards. In his 1983 book *Interpretations of Calamity: From the Viewpoint of Human Ecology* (Hewitt 1983), Hewitt demonstrated that natural hazards are always threatening but only develop into calamities when societies pursue development pathways that unwittingly increase their vulnerability. In his book *Regions of Risk: A Geographical Introduction to Hazards*, published in 1997, Hewitt elaborated on this claim by showing that calamity results from three factors: the natural hazard itself, the vulnerability of societies, and the degree to which societies are active in defending themselves.

In the early 1990s, British geographers Keith Smith and David N. Petley's book *Environmental Hazards: Assessing Risk and Reducing Disaster* (Smith and Petley 1991) and US and British geographers and scholars of international development Piers Blaikie, Ben Wisner, Terry Cannon, and Ian Davis's book *At Risk: Natural Hazards, People's Vulnerability and Disasters* (Blaikie et al. 1994) also played a leading role in shifting the attention of human geographers toward the role of socio-political factors in hazard events.

According to Smith and Petley, through time, hazard studies has developed four main paradigms, each paradigm viewing hazards and approaching their mitigation in particular ways. Before 1950, an *engineering paradigm* dominated. Here nature's extremes were perceived to be the root cause of natural disasters, and technological and engineering solutions were advocated to shield communities where possible. Between 1950 and 1970, a *behavioral paradigm* captured thought. Here the focus was upon hazards in more developed countries, the ways in which development was encroaching onto land exposed to hazards, and the role of spatial planning in steering future land-use development toward safer sites. The period of 1970 to 1990 witnessed the rise of a *development paradigm* in which underdevelopment and poverty in lesser developed countries were viewed as key sources of heightened vulnerability, and social, political, cultural, and economic change as the only route to redemption. Finally, according to Smith and Petley, from 1990 onward, a *complexity paradigm* has emerged, which has located hazards within coupled human and natural systems (CHANS) and sought to help local communities better manage interactions between society and nature so as to minimize their vulnerability to hazards.

At the center of Blaikie et al.'s (1994) approach to hazards is the view that because there exist marked inequalities within and between societies, between more-resourced and less-resourced social groups, it follows that vulnerability to hazards is unevenly distributed. Blaikie et al. developed a Pressure and Release (PAR) model to explain the ways in which deep structures in society (how society is organized politically, economically, socially, and culturally) lie at the "root" of vulnerability, generating "dynamic processes" (such as wars, poverty, and urbanization) that in turn lead to "unsafe conditions." The ability of populations in more impoverished regions of the world – the Global South – to withstand the effects of a natural hazard is likely to be less than that of populations in more affluent parts of the Western world – the Global North. Moreover, there exists a marked variation in vulnerability to hazards between social groups within the Western world, and indeed within the developing world. Hazards produce more disastrous consequences for communities already suffering from socio-structural disadvantage and exclusion. In contrast, those with more command over socio-political and economic resources and capital (for example, the wealthy, men, the ethnic majority, the better educated, the able-bodied, adults, those with strong social networks, and settled populations) benefit from enhanced resilience.

It is possible to identify six social, political, cultural, and economic conditions which increase the vulnerability of populations exposed to natural hazards: poor governance, poverty, social exclusion, war and violent conflict, megacities, and environmental degradation. Of course, all six feature in all world regions, rich and poor. But they tend to manifest themselves in particularly acute ways in the most destitute regions of the world.

Poverty: Poverty alone is perhaps the greatest progenitor of precarity. Poor people tend to live in overcrowded conditions and suffer from poorer nutrition and even malnutrition, both environments in which disease more easily incubates and spreads. Their means of subsistence and lack of savings ensure that they have limited access to resources in times of environmental upheaval. They are less likely to have political influence and representation in local, regional, and national governments. Poor people tend to live in homes that are poorly constructed and particularly fragile. They tend to live in areas of cities and regions which are especially exposed to hazards and where land prices and rents are in consequence lower. They are less likely to have life and health assurance, or insurance against loss of property owing to natural hazards.

Social exclusion: In addition to poverty, other forms of social exclusion can add to people's precarity, such as gender, age, ethnicity, disability, sexuality, and so on. Many societies are riven with inequalities in opportunity and welfare. For women, children, ethnic minorities, people with disabilities, those stigmatized on the basis of their sexuality, and, indeed, those marked out as different from mainstream society, natural hazards can often bring elevated vulnerability to an already highly precarious existence. In times of scarcity, it is an unfortunate fact that white middle-class men, adults, majority and mainstream populations, and the able-bodied capture more than their fair share of emergency resources and accordingly are better equipped to weather hardships.

Poor governance: Effective governance and strong and (st)able institutions help societies protect their most vulnerable people, develop long-term plans to mitigate against the effects of disasters, prepare for disasters, cope with disasters when they occur, and recover from disasters more speedily. Poor governance, in contrast, heightens the precarity of vulnerable groups, militates against long-term planning, and reduces the capacity of communities to prepare for, cope with, and recover from a hazard event. Countries marked by military dictatorships, repeated coup d'états and political instability, weak institutional capacity, and political corruption are likely to render their populations more vulnerable to nature's extremes.

War and violence: Violent conflicts increase vulnerability to natural hazards by making disaster preparation and planning impossible or less of a priority; displacing people from their homes and forcing them to live in makeshift and unsanitary refugee camps; disrupting emergency supplies and humanitarian relief; depleting resources and raw materials to the point of exhaustion, thereby increasing precarity; intentionally or unintentionally destroying hazard defense infrastructure; and increasing predatory practices (such as human trafficking and bonded labor) and crime (especially rape, looting, and murder).

Rapid urbanization: The rise of megacities is another factor that serves to deepen the precarity of already marginalized populations in the Global South. Migrants living in unregulated and underregulated shantytowns, *favelas*, and slums are among the most vulnerable of all populations. Megacities both draw upon the resources of the surrounding hinterland (for food and raw materials) and provide services to that hinterland (energy, employment, finance, water, and so on), and so any disruption caused by natural hazards can quickly become a disaster for populations apparently at a remove from the natural hazard itself. For good reason, many megacities form in coastal regions, and the effects of hazards can be heightened by the supplementary and associated risks of subsidence, salinization, liquefaction, flooding, and tsunamis.

Environmental degradation: For marginal communities in the Global South who eke out a living at the extremes of the natural environment, living sustainably is a particular challenge, and environmental degradation and resource depletion can in turn become problems that affect vulnerability to natural hazards. These problems can be further aggravated by climate change. Some ecosystems, such as forests, wetlands, mangroves, and coral reefs, can shield people from risks posed by landslides and tidal waves. Because ecosystems often provide nutrition, income (for example, tourism), building materials, and medicines, their destruction increases communities' susceptibility to hazards. Ecosystems can often help communities cope with natural hazards by providing emergency resources such as food and freshwater when normal supplies are interrupted. Finally, by increasing options and possibilities to diversify the available pool of local resources, biodiversity helps communities formulate long-term hazard mitigation plans; a reduction in biodiversity, therefore, can render populations more vulnerable.

Mapping the World at Risk

The United Nations University (UNU) in Bonn, Germany, provides a useful framework through which the social production of vulnerability might be better understood (UNU 2018). The UNU begins with the formula $R = H \times V$, but then breaks down vulnerability into three component parts: degree of susceptibility to hazards (likelihood of suffering harm), capacity to cope with hazards (capacity to mitigate the impact of hazards when they do occur), and ability to plan ahead to adapt to natural extremes (ability to minimize the degree to which exposure to hazards is increased by prior poor human decision making).

According to the UNU, economic and socio-institutional conditions determine a society's degree of susceptibility, coping capacity, and ability to adapt (and see Figure 12.4).

Susceptibility: Societies marginalize and impoverish some social groups to the extent that their existence is so precarious that small setbacks have significant consequences. They are likely to be predisposed to feel the full ferocity of hazards. In contrast, social, political, cultural, and economic processes also enrich and empower other social groups to the extent that they are fortified and inoculated to a degree against hazard events.

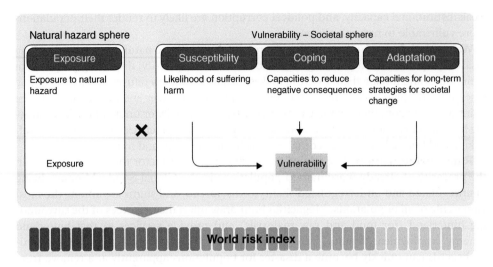

Figure 12.4   Factors in the calculation of the World Risk Index. Source: Redraw from UNU, Factors in the calculation of the World Risk Index.

Coping: The ability of a society to cope with a natural hazard when it does occur is a function of competencies in the areas of disaster preparation (the quality of forecasts and early warning systems); disaster management (the readiness of emergency and humanitarian services to evacuate, provide medical support, conduct search and rescue, provide temporary shelter, distribute food and supplies, and maintain law and order); and disaster recovery (the availability of resources to rebuild and repair communities and infrastructure, and social insurance schemes). Communities with resources are more likely to have superior systems of preparation, management, and recovery. Impoverished communities are liable, in contrast, to be underprepared, to suffer from poor governance, to have poorly co-ordinated emergency management systems, and to lack the resources to rebound quickly after disaster strikes.

Adaptation: Ideally, societies should formulate comprehensive long-term disaster management plans. In particular, spatial planning should be used where possible to steer human activities from areas exposed to hazards and to work to make communities, both poor and wealthy, more resilient. Wealthy societies generally have stronger institutions and superior systems of governance and are better able to engage in long-term planning. Lesser developed societies, in contrast, tend to suffer from weak and failing institutions and poorer governance, and as a consequence find it difficult to formulate and implement long-term plans.

Based upon these distinctions, the UNU has developed a set of indicators and produced maps showing global patterns of exposure, vulnerability (a composite of susceptibility, coping, and adaptation), and overall risk (a risk index comprising exposure and vulnerability) (see Map 12.1).

Although patterns of exposure express natural processes, it is clear that patterns of vulnerability reflect the history of Western hegemony and the legacies of uneven geographical development this history has bequeathed. Five hundred years of a European-centered world capitalist system and European empire has left an indelible geography of risk, with natural hazards more likely to translate into natural disasters in parts of Africa, Asia, and Latin America. In contrast, natural hazards pose less of a threat for rich advanced Western countries in the Global North. Specifically, it is possible to discern in the data at least five important trends.

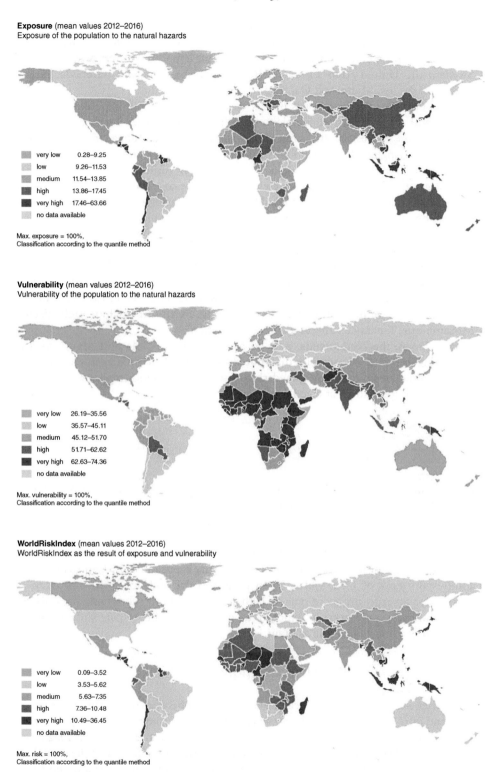

Exposure (mean values 2012–2016)
Exposure of the population to the natural hazards

very low	0.28–9.25	
low	9.26–11.53	
medium	11.54–13.85	
high	13.86–17.45	
very high	17.46–63.66	
no data available		

Max. exposure = 100%,
Classification according to the quantile method

Vulnerability (mean values 2012–2016)
Vulnerability of the population to the natural hazards

very low	26.19–35.56	
low	35.57–45.11	
medium	45.12–51.70	
high	51.71–62.62	
very high	62.63–74.36	
no data available		

Max. vulnerability = 100%,
Classification according to the quantile method

WorldRiskIndex (mean values 2012–2016)
WorldRiskIndex as the result of exposure and vulnerability

very low	0.09–3.52	
low	3.53–5.62	
medium	5.63–7.35	
high	7.36–10.48	
very high	10.49–36.45	
no data available		

Max. risk = 100%,
Classification according to the quantile method

Map 12.1 Mapping the world at risk. Source: Courtesy of World Risk Index, United Nations University Bonn.

- Some countries enjoy lower levels of risk because they are neither exposed to natural hazards nor especially vulnerable. Countries in this classification are among the safest in the world (for example, Germany, Estonia, Israel, Egypt, Norway, Finland, Sweden, the United Arab Emirates, Bahrain, Kiribati, Iceland, Grenada, Saudi Arabia, Barbados, Malta, and Qatar).
- Other countries suffer high levels of risk because they are both exposed to natural hazards and are especially vulnerable to these hazards. Countries in this classification are among the riskiest in the world (for example, Vanuatu, Tonga, the Philippines, Guatemala, Bangladesh, the Solomon Islands, Costa Rica, Cambodia, Timor-Leste, El Salvador, Brunei Darussalam, Papua New Guinea, Mauritius, Nicaragua, and Fiji).
- A third group of countries bear an elevated exposure to hazards but, because they are not vulnerable, their risk of a hazard becoming a disaster is lessened. These countries are better able to withstand hazards and diminish their lethality (good examples are Australia, New Zealand, Ireland, Italy, and, to a lesser extent, the United Kingdom, Greece, and the United States).
- Conversely, there exist countries that are not especially exposed to hazards but that, because they are exceptionally vulnerable, tend to amplify the effects of hazards and to experience hazard events as particularly devastating. These countries are liable to convert relatively minor disturbances into traumatic events (they include Afghanistan, Mozambique, Tanzania, Liberia, Eritrea, and Yemen).
- Finally, Japan, the Netherlands, and, to a lesser extent, Chile and Mauritius are interesting examples of countries that, notwithstanding their abilities to reduce their vulnerabilities, remain at heightened risk of natural disasters simply because their level of exposure is so great (see Deep Dive Box 12.2).

Deep Dive Box 12.2   A Tale of Two Earthquakes: Haiti and Japan

Insights into the ways in which economic and socio-institutional processes can increase or decrease countries' vulnerability to natural hazards can be gleaned from a comparison of the impacts of the Haiti earthquake in 2010 and the Tōhoku earthquake and tsunami in Japan in 2011.

According to the UNU's World Risk Index 2020, Haiti is ranked the 22nd most risky country in the world overall with an exposure score rank of 31 and vulnerability score rank of 13th, whilst Japan is ranked the 46th most risky country overall with an exposure score rank of 10th and vulnerability score rank of 144th. Both countries are especially exposed to earthquakes, but Japan suffers less because it is not as vulnerable as Haiti.

The Haiti Earthquake, 2010

The Caribbean island of Haiti is located at the boundary between the North American Plate and the Caribbean Plate, and as such is exposed to earthquakes. On 12 January 2010, at the Enriquillo–Plaintain Garden fault zone (EPGFZ), an earthquake of magnitude 7 on the Richter scale occurred, with an epicenter 25 km southwest of the capital city Port-Au-Prince and 13 km below the earth's surface (a relatively shallow depth) (see Map 12.2). Across the next fortnight, over 50 aftershocks measuring 4.5 or higher on the Richter scale, aggravated the consequences of the main earthquake.

Estimates of the damage caused by this earthquake continue to be disputed. The Haitian government initially estimated the death toll to be around 230 000 but later revised this upward to 316 000 (the island has a total population of 10 million, with 940 000 in the capital city). Others dispute this claim and suggest that between 46 000 and 85 000 perished. Approximately 250 000 homes collapsed or were severely damaged, resulting in 1.5 million people being displaced, around 550 000 of whom remained homeless two years after the event. A further 30 000 commercial buildings were said to have collapsed or were severely damaged. The social infrastructure on the island was also badly affected, with an estimated 1300 schools and 50 health care facilities destroyed and the island's main prison flattened, leading to 4000 inmates escaping (Plate 12.2).

Having secured independence from French colonial rule in 1804, Haiti has suffered from a series of dictatorships, corrupt political regimes, coup d'états, violence, and social unrest and remains one of the most impoverished countries in the world. Haiti's troubled political history, broken economy, inadequate infrastructure, poverty, and inept system of governance all contributed to a magnification of the lethality of the 2010 earthquake.

- The population of Haiti is among the poorest in the world. Haiti has a Gross Domestic Product (GDP) per capita of US$797 and a Human Development Index ranking of 169 out of 189 countries. Over 6 million Haitians live below the World Bank IPL of US$3.50 per day, and more than 2.5 million fall below the extreme poverty line of US$1.90 per day. As a consequence, poor nutrition, sanitation, and water supplies and high levels of disease meant people were already living in extreme precarity.

Plate 12.2 Aftermath of the Haiti earthquake, 2010. Source: US Navy, https://commons. wikimedia.org/wiki/File:LA_County_SAR_pulls_Haitian_woman_from_earthquake_debris_2010-01-17.jpg. Public Domain.

(Continued)

Box 12.2   *(Continued)*

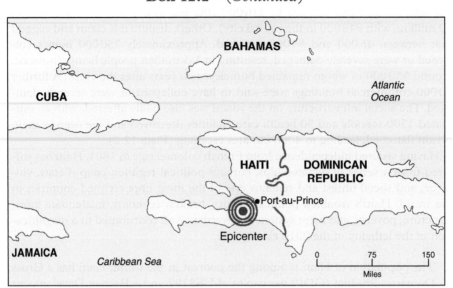

Map 12.2   The epicenter of the Haitian earthquake, 2010. Source: Compiled by the author.

- The earthquake's epicenter was in close proximity to Port-au-Prince, a city that has experienced rapid urbanization and where the majority of its inhabitants live in makeshift and illegal slums. Because construction on the island has been unregulated and underregulated, many buildings, including those in Port-au-Prince, were not designed or erected to be earthquake resistant.
- Already poor transportation connections exacerbated the capacity of relief efforts to ameliorate the plight of the most vulnerable populations immediately following the earthquake; the island only has a single airport (and a single runway) and a single principal port, both of which were damaged by the earthquake. Aid that had arrived at the airport was not distributed efficiently because of limited onward transportation options and personnel.
- Due to a shortage of trained local emergency teams and medical personnel, the search for and treatment of those buried in rubble and collapsed buildings was hampered, leading to unnecessary death, suffering, and injury.
- Desperation and inadequate governance led to looting, anarchy, and violence, exposing the population to new risks.

The Tōhoku Earthquake and Tsunami in Japan, 2011

Japan is located west of the boundary between the Eurasian and Pacific Plates. On 11 March 2011, an earthquake registering 9 on the Richter scale occurred, with an epicenter 72 km east of the Japanese coastal region of Tōhoku and 130 km northeast of the city of Sendai, and at a relatively shallow depth (32 km beneath the sea) (Map 12.3). Shortly after, a tsunami carrying waves in excess of 40 m in height crashed onto the northeastern coastline of Japan.

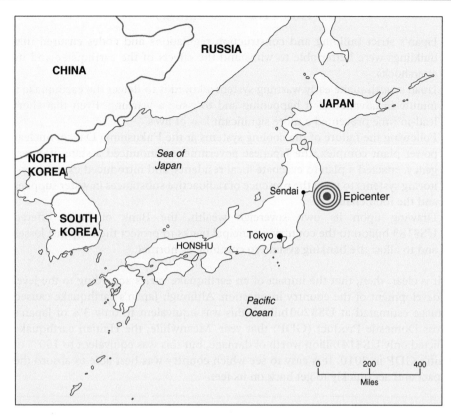

Map 12.3   The epicenter of the Japanese earthquake, 2011. Source: Compiled by the author.

According to the Japanese government, 15 883 people died in the disaster (the vast majority by drowning), 6145 people were injured, 129 225 buildings collapsed, 254 204 buildings partially collapsed, 691 766 buildings were damaged, approximately 4.4 million households had their electricity supplies cut, and 1.5 million households were left without clean water and sanitation. Moreover, the tsunami damaged the Fukushima Daiichi nuclear power plant complex, leading to a dangerous meltdown of reactors. In total, 340 000 people were temporarily displaced.

Although the Tōhoku earthquake and tsunami highlight the risks that natural hazard events pose even for wealthy countries, arguably Japan's privileged status, world-class infrastructure, strong institutions, and effective systems of governance ensured that it escaped even harsher punishment.

- The population of Japan is among the wealthiest in the world. Japan has a Gross Domestic Product (GDP) per capita of US$43,880 and a Human Development Index ranking of 19 out of 189 countries. Effectively no citizen lives beneath any of the World Bank's IPLs.
- Japan enjoys one of the highest life expectancies in the world, and its population boasted comparatively strong health prior to the disaster.

(Continued)

Box 12.2   *(Continued)*

- Japan's strict building and construction regulations and codes ensured that buildings were better able to withstand the effects of the earthquake and its aftershocks.
- Japan's earthquake early warning system allowed it to detect the earthquake a minute in advance of it happening and to issue a warning. Even that short lead-in time prevented a more significant loss of lives.
- Following the failure of the cooling systems at the Fukushima Daiichi nuclear power plant complex, the Japanese government announced a state of emergency, enacted a plan to evacuate local residents, and introduced expert monitoring systems to trace the presence of radioactive substances in water supplies and the food chain.
- Drawing upon its own sovereign wealth, the Bank of Japan offered US$183 billion to the country's principal banks to protect them against losses and to allow the banking system to continue as normal.

It is clear, then, that the impact of an earthquake varies according to the level of development of the country in question. Although Japan's earthquake caused damage estimated at US$200 billion, this was equivalent to only 3% of Japan's Gross Domestic Product (GDP) that year. Meanwhile, the Haitian earthquake inflicted only US$14 billion worth of damage, but this was equivalent to 160% of Haiti's GDP in 2010. It is easy to see which country was best able to absorb the impact and act quickly to get back on its feet.

Disaster Risk Reduction: What Stops Hazards from Becoming Disasters and Disasters from Becoming Catastrophes?

The Sendai Framework for Disaster Risk Reduction 2015–2030

Responsibility for promoting and coordinating international efforts to minimize the risk that natural hazards become disasters falls on the shoulders of the United Nations. Since its inception in 1945, the United Nations has adopted at least four different approaches to disaster risk reduction:

- **Phase 1 (1946–1970):** In the first phase, the focus was largely upon the coordination of stakeholders and partners following a natural hazard event to provide relief and to assist countries to return to normality.
- **Phase 2 (1971–1999):** In the second phase, a series of formal policy frameworks and institutional arrangements were pioneered to help improve countries' preparedness, ability to respond, and capacity to mitigate damage. During this phase, two important milestones were crossed. First, the United Nations established UNDRO (Office of the United Nations Disaster Relief Coordinator) in 1971, which evolved later into the present United Nations International Strategy for Disaster Reduction (UNISDR). Second, in 1999 the United Nations adopted a strategy called "A Safer World in the 21st Century: Disaster and Risk Reduction" (later labeled the International Strategy for Disaster Reduction).

- **Phase 3 (2000–present)**: In the third phase, further policy frameworks and institutional innovations were pioneered. Increasingly, the focus has been upon managing the risk presented by disasters. Reflecting this trend, in 2005, the United Nations adopted the Hyogo Framework for Action 2005–2015: Building the Resilience of Nations and Communities to Disasters, and is now embarking on The Sendai Framework for Disaster Risk Reduction 2015–2030.
- **Phase 4 (future)**: According to some, we now stand on the threshold of a new approach to disaster risk reduction, which is emphasizing the conscious incorporation of disaster risk management into social, political, and economic policies, for example in the areas of sustainable development, environmental management, and climate change.

Although it might be possible to reduce the magnitude and frequency of natural hazard events (for example, through geophysical intervention and engineering, and minimizing the contribution of human-induced climate change to extreme weather events), it is likely that future disaster management strategies will focus principally upon decreasing people's vulnerability to natural hazards. These strategies will work to fortify communities who, on account of their precarity, are most susceptible to hazard events; will strengthen the capacity of these and other communities to cope better when hazards do strike; and will work to enable all societies to devise and implement long-term strategies through which adaptation might be improved (see Deep Dive Box 12.3).

The United Nations Office for Disaster Risk Reduction (UNDRR) is the mandated focal point for disaster risk reduction in the UN system. The United Nations' Hyogo Framework for Action 2005–2015 provided an authoritative framework around which countries sought to meet the public policy challenges presented by natural hazards. At the heart of the HFA was the claim that there exist five key areas where action is needed:

- **Governance: Organizational, legal, and policy frameworks**. Greater effort is needed to ensure that comprehensive disaster planning is both a national and a local priority and is supported by strong institutions and effective governance.
- **Risk identification, assessment, monitoring, and early warning**. Notwithstanding recent improvements in understanding, greater competence is needed in the areas of disaster forecasting, early warning, and analyses of the vulnerabilities of different populations.
- **Knowledge management and education**. The management and distribution of existing knowledge needs to be improved, and a culture of sharing best practices and experiences needs to be built.
- **Reducing underlying risk factors**. There needs to be a renewed interest in tackling poverty, inept and corrupt governance, social inequalities, war, urban slums, and environmental degradation so as to reduce the precarity that turns hazards into disasters.
- **Preparedness for effective response and recovery**. Improvements in disaster management are required. Competency in the coordination and administration of preparation plans, emergency services, relief efforts, and recovery strategies needs to be scaled up.

The Sendai Framework for Disaster Risk Reduction 2015–2030 has continued this work and has championed four priorities – understanding disaster risk; strengthening disaster risk governance; investing in disaster risk reduction for resilience; and enhancing disaster preparedness for effective response and to "build back better" via recovery, rehabilitation, and reconstruction. The Sendai Framework also has seven targets:

- Substantially reduce global disaster mortality by 2030, aiming to lower average per 100 000 global mortality for 2020–2030 compared to 2005–2015.
- Substantially reduce the number of affected people globally by 2030, aiming to lower the average global figure per 100 000 for 2020–2030 compared to 2005–2015.
- Reduce direct disaster economic loss in relation to global GDP by 2030.
- Substantially reduce disaster damage to critical infrastructure and disruption of basic services, among them health and educational facilities, including through developing their resilience by 2030.
- Substantially increase the number of countries with national and local disaster risk reduction strategies by 2020.
- Substantially enhance international cooperation with developing countries through adequate and sustainable support to complement their national actions for implementation of the framework by 2030.
- Substantially scale the use of multi-hazard early warning systems and more effectively cascade and disseminate disaster risk information and assessments (not least to the public) by 2030.

The United Nations convenes a biennial Global Platform for Disaster Risk Reduction to take stock of progress in the implementation of the Sendai Framework and to share good practices. The next Global Platform is scheduled for 2021; it is to be expected that the Covid-19 global pandemic will have dramatically transformed approaches to risk management by that point, and it remains to be seen if the Sendai approach and targets will have to be fundamentally rethought before the 2030 end date.

Deep Dive Box 12.3   Case Study: Disaster Risk Reduction in Peru

Peru provides a good example of a country that responded positively to the United Nations' Hyogo Framework for Action 2005–2015 and that is currently implementing the principles enshrined in the Sendai Framework for Disaster Risk Reduction 2015–2030. Although it has undoubtedly made progress, the coastal El Niño event of 2017 reveals that work remains to be done.

Peru is a country especially exposed to a range of hazards, including earthquakes, avalanches, floods, mudslides, and the periodic weather events El Niño and La Niña. Peru is located close to the Peru–Chile Trench, where the Nazca Plate subducts under the South American Plate, creating a fault and consequently significant seismic activity. Perhaps Peru's most serious earthquakes in recent times were those that struck in 1970, 2001, and 2007. Of these three, the Ancash earthquake, which occurred on 31 May 1970, and registered 7.7 on the Richter scale was undoubtedly the most lethal (Map 12.4). Approximately 73 000 Peruvians perished in this quake, and a further 22 000 were presumed killed by an associated avalanche that occurred on Mount Huascarán. Over 140 000 people were injured, and over 500 000 people were left homeless. Peru has also experienced a long history of disasters linked to the El Niño–Southern Oscillation (ENSO), including during the global El Niño events of 1982–1983, 1997–1998, and 2017 (Table 12.1).

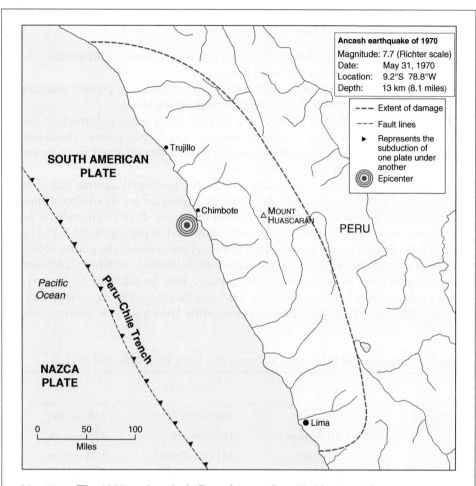

Map 12.4   The 1970 earthquake in Peru. Source: Compiled by the author.

Given the degree to which it is exposed to natural hazards, it is unsurprising that Peru has built up expertise in disaster planning. This is coordinated by Peru's Sistema Nacional de Defensa Civil (SINADECI) – the national civil defense system – and, in particular, the office termed El Instituto Nacional de Defensa Civil (INDECI). The purpose of SINADECI is to help prevent natural hazards from becoming human disasters and to manage disasters when they do occur. Although disaster planning is ultimately the responsibility of the presidency, the cabinet, and the national offices of civil defense, in reality it is INDECI that has primary responsibility on the ground. INDECI governs Civil Defense Committees that operate variously at the regional, provincial, and district levels. Their job is to:

a)   Promote disaster prevention (annual prevention plans are drawn up).
b)   Implement disaster prevention plans.
c)   Use science and technology to better understand which populations are most exposed and vulnerable, and when.
d)   Enhance the capacity of emergency services to handle disasters.
e)   Coordinate logistics and humanitarian aid in the event of crises.

(Continued)

Box 12.3   *(Continued)*

f)   Prepare health services to deal with crises and to coordinate services during crises.

g)   Ensure law and order is maintained during crises.

h)   Improve communications vis-à-vis promoting education to prevent disasters and to enable communication systems to work during crises.

Peru's approach to disaster management is famed for its command structure, fiscal commitments and budget allocations, and attempts to incorporate subnational organizations down to local communities into disaster planning and to empower these groups to deal with disasters when they occur.

Still, many claim that Peru remains vulnerable to natural hazards and that the principal cities of Lima and Arequipe in particular are ill equipped and poorly prepared to deal with hazards should they arise. Peru's vulnerability to natural hazards is aggravated by its poverty. With a GDP per capita of US$6470 (US$12 157 in purchasing power parity [PPP]), Peru is one of the poorer countries in Latin America. It has a population of 32 million, of whom 20% are officially classified as living below the poverty line. In addition to poverty, Peru's vulnerability to natural hazards has been heightened by the inappropriate development of settlements in regions with known seismic activity, the

Table 12.1   Impacts of El Niño events in Peru, 1982–1983, 1997–1998, and 2017.

	1982–1983	*1997–1998*	*2017*
Population	512 deaths	366 deaths	138 deaths
	1304 injuries	1040 injuries	459 injuries
	1.27 million affected	531 104 affected	1.74 million affected
Transportation	2600 km of highway damaged	3136 km of highway damaged	13 311 km of highway damaged
	47 bridges destroyed	370 bridges destroyed	449 bridges destroyed
Housing	98 000 homes destroyed	42 342 homes destroyed	63 802 homes destroyed
	111 000 homes damaged	108 000 homes damaged	350 181 homes damaged
Education	875 schools damaged	956 schools damaged	2870 schools damaged
Health	260 health posts damaged	580 health posts damaged	934 health posts damaged
Total monetary losses, USS	3.28 billion	3.5 billion	3.1 billion

Source: French et al. (2020). ©Elsevier.

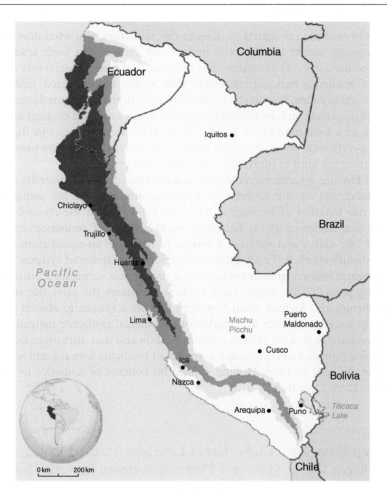

Map 12.5   Peru 2016–2017: the main areas affected during the coastal El Niño event (red/ areas: most impact; white areas: least impacted). Soure: Aspíllaga, "Niño costero", https:// commons.wikimedia.org/wiki/File:Niño_costero_(2016-2017).png. CC BY-SA 4.0.

rapidity with which cities have expanded (resulting in the shoddy construction of buildings), deforestation, and inadequate infrastructure. According to French et al. (2020), the coastal El Niño event of 2017 overwhelmed Peru's disaster response systems, despite recent innovations in disaster risk reduction (Map 12.5 shows the areas most impacted). They argue that inefficiencies in response and reconstruction efforts played a role, but the root causes of the inadequate response had to be traced to deeper structural factors, including still-limited emergency response capacity, infrastructure deficits, and poverty. The 2017 disaster revealed that, notwithstanding its "theoretical" framework, in reality disaster risk management in Peru continues to suffer from high levels of centralization, lack of functional articulation between government departments and lowers state tiers, and widespread corruption.

Resilience Politics: Robustness, Recovery, Reform, or Redesign?

The concept of resilience is central to disaster risk reduction. But what does "building resilience" actually mean? Resilience is understood variously in both academic and practitioner communities. This matters, because framings play a crucial role in shaping the kinds of resilience-building strategies that might be imagined and enacted. Geographer's use the term "resilience politics" to refer to the differential consequences of different perspectives on how to build resilience in the wake of a disaster and against the backdrop of a looming risk or hazard. When rebuilding societies in the name of strengthening resilience, political leaders need to recognize that they are making political choices about the kind of future they are working to create.

CS "Buzz" Holling, a Canadian ecological scientist based at the University of Florida, is widely recognized as the founder of resilience thinking. In his seminal article "Resilience and Stability of Ecological Systems" published in the *Annual Review of Ecology and Systematics* in 1973, he identifies two types of resilience: engineering resilience, or "the ability and speed of a system to return to an equilibrium or steady-state after a disturbance"; and ecological resilience, or "the ability of systems to reset to a new equilibrium when a stressor pressurizes it such that it surpasses a critical threshold or tipping point." While engineering resilience assumes the existence of a single, stable equilibrium and focuses upon how long it takes a system to absorb or resist a shock and "bounce back" after a perturbation, ecological resilience instead acknowledges the possibility that there exist multiple equilibria and that disruption can result in systems flipping into alternative stability domains, or bouncing forward and settling into a new regime. As early as 1986, Holling applied his concept of resilience to global climate change (Deep Dive Box 12.4).

Deep Dive Box 12.4   James Lovelock on the Resilience of Planet Earth: Gaia and Pushing Systems Beyond Their Equilibrium State

For nearly 50 years, British scientific inventor and public intellectual James Lovelock has served as a leading spokesperson for the environmental movement and a chief exponent of the Gaia hypothesis. First introduced in his seminal work, *Gaia: A New Look at Life on Earth*, published in 1979, the Gaia hypothesis holds that planet earth functions as a self-regulating single system that tends toward homeostasis. The totality of organic life (plants, animals, and humans) on earth works to regulate the more inorganic drivers in the system (soils, oceans, the atmosphere, hydrological cycles, and so on) to ensure that the earth remains habitable. When change occurs in one part of the system, a number of feedback loops are triggered, and the rest of the system seeks to compensate and restore conditions to the equilibrium position that prevailed at the outset.

In 2007 in *The Revenge of the Gaia: Why the Earth Is Fighting Back and How We Can Still Save Humanity*, Lovelock warned that through arrogance and complacency, the human race was now meddling with Gaia in a reckless and irresponsible way. In particular, by polluting the atmosphere, human beings were giving Gaia a fever; by 2100, this fever could lead to critical illness and even death. In

2009 in *The Vanishing Face of Gaia: A Final Warning*, Lovelock went further and suggested that if sufficient carbon concentrations exist in the atmosphere, it is possible that abrupt temperature jumps may occur in a matter of years or, at most, decades. Lovelock argued that Intergovernmental Panel on Climate Change (IPCC) estimates of global warming were conservative because they failed to allow for such jumps. He predicted that within a decade or two, the temperature of the earth may rise by as much as 9° centigrade.

In a 2012 television interview, however, Lovelock caused a furor by admitting that he had been "alarmist" in his earlier work; that instead of being "halfway towards frying," climate was changing very slowly; and that climate scientists had insufficient understanding of climate systems.

Subsequently, in a 2014 television interview, when asked why he had retracted his most alarming predictions, Lovelock replied: "Well, that's my privilege. You see, I'm an independent scientist. I'm not funded by some government department or commercial body or anything like that. If I make a mistake, then I can go public with it. And you have to, because it is only by making mistakes that you can move ahead.... They [scientists] all talk, they pass laws, they do things, as if they knew what was happening. I don't think anybody really knows what's happening. They just guess. And a whole group of them meet together and encourage each other's guesses."

Nevertheless, in his 2014 book *A Rough Ride to the Future*, Lovelock argues that human beings have changed the Gaia forever and that our era is indeed the era of the "Anthropocene." But instead of trying to hold back the tides of change, governments should recognize that the change that has occurred is irreversible and should develop a plan b that focuses upon living with the new Gaia. Lovelock suggests that instead of saving the planet, humans should cluster into large cities (lifeboat cities) where humane conditions can be artificially produced and protected. Furthermore, human beings can use new smart technologies to monitor (and therefore know) intimately Gaia's new trajectory. Armed with big data about Gaia and computer software to analyze these big data, they might even be able to shape the direction in which Gaia is moving.

On 26 July 2019, Lovelock celebrated his 100th birthday. That same month he published his latest book, *Novacene: The Coming Age of Hyperintelligence*. Lovelock now contends that human beings are reaching the end of the age of the Anthropocene and dwell on the brink of a new age – the Novacene. In the Novacene, artificial intelligence will develop and supercomputing machines operating with 10 000 times the intelligence and analytical capacity of humans will rise to displace the human race as the ruling species. But they too will rely on planetary resources and, threatened by an ever-warmer sun, will need Gaia to cool the planet. Being part of Gaia, they will need human beings, and therefore there is no need to fear for our future.

Inspired by Holling, there has emerged a scholarly tradition often referred to as social-ecological resilience studies, which focuses upon social-ecological systems (SES; also known as CHANS). SES are systems that rely upon interlinkages and co-dependencies between social systems (for example, culture and institutions), economic systems (technologies and preferences), and environmental and ecological systems (climate and habitat). In addition to engineering resilience and ecological resilience, also emerging are the concepts of adaptive resilience and evolutionary resilience. US economist Adam Rose distinguishes between inherent resilience (the resilience capacity already built into a system) and adaptive resilience (the active work of governments to build added buffers and supports around systems). Meanwhile, Canadian natural resources scientist Fikret Berkes and colleagues (2000) have popularized the tenets of evolutionary resilience that challenge the idea that systems are static and exist in a steady or equilibrium state, arguing instead that they are more often restless, dynamic, changing states and prone to constant churn. Some versions of evolutionary resilience assume that systems adapt so that they strengthen their resilience through time; survival of the fittest ensures that only robust systems or elements of systems survive through time.

According to Australian economist R. Quentin Grafton and colleagues (2019), resilience is best understood in the context of the three Rs; resistance, recovery, and robustness. Resistance refers to the ability of a system to maintain its structure, function, and capacity to perform following a disturbance. Recovery denotes the normalization of a system and its return to initial levels of performance following a shock. Finally, robustness captures the extent to which a system retains its basic identity and does not jolt beyond an undesirable (and possibly irreversible) threshold following an adverse event. For Grafton et al., resilience management can be understood as the planning, adaptation, and transformation activities undertaken by resilience managers to strengthen a system's resilience (its resistance, recovery, and robustness). They suggest that resilience management attend to seven key questions:

- Resilience of what objects (system, system component, or interaction) is being managed?
- For whom (stakeholders) is resilience being managed?
- What are the metrics of system performance for the identified stakeholders?
- What are the viability (or safety) goals of the stakeholders (and associated metrics) for key system variables that allow a system to retain its identity?
- What adverse events or causes, in relation to resilience, are being considered?
- How are the three Rs measured in relation to system performance and in response to adverse events?
- What are the expected net benefits, currently and over time and space, of resilience management actions?

In *Searching for Safety*, US political scientist Aaron Wildavsky (1988) identifies six principles that collectively define the capacity of a system to absorb the shock of a natural hazard and stay on course:

- Homeostasis principle: Promotes the view that systems regulate themselves through feedback loops that work to restore equilibria.
- Omnivore's principle: Holds that systems that procure resources from multiple sources are less vulnerable to system collapse.
- High-flux principle: Promotes the idea that the faster resources flow through a system, the more capable that system will be of recovering following a disaster.
- Flatness principle: Contends that systems that are governed from above in centralized, hierarchical, and top-down ways are less resilient than systems in which decision making is decentralized to grassroots communities with strong "on the ground" local capacity and knowledge.

- Buffering principle: Refers to the need for systems to operate in excess of actual need, so that any reduction in performance caused by a hazard does not lead to fatal shortfalls in capacity.
- Redundancy principle: Promotes doubling up system functions so that, in the event a function is lost, a backup can be easily deployed as a substitute.

We might say, then, that the most resilient systems are ones in which all actors in the system automatically respond to neutralize a disaster when any one actor is threatened; when systems procure inputs (food, water, medical supplies, and finance) through multiple channels and from multiple origins; when resources move through systems in a free-flowing way, efficiently and without impediment; when decision making is decentralized to communities and local stakeholders; when systems are built to perform at levels well in excess of actual need; and when systems can substitute functions that fail with a backup.

In his 1992 book *Balance of Nature? Ecological Issues in the Conservation of Species and Communities*, American-British biologist and theoretical ecologist Stuart Pimm contends that within literature on ecology, resilience is variously studied through the prism of the concepts of stability, scale of organization, and complexity. Each lacks clear meaning, but it is possible to discern five definitions of stability: mathematical stability (the extent to which variables within a system all return to equilibrium after a disturbance), resilience (how quickly a displaced variable returns to equilibrium), persistence (how long a variable lasts before it is reset by a shock to a new value), resistance (what happens to other variables in a system if one variable is disturbed), and variability (the number of times and extent to which a variable changes over time). Meanwhile, stabilization is understood to occur at three different levels of organization: at the species level, at the level of community composition (biodiversity), and with respect to community biomass (abundance biomass). Meanwhile, complexity too has been construed through three lens: number of species in a system, the degree of connectedness of the food web, and the relative abundance of a species in an ecological community. Pimm argues that research to date has only looked at some of the 45 permutations of resilience that these various categories generate, and in any case has failed to generate consistent conclusions. For example, for a while it appeared that simple ecosystems were less stable than complex ones, but more recent research suggests that the opposite might be the case.

Focusing specifically upon mitigating the impacts of earthquakes, American environmental engineer Michel Bruneau et al. (2003) suggest that resilience operates differently at different levels: technical, organizational, social, and economic. They define resilience in terms of four key ideas: robustness, redundancy, resourcefulness, and rapidity. For example, when applied to economic systems, these terms refer to:

- **Robustness**: Strength, or the ability of systems to withstand a given level of disturbance without suffering loss of function.
- **Redundancy**: The possibility of replacing a system with a substitute.
- **Resourcefulness**: The capacity to identify problems, establish priorities, and mobilize resources when conditions exist that threaten to disrupt a system.
- **Rapidity**: The capacity to implement remediating actions in a timely manner in order to limit losses and future disruption.

For Bruneau et al. (2003), a resilient system is one that delivers the following: reduced failure probabilities (reduced likelihood to collapse), reduced consequences from failures (death, infrastructure collapse, and economic and social decline), and reduced time to recovery (restoration of a system to its "normal" level of performance).

Deep Dive Box 12.5 Applying the Concept of Resilience to Better Understand the Geography of Food Security: The Food and Agricultural Organization (FAO) and RIMA/RIMA-II

Since 2008, FAO has been at the forefront of efforts to measure resilience to food insecurity and has pioneered the development and use of Resilience Index Measurement and Analysis (RIMA; from 2016, RIMA-II). RIMA-II works to better understand why some households appear to cope better with food shortages than others. Resilience is defined as "the capacity of a household to bounce back to a previous level of well-being (for instance food security) after a shock."

RIMA-II calculates household resilience using both direct (or descriptive) and indirect (or inferential) measures. The direct approach computes a Resilience Capacity Index (RCI), which is a measure of the capacity of a household to cope with a shock, and a Resilience Structure Matrix (RSM), which establishes the relative contribution of four critical factors implicated in household resilience. These are:

- **Adaptive Capacity (AC):** The ability of a household to adapt to a new situation and develop new livelihood strategies.
- **Social Safety Nets (SSN):** The ability of households to access help from relatives, friends, and governments, and to obtain timely and reliable assistance from international agencies, charities, and nongovernmental organizations (NGOs).
- **Assets (AST):** The productive (wealth creating) and non productive (pertaining to consumption) assets owned by members of a household.
- **Access to Basic Services (ABS):** The ability of a household to meet basic needs, and access and effectively use basic services (such as schools, health facilities, infrastructures, and markets).

The indirect approach looks at the determinants of food security loss and recovery, and predicts and accounts for any change in a household's resilience over time.

RIMA-II identifies three types of exogenous shocks that can impact households' resilience – idiosyncratic shocks (such as livestock death, job loss, and illness of a household member), covariate shocks (climate shocks, such as droughts, floods, temperature variations, storms, and other natural hazards), and conflict shocks (such as wars, terrorism, and public disorder).

To measure the impact of any aid package or development intervention on household food security, the FAO has also developed a RIMA-II tool, the Resilience Marker.

Currently, the FAO is using RIMA-II to support 10 countries (including Burkina Faso, Mali, Niger, Senegal, Somalia, South Sudan, and Sudan) to build household resilience and mitigate exposure to food shortages.

In his 2010 book *Adaptation to Climate Change: From Resilience to Transformation*, British geographer Mark Pelling cautions against promoting resilience as always and everywhere the goal of disaster management (Pelling 2010). Pelling calls for more attention to be given to the ways in which natural hazards play into the politics that prevail in countries. He deploys the term "disaster politics" to refer to the ways in which natural hazards interact with the existing political order, consolidating, destabilizing, and transforming this order in different circumstances. Societies that frame comprehensive disaster management in terms of the pursuit of greater resilience need to recognize that, in so doing, they are making a political choice as much as a scientific determination. Thinking in terms of resilience implies prioritizing bounce-back (to return to the status quo). In fact, disasters provide opportunities to bounce forward (to establish a new and better equilibrium state). According to Pelling, natural hazards can lead to at least three different political outcomes: in addition to "resilience," there can be "transition" and "transformation" outcomes.

- Resilience is the most conservative outcome of all. Here comprehensive disaster management works simply to help societies improve their capacities to return to their pre-disaster condition as quickly as possible after a hazard event. Pelling contends that, wittingly and unwittingly, resilience has become the default option in many countries.
- Transition, in contrast, refers to outcomes in which comprehensive disaster management works to help citizens exercise their existing rights more effectively, but within the prevailing political order. Here disasters force political regimes to introduce incremental improvements in the protection of citizens against harm.
- Transformation stands as the most radical of possible outcomes, promoting profound changes in the relationships that exist between governments and citizens. Pelling contends that hazard events alone are unlikely to trigger transformations in society but can tip political regimes already on the brink of change over the edge.

Mackinnon and Derikson (2013), meanwhile, distinguish resilience policy from the related but alternative idea of resourcefulness policy. Resilience policy simply works to help vulnerable communities improve their capacities to return to their original status as quickly as possible, and wittingly and unwittingly works to preserve the status quo. If the prevailing social and economic system knocks people down, it is the job of resilience policy to build people back up so that they can better survive within that system. Resourcefulness policy, in contrast, helps citizens exercise greater agency in part to prosper better within the existing political order but equally when relevant to challenge this order and strengthen their structural position. Clearly shielding vulnerable populations by improving their resilience is a worthy endeavor, but not if it merely serves to preserve the social, economic, cultural, and political processes that produce precarity in the first instance. Strengthening the capacity of beneficiaries so that people are better able to address the root causes of precarity provides a better option. Mackinnon and Derikson (2013) identify four pillars of resourcefulness, or arenas of capability, that merit particular attention:

- **Resources**: In the context of social inequality, increasing the resources (economic, cultural, political, and social capital) available to people can be the most direct way to build their resourcefulness.
- **Skill and competencies**: Increasing people's capacity for self-management and their skills and technical knowledge can result in increased resourcefulness.
- **Indigenous and folk knowledge**: Local cultural lore plays a vital role in tempering aspiration levels and personal goals; alternative cultural repertoires and local stories can reframe what is understood to be possible, achievable, and worthy of pursuit.

- **Recognition**: Listening authentically to people, respecting their analysis of where they are at and why, dignifying their concerns and ideas, entertaining the solutions they propose, and conferring recognition upon them can be affirming and empowering and can strengthen their resourcefulness.

So, to return to our original question: What does building resilience mean? Drawing together these various strands of work, it is clear that resilience is being understood variously and that these various framings are playing a crucial role in shaping the kinds of resilience-building strategies that are being imagined and enacted. We might identify four overall frameworks; although not mutually exclusive, each does focus attention and effort with different ends in mind.

Resilience as robustness: Focusing upon the amount of shock a system can absorb and continue to function effectively, and prioritizing strengthening the resistance of systems to external disturbances.

Resilience as recovery: Focusing upon the capacity of systems to return to a steady initial equilibrium steady state after a shock, and prioritizing solutions that help systems heal and repair faster.

Resilience as reform: Focusing upon the capacity of systems after a shock to adapt and evolve so that they are stronger than before, and prioritizing reform within the same politico-institutional "normal".

Resilience as redesign: Focusing upon the necessity of reconfiguring systems root and branch after a shock, and prioritizing politico-institutional transformation as the only enduring solution.

Clearly, robustness, recovery, reform, and redesign all have strengths and weaknesses in different contexts. Engineering systems so that they might increase their immunity to external disturbances affords reassuring protection, but there will be hazards that overwhelm even the strongest of vaccines, and in these instances resistance will be futile. Helping vulnerable populations recover from a disaster is a worthy endeavor, but not if it merely serves to preserve economic and socio-political processes that produced precarity in the first instance. Strengthening the rights of citizens by reforming the existing political order is obviously a welcome development, but not if it produces tokenistic transfers of power that only marginally reduce risk. Finally, redesigning societies to address the root causes of precarity may provide the only durable solution to human-induced vulnerability, but it is questionable whether deep-seated societal reconstruction is wise in times of existing upheaval or in the immediate aftermath.

Conclusion

This chapter has argued that, at least according to human geographers, there is nothing particularly natural about natural hazards. Although global variations in the risk posed by a natural hazard is in part a product of geological, meteorological, and hydrological mechanisms, it is also a function of the ways in which social, economic, cultural, and political processes render some populations more vulnerable to hazard events than others. The same is true for societal and technological hazards. It is little surprise, then, that hazards become disasters more readily in the Global South, and are more lethal to and impactful on populations living in greater precarity in both the Global North and the Global South, including the poor, women, children, people with disabilities, and ethnic minorities. As US geographer Susan L. Cutter has vividly shown, vulnerability turns hazards into disasters and disasters into catastrophes. If the risks associated with hazards are to be reduced, greater attention needs to be given to minimizing the susceptibility of vulnerable populations, increasing their capacities to cope with disasters, and fortifying their abilities to adapt in the long term. This will require that more resilient societies be built. But the idea of resilience comes freighted with many assumptions, and its different meanings (for example, system robustness, recovery, reform, or redesign) imply radically different policies and interventions. When working to stop hazards from becoming disasters, disaster risk strategies – such as the Sendai Framework for Disaster Risk Reduction 2015–2030 – need to attend to the question of resilience politics.

Checklist of Key Ideas

- Natural, societal, and technological hazards continue to threaten humanity, too often leading to death and injury, wreckage of vital infrastructure, social disruption, and economic carnage.
- Gilbert White believed that natural hazards only become natural disasters when societies expose people to unnecessary risks. White's public advocacy of the need for comprehensive floodplain management in the United States and elsewhere proved to be midwife to the birth of the interdisciplinary field of hazards research.
- Although exposure to natural, societal, and technological hazards remains a crucial factor, increased attention is now being given to the role of social, economic, and political forces as root causes of disasters. Geographers now use the formula Risk = Exposure × Vulnerability (R = E×V) to study disasters. Susan L. Cutter has been at the forefront of geographical scholarship on the social production of vulnerability to hazards. She has argued that capitalism, patriarchy, racism, ageism, ableism, and so on structure precarity unevenly, such that hazards exert a disproportionate toll on poor people, women, ethnic minorities, children, the disabled, and so on.
- Adopting the above formula, the United Nations University (UNU) has mapped our unequal world at risk. Although exposure is distributed chaotically across the earth's surface, it is clear that the Global South in particular suffers from enhanced vulnerability to hazard events.
- The UN Office for Disaster Risk Reduction's Sendai Framework for Disaster Risk Reduction 2015–2030 is working to stop hazards from becoming disasters and disasters from becoming catastrophes. Peru provides a good example of a country implementing the principles of disaster risk reduction; while it has made progress, the coastal El Niño event of 2017 reveals that work remains to be done.
- The concept of building resilience lies at the heart of disaster risk reduction, but it carries different meanings (for example, system robustness, recovery, reform, and redesign), and these meanings imply different ways of building safer communities. Resilience politics refers to the politics that underpin the choice of approach.

Chapter Essay Questions

1) Document the contributions of Gilbert F. White to hazard studies.
2) Using the UNU's World Risk Index, describe and explain global variations in risks from natural hazards.
3) Describe and comment upon the efficacy of the Sendai Framework for Disaster Risk Reduction 2015–2030.
4) Define what is meant by "resilience politics," and identify the political implications of responding to hazards by promoting system robustness, recovery, reform, and redesign.

References and Guidance for Further Reading

Gilbert White's final book with James Westcoat provides an excellent summation of his thinking on floods and flood hazard management:

Westcoat, J.L. and White, G.F. (2003). *Water for Life: Water Management and Environmental Policy*. Cambridge: Cambridge University Press.

Excellent overviews of the life and contributions of Gilbert White can be found in:

Cutter, S.L., Rutherford, H.P., Burton, I. et al. (2019). Reflections on Gilbert F. White: scholar, advocate, friend. *Environment: Science and Policy for Sustainable Development* 61: 4–21.

Hinshaw, R.E. (2006). *Living with Nature's Extremes: The Life of Gilbert Fowler White*. Boulder, CO: Johnson Books.

The works of US geographer Susan L. Cutter includes:

Cutter, S.L., Mitchell, J.T., and Scott, M.S. (2000). Revealing the vulnerability of people and places: a case study of Georgetown County, South Carolina. *Annals of the Association of American Geographers* 90: 713–737.

Cutter, S.L., Boruff, B.J., and Shirley, W.L. (2003). Social vulnerability to environmental hazards. *Social Science Quarterly* 84: 242–261.

Cutter, S.L., Barnes, L., Berry, M. et al. (2008). A place-based model for understanding community resilience to natural disasters. *Global Environmental Change* 18: 598–606.

Cutter, S.L., Emrich, C.T., Mitchell, J.T. et al. (2012). *Hurricane Katrina and the Forgotten Coast of Mississippi*. Cambridge: Cambridge University Press.

Cutter, S.L. (2020). The changing nature of hazard and disaster risk in the Anthropocene. *Annals of the American Association of Geographers* https://doi.org/10.1080/24694452.2020.1744423.

Rubin, C.B. and Cutter, S.L. (eds.) (2020). *Emergency Management in the 21st Century: From Disaster to Catastrophe*. Oxford: Routledge.

Other influential books that helped to recenter hazards research on the social, cultural, political, and economic roots of vulnerability are:

Blaikie, P., Wisner, B., Cannon, T., and Davis, I. (1994). *At Risk: Natural Hazards, People's Vulnerability and Disasters*. London: Routledge See also 2nd edition, 2013.

Hewitt, K. (1983). *Interpretations of Calamity: From the Viewpoint of Human Ecology*. London: Allen and Unwin.

Hewitt, K. (1997). *Regions of Risk: A Geographical Introduction to Hazards*. London: Longman.

Smith, K. and Petley, D.N. (1991). *Environmental Hazards: Assessing Risk and Reducing Disaster*. London: Routledge See also 6th edition, 2013.

A comprehensive mapping of risk, exposure, and vulnerability to natural hazards across the globe is provided in:

Centre for Research on the Epidemiology of Disasters (CRED) (2020). *EM-DAT: The International Disaster Database*. Brussels: CRED.

World Economic Forum (2020). *The Global Risks Report 2020*. Geneva: World Economic Forum.
United Nations University (UNU) (2019). *World Risk Report 2018*. Bonn: UNU.

An overview of the Hyogo Framework can be found at:
United Nations Office for Disaster Risk Reduction (UNDRR) (2005). *Hyogo Framework for Action 2005–2015*. New York: UNDRR.

An overview of the Sendai Framework can be found at:
United Nations Office for Disaster Risk Reduction (UNDRR) (2015). *Sendai Framework for Disaster Risk Reduction 2015–2030*. New York: UNDRR.

See also:
Christo, C. and Livhuwani, N. (2020). Implementing the Sendai Framework in Africa: progress against the targets (2015–2018). *International Journal of Disaster Risk Science* 11: 179–189.

The latest updates on progress in disaster risk reduction can be found at:
United Nations Office for Disaster Risk Reduction (UNDRR) (2020). *Global Assessment Report on Disaster Risk Reduction*. Geneva: UNDRR.

Important works on disaster risk management are:
Alexander, D. (2000). *Confronting Catastrophe: New Perspectives on Natural Disasters*. Harpenden, NY: Terra Publishing.
Alexis-Martin, B. (2019). *Disarming Doomsday: The Human Impact of Nuclear Weapons Since Hiroshima*. London: Pluto Press.
de Bruijn, K., Buurman, J., Mens, M. et al. (2017). Resilience in practice: five principles to enable societies to cope with extreme weather events. *Environmental Science & Policy* 70: 21–30.
Hein, C. and Schubert, D. (2020). Resilience, disaster, and rebuilding in modern port cities. *Journal of Urban History* doi: 10.1177/0096144220925097.
IPCC (2019). *The Routledge Handbook of Urban Disaster Resilience: Integrating Mitigation, Preparedness, and Recovery Planning* (eds. C.B. Field, V. Barros, V.F. Stocker, et al.). London: Routledge.
Keller, E. and DeVecchio, D. (2019). *Natural Hazards: Earth's Processes as Hazards, Disasters, and Catastrophes*. Sydney: Pearson Australia.
Kelman, I. (2020). *Disaster by Choice: How Our Actions Turn Natural Hazards Into Catastrophes*. Oxford: Oxford University Press.
Kelman, I., Mercer, J., and Gaillard, J.C. (eds.) (2017). *The Routledge Handbook of Disaster Risk Reduction Including Climate Change Adaptation*. London: Routledge.
Lupton, D. (ed.) (2013). *Risk*. London: Routledge.
Paul, B.K. (2011). *Environmental Hazards and Disasters*. Chichester: Wiley-Blackwell.
Pilkey, U.J. and Pilkey, K.C. (2019). *A Slow Tsunami on America's Shores*. Durham, NC: Duke University Press.
Pine, J.C. (2015). *Hazards Analysis: Reducing the Impact of Disasters*. Boca Raton, FL: CRC Press.
Plattner, G.K., Allen, S.K., Tignor, M., and Midgley, P.M. (eds.) (2012). *Managing the Risks of Extreme Events and Disasters to Advance Climate Change Adaptation*. Cambridge: Cambridge University Press.
Seo, S.N. (2019). *Natural and Man-Made Catastrophes: Theories, Economics, and Policy Designs*. Oxford: Wiley.
Tierney, K. (2018). *Disasters: A Sociological Approach*. Cambridge: Polity.

Important works exploring the ideas of resilience, resilience policies, and resilience politics include:
Adger, W.N. (2000). Social and ecological resilience: are they related? *Progress in Human Geography* 24 (3): 347–364.
Berkes, F., Folke, C., and Colding, J. (eds.) (2000). *Linking Social and Ecological Systems: Management Practices and Social Mechanisms for Building Resilience*. Cambridge: Cambridge University Press.

Bruneau, M., Chang, S.E., Eguchi, R.T. et al. (2003). A framework to quantitatively assess and enhance the science the seismic resilience of communities. *Earthquake Spectra* 19: 733–752.

Davoudi, A., Shaw, K., Haider, L.J. et al. (2012). Resilience: a bridging concept or a dead end? Reframing, resilience: challenges for planning theory and practice, interacting traps: resilience assessment of a pasture management system in Northern Afghanistan, urban resilience: what does it mean in planning practice? Resilience as a useful concept for climate change adaptation? The politics of resilience for planning: a cautionary note. *Planning Theory and Practice* 13: 299–333.

Food and Agriculture Organization (FAO) (2016). *Resilience Index Measurement and Analysis – II.* Rome: FAO.

Grafton, R.Q., Doyen, L., Béné, C. et al. (2019). Realizing resilience for decision-making. *Nature Sustainability* 2: 907–913.

Holling, C.S. (1973). Resilience and stability of ecological systems. *Annual Review of Ecology and Systematics* 4: 1–23.

Lindell, M.K. (ed.) (2019). *The Routledge Handbook of Urban Disaster Resilience: Integrating Mitigation, Preparedness, and Recovery Planning.* London: Routledge.

MacKinnon, D. and Derickson, K.D. (2013). From resilience to resourcefulness: a critique of resilience policy and activism. *Progress in Human Geography* 37: 253–270.

Nussbaum, M. (2011). *Creating Capabilities: The Human Development Approach.* Cambridge, MA: The Belknap Press of Harvard University Press.

Pelling, M. (2010). *Adaptation to Climate Change: From Resilience to Transformation.* London: Routledge.

Pimm, S.L. (1991). *The Balance of Nature? Ecological Issues in the Conservation of Species and Communities.* Chicago: University of Chicago Press.

Rose, A. (2004). Defining and measuring economic resilience to disasters. *Disaster Prevention and Management* 13 (4): 307–314.

Wildavsky, A. (1988). *Searching for Safety.* New Brunswick, NJ: Transaction Books.

James Lovelock's seminal work is:

Lovelock, J. (1979). *Gaia: A New Look at Life on Earth.* Oxford: Oxford University Press.

Lovelock, J. (2007). *The Revenge of Gaia: Why the Earth Is Fighting Back and How We Can Still Save Humanity*, vol. 36. London: Penguin.

Lovelock, J. (2020). *Novacene: The Coming Age of Hyperintelligence.* London: Penguin.

A good policy-centered introduction to disaster risk reduction in Peru is provided in:

World Bank (2016). *Peru: A Comprehensive Strategy for Financial Protection Against Natural Disasters.* Washington, DC: World Bank.

A good critical introduction to Peru's approach to disaster risk management can be found at:

French, A., Mechler, R., Arestegui, M. et al. (2020). Root causes of recurrent catastrophe: the political ecology of El Niño-related disasters in Peru. *International Journal of Disaster Risk Reduction* 47: 101539. https://doi.org/10.1016/j.ijdrr.2020.101539.

The El Niño phenomenon is explained in:

Grove, R. and Adamson, G. (2018). *El Niño in World History.* Basingstoke: Palgrave Macmillan.

Website Support Material

A range of useful resources to support your reading of this chapter are available from the Wiley Human Geography: An Essential Introduction Companion Site http://www.wiley.com/go/boyle.

Chapter 13

Remaking the West, Remaking Human Geography

Chapter Table of Contents

Chapter Learning Objectives

By the end of this chapter you should be able to:
- Summarize the principal arguments advanced in this book, and explain why it makes sense to introduce the story of human geography in and through the story of the rise, reign, and faltering of the West from the fifteenth century.
- Reach a judgment about the achievements and failures of the West's politico-economic-institutional model. Explain what is meant by the pursuit of "a mission civilisatrice for the metropole," and discuss the merits and demerits of using such terminology and pursuing such a mission.
- Explain why human geography needs to become a *postcolonial* academic subject and why the idea of human geography "without a center" can help it to make this transition. Discuss the extent to which – and the ways through which – it might be

Human Geography: An Essential Introduction, Second Edition. Mark Boyle.
© 2021 John Wiley & Sons Ltd. Published 2021 by John Wiley & Sons Ltd.
Companion website: www.wiley.com/go/boyle

> possible to *both* situate, critique, provincialize, and decolonize the Eurocentrism inherent in human geography *and* insist upon rigor, standards, and objectivity in human geographical inquiry.
> - Picture the West in 2050. Guided by the idea of "utopia for realists", imagine from your perspective what a rejuvenated West might look like. Identify what would need to be done to build such a West?
> - Discuss the ways in which European-Anglo-American human geography might become a better version of itself in the twenty-first century by supporting the West to become a better version of itself.

Introduction

European colonial imaginaries linger deep in the Western psyche. Human geography is now awake and conscious of the privileges and debilitations bestowed by its imbrication in European imperialism and hegemony and its partial perspective. But frames of reference that have been in the making for centuries will not be unthought overnight, not least because there is a politics to their maintenance.

Just like Phileas Fogg, we have now completed our circumnavigation of the globe (Plate 13.1). Perhaps it has taken us a little longer than 80 days, but in our defense we have traveled more expansively and permitted ourselves more time to get to know the places on our itinerary. As he made his way back to the London Reform Club on 21 December 1872, one wonders if Fogg's Britannocentric worldview had survived his wanderings. Travel broadens the mind, and his certainly needed broadening. Alas, one suspects that Fogg's principal preoccupation was collecting the wager he had agreed for making it back on time; trifling matters such as cultivating the intellect would not appear to be foremost in his mind. Living at the height of the European colonial adventure, Fogg's French creator Jules Verne (1828–1905) had neither cause nor capacity to gift to Fogg a more cosmopolitan worldview. Verne attributed the success of his corpus of work – gathered together under the title *Extraordinary Voyages* and published as 54 novels (including *Around the World in Eighty Days*, 1873) between 1863 and 1905 – to his love of geography. From 1865 to 1898, he was an active member of the French Société de Géographie. His geographical imagination was sculpted in and through a Paris that then served as the capital of modernity. According to French geographer Lionel Dupuy (2013), Verne himself was overly invested in phantasmagorical tales of exploration and adventure and suffered from a colonial penchant for "geographical exoticism." He delighted in writing about "elsewheres" that he often "idealized, fantasized, and stereotyped." His disposition to fetishize technology and fondness for science fiction were to render him a pioneer of Steampunk, a literary genre that aligned well with Western capitalist modernity.

In this closing chapter, I will take stock of what we have covered in the book, reflecting upon the ways in which the making of the West and the making of human geography have been deeply entangled. I will then argue that there is merit in interlacing the projects of remaking human geography and remaking the West for mutual gain, and will make a case for further developing a *postcolonial* human geography and putting it in the service of what I will call a *mission civilisatrice* for the metropole.

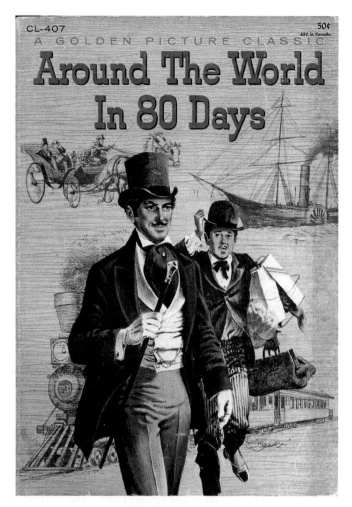

Plate 13.1 From the cover of Jules Verne's book *Around the World in Eighty Days*. Source: From cover of Jules Verne's book Around the world in eighty days (1958 Golden Picture Classic reprint).

Summary: Making the West, Making Human Geography

Human geographers work to describe the ways in which human beings have occupied the surface of the earth and to explain the mosaic of areal differentiation that has arisen as a result. In Chapter 1, we noted that the distinguishing scholarly concerns of human geography are: (i) the mutual constitution of society and space, (ii) place and the cultural landscape, and (iii) human–environment interactions. Everyone has a geographical imagination, that is, a mental map of the various ways in which human beings have occupied the earth's surface in different parts of the world. This book has challenged you to become more conscious of your geographical imagination and to strengthen and cultivate that imagination through formal education. Its core supposition has been that to build an informed geographical imagination, it is first necessary to understand the history that lies behind the human geographies that now mark the earth's surface. More particularly, to understand variations from place to place in the ways in which humans have made earth home, it is vital to understand the story of the rise, reign, and faltering of the West from 1500 CE and the the extent to which this threshold episode in "big human history" has shaped the trajectories of different world regions.

The ascent and dominance of the West as the world's first hegemon have left in its wake a deeply unequal and socially differentiated world. For nearly five centuries, countries in the Global North have grown affluent and developed, while countries in the Global South have been left underdeveloped and in too many cases in acute poverty. Of course, this is not to imply that human history began only in 1492, nor that alternative non-European models of political economy, society, and culture have been inconsequential since 1492. Europe did not jump-start human history, and Africa, Asia, and the Americas witnessed the rise and fall of many indigenous societies, cultures, polities, and economies both before and after Europe's first conquistadors arrived. And in any case, the West itself is a mongrel civilization, constituted by the coming together of many of these antecedent cultures and developments. But it is to say that it is impossible to understand the historical development of any world region, including its principal demographic, social, cultural, economic, technological, political, and environmental features, without first understanding how that region figured in the story of the rise, reign, and stumbling of the West.

Specifically, in this book you have seen how the rise of the West as a global hegemon has both triggered and been defined by:

- The establishment of capitalism as an economic system, the formation of a world capitalist economy, and the engraving of uneven development across the surface of the earth (Chapter 4).
- The innovation of the sovereign nation-state and liberal democracy; the violent colonization and command by European nation-states over large parts of the Americas, Asia, Africa, Oceania, and the Polar regions; and the postcolonial new geopolitical order that is emerging today (Chapter 5).
- Civilizing missions evangelizing the myth that "West is best," symbolic landscapes marked by the "imperialism of the straight line," and culture wars over the story of the West and what constitutes civilized spaces and unruly places (Chapter 6).
- The globalization of Eurocentric development agendas, manifest most clearly in the imposition of the Washington Consensus and neoliberal development models upon countries in the Global South, and the emergence of new neocolonial relations between the Global North and Global South (Chapter 7).
- A dramatic rise in world population and new geographies of population across the earth's surface (Chapter 8).
- The super-exploitation by human beings of the earth's resources and the development of a new geological period, the Anthropocene or, more accurately, the Capitaloecene (Chapter 9).
- The urbanization of the face of the earth and the rise of a new age of planetary urbanization (Chapter 10).
- Structured patterns of migration within and between countries in the Global South and the Global North (Chapter 11).
- Notwithstanding technological innovation and historically unprecedented levels of control over the natural environment, the heightened exposure of whole new swathes of humanity to natural hazards (Chapter 12).

There remains disagreement within the human geographical community as to how the story of the rise of the West might best be narrated. In Chapter 1, we noted the existence of at least four ways in which this story has been recounted. We labeled these: (i) only in the West because of favorable environmental endowments; (ii) first in the West, then elsewhere; (iii) because in the West not elsewhere; and (iv) the West versus the rest. You have encountered each story in various incarnations throughout this book. Of course, there is no need to prefer one over the other; eclecticism is not a sin. Ours is a complex world, and the layering and complexifying of arguments are often more productive than the imposition of a single muscular narrative thread. But it does matter which story our political leaders believe to be most compelling, because each speaks to a different kind of regenerative project for the West.

In Chapter 2, we noted that the rise of the West has also bequeathed a geography of knowledge production. As a politico-institutional project, human geography is in its outlook a quintessential European intellectual project; its theories, concepts, methods, and substantive research studies are inescapably situated and parochial. It is no accident that human geography flourished at precisely the same moment that the West began its ascent to the summit of world history and, in particular, in the period spanning the European Enlightenment, Age of European Exploration, and Age of European Empire. According to some, human geography's ways of looking at the world are also institutionally white, middle-class, able-bodied, masculine, adult-centered, and heteronormative. Human geography does not simply put a mirror up to the world and reflect back what it sees. The view (of the world) from nowhere is a myth; there can only be a view from somewhere and, in the case of human geography, somewhere very specific.

This is not to imply that the production of human geographical knowledge began only in 1492, nor that alternative non-European human geographical traditions have been inconsequential since 1492. But it is to recognize that European-Anglo-American human geography has been as hegemonic in the world of human geographical scholarship as the West has been in the world itself. And it is to heighten awareness that alongside bequeathing powerful understandings of the world and its tapestry of human cultures, societies, economies, polities, and environments, wittingly and unwittingly, European-Anglo-American human geography has supported a wealth of deep-seated misunderstandings of the prowess of the civilization that has given it life and the culture and character of the non-Western peoples (in particular, the peoples of the Global South and indigenous First Nations peoples) that this civilization has sought to dominate.

Our tale has a twist. A product of the rise and reign of the West, it is not surprising that the faltering of the West as a global hegemon is etching itself in the stratigraphic record of human geographical thought and practice. Its most serious manifestation is a certain paralysis of purpose wrought by the ascent from the early 1980s of the philosophy of postmodernism.

In the shadow of British historian Arnold J. Toynbee, Chapter 3 noted that the story of "big human history" is the story of the rise and fall of great civilizations; all believed they were impregnable, but all fell. The West is unlikely to be an exception. Indeed, with feverish intensity, across the past century, a chorus of eschatologies have predicted the imminent collapse of Western civilization and its core institutional pillar, capitalist modernity. Few have matched Toynbee's sage capstone that the biggest threat to the West is the West itself: civilizations are rarely murdered and more often collapse through suicide and self-harm. In a world burning from economic crises, austerity, environmental pillage, pollution and climate change, widening social and spatial inequalities, and a mental health epidemic, it is unsurprising that the West's politico-institutional model has steadily appeared less self-assured. Today this faltering in self-belief has deepened into an existential crisis. The loss by Europe of its empires across the twentieth century proved to be a crucial blow. If the Suez Crisis of 1956 shook the complacency of Europe, so too the Vietnam War (1955–1975) vexed the United States. Crisis has escalated since the collapse of Fordist-Keynesianism in the mid-1970s and in particular since the global financial crash in 2007. The recent rise of nationalist and populist movements and leaders appears to be more a defensive reaction reflecting weakness than a bold statement announcing a resurgent West. And the 2020 Covid-19 outbreak and its aftershocks are testing the underlying frailties of the West's model of political economy to the full.

The accumulation of the problems it itself has created has brought the West to the brink: but to the brink of what?

Rooted as it is in the fate of the West, it comes as little surprise that European-Anglo-American human geography too has become ever more troubled with self-doubt as to the validity and veracity of its theories: Have we got it right? Does the world really work that way? Contemporaneously, human geographers have come to understand more keenly that their theories are situated in time and space and therefore endemically vulnerable to a lack of objectivity and neutrality. In the 1980s and 1990s, a tradition of postmodern human geography announced the death of the meta-narrative and placed doubt on all claims to objective knowledge. Postmodernism ridiculed science, ethics, and political ideology and was ambivalent about solving societal problems. Although its worst excesses have waned, its intellectual challenge has yet to be properly dealt with. At times, human geography seems to be somehow just muddling on. A key question persists: what are human geographers to do with human geographical knowledge if it is accepted that all knowledge is produced within history and brought to existence only from the vantage point of particular locations? More particularly, what to do when this viewing platform is crumbling?

And so too, then, the accumulation of the problems that human geography has created for itself has brought it to the brink: but again, to the brink of what?

Human Geography: An Essential Introduction has been written at a particular moment in the history of the entanglement between the West and human geographical enquiry. The confluence of the faltering of the West as a hegemonic political and economic project and growing skepticism over the universality and authority of the West's knowledge formations, including the intellectual enterprise of human geography, provides the essential context. This book is both a product of and a response to the crisis in and of West and the structuration of human geographical thought that this crisis is inducing.

So how to conclude? What is to be done next?

In closing this book, I advance the argument that there is a need for a twenty-first-century *mission civilisatrice* for the metropole and that human geography has a role to play in remaking the West as both a home maker and a neighbor. Given its historical resonances, to use such language is of course to risk provoking the ire of some. I take this risk so as to explicitly trouble and invert colonial geographical imaginaries and hierarchies that construe the West as an agent doing the civilizing and not as an object to be civilized. Just as the rise of the West and the intellectual project that is human geography were made together for mutual advantage but oftentimes with regrettable results, so too the future of the West and of European-Anglo-American human geography might be remade together, on this occasion for mutual advantage and welcome outcomes.

Remaking the West

Evangelists continue to believe that the West's central institutions are essentially omnipotent; indeed, there are some who continue to believe that the West rose from the wilderness by divine providential design. US political scientist Francis Fukuyama (2006) once famously declared that the birth of liberal capitalist democracy signaled an end to history; humans should give up trying to invent a better model of society because there was none to be found. This kind of thinking informs the work of the neoconservative US-based historian Niall Ferguson (2014) and British neoconservative political commentator Douglas Murray (2017), both of whom argue that the world needs an effective liberal empire and that the United States is well placed to build that empire. Both recognize that the West is under threat but believe that the enemy within – pusillanimity – is the real danger. Illiberal liberalism and cancel culture have disarmed Europe's capacity to wage a cultural war on behalf of its own story. The West needs to boldly reaffirm the superiority of its core institutions. (See Deep Dive Box 13.1.)

Deep Dive Box 13.1   Culture Wars over the American Story:
A Tale of Two Speeches

There remains a lot at stake in the ongoing preservation of the stories that the West likes to tell the world about itself. In particular, who controls the myth that "West is best" controls the future of the West. Consider the following two US Independence Day (July 4th) speeches. Which do you think captures the American story best? Is there a single story to tell at all?

On 4 July 1965, during a sermon delivered at the Ebenezer Baptist Church in Atlanta, Martin Luther King Jr. famously described as sublime Thomas Jefferson's words in the United States Declaration of Independence in 1776 (Plate 13.2) and in particular the passage:

"We hold these truths to be self-evident, that all men are created equal, that they are endowed by their Creator with certain unalienable Rights, that among these are Life, Liberty and the pursuit of Happiness."

Astonished by the "amazing universalism" expressed in this passage, King reflected:

"It says that each individual has certain inherent rights that are neither derived from or conferred by the state. They are gifts from the hands of the almighty God. Very seldom, if ever, in the history of the world has a socio-political document expressed in such profound, eloquent, and unequivocal language, the dignity and the worth of human personality."

Here was a "promissory note," issued to future generations and promising liberty and equality for all. But in the very same sermon, King went on to say:

"Ever since the Founding Fathers of our nation dreamed this dream, America has been something of a schizophrenic personality. On the one hand, we have proudly professed the noble principles of democracy. On the other hand, we have sadly practiced the very antithesis of those principles. Indeed, slavery and segregation have been strange paradoxes in a nation founded on the principle that all men are created equal."

(Continued)

Box 13.1 *(Continued)*

Plate 13.2 Writing the Declaration of Independence – Thomas Jefferson (right), Benjamin Franklin (left), and John Adams (center) meet at Jefferson's lodgings, 1776. A painting by Jean Leon Gerome Ferris (Virginia Historical Society). Source: Library of Congress.

For King, Jefferson's promissory note would be rendered meaningless so long as the United States presided over the triple evils of racism, war mongering, and poverty.

On 4 July 2020, then US President Donald Trump delivered a speech at Mount Rushmore National Memorial in South Dakota. Carved into the mountain are 60-ft-tall faces of Presidents George Washington (1732–1799), Thomas Jefferson (1743–1826), Theodore Roosevelt (1858–1919), and Abraham Lincoln (1809–1865), selected to symbolize the United States' birth, growth, development, and preservation, respectively. Trump had chosen to give his Independence Day speech at Mount Rushmore as a direct response to the actions of the Black Lives Matter movement who had taken to the streets in many US cities – and, indeed, in cities around the world – to protest police brutality against the black community, including by tearing down place names, flags, monuments, and statues associated with slavery. Trump perceived their actions to be undemocratic, violent, and illegal. Condemning what he called "cancel culture," in a defiant and fiery tone he declared:

> "Make no mistake: this left-wing cultural revolution is designed to overthrow the American Revolution. In so doing, they would destroy the very civilization that res-
> cued billions from poverty, disease, violence, and hunger, and that lifted humanity to
> new heights of achievement, discovery, and progress. There is a new far-left fascism
> that demands absolute allegiance if you do not speak its language, perform its rituals,
> recite its mantras and follow its commandments, then you will be censored, banished,
> blacklisted, persecuted and punished. It is not going to happen to us. We will not be
> silenced. . .. They think the American people are weak and soft and submissive, but
> no, the American people are strong and proud and they will not allow our country

and all of its values, history, and culture to be taken from them. This monument will never be desecrated. These heroes will never be defaced. Their legacy will never, ever be destroyed. Their achievements will never be forgotten. And Mount Rushmore will stand forever as an eternal tribute to our forefathers and to our freedom."

Culture wars over ownership of the American story have become partisan and febrile. In January 2021 outgoing President Donald Trump became the first President to be impeached twice on the grounds that he "engaged in high Crimes and Misdemeanors by inciting violence against the Government of the United States," Trump was accused of encouraging Far Right supporters (including QAnon conspiracy theorists) to attack the Capitol building (the "citadel" of democracy) in Washington DC in a bid to overturn the results of an election he lost to the Democratic incumbent Joseph Biden. Although the Senate voted 57-43 to convict Trump of inciting insurrection, it did not secure the two-thirds majority vote required by the Constitution and Trump was acquitted for a second time. Meanwhile, at least for some Republican "Grand Old Party" (GOP) supporters, Antifa (a left-wing anti-fascist and anti-racist political movement) protests across United States cities in support of the "Black Lives Matter" cause have served only to divide the nation.

Anger and untruths have to be removed from the culture wars now raging across the West so that a rigorous and balanced conversation can be had about the most appropriate way to narrate the United States' story and its place in the grander sweep of big human history. Getting this story right will be vital if we are to embark on the kind of farsighted transformation of the West's politico-institutional model which will necessary if it is to continue to command a democratic license in the twenty first century.

The Western model has delivered in a short period of 500 years unprecedented human progress. The human species has never lived longer, enjoyed better health, or been better educated. Market economies have lifted billions out of poverty. Democracy and mass enfranchisement have empowered those hitherto excluded from choosing and holding to account their political leaders. The fourth estate has never been as able to call the political classes to account. People have never been as free to speak their mind, assemble in public, and pursue their conscience. The rule of law has enabled people to coexist peaceably. Western technology has vastly enriched the quality of human life and helped those with disabilities to live well and more fully. Ever-expanding consumption has created worlds of luxury and abundance, at least for some. Innovations in transportation have enabled people to explore the world, from the tip of the highest mountain to the bottom of the deepest ocean, from the West to the East and the North to the South. In short, the peoples of Europe and the settler colonies Europe birthed would undoubtedly be living today significantly less agreeable lives had the European world capitalist economy not appeared on the historical stage 500 years ago. In his final book, Factfulness, health statistician Hans Rosling et al. (2018) condemned the pervasiveness within the West of excessively pessimistic stories about the state of the world. In fact, nearly every significant indicator points to a world that is constantly improving: critics need to temper their inherent biases and make judgments only on the basis of facts.

But equally, at the same time as they were espousing the noble Enlightenment ideal that each human life is infinitely precious, endowed with freedom and dignity, and of equal value, European states were also building rapacious empires, colonizing and subjugating other peoples in other places, enslaving millions of Africans, presiding over vicious regimes of institutional racism, cruel famines, enacting genocide against indigenous peoples, waging wars to hold onto colonial lands, and pillaging resources that they had no right to "own."

Western greed has led to more wars (most often between Western powers rather with non-Western powers) and more lives lost to war per capita than in any prior historical period. The West has birthed a schizophrenic global legal and ethical machinery that, according to US philosopher Judith Butler (2016), seems somehow able to live with Western atrocities while prosecuting crimes against humanity committed by those at a distance. How on earth has it become possible to divide the world into grievable and ungrievable lives? Nuclear technology has provided a handful of human beings with the power to render the entire human species extinct. The capitalist economy has proven difficult to regulate over the *longue durée* and has lurched from one crisis to the next, creating endless and socially painful cycles of boom and bust. Capitalism has brought humanity to the precipice of a fourth industrial revolution that risks unleashing a dangerously under-regulated body of artificial intelligence (AI). Unchecked human-induced climate change, deteriorating air quality, the generation of toxic mountains of waste, and the loss of biodiversity are taking the planet to the edge of ecocide. Gross inequalities are unleashing dark neo-fascistic forces and leading to the election of divisive populist governments. A product of geopolitical failures and climate change, it is anticipated that the Global North will witness a significant growth in asylum claims and refugees, at a time when populist revolt against "whiteshift" is heightening racial tensions. Meanwhile, unsustainable human ecologies (including the building of ever-larger cities whose chaotic expansion, poor-quality built environments, and frantic and frenetic rhythms are taxing the structure and functioning of the human central nervous system) are becoming manifest in a global mental health crisis – arguably the emperor of all maladies. (Re)emerging infectious diseases, including but not limited to Covid-19, are presenting as a fresh threat to the West, its people, and its model of political economy. And the rise of corporate media and social media has led to a post-truth "digital" public realm, obfuscated public understanding of the policies and performance of political leaders, and diminished the democratic process.

So, what are we to conclude? My own view is that the idea that human history has ended with the ascent of the West is surely a preposterous one. History will continue to be alive for as long as human beings exist. In time the West will die as all prior civilizations have, and there will arise a new politico-economic-institutional model to take its place. Until such times, however, I also believe that: (i) there is much to admire and much to lament about the West's model of politics and economy; (ii) the histories of many alternative politico-economic-institutional models, including those that prevail in the Global South, are littered with actions and inactions that are inglorious, abhorrent, indefensible, reprehensible, and no better alternative is readily apparent today; (iii) the real politic is that the West will continue to be a global power for the remainder of this century; and (iv) there is still mileage left in Whiggish histories of the West, that is, there remains scope within the Western model for further self-reflection, enlightenment, and deep structural improvement.

For sure, the West has some very substantial work to do on itself. Its day of reckoning has come. Nothing less than a disruptive intervention will do. But a radically transformed West has the potential to change the course of human history for all of humankind for the remainder of this century. There are likely to be as many as 10 billion reasons why the successful prosecution of such a project matters. It is a mission worthy of the attention of human geographers.

Remaking Human Geography

The Royal Geographical Society (with the Institute of British Geographers, or RGS[IBG]) is a learned society and professional body dedicated to the advancement of geographical understanding. For over a century Lowther Lodge, an iconic Victorian building bearing the equally iconic address 1 Kensington Gore, London, has served as the Society's headquarters. At the heart of Lowther Lodge is the magnificent 750-seat Ondaatje Theatre – named after Christopher Ondaatje, born into one of Sri Lanka's most powerful colonial families and later to become a wealthy Canadian financial and publishing mogul (Plate 13.3). While steeped in colonial history, the RGS(IBG) Annual Conference provides a democratic space where nearly 2000 geographers meet to exchange their latest research findings and to discuss the future of the discipline. Although Lowther Lodge remains adorned with symbols of British exploration and imperial expeditions, the Ondaatje Theatre plays host to vigorous debates over how geography might confront its complicity with the European imperial adventure, challenge the echoes of empire that reverberate in our colonial present, provincialize and decolonize its theories, and become a truly global subject. The juxtaposition is emblematic of where the discipline now finds itself.

In 2017 the RGS(IBG) Annual Conference was chaired by Sarah Radcliffe, a geographer working at Cambridge University. An expert on Latin America, Radcliffe's wider research program explores "the ways in which we might learn about, write, and do geography in ways that open it out to other ways of seeing, living in and knowing the world" (Radcliffe 2017). Specifically, she has published widely on the themes of the decolonization

Plate 13.3   The magnificent 750-seat Ondaatje Theatre, Lowther Lodge, in Kensington, London, where the Annual Conference of the Royal Geographical Society with the Institute of British Geographers (RGS[IBG]) takes place. Source: Panel 8 Photography, Courtesy of the RGS(IBG).

of theories of development, new sites of knowledge production in the Global South, and indigenous geographical knowledges. Radcliffe was acutely aware of the role of the Royal Geographical Society in the production of colonial authority. Her choice of conference theme was unsurprising but widely applauded: "Decolonising Geographical Knowledges: Opening Geography to the World." For centuries, Europe's metropolitan heartlands had produced knowledge of the colonized world that had crowded out other knowledge formations such as "indigenous knowledge, alternative universalisms, locally-based research agendas, and the Southern theory arising from the colonial encounter itself." It was now time to "decolonize curricula, research and theory." In her opening welcome to delegates, held in the Ondaatje Theatre, Radcliffe noted, "decolonising geographical knowledges remains polemical and rightly so, representing as it does a problem space for geography." But, she added, "what better place to start than at the major British geography conference among its thousands of (international) participants?"

In fact, for some time now, a postcolonial human geography has been a work in progress, developing variously along four complementary tracks. A first track has sought to undertake a ***genealogy of human geography's development as a colonial subject***. European-Anglo-American human geography is asked to take stock of its imbrication in colonial projects and to better understand how its past theories, methods, and substantive studies helped to enable these projects. How might it atone? What reparations are due? *A postcolonial human geography is a human geography that "owns" its past complicity with empire.* A second track seeks to ***provincialize human geography***. European–Anglo–American human geography needs to be put in its place but not dismissed. The West's knowledge formations may be more parochial than hitherto believed, but if handled critically they continue to offer powerful understandings of the world and its workings. *A postcolonial human geography is a human geography that recognizes the importance of working critically with a partial perspective.* A third track focuses upon ***decolonizing human geography***. Alongside European-Anglo-American regimes of truth, there is a need to radically valorize ways of being in and knowing the world whose roots lie in the overlooked intellectual traditions of the Global South and in dispossessed indigenous communities in settler colonies. *A postcolonial human geography is a human geography without a single hegemonic intellectual heartland.* A final track dedicates itself to ***anticolonial human geography***, which uses human geographical knowledge from the European-Anglo-American tradition and beyond to critique the colonial present. *A postcolonial human geography is a human geography that deploys its expertise and knowledge to critique today's imperialists and neo-imperialists.*

Recognizing that there are multiple knowledge formations and competing regimes of truth does not license scholars to lapse into a debilitating relativism or to practice bad social science. According to US geographer Farhana Sultana (2018), progress can be made if we first distinguish between free speech and academic freedom. While free speech is speech that is protected irrespective of its veracity (even if it is nonsensical, prejudiced, or inciteful speech), academic freedom denotes free speech that arises from scholarly intensity and that meets rigorous and respected intellectual standards and impartial criticality.

In an influential article entitled "Situated Knowledges: The Science Question in Feminism and the Privilege of Partial Perspective," published in 1988 in the journal *Feminist Studies*, US feminist philosopher Donna Haraway argues that provincializing Western knowledge and entertaining multiple ways of knowing the world provides a superior route to objectivity than that proffered by traditional science. That there may exist a variety of regimes of truth need not dishearten us or weaken our resolve to affirm our

Deep Dive Box 13.2   Culture Wars Over Stewardship of the Western University

The project of building a postcolonial human geography has become caught up in culture wars over the story of the West, which have spilled over to the university campus.

If university leaders were to take a Hippocratic Oath, it would surely center upon a commitment to defend academic freedom at all costs; if scholars were to have one, it would pivot upon their sacred duty to dignify this freedom by ensuring that the knowledge they issue to the world exceeds minimum benchmarks of intellectual rigor and objectivity. It is the combination of both that imbues academic knowledge with privileged authority. For some critics, threats to both are leading to an erosion of this authority. To gain credibility, a postcolonial human geography will need to find a way through the minefields currently being laid by warring factions.

The (Neo)conservative Critique

In the first instance, (neo)conservative commentators lament what they consider to be a left-liberal (Hollywood-esque) capture of the university. Far from broadening the intellect, universities have presided over a "closing of the American mind": radical "social justice warriors" have shut down genuine agonistic debate on campus, de-platformed those espousing conservative points of view, and licensed (as they see it) "fashionable nonsense" and "higher superstition" to parade as objective knowledge. Those who believe in free speech and academic freedom and who dare to defend the university as a place where all ideas (good and bad) are allowed to surface and be interrogated are being subjected to a "Maoist" purge. Political correctness is killing the pursuit of truth. The "snowflake" generation have no business in the hallowed halls of the university - where contesting and falsifying knowledge claims is the entire purpose - if they are not prepared to debate those who share different points of view and contemplate the possibility that there are other equally valid ways of looking at the world. A favorite neoconservative trope declares: "facts do not care about feelings".

The Left Critique

In the second instance, left scholars question (as they see it) the extent to which a creeping corporatization of the university is stifling academic freedom and forcing adherence to applied bourgeois analytics. Academics have lost control over their intellectual labor, and the precarity of their tenure has been weaponized to constrain their capacity to undertake studies that are critical of powerful actors. With the election of populist governments, and especially the outgoing Trump administration in the United States, the university is also now coming under threat from authoritarian "post-truth" warriors pedaling anti-intellectualism, white supremacy, and neocolonial violence. How can one speak truth to power when those in power control the allocation of resources? To survive, scholars now need to prostrate themselves before "national priorities" and justify their contribution to the world as it is, not as they would like it to be.

(Continued)

Box 13.2 *(Continued)*

The Equality, Diversity, and Inclusion (EDI) Critique

Even though the mission to un-think Eurocentrism is laudable, some indigenous and subaltern communities question (as they see it) the limits of its enactment in practice. Some have gone as far as saying that it is now necessary to "trouble good intentions." Efforts to provincialize and decolonize knowledge are being undertaken from a limited number of locations, mainly in the United States, the United Kingdom, and more developed (settler) colonies such as Ireland, Singapore, Australia, South Africa, and Canada. Most scholars within the Western academy are Western by birth and/or work in Western universities and/or perform to Western registers of theory generation, and it is dangerous that they presume to speak on behalf of colons and indigenes. The risk is that they unwittingly commit epistemicide, defined as the destruction by one knowledge formation (say, in the Global North) of knowledge formations elsewhere (say, in the Global South).

All three camps raise questions that cannot be ignored; the stakes are simply too high. So how to proceed?

In the United States, in reaction to the New York Times Magazine's "1619 Project" which has sought to locate slavery and Black Americans at the heart of American historiography, former President Donald Trump established a "1776 Commission" to valorise what he termed "patriotic education". This Commission was disbanded by President Joe Biden as one of his first acts in office. Meanwhile, in part a response to pervasive anti-Brexit sentiment among faculty members, the UK government has empowered the Office for Students (OfS) (the sector regulator), to sanction universities judged to have stifled free speech and has created a new board level role - "free speech and academic champion" - to investigate incidents of censorship. At the same time the UK Department for Education (DfE) issued guidance to school leaders and teachers forbidding the inclusion in curricula of "resources produced by organisations that take extreme political stances on matters". These include positions such as "a publicly stated desire to abolish or overthrow democracy, capitalism, or to end free and fair elections"; opposition to freedom of speech; the use of racist, including antisemitic, language; the endorsement of illegal activity; and a failure to condemn illegal activities perpetrated in support of their cause. Whilst seemingly uncontroversial, such efforts to promote free speech themselves threaten to become chilling instruments of surveillance and silencing. History instructs that in spite of good intentions, governments can all to easily default to totalitarian and authoritarian censorship.

At a moment when the very concept of truth is in crisis, the future vitality of a postcolonial human geography will pivot upon the veracity and persuasiveness of its claim to speak 'truth to power'. A postcolonial human geography must serve as a better custodian of truth – or perhaps more modestly a better steward of verisimilitude – than uncritical human geography, including uncritical critical human geography. The claim that all observation is theory laden, all knowledge is situated, all scholarship is ideological, and all research underpinned by values does not provide a license to practice bad social science. A postcolonial human geography must establish its credentials as a truly liberal, self-critical, and compelling situated science: if its truths are to be accepted as truthful truths or at least 'truthy truths' , its theories and concepts must be open to relentless and uncompromising interrogation and critique.

commitment to objectivity. To consider all knowledge equal is indeed to fall prey to the tyranny of relativism. Relativism was as totalitarian as scientific ideologies of objectivity; to Haraway, "both deny the stakes in location, embodiment, and partial perspective; both make it impossible to see well. Relativism and totalization are both 'god tricks' promising vision from everywhere and nowhere equally and fully, common myths." All scholars are obliged to take seriously their duty to pursue "passionate detachment" and self-critical, rational enquiry. Only by attending to the "politics and epistemology of partial perspectives" can scholars practice objectivity. Ruthless self-critical partiality, and not pretensions to neutral observation, is the most rigorous route to trustworthy knowledge.

If it is to prosper, a postcolonial human geography will need to avoid practicing both uncritical critical geography and uncritical applied geography and will need to find a balance between giving western epistemologies their proper due whilst challenging their proclivity to commit epistemicide. A postcolonial human geography must place a premium on the quality of the knowledge it produces. It must be humble enough to recognize its situated, modest, nonfoundational, and always provisional insights. It must practice vigorous self-criticality and open itself up fully to interrogation from all sides of the political spectrum. It must apply uncompromising and passionate objectivity and always aspire to verisimilitude. And it must test its conjectures with an inclination to falsify rather than verify them. In this way, it will elevate the status of its claims. A postcolonial human geography should aim to become nothing less than a "situated science." By its scholarly standards, it shall be known – and judged.

Remaking the West, Remaking Human Geography

Introducing the novel idea "utopia as method", British sociologist Ruth Levitas (2013) champions the merits of (re)making social science disciplines in and through the process of proposing and testing visions for a better world. One way to decolonise European-Anglo-American human geography is to put it in the service of what I will call a *mission civilisatrice for the metropole*. By supporting the West to repurpose itself, human geography will create an intellectual climate conducive to the remaking of Western human geography itself, as a globalized human geography "without a center." Equally, decolonising European-Anglo-American human geographical scholarship and catalysing and complexifying conversations between Global North and Global South human geographical traditions will inject much needed fresh thinking into the project of making the West anew. A virtual cycle will result.

What ideas and resources are available to help us begin our journey? For British born-Singaporean based geographer Tim Bunnell and Singaporean sociologist Daniel Goh (2018), the capacity to imagine different futures varies over time and space: a culture rich in hope and aspiration and invigorated by an open public realm has at least a chance of sponsoring significant social change; a culture suffering from a diminished public square and bereft of resources to engage in vibrant futurity talk is condemned to accept the mantra 'there is no alternative'. Alas, for a long time the West has suffered from a famine of hopeful voices. Fortunately we find ourselves today in a world consumed by futurity chatter and in the company of a veritable library of thought experiments setting out what a post imperial, post neoliberal and post growth-centered liberal capitalist democracy might look like.

In the wake of the global financial crash of 2007, in February 2013 the journal *Soundings* convened a conference to explore how those most impacted by the ensuing period of austerity might respond. The conference organizers, British cultural theorist Stuart Hall, geographer Doreen Massey and sociologist Michael Rustin, all with attachments to Kilburn in North London, believed that: 'Although the neoliberal economic settlement is unravelling, its political underpinning remains largely unchallenged'. It was time neoliberal London sponsored a vibrant new tradition of futurity speech. Over the next eighteen months follow up evening seminars explored radical alternatives to the neoliberal order. In 2015, these seminars were gathered together in a book edited by Hall, Massey and Rustin and bearing the title *After Neoliberalism? The Kilburn Manifesto* (Hall et al. 2015).

Inspired by the *Kilburn Manifesto*, I will now fashion my own personal vision of what a better West might look like. Decoupling the West from rampant unbridled capitalism and relentless economic growth at any cost is my central theme. Whilst recognizing the value of thinking seriously about 'de-growth', I believe that securing economic growth should continue to be a priority. We do not exist in a world of surplus or abundance; our world is no Garden of Eden. But this aspiration should now be radically tempered: it is the quality of economic growth that matters most; growth is only virtuous if it contributes to human flourishing and wellbeing and the maintenance of the planetary Gaia. And also of course, if it does no harm to others at a distance. Good quality growth – resilient, sustainable, inclusive, ethical and in support of life itself - must replace pro-growth as the new North Star. My task is to identify the changes I believe we will need to make to give effect to this vision. Of course, you are not obliged to agree with my ideas; infact I would be disappointed if you did. My blueprint is built upon provocations more so than prescriptions. But as an educated and concerned citizen, you do have an obligation given the times in which we live to cultivate your own vision for a hopeful future and be prepared to fight for your nirvana in the market place of ideas.

In starting with the idea of good quality inclusive and clean growth, I find myself immediately in the company of giants. According to US economist Joseph Stiglitz's (2019) it is surely time for world leaders to respond to the emerging "global social movement for well-being," shift from measuring GDP growth to "measuring what actually counts," and develop "a market economy that works for people and not the other way around." At least in the Global North, GDP growth is no longer an appropriate measure of the efficacy of an economy and must be replaced with metrics pertaining to improved human health, well-being, prosperity, and quality of life. In his 2017 book *Building the New American Economy: Smart, Fair, and Sustainable*, US economist Jeffrey D. Sachs likewise attacks what he considers to a corporate takeover of American democracy (by the military–industrial complex, the Wall Street–Washington complex, the Big Oil–transport–military complex, and the health care industry), and calls upon the United States to confront this "Corporatocracy" and refocus its attention away from relentless GDP growth and toward sustainable development, social justice and decarbonization. In *Doughnut Economics: Seven Ways to Think Like a 21st Century Economist*, British economist Kate Raworth (2017) also provides a framework in support of a recalibration of the mission of the economy in favor of human well-being and prosperity. A certain scale of economic activity is required to ensure that all members of society enjoy a decent standard of living but the key is to manage the economy so as to "land" within safe operating zones and planetary boundaries. Meanwhile, drawing upon a wealth of data, British geographer Danny Dorling (2020) argues that the age of turbo-capitalism – the great acceleration – is now ceding to a new period of *Slowdown*, a great deceleration in economic growth, and perhaps even the start of a period of de-growth. For Dorling, slowdown should be interpreted as a signal that our civilizational model is exhausted and embraced as a prelude to a sustainability transition.

Who will lead us to a better tomorrow? United States President Ronald Reagan famously stated that the nine most terrifying words in the English language are *'I'm from the government, and I'm here to help.'* But arguably, a different nine words have proven to be more terrifying: *Trust the market for private enterprise delivers public good.* If the hegemony of unfettered free market capitalism has taught us anything, it is that if left to its own devices the market will not deliver sustainable, fair, or green development. In fact, it is anatomically wedded to boom and bust economics and unsustainable and unequal development and what British-United States based geographer David Harvey has called "accumulation by dispossession". Accordingly, we need to consider establishing what Australian urbanist Brendan Gleeson (2010) refers to as a pioneering 'guardian state'. My own preferred label is a "social democratic state for the twenty first century" or SDS c.21.

What does a SDS c.21 need to fix to get us to where we want to go? SDSs tend to enjoy a strong social license because they democratize economies and humanise capitalism by ensuring that markets serve the commonweal and not the commonweal markets. They do so by: a) emphasizing economic democracy and the democratization of the management of the capitalist economy, b) incentivising and regulating markets so that they unleash creativity and energy but support private gain that does public good, c) building a mixed economy to compliment markets and valorizing alternative economic logics, and d) harnessing wealth created by markets to deliver comprehensive and high quality universal basic services and social protections. SDSs have long existed, but a SDS c.21 must be like no other: tired old recipes will not be up to the task at hand and radically disruptive entrepreneurship and innovation must be written into its source code. United States economist Joseph Stiglitz (2019) gets to the heart of the matter when he asks: "Are there ways of designing governmental institutions which enhance the likelihood of, if not ensuring that, public interventions are welfare enhancing?" The ultimate objective of a SDS c.21 should be to champion and oversee a great reboot and reset of the West's central institutional pillars; "liberalism", "capitalism", and "democracy" and to imagine how its core (French inspired) motifs of "Liberté" (freedom), "Égalité" (equality), and "Fraternité" (solidarity) might be put to work more productively.

Therein, a SDS c.21 might attend to seven particular priority actions: regaining a social license for democracy, reasserting liberal citizenship (in the digital age), redevising market logics, compelling equality of opportunity, valorizing and scaling alternative non-commoditised economies, championing tech for public good, and becoming a better neighbor.

Regaining a Social License for Democracy
The first job facing a SDS c.21 will be to save democracy from elections (at least in their current form) and reconnect governments with the citizens they are supposed to serve. According to British political scientists Roger Eatwell and Matthew Goodwin (2019) Western democracy is suffering on account of 4d's; distrust of politicians, destruction of national identities, deprivation and a sense of being left behind, and dis-alignment of political parties from their bases. Representative democracy and popular sovereignty have drifted apart, dangerously so. Remediation will require new experiments in, and a blending of, instruments of representative and direct democracy (participative and deliberative democracy). But we surely also must counter the mutation of an already flawed liberal democracy into something worse - an illiberal democracy. As United States political scientist Kevin O'Leary (2020) has shown, in many ways political populisms stand in direct opposition to liberal democratic enfranchisement and equality. We need polities which are owned by all of the people for all of the people and not simply by a majority for a majority or worse still a minority for a minority

Deep Dive Box 13.3   Putting Human Geography in the Service of a Mission Civilisatrice for the Metropole

What might a specifically human geographical contribution to an alternative West look like?

In his book *Spaces of hope*, British-United States Marxist Geographer David Harvey (2000) called upon human geographers to rescue and rehabilitate the concept of utopian geographies. For Harvey, to imagine a better future by imagining only a better society is to fall short. Any utopian project must recognize that to build a new politico-economic-institutional model it will be necessary to redevise the spatiality of social life and people's relationships with the natural environment. It is for this reason that a social democratic state for the twenty first century (SDS c.21) must acquire spatial analytic literacy – variously defined and deployed at a variety of scales. History makes geography but so too geography makes history; social justice can only be achieved alongside spatial justice and spatial justice alongside social justice, sustainable and resilient development alongside new relationships between people and the natural environment and new relationships between people and the environment alongside new models of sustainable and resilient development. There can be no "imaginary reconstitution of society" without a parallel "imaginary reconstitution of space, place, landscape, and nature".

And so we must ask:

Regaining a social license for democracy In what ways are spatialities of political power implicated in the present crisis in democracy? If they are to be part of the solution, how do political geographies of power need to change?

Redevising market logics In what ways can uneven geographical development and the global climate and ecological crisis be said to be expressions of market failure? How might spatial planning and climate action planning instruct the project of redesigning markets?

Reasserting liberal citizenship (in the digital age) In what ways does a denuded public square contribute to the erosion of liberal citizenship? How might we defend the public character of public space (freedom of assembly, movement, association, expression, speech etc) in support of a healthy liberal public realm?

Compelling equality of opportunity How does who gets what, where, why and how influence what people can be and become? How might fairer welfare geographies contribute to the project of equalization opportunity and unlocking the potential of all?

Valorizing and scaling alternative non-commoditised economies In what ways has the neoliberal economy and the spaces it has left behind impacted the geography of alternative non-commoditised economies? What challenges and opportunities do geographies of the foundational economy, social economy and caring economy present for those who want to support and scale all three?

Championing tech for public good How might uneven geographies in the innovation and adoption of AI/digital innovation/big data technology enable and constrain plans to accelerate the harnessing of tech for public good?

> **Becoming a better neighbor**: In what ways are regressive concepts of place implicated in the rise of populist isolationism and the emergence of a new politics of hospitality. How might the ideas of "global sense of place", "power geometries of relational space" and "geographies of responsibility" help the West become a better neighbor to the world?

Reasserting liberal citizenship (in the digital age)

We have now come to understand that Western in origin and design, the 1948 United Nations Universal Declaration of Human Rights (UNDHR) codifies the rights that political liberalism bestows upon the western citizen more so than it codifies universal human rights per se. Still, the UNHDR gives voice to ideals that are infinitely precious. The liberal citizen is dignified by basic freedoms (of speech, movement, from arbitrary arrest, of assembly, of association, to vote, to labor, to worship, and so on). These ideals are worth defending. The West is at its best when it upholds these freedoms. But it is clear that they are under threat – everywhere, including in the West. They are been denied, exploited and abused, not least by those who govern and those who poison the ethernet. The (virtual/digital) public square is today significantly denuded. Fundamentalists from each of the political right, left and centre have been accused of practicing illiberal liberalism – standards need to be introduced to adjudicate on these accusations. A SDS c21 must regain custody of political liberalism and

Plate 13.4 The West at its best? Eleanor Roosevelt with the English language version of the Universal Declaration of Human Rights, December 1948. Source: FDR Presidential Library & Museum. https://commons.wikimedia.org/wiki/File:Eleanor_Roosevelt_UDHR.jpg, CC BY 2.0

defend the rights and freedoms enjoyed by the liberal citizen, even (and especially) when this leads to agonistic politics. But it must also serve as a steward of the public realm and insist that the right to be part of such a community brings with it an obligation to meet a number of basic standards of conduct – such as doing no harm to others, speaking the truth, trusting science, valuing expertise, respecting facts and acting responsibly.

Redevising market logics

The market will surely have to play a central role in solving public problems; it has enormous resources, talent, dynamism, expertise and innovative capacity that needs to be harnessed. But it is clear that the market will require regulation, incentives and discipline if it is to serve the public good. A SDS c.21 must reconstitute, redesign and reframe markets so that they serve the commonweal.

Conscious that capitalism will need to up its game if it is to avoid seismic shifts in market regulation, in 2020 Executive Chairman of the World Economic Forum Klaus Schwab crafted a prospectus titled: *Davos Manifesto: The Universal Purpose of a Company in the Fourth Industrial Revolution*. From the 1980s, companies have been driven by the imperative of maximizing shareholder value (MSV). A risky and volatile economy, unsustainable social and spatial inequalities in wealth and income and a global climate and ecological crisis had been the result. A turn towards an ethical, inclusive and green capitalism – profit from purpose, public good through private means, moral markets, and tech for good - is now required. An old mid-twentieth century model of capitalism – stakeholder capitalism – needed to be dusted down and given new life for the twenty-first century. Standard financial metrics of performance needed to be replaced with environmental, social, and governance (ESG) metrics. Responding to the same set of pressures, according to former Managing Director with JPMorgan and President of the Capital Institute, John Fullerton, whilst the current capitalist economy is unsustainable and heading for collapse, this does not mean that the market mechanism is flawed per se, only that turbo-charged neoliberal capitalism is pathological. Drawing insights from the Eastern wisdom traditions and from the science of biomimetics, a new 'regenerative' wellbeing centred market economy should be built modelled upon high performing natural ecosystems which have prospered over millennia.

But will preemptive action by market actors and voluntary self-reform be enough?

For Irish marketing scholar Susi Geiger, key to rectifying 'market misfires' and to securing for markets a social legitimacy is the relocation of market design, regulation and governance from the sphere of 'technical' economic planning to that of society and politics. Geiger asks: how can markets be made to be more inclusive and open to the concerns of those who are let down by them? She has sought to bring new instruments of economic democracy to bear on the social and political making of markets. Markets might fail less often if they are configured, disciplined and incentivized by the people and for the people. Likewise, British geographer Andrew Cumbers (2020) argues that in the context of the rising tide of populism and a crisis in liberal democracy, there is a need to invoke the idea of "economic democracy." It is time for people to "take back control" of the economy. Workers have a right to participate in the design of the economy as a whole: how and what to tax, where investment is to go, what is to be subsidized, workers' rights, regulatory standards and restrictions, and so on. While not entirely redundant, older traditions of economic democracy based upon outdated ideas of public ownership and union control of the factory floor need to cede to three new critical interlocking foci: individual economic rights, diverse forms of democratic collective ownership of companies, and greater public participation in economic decision making.

Compelling equality of opportunity

In 1967, in an interview with Sander Vanocur at the Ebenezer Baptist Church in Atlanta, Georgia, and aired by NBC News, Martin Luther King Jr. spoke in his own uniquely eloquent way about the importance of ensuring that everyone who enters the capitalist race begins from the same start line. Comparing the gift of land to "white peasants from Europe" to the lack of provision of any kind of economic base to "black peasants from Africa," who came to America "involuntarily in chains and worked free for two hundred and forty-four years," King Jr. reflected: when white Americans tell the Negro to "lift himself by his own bootstraps," they don't look over the legacy of slavery and segregation. I believe we ought to do all we can and seek to lift ourselves by our own boot straps, but it's a cruel jest to say to a bootless man that he ought to lift himself by his own bootstraps. No one – no matter their class, ethnicity, race, religion, gender, age, sexuality, or disability– should enter or dwell in a civilized world bootless.

The juxtaposition of the role played by wealth inequality in income inequality on the one hand, and the promise of equality of opportunity on the other, stands as one of the greatest contradictions which liberal capitalist democracies somehow manage to live with. It is also one of the greatest source of market failure and lost human potential. A SDS c.21 must disrupt the link between wealth and income generation and better equalize the starting capacity of all new entrants to the market economy. French economist Thomas Piketty (2019) has done more than most to track the impact of wealth on growing income inequalities in the advanced capitalist world.. Piketty acknowledges that although some inequalities will be inevitable and perhaps even desirable in a capitalist economy, the scale of present inequalities is harming the economy, society, the environment, and democracy. For Piketty, there needs to emerge a new "participatory socialism" at the center of which is a new tax on wealth; the state should levy wealth taxes of no less than 90% on assets over $1 billion, and marginal income tax rates could legitimately rise to as high as 80%. A bigger state budget could then be used help capitalism win back its social license by providing those with no inheritance a lump sum to afford them a more equal shot at life.

But there is more to equality of opportunity than levelling up inherited wealth. There has emerged within some Western countries a widely held belief that public services are broken. In particular, the welfare state is in crisis and has proven ineffective in addressing the complex needs of communities debilitated by low aspiration, poverty, unemployment, family breakdown, poor housing, ill health, and low educational attainment. An emerging consensus holds that "universal", "anticipatory" "capabilities focused", "user centered" and "preventative" public services ought to replace dated transactional models of welfare provision which seek to "fix" failing communities. And so Universal Basic Services (UBS – unconditional access by all to essential services which are made available to all free at the point of delivery) and Universal Basic Income (UBI – a universal and unconditional basic income paid to everybody) are now being widely touted.

For British development studies scholar Guy Standing UBI is the weapon best equipped to slay what he identifies to be the eight giant ills that bedevil society: inequality, insecurity, debt, stress, precarity, automation, populism, and extinction. Likewise, for Dutch popular historian Rutger Bregman (2016), at the heart of a "utopia for realists" should be UBI, a short working week of 15 hours, and open borders and free movement worldwide. For British-American anthropologist David Graeber (2018), UBI presents one decisive response to economies increasingly dominated by "bullshit jobs," or work that is essentially pointless and whose meaningless has become psychologically destructive and harmful for large swathes of the population. For British public policy expert Andrew Simms (2013), in support of those who seek to perfect the "art of living," governments will need to introduce radical policies that sponsor a shorter working week, rewarding jobs, a zero carbon economy, locally based food production and consumption systems, local currencies, a new generation of community services, and advertising-free zones. In calling for an "economy of belonging" Norwegian-British economics commentator for the Financial Times Martin Sandbu (2020) notes the importance of UBS and UBI and the need to invest in education, skills, and labor activation programs, cap pay ceilings and increase the minimum wage, strengthen workers' rights; and regulate ownership rights for personal data. Meanwhile for British social entrepreneur Hilary Cottam (2018) UBS and "user centered", "capablities focused" and "relationship rich asset based community development (abcd)" provide a template for a new welfare state.

Valorizing and scaling alternative non-commoditised economic logics

British Geographer Doreen Massey's (Hall et al. 2015) particular contribution to the *Kilburn Manifesto* was to call into question the vocabulary we use when talking about the economy and in particular the special status we afford markets. What about all the other work which we need to do – outside the market economy and often unpaid - to keep our world going? In his book *What Money Can't Buy* United States political philosopher Michael Sandel (2013) likewise argues that even though they often try, markets are singularly ill equipped to deliver the moral and civic goods and services that human beings need to prosper. A SDS c.21 must recognize the existence of a "more than capitalist" economy and value the unique contributions that non-commoditized alternative economies make in the service of economic production and social reproduction.

The failure of centralized command economies built around extensive state ownership of enterprises allied with neoliberalism's quasi-religious devotion to private property rights and privatization has compromised a far sighted debate on the role of the state as an economic actor. This lacunae is now being filled.

According United States Economist Stephanie Kelton (2020) the capitalist state has developed an unhealthy aversion to government borrowing believing budgetary deficits to be irresponsible and an albatross around the neck of economic growth. This belief has led to prolonged periods of fiscal austerity and formidable social and economic pain. But it is simply wrong. In her book *The Deficit Myth: Modern Monetary Theory and the Birth of the People's Economy*, Kelton outlines a case for supporting the Keynesian argument that governments can and should spend themselves out of recession. In their *Manifesto for the foundational economy*, the British based Centre for Research on Socio-Cultural Change (CRESC) (2013) observes that governments preside over vast public expenditure and own significant infrastructure but both are overlooked because they appear banal. The 'foundational economy' (for example publicly funded and/or subsidized transport, energy, digital capacity, (air)ports, schools, hospitals, food markets and housing) is vital to the reproduction of everyday life and uniquely has a greater presence in socially economic deprived cities and towns. CRESC asks: how can anchor institutions help the foundational economy better support a new tradition of local economic development predicated upon community wealth building Meanwhile in her book *On Fire: The Burning Case for a Green New Deal*, Naomi Klein (2019) draws upon this logic to call for a Green New Deal. Inspired of Franklin D. Roosevelt's 1933 New Deal dedicated to lifting the United States economy out of deep depression, a Green New Deal could work to decarbonize the economy and in the process, create jobs, ensure fair wages are paid, secure adequate social insurance, guarantee high quality health care, protect paid vacations and furnish retirement security.

Taking a different tack, Australian and US human geographers J. K. Gibson-Graham (2006) have developed the idea of the postcapitalist economy – perhaps more accurately expressed as the diverse economy. Gibson-Graham critiques economic policy for being "capitalistocentric," referring to the tendency of governments and economic planners to see capitalism only and everywhere, and as a ubiquitous, inevitable, and all-conquering economic system. In fact, capitalism coexists with many other forms of economy. For example, an important "social economy" exists, built around economic actors with social, ecological, and ethical motives (charities, social enterprises, cooperatives, self-help initiatives, housing associations, etc.). Social enterprises often make profit (or surplus); they are part of the economy like any other business. But they reinvest this surplus in producing further social value rather than returning it to private shareholders. Meanwhile, in her book *Caring Democracy* (2013), US political scientist Joan Tronto argues that it is mainly women, the poor, or ethnic minorities that carry out the care work necessary for societies to function, and yet this work is rarely recognized or rewarded. Market economies further redistribute and reallocate care work by facilitating "privileged iresponsibility" – the tendency of the wealthy to pay others to do their care work. Moreover, liberal democracies and market economies hand out "charity passes" (charitable assistance) and "bootstrap passes" (help to kin) as an alternative to providing structural and systematic supports to care providers. Tronto calls for a new rapprochement between politics and care: care work should not provide the state with an opportunity to abrogate on its responsibilities; instead, care should be registered, rewarded, and supported.

Championing tech for public good

A scaling, and accelerating of the innovation and adoption of artificial intelligence (AI) and computerised data analytics has led many to claim the onset of a fourth industrial revolution. In his 2018 book *Big Mind*, British public policy scholar Geoff Mulgan draws attention to the potential of Collective Intelligence (hybrid socio-technical networks combining human and machine intelligence) to secure solutions to complex societal problems. But critical market failures (such as insufficient use of data, unaccountable corporate monopolies, unconscious bias and lack of transparency in algorithms, the robotisation of the workforce and large-scale job displacement, and digital inequalities, poverty and addiction) underscore the extent to which Collective Intelligence has already accelerated ahead of ethics, law, public policy, regulation, and governance. According to United States Business Studies scholar Shoshana Zuboff (2019) without a clear ethical, legal, and stable regulatory compact the risk is that a techno-dystopian and pernicious surveillance capitalism will emerge. More provocatively again, Australian-American scholar of media and culture McKenzie Wark argues that any fourth industrial revolution will not be a capitalist revolution; infact it will be something worse. 'Capital is dead' Wark declares, a tiny number of social media and digital platform organisations now control our data and rule the world as a neo-feudal overlord demanding mass serfdom. It is crucial that open data, ethical data governance (including citizen juries, data cooperatives, and civic data trusts), digital equality, 'tech for public good' and 'algor-ethics' wins out over a deep state and predatory corporate take-over of AI, digital infrastructure and data markets.

Becoming a better neighbor

In a speech to the World Economic Forum in Davos in 2019 (before the global Covid-19 pandemic) António Guterres, Secretary-General, United Nations stated: *"If I had to select one sentence to describe the state of the world, I would say we are in a world in which global challenges are more and more integrated, and the responses are more and more fragmented, and if this is not reversed, it's a recipe for disaster"* In contradistinction to those who believe that places have unique souls, innate identities, and monolithic cultures what makes places unique is infact the specific bundle of relationships they have with other places. Places always have permeable borders, multiple identities, and complex relations with other places. Doreen Massey observed that our world is a world of power-laden and asymmetrical relationships and connectivities between places; the story of Europe's subordination of the Rest. She referred to these hierarchical relationships as the 'power geometries of relational space'. She saw human geography as uniquely well placed to unearth the multiplicity and webs of connective lines and tissues which connect peoples and places across the world today and to promote thereafter peaceable and mutually enriching relations between peoples and places based upon principles of reciprocity and relationality rather than exploitation, dependency, and extraction (Massey 2004). For Massey obligations to be a good neighbor to the world develop unevenly over space because places are differentially imbricated in the fate and fortunes of other places. We need to think in terms of a global sense of place. What it means to act ethically toward those from "elsewhere" and to extend hospitality to the "stranger" in an unequal world varies from one place to the next because each place has different interconnections with different distant others.

So there you have it! My vision for a better West and for a human geography implicated in building back better. The floor is now yours.

Conclusion

Alas we now reach the end of our journey together. We are privileged to be living through an especially restless, turbulent, and volatile but equally exciting, fascinating, and potentially promising moment in human history. I hope you have found in these pages a concise and essential introduction to human geography that has provoked you to awaken, test, grow, refine, and gain discipline over your geographical imagination. You will by now have realized that to succeed as a human geographer, it is first necessary to develop as a Renaissance scholar; disposed always to think and act expansively, catholic in your interests, cosmopolitan in your outlook, intellectually curious, inquisitive, alert to the ennoblement that learning brings, and awake to the grand challenges facing humanity in the twenty-first century and what you might do about these challenges as a concerned and educated citizen. Human geography is one of the academy's pre-eminent lighthouse subjects; its gift is to furnish you with the bandwidth you will need to play your part in guiding a chastened human species to safer shores.

Checklist of Key Ideas

- To understand the story of the ascent, dominance, and stumbling of European-Anglo-American geography, it is necessary to understand how this story is intertwined with the story of the rise, reign, and faltering of the West from the fifteenth century.
- After 500 years of global hegemony, the West is faltering. Who controls the story of the West controls the future of the West. The West may be structurally compromised as a civilization; its survival may depend upon what might be termed "a mission civilisatrice for the metropole."

- Scholars who wish to develop a postcolonial human geography "without a center" are working to provincialize and decolonize Eurocentric knowledge formations. To prosper, any postcolonial human geography will need to establish itself as a "situated science," at once anti-foundational and anti-relativist in its approach to claims of objectivity and verisimilitude.
- European-Anglo-American human geography might become a better version of itself in the twenty-first century by supporting the West to become a better version of itself. Human geography will play an indispensable role if it works to help the West become a better home maker and better neighbor to the rest of the world. And so a worthy mission for a postcolonial human geography fit for purpose for the twenty-first century might be: for the advancement of geographical understanding and the ennoblement of human life - everywhere.

Chapter Essay Questions

1) Write an essay entitled "my geographical imagination." Include in this essay a commentary on the ways in which this book has helped you to (re)form your geographical imagination.
2) Write an essay entitled: My thoughts on the case for a postcolonial human geography.
3) European-Anglo-American human geography might become a better version of itself in the twenty-first century by supporting the West to become a better version of itself. Discuss.
4) Guided by the idea of "utopia for realists" sketch a picture of the human geography of the United States at 2050 that you would hope to see.

References and Guidance for Further Reading

Lionel Dupuy's studies examining author Jules Verne's geographical imagination include:

Dupuy, L. (2013). Jules Verne's extraordinary voyages, or the geographical novel of the 19th century. *Annales de Géographie* 690 (2): 131–150.

Recent works by Sarah Radcliff espousing a decolonizing agenda for geography include:

Radcliffe, S.A. (2017). Decolonising geographical knowledges. *Transactions of the Institute of British Geographers* 42 (3): 329–333.

Radcliffe, S.A. (2020). Geography and indigeneity III: co-articulation of colonialism and capitalism in indigeneity's economies. *Progress in Human Geography* 44 (2): 374–388.

Radcliffe, S.A. and Radhuber, I.M. (2020). The political geographies of D/decolonization: variegation and decolonial challenges of/in geography. *Political Geography* 78 (2020): 102–128.

Important works charting a new set of "postcolonial" concerns within human geography include:

Blunt, A. and McEwan, C. (2002). *Postcolonial Geography*. London: Continuum.

Chakrabarty, D. (2000). *Provincializing Europe: Postcolonial Thought and Historical Difference*. Princeton, NJ: Princeton University Press.

de Leeuw, S. and Hunt, S. (2018). Unsettling decolonizing geographies. *Geography Compass* 12 (7): e12376.

Ferretti, F. (2019). Rediscovering other geographical traditions. *Geography Compass* 13 (3): e12421.

Gregory, D. (2004). *The Colonial Present: Afghanistan. Palestine. Iraq*. Malden, MA: Wiley.

Jazeel, T. (2019). *Postcolonialism*. London: Routledge.

Jazeel, T. and Legg, S. (eds.) (2019). *Subaltern Geographies*. Athens: University of Georgia Press.

Jensen, L. (2020). *Postcolonial Europe*. London: Routledge.

McEwan, C. (2018). *Postcolonialism, Decoloniality and Development*. London: Routledge.

Mohan, G., Shurmer-Smith, P., and Stokke, K. (2020). *Postcolonial Geography*. Thousand Oaks, CA: Sage.

Noxolo, P., Raghuram, P., and Madge, C. (2012). Unsettling responsibility: postcolonial interventions. *Transactions of the Institute of British Geographers* 3 (37): 418–429.

Noxolo, P. (2017). Introduction: decolonising geographical knowledge in a colonised and re-colonising postcolonial world. *Area* 49 (3): 317–319.

Oswin, N. (2020). An other geography. *Dialogues in Human Geography* 10 (1): 9–18.

Roy, A. (2016). Who's afraid of postcolonial theory? *International Journal of Urban and Regional Research* 40 (1): 200–209.

Sharp, J. (2009). *Geographies of Postcolonialism: Spaces of Power and Representation*. London: Sage.

Sidaway, J.D. (2000). Postcolonial geographies: an exploratory essay. *Progress in Human Geography* 24: 573–594.

A neoconservative critique of the liberal-left "capture" of the university is:
Bloom, A. (1987). *The Closing of the American Mind: How Higher Education Has Failed Democracy and Impoverished the Souls of Today's Students*. New York: Simon and Schuster.

A left critique of the neoliberal and neoconservative "capture" of the university is:
Giroux, H.A. (2015). *University in Chains: Confronting the Military-Industrial-Academic Complex*. London: Routledge.

Good overviews of the sites from which critical human geographical scholarship has been produced:
Barnes, T.J. and Sheppard, E. (eds.) (2019). *Spatial Histories of Radical Geography: North America and Beyond*. Hoboken, NJ: Wiley.

Santos, B. (2016). *Epistemologies of the South: Justice Against Epistemicide*. London: Routledge.

The ongoing quest for equality, diversity, and inclusion (EDI) in the university is explored in:
Henry, F., Dua, E., James, C.E. et al. (2017). *The Equity Myth: Racialization and Indigeneity at Canadian Universities*. Vancouver: UBC Press.

An influential article setting out a case for anti-foundational but also anti-relativist knowledge is:
Haraway, D. (1988). Situated knowledges: the science question in feminism and the privilege of partial perspective. *Feminist Studies* 14: 575–599.

The importance of upholding standards in critical human geographical scholarship is discussed in:
Blomley, N. (2006). Uncritical critical geography? *Progress in Human Geography* 30 (1): 87–94.

Sultana, F. (2018). The false equivalence of academic freedom and free speech. *ACME: An International Journal for Critical Geographies* 17: 228–257.

The vast literature (popular and scholarly) that is attempting to reimagine and reconstitute the institutional architecture of the great Western liberal capitalist democracy experiment includes:
Bregman, R. (2016). *Utopia for Realists: The Case for a Universal Basic Income, Open Borders, and a 15-Hour Workweek*. London: Bloomsbury.

Bunnell, T. and Goh, D.P.S. (2018). *Urban Asias: Essays on Futurity Past and Present*. Berlin: Jovis.

Butler, J. (2016). *Frames of War: When Is Life Grievable?* London: Verso Books.

Centre for Research on Social and Cultural Change (CRESC) (2013). *Manifesto for the Foundational Economy*. Manchester: CRESC.

Cottam, H. (2018). *Radical Help: How We Can Remake the Relationships Between Us and Revolutionise the Welfare State*. London: Hachette.

Cumbers, A. (2020). *The Case for Economic Democracy*. Oxford: Wiley.

Dorling, D. (2020). *Slowdown: The End of the Great Acceleration and Why It's Good for the Planet, the Economy, and Our Lives*. New Haven, CT: Yale University Press.

Eatwell, R. and Goodwin, M. (2018). *National Populism: The Revolt Against Liberal Democracy*. London: Penguin.

Ferguson, N. (2014). *The Great Degeneration: How Institutions Decay and Economies Die*. New York: Penguin.

Fukuyama, F. (2006). *The End of History and the Last Man*. London: Simon & Schuster.

Fullerton, J. (2015). *Regenerative Capitalism: How Universal Principles and Patterns Will Shape Our New Economy*. New York: Capital Institute.

Geiger, S. and Gross, N. (2018). Market failures and market framings: can a market be transformed from the inside? *Organization Studies* 39 (10): 1357–1376.

Gleeson, B. (2010). *The Urban Condition*. London: Routledge.

Gibson-Graham, J.K. (2006). *A Post-capitalist Politics*. Minneapolis: University of Minnesota Press.

Graeber, D. (2018). *Bullshit Jobs: A Theory*. New York: Simon & Schuster.

Harvey, D. (2000). *Spaces of Hope*. Edinburgh: Edinburgh University Press.

Homer-Dixon, T. (2020). *Commanding Hope: The Power We Have To Renew A World In Peril*. London: Penguin.

Hunter-Lovins, L., Wallis, S., Wijkman, A., and Fullerton, J. (2018). *A Finer Future: Creating an Economy in Service to Life*. New York: New Society Publishers.

Kelton, S. (2020). *The Deficit Myth: Modern Monetary Theory and the Birth of the People's Economy*. London: Hachette.

Klein, N. (2019). *On Fire: The (Burning) Case for a Green New Deal*. London: Simon & Schuster.

Levitas, R. (2013). *Utopia as Method: The Imaginary Reconstitution of Society*. Amsterdam: Springer.

Mulgan, G. (2018). *Big Mind: How Collective Intelligence Can Change Our World*. Princeton, NJ: Princeton University Press.

Murray, D. (2017). *The Strange Death of Europe: Immigration, Identity, Islam*. London: Bloomsbury.

O'Leary, K.C. (2020). *Madison's Sorrow: Today's War on the Founders and America's Liberal Ideal*. New York: Pegasus Books.

Piketty, T. (2014). *Capital in the 21st Century*. Cambridge, MA: Harvard University Press.

Piketty, T. (2019). *Capital and Ideology*. Cambridge, MA: Harvard University Press.

Raworth, K. (2017). *Doughnut Economics: Seven Ways to Think Like a 21st-Century Economist*. London: Chelsea Green Publishing.

Rosling, H., Rosling, O., and Rosling, R.A. (2018). *Factfulness: Ten Reasons We're Wrong About the World – and Why Things Are Better Than You Think*. Stockholm: Flatiron Books.

Sachs, J. (2017). *Building the New American Economy: Smart, Fair, & Sustainable*. New York: Columbia University Press.

Sandel, M.J. (2013). *What Money Can't Buy: The Moral Limits of Markets*. New York: Penguin.

Sandbu, M. (2020). *The Economics of Belonging: A Radical Plan to Win Back the Left Behind and Achieve Prosperity for All*. Princeton, NJ: Princeton University Press.

Schwab, K. (2019). *Davos Manifesto 2020: The Universal Purpose of a Company in the Fourth Industrial Revolution*. Geneva: World Economic Forum.

Simms, A. (2013). *Cancel the Apocalypse: The New Path to Prosperity*. London: Hachette.

Standing, G. (2020). *Battling Eight Giants Basic Income Now*. London: Bloomsbury.

Stiglitz, J.E. (2019). *Measuring What Counts: The Global Movement for Well-Being*. New York: The New Press.

Tronto, J.C. (2013). *Caring Democracy: Markets, Equality, and Justice*. New York: University Press.

Vollrat, D. (2020). *Fully Grown: Why a Stagnant Economy Is a Sign of Success*. Chicago: University of Chicago Press.

Wark, M. (2019). *Capital Is Dead: Is This Something Worse?* London: Verso Books.

Wilkinson, R. and Pickett, K. (2009). *The Spirit Level: Why More Equal Societies Almost Always Do Better*. London: Penguin.
Zuboff, S. (2019). *The Age of Surveillance Capitalism: The Fight for a Human Future at the New Frontier of Power*. New York: Profile Books.

Doreen Massey's thinking on geographies of the economy, place, and responsibility is outlined in:
Hall, S., Massey, D., and Rustin, M. (2015). *After Neoliberalism? The Kilburn Manifesto*. London: Lawrence & Wishart.
Massey, D. (1991). A global sense of place. *Marxism Today* 38: 24–29.
Massey, D. (2004). Geographies of responsibility. *Geografiska Annaler Series B* 86: 5–18.

Website Support Material

A range of useful resources to support your reading of this chapter are available from the Wiley Human Geography: An Essential Introduction Companion Site http://www.wiley.com/go/boyle.

Plate C19.1 Sign of the times? Leonardo da Vinci's Mona Lisa wearing a protective mask. Source: Sumanley on Pixabay.

Coda on Covid-19

Coda Table of Contents

Coda Learning Objectives

By the end of this chapter you should be able to:
- With reference to cases and deaths (absolute and per capita), map Covid-19's geographies, noting the spatial diffusion of the pandemic from its point of origin (patient zero) – most likely in Wuhan, China, in Autumn/Winter 2019.
- Deploying the formula Risk = Hazard × Vulnerability (R = E × V) and disaggregating vulnerability in turn into susceptibility (prior existing openness to harm), adaptation (institutional capacity and degree of long-term planning to minimize the impacts of disaster events), and coping (capacity to manage and recover from a

Human Geography: An Essential Introduction, Second Edition. Mark Boyle.
© 2021 John Wiley & Sons Ltd. Published 2021 by John Wiley & Sons Ltd.
Companion website: www.wiley.com/go/boyle

> hazard event), discuss critically a range of explanations (twenty three are explored herein) that have been convened to account for Covid-19's geographies. Weighing these explanations, discuss the sense in which Covid-19 might be construed as a "barium meal" for the West.
> - Identify the implications of Covid-19 and the "build back better" agenda it has spawned for the project of remaking the West (see Table C19.2).

Covid-19: The West's Barium Meal?

On December 31, 2019, the Wuhan Municipal Health Commission reported to the World Health Organization (WHO) the existence of a cluster of pneumonia cases in Wuhan, Hubei Province, China. Eventually, the novel coronavirus (severe acute respiratory syndrome coronavirus 2, or SARS- nCoV-2 2019), a highly contagious airborne communicable disease with an estimated case fatality rate of 1–2%, was identified as the pathogen. By January 30, 2020, WHO reported that they were aware of a total of 7818 cases across 18 countries (but almost all in China) and 43 recorded deaths (again, almost all in China). That same day, WHO declared Covid-19 a Public Health Emergency of International Concern (PHEIC). By March 11, 2020, 118 000 cases had been recorded in 110 countries, resulting in over 19 000 deaths: on that day, WHO updated and upgraded Covid-19 to the status of a global pandemic. Events thereafter have served to validate this alert level. At the time of writing (January 7th 2021), nearly 88.1 million cases of Covid-19 had been reported in 213 countries, resulting in 1.9 million deaths (see Figures C19.1 and C19.2).

Covid-19 entered the fray as this book was being prepared for final publication. It is clear that the pandemic is rapidly becoming an event of world historical consequence. Covid-19 has added to the catalog of problems that humanity was grappling with up until yesterday – which were and remain momentous in themselves – a fresh existential threat without historical precedent. Rarely before has humankind in its entirety – everyone and everywhere – been confronted with a public health crisis of this magnitude. Rarely has the world's population as a whole (everyone, 7.7 billion, without exception) been so comprehensively locked down and had their everyday lives so tumultuously disrupted. Rarely before has every politico-economic-institutional model been confronted by a common enemy and had its efficacy so severely tested. Rarely before have all global citizens been asked to embark upon a whole-of-society, whole-of-government, and whole-of-the-world project of rebuilding for a more resilient tomorrow. And never before have all four combined into a perfect storm at a given historical moment.

A story about Covid-19's geographies – which cuts to the heart of the organizing argument underpinning this book – is fomenting.

The formula Risk = Hazard × Vulnerability (R = H×V) stands as an article of faith within hazards research. On this basis, at the outset of the pandemic, commentators feared that it would be poorer countries in the Global South who would be most at risk. These places suffer from greater vulnerability and are less able to withstand shock due to their underlying socio-structural weaknesses, weaker institutional capacities, limited disaster risk management infrastructures, and impoverished leadership. It has come as something of a surprise, then, that to date the pandemic has exacted its heaviest toll on the superpowers of the United States and Europe, the oil-rich economies of the Arabian

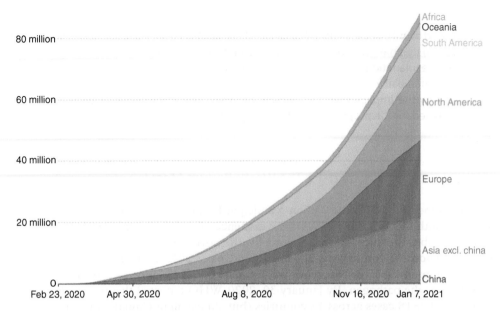

Figure C19.1   Total confirmed Covid-19 cases. The number of confirmed cases is lower than the number of total cases. The main reason for this is limited testing. Source: European CDC, "Situation Update Worldwide," January 7, 2021, http://OurWorldInData.org/coronavirus.

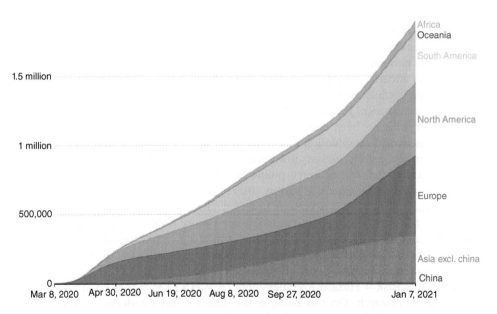

Figure C19.2   Total confirmed Covid-19 deaths. Limited testing and challenges in the attribution of the cause of death mean that the number of confirmed deaths may not be an accurate count of the true number of deaths from Covid-19. Source: European CDC, "Situation Update Worldwide," January 7, 2021, http://OurWorldInData.org/coronavirus.

Peninsula, and middle-income and emerging economies. Although SARS- nCoV-2 2019 originated in China, (Sinic) East Asia appears to have suppressed the pandemic much more effectively than North America and Europe, who have lurched from one crisis to the next, seemingly at a loss to know how to contain the virus and reopen the economy safely. Moreover, at least so far, the poorest countries in the world and especially those on the African continent appear to have escaped relatively unscathed.

How can we explain this unexpected turn of events? What significance might be attached to the fact that the Organisation for Economic Co-operation and Development (OECD) world appears to be bearing the brunt of Covid-19 and its aftershocks? Does it signal anything fundamental about the underlying health of the West's politico-economic-institutional model?

The European relationship with pandemics has been a complex one. In 1348, approximately one-third of Europe's population died from the Black Death, but European civilization went on to enjoy 500 years of unrivaled prosperity. Indeed, later, smallpox helped the Spanish conquer the Aztec and Incan Empires. And in the late nineteenth and early twentieth centuries, epidemics triggered a public health revolution in European cities, paving the way for a century of historically unprecedented levels of health. Nevertheless, pandemics of plague, smallpox, cholera, influenza, polio, measles, malaria, and typhus have played a crucial role in shaping the fortunes and fate of many past human civilizations. It is not altogether fanciful to suppose that Covid-19 could in time join the pantheon of episodes when communicable diseases have served as hinge points in the unfolding of Big History. Covid-19 could be a barium meal of sorts for the West, flashing red its vulnerabilities and surfacing ongoing weaknesses and deficiencies in its already-faltering civilizational model. Perhaps it might even prove in the end to be one crisis too many, the nudge that will finally topple the house of cards. In Covid-19, we could be witnessing a speeding-up of the collapse of the past.

Certainly, on the basis of this logic, gaining control over the story of Covid-19 and its geographies is becoming a decisive battleground in culture wars over the status of the claim that "West is best." Covid-19 braggadocio is assuming crucial geopolitical significance. Competency in responding effectively to the pandemic has become a *de facto* barometer of the vitality of rival hegemons, signaling who is rising and who is falling. As judged by the geopolitical conflicts that are brewing over the origins of SARS- nCoV-2 2019 and who might be responsible for its diffusion from Wuhan; the theatrical performances enacted in government-orchestrated daily press "progress" briefings; the positions on Covid-19 adopted by media moguls, pioneering new cartographies and visualizations of Covid-19 geographies; and nationalistic charged depictions of the race to invent a safe and effective vaccine, Covid-19 is as much an infodemic as it is a pandemic. Hijacked by social media warriors, truth has fallen prey to a surging industry of conspiracy theories and lies.

But what truth lies beneath the propaganda wars that are now frothing? What are we to make of the fact that against all prior assumptions and notwithstanding their obvious wealth, education, and health and health care advantages, Covid-19 is troubling advanced OECD societies and middle-income and emerging economies more so than the world's poorest economies?

In searching for an answer to this question, we must proceed with caution:

Data problems: Immediately, we must note that the mission to track and map Covid-19 is compromised, perhaps irrevocably, by severe data limitations. WHO, the Johns Hopkins Coronavirus Resource Center (CRC), the US Centers for Disease Control and Prevention (CDC), and the European Centre for Disease Prevention and Control (ECDC) are the most authoritative sources of up-to-date data on Covid-19 cases and deaths. But each data steward warns that because reporting criteria and testing capacity vary between countries, it is highly likely that countries with poor data infrastructures or particularly politicized census offices will be underreporting the extent to which they have been affected by the outbreak. It is entirely possible that commentaries on Covid-19 geographies may end up erroneously reporting and ruminating over spatial disparities that are little more than an artifact of variable data collection and publishing practices.

A hazard event unfolding in real time: Acute problems attend to the analysis of what is a dynamic and rapidly evolving situation. What needs emphasizing is that this Coda has been written while the pandemic has been unfolding. Ours is a restless, dynamic, and incomplete subject matter; the stocktaking, interpretations, and commentaries offered herein are necessarily tentative, provisional, and conjectural. It will fall to you to update and, hopefully at some point in the not-too-distant future, complete the story of Covid-19 and its uneven geographies. Only when the pandemic has been permanently suppressed will we be in a position to reach a final conclusion on the meaning and implications of its geographies for the future of the West.

These qualifications noted, the world historical import of the Covid-19 pandemic dictates that such analysis be undertaken, even if the conclusions reached are necessarily qualified by data concerns and even at the peril of it being overtaken by events and potentially rendered obsolete.

Covid-19's Unexpected Geographies

Maps C19.1 and C19.2 provide a profile of total and per million Covid-19 cases and deaths for countries as of January 7, 2021.

At least to date, Covid-19 appears to have been more of a communicable disease of the OECD world than one of the Global South. It is truly astonishing to witness wealthy countries with strong institutional capacity that hitherto had been understood to have reached the finishing line in the epidemiological transition – the United States and Canada, Europe's most developed countries (the United Kingdom, Spain, Italy, Sweden, Belgium, France, and Germany), and Israel – consistently rank at the top of the league table of the world's most impacted states, in terms of both absolute cases and deaths and cases and deaths per million. There can be no denying that the West has shown itself to be more vulnerable to the global pandemic than many other regions of the world.

But equally, there are other stories at play that cannot be ignored.

- **Differences exist within the OECD world**. Some Western countries have done better than others (in particular Japan, New Zealand, Australia, South Korea, Greece, and the Nordic countries of Denmark, Norway, and Finland (Sweden is the exception, having adopted a high-risk "herd immunity" strategy). Meanwhile, when compared to their wealthier Western European counterparts, the transition and emerging economies of Eastern Europe (especially Lithuania, Croatia, Slovenia, Latvia, Estonia, Georgia, Poland, and Belarus) have secured favorable outcomes.
- **The West has not been alone in facing the worst of the storm**. The virus has exacted a very heavy toll in Latin America, historically the richest region within the developing world, impacting in particular Peru, Brazil, Mexico, Colombia, Chile, Argentina, Panama, and Bolivia. Moreover, it has burdened countries of significant standing (middle-income and emerging economies) in other continents, in particular Iran, India, South Africa, and Russia. In addition, when expressed in per capita terms,

it is clear that the oil-rich countries of the Arabian Peninsula constitute an additional hub for Covid-19 cases and deaths (especially Qatar, Oman, Kuwait, and Saudi Arabia).

- **(Sinic) East Asian exceptionalism**. Although initially at the epicenter of the pandemic, (Sinic) East Asia (China, Japan, South Korea, Vietnam, Thailand, Taiwan, and Malaysia) has managed to suppress Covid-19 particularly effectively. Given that the countries straddling this region encompass a wide range of politico-economic-institution models, the implication may be that a specifically "Asian" cultural factor is at work.
- **The surprising African story**. South Africa (and to a much lesser extent Libya) presents as an exception in the wider African narrative, being among the most severely impacted countries in the world. Otherwise, it comes as a welcome surprise that, against all odds, it is the countries that hitherto have been perceived to be especially vulnerable to communicable disease – sub-Saharan African countries (including the very poor and very populous countries of the Democratic Republic of Congo, Malawi, and Nigeria) – which at least to this point have escaped the worst.

Explaining Covid-19's Geographies: Risk = Hazard × Vulnerability

Let us now make use of the formula Risk = Hazard × Vulnerability (R = E × V), disaggregating vulnerability in turn into susceptibility (prior existing openness to harm), adaptation (degree of long-term planning to minimize the impacts of disaster events) and coping (capacity to manage, and recover from a hazard event), to identify and test a variety of possible explanations (twenty three in total are considered here) for Covid-19's surprising geographies. We will ask, in what sense can advanced Western OECD countries be said to be especially *exposed* to the SARS- nCoV-2 2019 virus and especially *vulnerable* to being harmed by Covid-19.

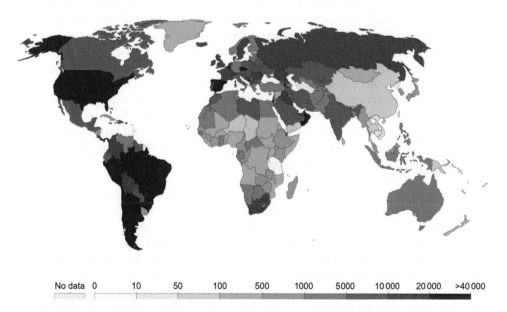

No data 0   10   50   100   500   1000   5000   10 000   20 000   >40 000

Map C19.1   Total confirmed Covid-19 cases per million people, January 7th 2021. The number of confirmed cases is lower than the number of actual cases. The main reason for this is limited testing. Source: European CDC, "Situation Update Worldwide," January 7th 2021, http://OurWorldInData.org/coronavirus.

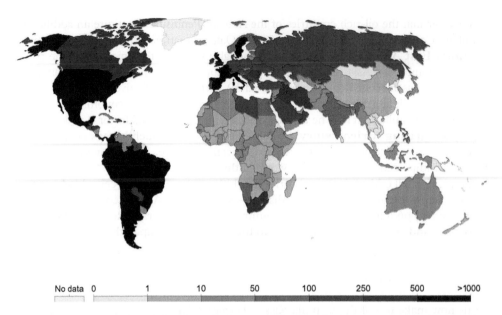

No data 0   1   10   50   100   250   500   >1000

Map C19.2   Total confirmed Covid-19 deaths per million people, January 7th 2021. The number of confirmed cases is lower than the number of actual cases. The main reason for this is limited testing. Source: European CDC, "Situation Update Worldwide," January 7th 2021, http:// OurWorldInData.org/coronavirus.

Exposure to Covid-19: Uneven Geographies of Viral Load

Perhaps Covid-19's geographies simply reflect uneven geographies of exposure to the hazard that is SARS- nCoV-2 2019. Unlike earthquakes or volcanoes or tropical storms, which are relatively easy to locate, infectious diseases spread through time and present as a threat to a wider range of populations. One must not forget that (it is most likely that) the virus has zoonotic origins (conventional wisdom holds that it is a bat-borne virus transmitted to humans possibly through an intermediate animal reservoir such as the pangolin community), and like other viruses is engaged in a (Darwinian) struggle for survival. It originated in a specific location and has spread thereafter to other locations in the world as a by-product of its proclivity to propagate. Of course, alongside epidemiological progenitors, social, economic, political, and cultural factors have combined to determine the spatial diffusion of SARS- nCoV-2 2019 and its directions of travel. Its propagation depends on its capacity to jump on the coattails of socially produced dynamics of human mobility, interaction, and crowding. But the important point is that not every place has found itself on the hurricane track or in the eye of the storm; not every place has the same – to coin a phrase – *geographical viral load* (proportion of the population exposed to particularly virulent strains of the virus).

Explanation 1: Societies whose position in the world economy demands that they function as critical nodes and hubs in global flows of people will be exposed to a greater number and variety of corridors of transmission. OECD countries are at the heart of the global economy; 12 are ranked in the top 20 most globalized economies (according to the DHL Connectedness Index). Geographical centrality may be playing a role by magnifying exposure to SARS- nCoV-2 2019 in countries such as the United States, the United Kingdom, the Netherlands, Belgium, and Switzerland, while economic peripherality and geographical isolation may be reducing outbreaks in some African countries and in Australia and New Zealand. But Singapore, the second most globalized economy in the world, and the globally connected economies of Taiwan, Israel, and South Korea all have substantially lower death rates than the much less globalized countries of Peru, Chile, Mexico, and Colombia.

Explanation 2: Given its level of contagiousness, Covid-19 will thrive in countries with higher population densities and less space per capita. Perhaps SARS-nCoV-2 2019 is more likely to spread in countries with higher population densities (defined by the UN World Population Division as people per square kilometer). Certainly, the relative "emptiness" and "spaciousness" of Australia, New Zealand, the African continent, and the Nordic countries have been cited as possible explanations for their lower numbers of cases and deaths, while the higher population densities that prevail in India, the United Kingdom, and Italy have been invoked as a contributor to these countries' poorer outcomes. But other countries with very low populations densities, including the vast majority of states in Latin America, have presided over very significant outbreaks, while some countries with very high population densities such as China, Japan, South Korea, and Nigeria have managed to avoid the worst of the pandemic. Population density per se does not appear to be a strong determinant of the scale of Covid-19's propagation.

Explanation 3: Given the contagiousness of Covid-19, urban density is the enemy of public health; therefore, Covid-19 will thrive in more urbanized countries. Globally, OECD and Latin American countries are among the most urbanized (as documented in the UN World Urbanization Prospects reports). Recognizing that large urban agglomerations represent a perfect petri dish for Covid-19 contagion, cities are working to adapt to the new realities. In both cases, Covid-19 proofing of cities is in its infancy, and agglomeration is likely to be a significant progenitor of elevated transmission. But (Sinic) East Asia has suppressed the virus in spite of the fact that it is a rapidly urbanizing global region and home to the largest number of the largest cities in the world. Moreover, the problem of containing the virus in Global South cities is intensified by a dangerous mix of severe overcrowding, poor sanitation, economic precarity, and poverty. Indeed, urbanization in the Global South has birthed vast overcrowded informal housing slums, places where virus containment is especially challenging. The virus is more likely to prosper in more urbanized societies, but it seems unlikely that urban geography is disproportionately heightening the vulnerability of OECD relative to Global South countries.

Explanation 4: Climatic cycles have increased the intensity of outbreaks of Covid-19 in the Global North. Some epidemiologists claim that Covid-19 spreads more easily in colder climates, and climatic factors have shaped the geography of its contagion. Covid-19 struck the northern hemisphere during its winter; cases and death rates were brought under a degree of control only with the arrival of summer. Meanwhile, as Australia and New Zealand moved into their winter, cases and deaths there (to a degree) increased. Furthermore, the Global North has witnessed second and third waves as it has lurched back to winter whilst rates of infection in Oceania have declined with the arrival of summer. But this claim seems at odds with the intensity of Covid-19 outbreaks in equatorial and desert zone countries and regions, such as Brazil (and Amazonian countries more broadly), southern India, Mexico, the southern United States (especially Florida), and the Middle East.

Explanation 5: Uneven geographies of Covid-19 reflect mutations in SARS-nCoV-2 2019. There is a claim that, after entering Europe in January–February 2020, SARS- nCoV-2 2019 mutated and became more potent. This more infectious and lethal strain swept across Europe and was then exported to the United States, where likewise it spread with heightened lethality. Moreover subsequent mutations have led to more contagious South African and British strains which have circulated speedily across the West. Uneven geographies of Covid-19 cases and deaths simply reflect uneven geographies of pathogen mutation, transportation, and potency. But within Europe and North America, there is a marked variation between countries and regions as measured by case, death, and case fatality rates, and death rates (although not necessarily absolute deaths) have fallen in most countries and regions from their peak in May–June 2020: in combination, these realities imply that something other than pathogen virility must be at work.

Explanation 6: Uneven geographies of Covid-19 reflect historical and racial differences in immunity to SARS(-like) viruses. A claim has emerged that SARS-nCoV-2 2019 is part of a family of viruses that have circulated historically in Asia but not (or less so) in Europe and North and Latin America, resulting in varying degrees of immunity and vulnerability. Nevertheless, Anglo-Saxon countries have witnessed different outcomes: for example, the United Kingdom and the United States have been highly impacted, Canada has been impacted but to a lesser degree, and Australia and New Zealand have been comparatively lightly impacted. Moreover, as noted, Eastern European transition populations have avoided the levels of cases and deaths witnessed in Western European countries, in spite of their close racial heritage.

Vulnerability to Covid-19: Susceptibility, Adaptation, and Coping

Although variegated exposure to hazards remains a crucial risk factor, increasingly it is recognized that it is primarily social, economic, and political forces that turn natural hazards into disasters and disasters into catastrophes. Covid-19 is less a freak of nature or act of god and more a socially produced hazard event – that has been made to be a disaster and increasingly a catastrophe only in certain places.

In his book *The Covid-19 Catastrophe: What's Gone Wrong and How to Stop It Happening Again*, editor-in-chief of leading medical journal *The Lancet* Richard Horton (2020) argues that the elevated impact of Covid-19 in the United Kingdom and the United States reflects at root government incompetence. Curiously, some populist governments have presided over particularly poor outcomes: Trump in the United States, Johnson in the United Kingdom, Bolsonaro in Brazil, Modi in India, and Erdoğan in Turkey. For too long, the pandemic was written off as no worse than the flu; supplies of Personal Protective Equipment (PPE) for frontline health workers and ventilators for patients were in short supply; test, track, and trace procedures were introduced too late and remain inadequate; poorly conceptualized ideas of "herd immunity" guided responses; international air travel continued; and mask wearing was optional. Horton argues that it is erroneous to imply that Covid-19 was unexpected: in fact, epidemiologists have been warning governments for years about the imminent threat of airborne communicable disease. For Horton, Covid-19 stands as the greatest science-policy failure in a generation.

As we will see, government mismanagement has undoubtedly played its part in shaping geographical variations in vulnerability to Covid-19 and its aftershocks. But there is more to this story than simple incompetence.

Vulnerability Wrought by Socio-structural Disadvantages
and Heightened Susceptibility

Explanation 7: Given their aging demographic structures, North American and European populations have been more vulnerable to Covid-19. OECD countries have an older population profile (as measured by median age and the percentage of the population over 65 years of age) and are therefore at greater risk from Covid-19. Italy has the second oldest population in the world; it is little surprise, then, that it has been particularly challenged by Covid-19. But Japan, with the oldest population profile in the world, has evaded the worst of the pandemic, while Mexico's youthful population structure has not afforded it significant protection.

Explanation 8: Because Covid-19-linked comorbidities vary between populations, so too there exists an uneven geography of vulnerability to Covid-19. Generally, people in OECD countries enjoy longer and healthier lives (they have much higher life expectancies and higher Health-Adjusted Life Expectancies, or HALEs) and ought to be at reduced and not elevated risk. But years of life lost to disease (YLDs) among people over 70 years of age living in both OECD and non-OECD countries appear to be broadly similar, giving neither an obvious advantage. Moreover, specific Covid-19-linked comorbidities are consistently present at high levels in OECD countries, placing them at a disadvantage. Within the OECD and beyond, countries particularly burdened by Covid-19-linked comorbidities (especially diabetes; kidney, chronic respiratory, and cardiovascular disease; obesity; and asthma) do tend to suffer from elevated Covid-19 case and death rates. Australia, Greece, New Zealand, Israel, China, and Taiwan, however, present as exceptions, performing better than the prevalence of comorbidities might imply.

Explanation 9: The uneven impact of Covid-19 is rooted in growing socio-spatial income and wealth inequalities. Covid-19 geographies cannot be explained by poverty (UN HDI scores, UNU World at Risk vulnerability scores, or World Bank Poverty Rates); a majority of the poorest countries in the Global South have had considerably fewer cases and deaths than the richest countries within the Global North. But Covid-19 has preyed on growing "within-country" inequalities (as measured by the Palma ratio and Gini coefficient), disproportionately impacting poor and Black, Asian, and Minority Ethnic (BAME) communities (those already furthest from the labor market, living in high-population-density neighborhoods, suffering digital poverty, engaged in the most hazardous occupations, and most reliant upon public transport). Countries with the highest levels of inequality in the OECD (the United States, the United Kingdom, Chile, and Colombia) and in the Global South (across the whole of Latin America, India, Russia, and South Africa) have endured the worst of the pandemic. Counter-examples, however, are to be found in (Sinic) East Asia (especially China and South Korea) and elsewhere (New Zealand), where growing inequalities have not tempered efforts to suppress Covid-19. Meanwhile, a significant number of African countries have avoided the worst of Covid-19, notwithstanding their very high levels of inequality.

Explanation 10: The uneven impact of Covid-19 is rooted in the demise of social cohesion; countries where social capital, solidarity, mutuality, and reciprocity have been eroded and depleted most will suffer disproportionate harm. Historically, the work that place-based communities did in protecting vulnerable people in times of need and strengthening resilience for all was critical. Alas, forty years of neoliberalism and a decade of austerity appear to have dismantled the complex webs of connective lines and tissues braided within and between (place-based) communities at myriad scales. Much has been made of the decline of community in Western societies in particular: the erosion of trust between people and depletion of social capital (a finding repeatedly confirmed by the World Values Survey (WVS) series). Responsibility for the provision of social care has shifted from families and communities to governments. When enacted without a sense of solidarity, such "caring from a distance" can be overly transactional and can present as impoverished, disembodied, cool, and detached care. Countries that have both preserved strong communities and endured Covid-19 better include the Nordic countries, Australia, China, Vietnam, and New Zealand. In contrast, countries that have both witnessed a decline in social trust and been especially vulnerable to Covid-19 include the United States, the United Kingdom, France, Spain, and the majority of Latin American states. But it is asserted that African countries are bereft of social capital and yet most have weathered the pandemic better than a number of countries that report relatively healthy stocks of social capital, such as India, Russia, South Africa, Saudi Arabia, and to a degree Canada and Germany. Alternatively, perhaps it is changing levels rather than absolute levels of social trust and community solidarity that do most damage.

Explanation 11: Covid 19's geographies reflect socio-economic differences in immunity to SARS (like) viruses. Notwithstanding variations in the quality of data, there exists a claim that case fatality rates and infection fatality rates have been higher in high income countries and lower in low income countries. A so called 'hygiene hypothesis' posits that people in poorer countries have developed strong immune systems by dint of their life circumstances (poverty, overcrowded housing, slum housing, poor sanitation, lack of access to clean water, work as waste pickers and so on). Microbiomes (bacteria, viruses, fungi and single-celled archaea microbiome) inside human bodies play a vital role in human health improving digestion, repelling disease-causing bacteria, regulating people's immune systems and producing vitamins. Whilst potentially harmful, 'gram-negative bacteria' in particular can produce an antiviral cytokine - molecules which help suppress pathogens - called interferon which shield cells against SARS- nCoV-2 2019. Global North countries in contrast, by sterilising their environments have effectively starved the immune systems of their populations of the training required to fortify personal resilience. Whilst the hygiene hypothesis has been deployed to account for lower death rates in India, its status remains undetermined. Arguably, people living in Europe's slums in the nineteenth and twentieth centuries found themselves immunocompromised, not immunocompetent. Moreover whilst bacterial infection cannot be compared with viral infections, it is becoming clearer that both interact, and that antimicrobial resistance can lead to secondary infections in Covid-19 patients, aggravating immunocompromisation. But whilst of concern, the United States and Europe have much lower AMR mortality rates than Africa and South-East Asia.

Explanation 12: Authoritarian regimes have been more able to mobilize and give effect to stringent public health controls than democratic governments. Authoritarian governance models go some way to explaining China and Vietnam's success in suppressing the virus. To be as effective, democracies need to command a very high level of public confidence; surveys on trust (such as those undertaken by the OECD, Edelman Trust, and WVS) suggest that the democratic governments are failing miserably to inspire trust and loyalty from citizens. A schism between representative democratic institutions and popular sovereignty has almost certainly mitigated against the capacity of democratic polities to implement stringent containment measures. Asian democracies such as South Korea, Taiwan, and Japan are exceptions to the rule;

enhanced reverence to the state and citizen attitudes toward authority and compliance may explain the better performance of these polities. Meanwhile, countries like Australia and New Zealand have demonstrated a capacity to gain a democratic license to limit personal liberties at least for a period.

Explanation 13: Centralized political systems that govern regions and cities from a distance preside over poorer outcomes than federalized states with decentralized or devolved powers and bespoke localized responses. Some political geographers have hypothesized that the proximity of governments and their disaster risk management institutions and infrastructures to the people they serve has been an important determinant of the efficacy of responses. Larger countries and countries with highly centralized government structures and powers are more likely to suffer significant outbreaks and elevated harm. By comparison, smaller countries and countries with more decentralized federal structures and more powerful regional and local authorities are better placed to understand local nuances, mobilize local assets, and suppress the virus locally. Certainly, some countries with enhanced capacity to design and implement localized bespoke place-based responses have done comparatively well, for example Germany and Australia. But many have not, including Belgium, India, Brazil, Argentina, Spain, the United States, and South Africa. Likewise, while some highly centralized countries have presided over poor outcomes (for example, the United Kingdom and Ireland), others have enjoyed relatively better results (such as China, Greece, and South Korea). Perhaps, then, it is not the degree of centralization or decentralization that matters but the extent of coordination, the balance struck between centralized and decentralized strategies, and the unlocking of local leadership, assets, and capacities.

Vulnerability Wrought by Weak Institutional Capacity for Advanced Adaptation and Preparation

Explanation 14: Countries with institutional capacity to give effect to disaster risk reduction plans and with effective co-ordinated emergency management have escaped the worst of Covid-19. The United Nations University's (UNU) World Risk Index identifies countries have robust disaster management capacity; they include many countries which have borne the brunt of Covid-19 (for example Spain, Belgium, Germany, Israel, Sweden, the United Arab Emirates, Bahrain, Saudi Arabia, Ireland, Italy, United Kingdom, and the United States), Conversely, there exist countries with poor hazard management infrastructure which ought to amplify the effects of hazards (they include Afghanistan, Nigeria, Malawi, Bangladesh, Mozambique, Tanzania, Liberia, Eritrea, Vietnam, and the Democratic Republic of Congo) but which have avoided Covid-19 catastrophe. Surprisingly, institutional capacity and disaster management capability have not shaped Covid-19 geographies.

Explanation 15: Countries with well-established and high performing medical and public health services will be better able to suppress the Covid-19 pandemic; those with inadequate health care systems will suffer most. According to the WHO Global Health Care Index, the overall quality of any health care system is a product of its health care infrastructure; the number and competencies of health care professionals (doctors, nursing staff, and other health workers); the cost of accessing health care; the quality of the medicines available, and government readiness. Poor institutional capacity in Russia, South Africa, Brazil, and Peru and to a degree the United States – helps to account for the poor outcomes endured by these countries whilst the better outcomes witnessed in Australia, New Zealand, South Korea, Japan, Taiwan, and the Nordic countries coincide with their high performing heath care infrastructures. But countries with exceptional health care systems, such as Italy, Spain, France, the United Kingdom, and to a degree India and the United Arab Emirates, have found themselves surprisingly incapacitated, while countries with weak health care systems such as China, Vietnam, and Nigeria have exceeded expectations.

Explanation 16: Health care and public health systems in Western OECD countries are designed to remediate degenerative disease and lack the institutional capacity and disaster risk management infrastructure needed to tackle airborne infectious disease. For too long, Global North governments have assumed that they have, by and large, completed the epidemiological transition, and have built capacity primarily around degenerative disease (such as cancers, strokes, and cardiovascular disease). These systems are ill equipped to respond to (re)emerging infectious disease and disease outbreaks. After waves of epidemic outbreaks in rapidly growing cities during the nineteenth century, public health concerns and the containment of epidemics became a key informant for housing provision, urban planning, and urban design. The tight connections between medical and epidemiological concerns, on the one hand, and housing provision and urban planning, on the other hand – once so prominent – are now lost. Likewise, health care and medical resources and expertise have pivoted from remediating infectious diseases to prolonging life to its biological maximum by delaying the onset of degenerative disease. The result has been that Western countries have lacked the infrastructure necessary to tackle Covid-19 and have had to (re)learn how to handle a pandemic – and especially one caused by the transmission of an airborne virus – in real time.

Explanation 17: Societies with more experience in handling communicable disease and disease outbreaks have responded more effectively. Relatedly, even if the required systems were present, Global North health care systems lack recent experience in handling pandemics. In contrast, China, Korea, Hong Kong, and Taiwan have had recent experience with serious acute respiratory syndrome (SARS) (2002–2003) and Middle East respiratory syndrome (MERS) (2012), and West Africa with Ebola (2013–2016). Declining case fatality rates in the Global North signal the growing competence of OECD health care systems from a low base in January 2020 and confirm the importance of system memory and learning. But MERS impacted many OECD countries, and many OECD countries were exposed to the swine flu pandemic (2009–2010), suggesting that both past experience and a willingness to learn from past experience may be of equal importance.

Explanation 18: Government policies toward Bacillus Calmette-Guérin (BCG) help to explain Covid-19 geographies. It has been hypothesized that the BCG vaccination for tuberculosis provides a degree of spillover protection against Covid-19. Many OECD countries have discontinued or never had mandatory and universal vaccination programs, while the vast majority of non-OECD countries continue to administer mass vaccination. People living in OECD countries may, then, lack a degree of immunity. This could help explain why Eastern European countries have been less impacted than Western European countries and why Spain has had significantly more cases and deaths per capita than Portugal.

Explanation 19: Transition to a service-based economy and offshoring of manufacturing alongside private ownership of the means of production have reduced industrial capacity in OECD countries and increased the

difficulty of speedily pivoting factories toward the production of virus-related products. Massive manufacturing capacity in (Sinic) East Asia has enabled the rapid production of virus-related products. The prevalence of state-owned enterprises, in China in particular, has enabled a rapid shift in production to virus-related products. And so China, South Korea, Taiwan, and Vietnam have become net exporters of these products. In contrast, in North America and Europe (notwithstanding enhanced state powers in times of emergency), a paucity of capacity combined with private ownership of manufacturing plants have mitigated against such agility and resulted in endemic equipment shortages (and, in particular, shortages of personal protective equipment for frontline health workers). A lack of manufacturing capacity is now hindering the production of vaccine doses at speed and scale.

Vulnerability Wrought by Poor Coping and the Speed, Quality, and Efficacy of Government Responses

Explanation 20: Covid-19 geographies arise from variations in the efficacy of government responses; those that have gone hard and gone early have enjoyed greater success in the suppression of the virus. In spite of their exceptional resources, infrastructures, and capacities, European and North American countries have failed to match their much poorer counterparts in terms of proactive and proportionate remediating actions. According to the Oxford Covid-19 Government Response Tracker (OxCGRT), at the peak of the first wave of the pandemic (May 1, 2020), most OECD countries scored relatively poorly on "containment and health" measures and especially on "stringency of lockdown" measures; their overall performance was rescued to some degree only by much stronger economic and social insurance supports. While they have improved their overall positive test rates (tests that returned positive, indicating the presence of the disease), at the peak of the first wave, many OECD countries exceeded the WHO recommended rate of <5%. Indeed, at that point, countries like the Netherlands, the United Kingdom, Italy, and the United States had higher positive test rates than Vietnam, Pakistan, Iran, and South Africa. Of the 3Ts (testing, tracing, and treatment), OECD countries have demonstrated strength only in treatment, but even here they have suffered from equipment shortages. Democratic East Asian nations are best in class. Taiwan is the gold standard for contact tracing, South Korea has set the global benchmark for testing methodologies, and Japan is leading the world in the treatment for pneumonia. Beyond these democratic polities, China and Vietnam, meanwhile, have presided over very stringent but effective lockdowns.

Explanation 21: The extent of Covid-19 deaths in long-term-care homes (LTCHs) in European and North American countries points to their moral failure to protect vulnerable elderly groups. In OECD countries, at the peak of the first wave (May 2020), Covid-19-linked deaths in LTCHs or of care home residents constituted 50% or more of all Covid-19 deaths, with Canada, Belgium, and Ireland recording the most concerning levels. Recognizing the death of LTCH residents as one of the great "policy disasters" of the pandemic, WHO Director-General Tedros Adhanom Ghebreyesus has called attention to the moral bankruptcy that appears to have led some governments to deprioritize the elderly. There is insufficient data to assess the plight of care home residents in non-OECD countries; this does not stop us from concluding that neglect of the elderly and a government failure in their duty of care to those relying upon state protection as they enter the final years of their lives have played a central role in the high rates of Covid-19 cases and deaths recorded in the West.

Explanation 22: Weakened by populist governments, the West has failed to show global leadership, and this failure to step up has boomeranged back and caused self-harm. The United States' America First policy and its drift toward populism and isolationism have, rightly or wrongly, deprived the world of global leadership, partnership, and cooperation at a critical moment. The United Nations and WHO have lacked the support necessary to step into the void. In the absence of US stewardship, China is posturing as a global leader in waiting. But tensions between the United States and China have threatened to escalate into a trade war, and neither are likely to emerge from the pandemic especially emboldened or in a way that will shift the balance of world power. Countries cannot tackle global problems with local actions; coordinated global solutions are required for hazards brewing elsewhere today that will inexorably become hazards for us tomorrow. By withdrawing from the international community and persisting with isolationism, the United States has rendered itself more vulnerable. Of course with the election of President Joe Biden this might change.

Explanation 23: As we reach the end of the pandemic cycle, new Covid-19 geographies will emerge as a reflection of the ownership and distribution of safe and effective vaccines. Remarkably, vaccines are now being approved (on emergency license) and deployed and there is hope that 2021 will witness a decline in the Covid-19 threat. Fueled in part by Operation Warp Speed, the United States is leading the way- in addition to already approved vaccines it has by far the greatest number of candidate vaccines and active trials; the United Kingdom has made significant investments in vaccine development and Canada, Australia, New Zealand, and South Korea are actively testing other variants of vaccines. Overall, the West enjoys a decisive competitive advantage. China, India, and Russia are the only non-Western countries to sponsor significant searches for a vaccine. Vaccine development could enable the West to recover lost ground in an instant, but we should note that a central tenet the Sendai Framework for Disaster Risk management (2015–2030) is that societies which seek to engineer themselves out of trouble rarely build resilience over the longue durée. Moreover, whether an ethical and equitable distribution of Covid-19 vaccines globally will be enacted remains to be seen.

Commentary: Disentangling Covid-19's Complex Causality

It is clear that a complex brew of epidemiological, demographic, health, social, economic, political, and environmental progenitors have combined to produce wholly unexpected Covid-19 geographies.

How are we to conclude?

By way of taking stock of our investigation, Table C19.1 employs a traffic light system: green is used to colour code conjectures we judge to be most compelling, amber for those we consider to suggestive but in need of clearer supporting evidence and red for those we find to most wanting. This traffic light system is a heuristic device for orientation only - as and when further data and evidence emerges the actual importance of each of the explanations identified will undoubtedly become more apparent.

It is remarkable that significantly higher GDP per capita and HDI and HALE scores, significantly lower poverty levels, and significantly greater institutional capacities to respond to hazard events and systemwide shocks have not protected OECD countries from the worst of the Covid-19 pandemic. The West's politico-economic-institutional model – or at least the variant that has been ascendant for over 40 years now – has been severely tested and its weaknesses ruthlessly exposed. Globalizing economies, hardwired

Table C19.1   Explaining Covid-19 geographies: Conjectures, refutations and provocations.

Risk of being harmed by Covid-19 = Exposure x Vulnerability	Author's provocation
EXPOSURE	
Explanation 1: Global connectivity	
Explanation 2: Population density at country level	
Explanation 3: Urbanisation and urban form	
Explanation 4: Climatic cycles	
Explanation 5: Mutations in SARS- nCoV-2 2019	
Explanation 6: Racial differences in immunity to SARS (like) viruses	
VULNERABILITY	
SUSCEPTIBILITY	
Explanation 7: Ageing population structures	
Explanation 8: Covid-19 linked co-morbidities	
Explanation 9: Socio-spatial wealth and income inequalities	
Explanation 10: Social solidarity and social cohesion	
Explanation 11: Socio-economic differences in immunity to SARS (like) viruses	
Explanation 12: Type of political system and public trust in government	
Explanation 13: Centralised versus federalised and devolved states	
PREPAREDNESS	
Explanation 14: Disaster risk reduction capacity	
Explanation 15: Strength of the health care system	
Explanation 16: Epidemiological transition and health care orientation, foci and expertise	
Explanation 17: Institutional (muscle) memory and recent experience in responding to pandemics	
Explanation 18: The Bacillus Calmette-Guérin (BCG) vaccine	
Explanation 19: Flexible manufacturing/production capacity	
COPING	
Explanation 20: Overall efficacy of government approaches/policies/responses	
Explanation 21: (In)Adequacy of long term care homes (LTCH) sector	
Explanation 22: Global leadership and partnership	
Explanation 23: Capacity to innovate and deploy vaccine	

Source: Compiled by the author. Guide to provocations: Green (judged by the author to be a strong cause of Covid-19's geographies), Amber (judged by the author to be a possible cause) and Red (judged by the author to a weak cause).

into global circuits of capital and mobility, adopting increasingly isolating foreign policies, bearing the scars of forty years of neoliberal rule, and debilitated by growing socio-spatial inequalities, diminishing social coherence, faltering democratic polities, increasingly polarized and polarizing politics, rising populist and nationalist movements, a more inhospitable climate for those deemed "other" and "foreign," and a broken LTCH sector for the elderly, have offered a perfect petri dish for Covid-19. But socio-structural factors have been aggravated by government mismanagement: too many OECD countries responded too late and have presided over inept lockdown, social distancing, and test, track, and tracing programs. Too many have adopted a morally unjustifiable triage system in which the elderly have been placed at the end of the queue. Moreover, most OECD countries also have older populations, large populations who are burdened by health inequalities and significant Covid-19-linked comorbidities, and populations disadvantaged by discontinued mass BCG vaccination.

But such a reading comes with caveats, qualifications, and conditional clauses.

While many democratic liberal market economies (LMEs – those that are neoliberal and market oriented) have witnessed poor outcomes both in the OECD (United States, United Kingdom, Chile) and indeed in the Global South (India), some have enjoyed better results (Australia, New Zealand). Equally, while many coordinated market economies (CMEs – greater social democratic oversight of markets) have performed relatively poorly (Belgium and the Netherlands), others have achieved better outcomes (for example, Japan, the Nordic countries, and to an extent Germany). Meanwhile, while the hybrid market economies of Eastern Europe appear to have been reasonably successful in suppressing the virus, the hybrid market economies of Latin America and South Africa's apartheid capitalism have been less successful. Among totalitarian regimes and severely flawed democracies, Vietnam's communist capitalism and China's breed of authoritarian capitalism have proven efficacious in suppresing the virus, but the Russian model of compromised democracy and command capitalism and the autocratic patriarchal monarchies and theocracies that prevail in the Arabian Peninsula and Near East (Saudi Arabia, Qatar, Oman, Iran, and Yemen) less so. Finally, while the pious-liberal democracies of Malaysia and Indonesia have achieved comparatively good outcomes, so too have Thailand and the Philippines. Meanwhile, although all Islamic, the theocracies of Pakistan, Saudi Arabia, and Iran appear to be on separate courses as a reflection of their varied HDI scores.

Perhaps the most important distinction to draw is not between the West and the rest, but instead between (Sinic) East Asia on the one hand and North America and Europe on the other. Certainly, autocratic governance has enabled better containment and enforcement. But alongside East Asian autocracies, democratic South Korea, Taiwan, and Japan and the developmental city-state of Singapore have all achieved good outcomes, suggesting the potential existence of a broader Asian cultural factor. Perhaps Asian Confucianism has nurtured a sense of community solidarity and mutuality that is lacking in the more individual-centered West? Perhaps social etiquette, reverential customs, and mask wearing in Asia have afforded it comparative advantage? But no matter their capacity to suppress the virus, there is a fundamental question to be asked of Covid-19 government responses that trade on coercion, norms of community approval, and citizen compliance. At the heart of the West's model is a desire to keep in check the capacity of the state to erode individual freedoms. This has made it difficult for governments to contemplate restricting and rethinking basic civil liberties and to discipline public behavior. Indeed, the West has witnessed freedom marches demanding "No New Normal" and an end to "Covid-1984," and calling for a reversal of mandatory lockdowns and mask wearing, strict social distancing, contact-tracing mobile phone apps and data harvesting, and enforced vaccination. If sacrificing individual liberties is the price for suppressing the virus, is this not too high a price to pay?

African exceptionalism is perhaps the most surprising feature of Covid-19's geographies. It is remarkable that in spite of very low GDP per capita, low to medium HDI and HALE scores, significantly higher poverty levels, and dramatically weaker institutional and governance capacities to respond to hazard events, sub-Saharan African countries (with the exception of South Africa) have somehow managed to avoid the worst of the Covid-19 pandemic. Of course, African economies reside at the margins of global circuits of capital and mobility, African populations are considerably more youthful, mass and mandatory BCG vaccination programs are almost universal in their coverage, and health care systems are attuned to working with populations burdened by communicable disease. But many African countries are poorly governed and do not command the confidence of citizens, suffer from very significant socio-spatial inequalities, are rapidly urbanizing and witnessing the growth of fast-expanding and poorly planned megacities, score poorly in government health and containment policies, and lack resources to protect citizens in furlough. Conditions do not inspire confidence that Africa will emerge from the pandemic unscathed; the mystery is how it has managed to outperform OECD countries to this point.

As ever, cornucopian beliefs and confidence in technological solutions lie at the heart of the West's response; in this case, hope is resting on the rapid production of a safe and effective vaccine. Significant strides have been made and a number of vaccines have now been approved for emergency use. To produce, distribute, and administer a safe and effective vaccine to 7.7 billion people in a short period of time will require an extraordinary partnership between governments, researchers, manufacturers, international organizations, and multilateral stakeholders. WHO and the Gavi Vaccine alliance have established Covid-19 Vaccines Global Access (COVAX), an initiative that is working to ensure all countries, rich and poor, have equitable access to vaccines. When in office, Donald Trump refused to participate in the initiative. Whilst promising to join COVAX, the Biden administration has likewise promised to pursue an America First Vaccine program. An in truth most Western governments are doing likewie. A geopolitics of vaccine distribution looms; countries that invent, patent, and govern access to effective vaccines against Covid-19 will be in a very powerful position to determine Covid-19 geographies in the end stage. In this sense, the West could make up all its lost ground instantly.

Any assessment of the intersection between Covid-19 and the future status of the West must recognize that, while at the center of the storm, it will not only or even principally be OECD countries that will face the most serious economic, social, and environmental aftershocks of the pandemic.

The world economy has experienced 14 global recessions since 1870: in 1876, 1885, 1893, 1908, 1914, 1917–1921, 1930–1932, 1938, 1945–1946, 1975, 1982, 1991, 2007–2009, and now in the wake of Covid-19 in 2020 (Figure C19.3). According to the World Bank, the global economy will shrink by 6.2% in 2020, making this recession the fourth deepest since 1871, the most severe since World War II, and twice as severe as the 2007–2009 financial crash and subsequent recession. In absolute terms, it will be the advanced capitalist economies that will be impacted most: the EU is expected to contract by 9.1%, the United States by 6.1%, and Japan by 6.1%. But economic output is expected to fall significantly in almost every country. Elsewhere, the economies of non-EU Europe and of Central Asia will contract by 4.7% in 2020, the Middle East and North Africa by 4.2%, Latin America and the Caribbean by 7.2%, sub-Saharan Africa by 2.8%, and Southeast Asia by 2.7%. The fast-growing Asia and Pacific region, meanwhile, is expected to grow but only by +0.5%. The extent to which this economic shock will continue into 2021 and beyond depends on how quickly the world can suppress the virus and globally roll out an effective vaccine. Optimists predict a V- or U-shaped economic recovery curve; pessimists fear a W- or L-shaped curve. Perhaps a tick-shaped recovery – with a long shallow tale – is most likely.

According to the United Nations, Covid-19 is likely to have:

- "a highly negative impact" upon the capacity of all governments, but particularly Global South governments, to meet Sustainable Development Goals (SDGs) 1 (No Poverty), 2 (Zero Hunger), 3 (Good Health and Well-Being), 8 (Decent Work and Economic Growth), and 10 (Reducing Inequality).
- a "mixed or moderately negative impact" on the achievement of SDGs 4 (Quality Education), 5 (Gender Equality), 6 (Clean Water and Sanitation), 7 (Affordable and Clean Energy), 9 (Industry, Innovation, and Infrastructure), 11 (Sustainable Cities and Communities), 16 (Peace, Justice, and Strong Institutions), and 17 (Partnerships for the Goals).
- an as-yet-unclear impact on SDGs 12 (Responsible Consumption and Production), 13 (Climate Action), 14 (Life Below Water), and 15 (Life on Land).

Covid-19 will aggravate global inequalities and poverty. According to the World Bank, the Covid-19 pandemic will push between 71 million and 100 million (others suggest the figure will be closer to 150 million) people into extreme poverty at the IPL of $1.90 per day, 176 million into poverty at the $3.20 IPL, and 177 million at the $5.50 IPL. Covid-19 will increase between-country inequality as Global North governments redirect international aid to fund their own domestic responses. Meanwhile, economic recession in destination countries is likely to seriously impact migrant remittances (a critical source of income in places such as the Philippines, Bangladesh, Ghana, and Honduras), further intensifying global poverty and inequality.

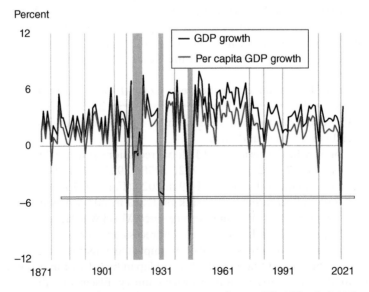

Figure C19.3   Global economic recessions, 1870–2021. Source: World Bank (2020).

Covid-19 will simultaneously intensify the conditions that foster movement from the Global South to the Global North while limiting safe and legal migration channels. It will also increase the vulnerability of refugees, asylum seekers, and internally displaced persons and make more challenging again the process of migrant settlement in host countries. Forced migrants living in refugee camps and temporary accommodation centers face not only heightened exposure to the virus but also deferrals and delays in due process and diminished institutional support.

Covid-19 will also have consequences for the global climate and ecological crisis due to its impacts on oil prices, carbon emissions, social attitudes and behavioral change, and government policy. Positive benefits have arisen: massive state investment has been required to keep economies alive, there is renewed interest in the idea of a Green New Deal, more people are walking and cycling, energy consumption and carbon emissions have dropped significantly, air pollution has improved dramatically, international air travel has declined, and people have awakened to the possibilities of living a more sustainable life. But Covid-19 has also negatively impacted the global climate and ecological emergency by delaying international meetings on climate, distracting efforts to curb illegal loggers and miners in Amazonia, deferring and canceling planned investments in renewable energy infrastructure, decreasing oil prices, increasing pressure to relax environmental regulations on development, increasing the production of single-use plastics (not least for personal protective equipment), and encouraging people to choose car journeys over public transit. Moreover, it has constrained the bandwidth of global leaders and the appetite of governments to prioritize global climate justice.

It is not yet entirely clear how, in the end, the burdens of the Covid-19 global pandemic will be distributed geographically and who will be most impacted, where, why, when, in what ways, and with what consequences. Emerging Covid-19 geographies at all scales – geographies of both the public health crisis and its social, economic, political, cultural, and ecological aftershocks – already signal the likelihood of a highly variegated and complex outcome. No world region is liable to exit Covid-19 unscathed; alongside the West, politico-economic-institutional models predicated upon alternative ideologies and systems (flawed democracies, hybrid market-central command capitalisms, kleptocracies, centrally planned command economies, theocracies, totalitarian regimes, autocracies, and so on), too, will be to varying degrees humbled. But there is a strong case to be made that Covid-19 is indeed becoming akin to a barium meal for the West. The pandemic is without question intersecting with the West's already precarious and fragile global hegemony and is surfacing its socio-structural weaknesses and fissures. That the OECD world is bearing the brunt of Covid-19 is surely instructive and underscores the urgency of the mission to remake the West.

Covid-19: A Portal to Another World?

Covid-19 provides further impetus for the liberal capitalist democratic politico-economic-institutional model to transition to a better version of itself. So-called one-in-100-year disasters are now occurring, it seems, once a decade or even more frequently! To endure, if not prosper, at the very least the West will need to convince doubters that it is committed to preparing the world for a Covid-20 or Covid-21 and that it is up to the job.

In Chapter 12, it was observed that the concept of resilience is deployed variously in disaster risk management:

- Resilience as robustness scrutinizes the amount of shock a system can absorb and continue to function effectively, and works to strengthen the resistance of systems to external disturbances.
- Resilience as recovery focuses upon the capacity of systems to return to a steady initial equilibrium steady state after a shock, and prioritizes solutions that help systems heal and repair faster.
- Resilience as reform re-centers attention upon the capacity of systems after a shock to adapt and evolve so that they are stronger than before, and emphasizes reform within the same politico-institutional norm.
- Resilience as redesign brings to the fore the necessity of reconfiguring systems root and branch after a shock, and affords priority to economic-politico-institutional transformation as the only lasting solution.

Covid-19 points to the importance of resilience as reform and perhaps even redesign. The West could emerge from Covid-19 emboldened if it embraces a more inclusive, just, and compassionate market economic model; restores social cohesion; reinvigorates democratic institutions; takes more seriously healthy aging, health inequalities, and communicable diseases; and provides wise leadership by making and remaking generous and effective global partnerships. But regressive actions, such as embracing neoliberalism redux and introducing postrecession austerity and stringent fiscal discipline, could further diminish public trust and confidence in democratic institutions and push the West to and beyond the brink.

Table C19.2   Building back better after the Covid-19 global pandemic.

Disaster Risk Reduction Pillar	Disaster Risk Reduction Actions
EXPOSURE	
Actions pertaining to epidemiological mechanisms of transmission, pathogen mutation, and infection	**Mastering the craft of circuit breaker lockdowns. Exploit innovations in ICT and promote new attitudes to international travel.** • Establish mechanisms to break connections between globalized and localised flows of people, capital, and goods and the transmission, circulation, and diffusion of pathogens. • Restore links between public health and urban governance/planning/place making (for example by adopting innovations such as 2m planning and the 15 minute city). • Accelerate moves towards Universal Basic Income and a future world of work where home working and a three or four day working week becomes the norm. • Leverage the benefits of online shopping. • Reconfigure Global Production Networks (GPNs) and in particular rethink TNC logistics and procurement practices.

Disaster Risk Reduction Pillar	Disaster Risk Reduction Actions
VULNERABILITY **Susceptibility** actions pertaining to the general efficacy of the prevailing socio-structural model	**Fortify the resilience of those placed in harm's way disproportionately due to prior underlying socio-structural disadvantages** • Reduce wealth and income inequalities. • Promote healthy aging and delay until later in life the onset of chronic degenerative disease. • Tackle health inequalities, promote universal and preventative health, and address the problem of multiple comorbidities. • Build communities, revalorize social capital, prioritize social inclusion, and enhance social coherence. • Restore public trust and confidence, and reform the institutions of democracy. • Align and strengthen multiscalar governance arrangements, fortifying the powers and resources of regional and local authorities who are closest to the people they serve.
Adaptation actions pertaining to the institutional capacity and efficacy disaster planning and management	**Fortify the resilience of those placed in harm's way disproportionately due to poor prior disaster planning and preparation** • Invest a greater percentage of GDP in health care, and ensure access to health care for all. • (Re)prioritize re-emerging communicable disease (especially airborne infectious disease) in public health and health care strategies and systems. • Rethink isolationist policies, restore global leadership, and work in partnership with the international community to tackle future global shocks • Consider again the merits and demerits of BCG vaccination programs. • Build and stock large warehouses of essential medical supplies, including supplies of personal and protective equipment. • Codify a social contract that commands a social science which enables compliance, surveillance and protocols to be reconciled with freedoms and liberties and data sovereignty during pandemics. • create formal institutional feedback loops to bank and embed learning.
Coping actions pertaining to the efficacy of governments' responses to the pandemic	**Fortify the resilience of those placed in harm's way disproportionately due to ineffective and incompetent handling of the pandemic.** • Remediate broken long-term care services for the elderly. • Strengthen institutional checks on government incompetence, and call to account governments who have mismanaged responses to the pandemic. • Oversee an efficacious, ethical and equitable distribution of Covid-19 vaccines globally.

Source - compiled by the author.

Checklist of Key Ideas

- Covid-19 is bringing to the fore weaknesses in the West's politico-economic-institutional model. In spite of their health, wealth, and institutional capacities, OECD countries have struggled to suppress the virus and have witnessed an exceptional level of Covid-19 cases and deaths. They occupy the top positions in league tables. But they are not alone: Covid-19 has exacted a particularly heavy toll on Latin America (especially Brazil and Peru) and also the Middle East, Russia, South Africa, and India. Moreover, Covid-19's cascading political, economic, social, and environmental aftershocks will be felt most acutely in the Global South and among the world's poor.
- The risk of suffering harm from Covid-19 is a function of *geographical viral load* (exposure to epidemiological mechanisms of transmission, pathogen mutation, and infection) and socially produced vulnerability (the general efficacy of prevailing politico-economic-institutional models – susceptibility; the extent of advanced disaster planning and preparation – adaptation; and the speed, quality, and efficacy of government responses to the pandemic – coping).
- To build back better, OECD countries will need to attend to weaknesses at the heart of the West's current variant of liberal capitalist democracy. Covid-19 provides another opportunity for the West to rediscover its strengths by admitting its weaknesses.

Chapter Essay Questions

1) To what extent, and in what ways, have variations in population age structures affected Covid-19 geographies?
2) To what extent, and in what ways, have variations in wealth and income inequalities affected Covid-19 geographies?
3) Identify and comment upon the factors that have enabled East Asian countries to suppress Covid-19.
4) Identify and comment upon lessons that OECD countries might learn from the Covid-19 global pandemic.

References and Guidance for Further Reading

A good introduction to geographical writings on Covid-19 is
Special Issue: Geographies of the Covid-19 Pandemic. (2020) *Dialogues in Human Geography* 10: 97–295.

The World Bank provides an analysis of the impact of Covid-19 on the global economy in:
World Bank (2020). *Pandemic, Recession: The Global Economy in Crisis.* Washington, DC: World Bank.

An oversight of the United Nations' efforts to remediate the Covid-19 global pandemic can be found in:
United Nations (2020). *Comprehensive Response to Covid-19: Saving Lives, Protecting Societies, Recovering Better.* New York: United Nations.

The impact of Covid-19 on the capacity of countries to meet the United Nations' SDG 2030 goals is discussed in:

Sachs, J., Schmidt-Traub, G., Kroll, C. et al. (2020). *The Sustainable Development Goals and Covid-19. Sustainable Development Report 2020.* Cambridge: Cambridge University Press.

In important book which places Covid-19 in the long historical context is:

Christakis N A (2020) *Apollo's Arrow: The Profound and Enduring Impact of Coronavirus on the Way We Live* (Little, Brown, New York)

A good general introduction to the pandemic as a public health crisis is provided in:

Brown A and Horton R (2020) A planetary health perspective on Covid-19: a call for papers *The Lancet* 395(10230) 1099

Important websites providing authoritative data and analysis of Covid-19 and its geographies include:

World Heath Organization: https://www.who.int
Association of American Geographers: http://www.aag.org/Covid-19TaskForce
World Bank: https://www.worldbank.org/en/topic/health/coronavirus
International Monetary Fund: https://www.imf.org/en/Topics/imf-and-covid19
European Union: https://europa.eu/european-union/coronavirus-response_en
US Centers for Disease Control and Prevention (CDC): https://www.cdc.gov/coronavirus/2019-nCoV/index.html
The Lancet journal: https://www.thelancet.com/coronavirus
The British Medical Journal: https://www.bmj.com/coronavirus
The British Office of National Statistics: https://www.ons.org/coronavirus
John Hopkins University: https://coronavirus.jhu.edu
European Centre for Disease Control and Prevention (ECDC): https://www.ecdc.europa.eu/en/coronavirus

Website Support Material

A range of useful resources to support your reading of this chapter are available from the Wiley Human Geography: An Essential Introduction Companion Site http://www.wiley.com/go/boyle.

Glossary

Anthropocene   A term used to capture the idea that we are now living in a new geological time period. The 12 000-year-old Holocene period is giving way to the Anthropocene period. Today planetary ecosystems have been so modified by humanity that there is no such thing as pristine or first nature, but only historical nature or socio-nature.

anthropocentric   The belief that the human species is at the center of the universe and is sufficiently enlightened to dominate over the rest of the natural world.

arithmetic rate   An arithmetic rate of growth is a rate of growth that occurs in a step-like fashion; for example, 1, 2, 3, 4, 5, 6, and so on.

artificial intelligence (AI)   AI is intelligence demonstrated by machines, in contrast to the intelligence displayed by humans. AI is based on algorithms: in mathematics and computer science, an algorithm is a set of instructions, typically to solve a class of problems or perform a computation.

big data   A term used to denote the explosion in the *volume* (quantities), *velocity* (created in real time), *exhaustivity* (full population coverage rather than sampling, so $n = $ all), and *relationality* (interoperability) of data being generated today. Data is now a crucial raw material fueling economic development, and as such is referred to as the new oil.

Big History   A grand and sweeping approach to the history of the earth, from the very beginning of time ("Big Bang") to the present, bringing to the fore the principal watersheds in natural history and human history.

Brexit   British exit (in 2020/1) from the European Union (EU).

Brici countries   A group of emerging economies (Brazil, Russia, India, China, and Indonesia) that, due to their pace of growth and size, could rise to become future economic and global superpowers. Some commentators use the phrase Brics, replacing Indonesia with South Africa. Others speak in terms of Brick countries, replacing Indonesia with Kazakhstan.

Human Geography: An Essential Introduction, Second Edition. Mark Boyle.
© 2021 John Wiley & Sons Ltd. Published 2021 by John Wiley & Sons Ltd.
Companion website: www.wiley.com/go/boyle

capabilities approach   An approach to international, national, regional, and urban development that emphasizes strengthening human development and not just human capital; building people up as people and not just as market actors, active citizens, or consumers; and starting with the skill sets and assets that people already have and helping them build capabilities and flourish as human beings.

capitalism   A political and economic system based upon private ownership of the means of production and the distribution of wealth, property, and income through the mechanism of the free market. This system is regulated to varying degrees by liberal democratic states.

capitalocene   A concept used alongside and sometimes as an alternative to that of the Anthropocene, to call attention to the central role that capitalism has played (not just humanity generally but the capitalist organization of humanity specifically) in generating the present climate and ecological crisis.

carrying capacity   The maximum number of people the earth (or any other defined territory) is able to sustain given known resources and current technology.

cartography   The art and science of making maps and representing the world spatially.

centripetal forces   Forces that cluster human activities around a common center.

centrifugal forces   Forces that disperse human activities across a wider area.

circular economy   A highly efficient zero-waste economy in which waste becomes a source of raw material for other downstream industries and waste pollution ceases.

classical (and neoclassical) economics   An approach to economics pioneered (most notably in his 1776 book The Wealth of Nations) by Adam Smith in which building economies around liberal laissez-faire capitalist markets is presumed to maximize economic efficiency and optimize human welfare. Neoclassical economists elaborate upon the work of classical economists and are likewise market fundamentalists.

Cold War   A period from 1945 to the early 1990s when the world became polarized into two camps, those who supported the United States and cherished liberal democratic capitalist systems and those who supported the USSR and wished to see a communist world predicated upon state control and central planning.

collective consumption   The provision of mass public services, such as health care, education, and housing, by governments to citizens.

colonialism   The systematic occupation and governing by one nation-state of lands already settled and claimed by other peoples.

command capitalism   A form of capitalism most associated with rising Asian economies in which strong "developmental" states, using sovereign wealth and at times the vehicle of state-owned enterprises (SOEs), orchestrate the integration of economies into the global capitalist economy.

communism   A political and economic system inspired by the writings of German economist and philosopher Karl Marx, which is based upon centralized state planning of the economy, state control over the means of production, and state control over the distribution of national wealth, property, and income.

cornucopians   Adherents to the view that the earth has limitless supplies of resources and is more than capable of sustaining and nourishing the human population irrespective of the size to which it might grow.

cultural landscape   The etching, engravings, deposits, and imprint of human activities on the surface of the earth. Cultural landscapes are built environments that are inscribed onto natural environments and which are invested with meaning.

cultural politics/culture wars   Struggles between social groups who represent or depict or caricature people, places, and things in different ways.

decolonization   The withdrawal by a colonial power (by choice or under threat of force) from lands formerly held as part of an empire, and the return of those lands to indigenous peoples.

decolonizing human geography   An agenda for human geography that recognizes the complicity of the discipline in the European imperial adventure and the ongoing saturation of the geographical imagination with Euro-centric theories and concepts. To decolonize human geography is to heighten awareness of the historical and political origins of the subject's knowledge formations and to strive to vent and ventilate alternative geographical imaginations that approach the world from the point of view of the Global South, colonized populations, and indigenous groups.

division of labor   The fragmentation of a given production process into a number of discrete tasks.

DNA mapping   A technique that allows the unique codes or signatures contained in genetic and familial lines to be read.

domesticable species   Species that are capable of being brought under the stewardship of farmers and formally cultivated and engineered.

dystopia   A world that arises when bold experiments to change the world for the better end up changing it for the worse, aggravating rather than ameliorating human suffering, poverty, oppression, and misery.

ecological footprint   The amount of useful land and sea area necessary to supply the resources a human population consumes and to assimilate human waste. The ecological footprint is a measure of the scale of the human impact on the natural environment through resource exploitation and pollution.

empire   The total bundle of territories that a country amasses during its colonial expansion and that now come under the rule of the colonizing power.

environmental justice   The doctrine that responsibility to ameliorate the environmental challenges facing humankind today should be borne in proportion to the ecological footprint particular societies create *and* their capacities to change without further impoverishing vulnerable groups.

eugenics   Born as a political movement in the late nineteenth and early twentieth centuries, eugenics promotes the use of scientific knowledge to socially engineer human reproduction and to "improve" the genetic composition of the human race. It is now a discredited enterprise.

Eurasia   The single landmass connecting the continents of Europe, the Middle East, and Asia. In some definitions, the Maghreb in Northern Africa is also included.

exopolis   Cities that are not organized around an urban core and where polynucleated urban development occurs.

exposure   The proportion of a population likely to be effected by a natural hazard as a function of the magnitude and frequency of hazards in any particular location.

feudalism   A rigid and hierarchical social structure that dominated across Europe prior to the rise of the European-led capitalist economy in the fifteenth century.

First Americans   The term "First Americans" refers to both the first humans to settle the Americas and the descendants of these Native Americans.

floodplain   Lands that flood under particular meteorological and hydrological conditions. Floodplains include, but are not limited to, low-lying estuaries where major river systems meet the sea.

Fordism   An approach to production that seeks to use rigid divisions of labor and assembly lines to improve productivity. Fordism lay behind the rise of mass production in the twentieth century.

Fordist-Keynesian compromise   A pact made between governments, Fordist firms, and organized labor designed to support growth, profits, and productivity on the one hand and wages and improvements in standards of living on the other. This compromise undergirded the thirty glory years of capitalism (1945–1975).

Fourth Industrial Revolution (aka IR4, cyber-physical revolution, Society 5.0, Industry 4.0)   This refers to the claim that we live at the start of a new generation of economic growth based upon smart technology encompassing AI/autonomous systems, the Internet of Things (IoT)/5G, and quantum leaps in data science, infrastructure, and capacities.

gentrification   The process through which former deindustrialized sites in inner cities (often along abandoned docks) are regenerated by or on behalf of young professionals who choose to dwell in the heart of cities for cultural and lifestyle as well as professional reasons.

Geocomputation   Applying computerized data analytics and spatial statistics to better understand and visualize geographical patterns and to solve geographical problems.

geographical imagination   The world as imagined by people. People's mental maps, perceptions, and images of the world's mosaic of cultures and environments.

geography journal   A magazine that publishes formal academic articles on topics of interest to geographers. These articles are normally peer reviewed and therefore of significant quality. There are over one hundred journals dedicated to geographical topics. Geography journals normally publish between two and twelve editions each year.

geometric rate   A geometric rate of growth is a rate of growth that occurs in an exponential fashion; for example, 1, 2, 4, 8, 16, 32, and so on.

Global Commodity Chains   The stages involved in the production of a good or a service, from the gathering of raw materials to final distribution to the market.

Global North   An imagined region of the world comprising the more developed countries (most of which reside in the northern hemisphere, hence the reference to "North").

Global Production Networks   The political, economic, social, environmental, and cultural conditions in interconnected regions that enable a specific product or service to be produced, distributed, and consumed.

Global South   An imagined region of the world comprising the more underdeveloped countries (most of which reside in the southern hemisphere, hence the reference to "South").

Global Value Chains   The level of technology and the value that is added incrementally at various stages in the production and delivery to market of a good or service.

Greco-Roman world   Those regions of the world that formed part of first the Greek Empire, and then later the Roman Empire. The Greco-Roman world provided important foundations for, and made possible the later rise of, European civilisation.

Green New Deal   Taking its name from Franklin D. Roosevelt's "New Deal" response to the Great Depression of the 1930s, the idea of a Green New Deal is being invoked today to refer to the role that political reform and a massive public works program (investing especially in low-carbon energy sources) might play in tackling (post Covid-19) economic recession, climate change, and economic inequality.

Gross Domestic Product (GDP)   The total domestic output claimed by residents of a country, excluding incomes earned by nationals living overseas but including income earned in the home economy by non-nationals. It is normally expressed in per capita terms.

Gross National Income (GNI)   The total domestic and foreign output claimed by residents of a country, consisting of Gross Domestic Product (GDP) plus incomes earned by foreign nationals minus income earned in the home economy by non-nationals. It is normally expressed in per capita terms.

hegemony   The dominance of one social group over another by presenting this dominance as natural and inevitable.

Holocene   A geological time period dating from the last ice age (around 13 000 YBP) to the present and marked by comparatively warmer and milder climates and a retreat of glaciers toward the polar North and South.

hominin   A term used to describe modern humans, human species that once lived but have now become extinct, and all immediate ancestors of modern humans. The term "hominin" does not include more primitive species such as chimpanzees, gorillas, and orangutans.

Homo erectus   An extinct ancestor in the hominin family, given first life some 1.8 to 1.3 MYBP and who migrated from its place of origin (Africa) to Asia (as far as India, China, and Java).

Homo ergaster   An extinct ancestor in the hominin family, given first life some 1.9–1.4 MYBP in central and east Africa.

Homo neanderthalensis   An extinct ancestor in the hominin family (the closest ancestor to modern humans), which migrated perhaps as long ago as 350 000 YBP from Africa to western Europe and which lived for a time among modern humans, ebbing to extinction around 30 000 YBP.

Human Development Index (HDI)   A composite measure employed by the United Nations to gauge the level of development of a country.

human genes   Biological codes that are transmitted to siblings, thus allowing family blood lines to be traced.

humanities   The set of academic disciplines that attempt to explore the human condition in all its complexity through means that permit and encourage subjective speculations.

inequality The uneven distribution of wealth and income across a particular population. Although limited inequality can be productive in an economy, too much inequality can be detrimental to social cohesion, economic growth, and political stability.

international development aid (IDA) IDA (often called official development assistance [ODA]) refers to the funding, finance, and expertise provided by governments, philanthropists, and charities from (principally) Global North countries to improve living conditions in the world's least well-off countries.

International Poverty Line (IPL) A term used by the World Bank to denote the threshold beneath which it considers people to be living in absolute or extreme poverty. Currently, the World Bank have set the IPL at US$1.90 per person per day.

Internet of Things (IoT) Devices, sensors and objects which can communicate and interact with other devices, sensors and objects via the internet, and which can be remotely monitored and controlled.

Judeo-Christian A religious tradition based upon writings, beliefs, values, and doctrines shared jointly by Judaism and Christianity.

megacity A city with more than 10 million inhabitants.

Millennium Development Goals (MDGs) Targeted actions, specified by the United Nations and agreed in the year 2000, designed to transform the plight of the Global South by 2015.

mission civilisatrice The concept that European colonization of the Americas, Asia, and Africa was a benevolent act by Europe aimed at bringing civilization and progress to foreign cultures perceived to be comparatively more primitive, backward, and irrational.

mode of production The way in which society organizes itself (creates institutions) to produce, distribute, and consume material goods hewn from raw materials procured from the surface of the earth.

multi-level governance A form of governance in which no actor has effective sovereign authority over a territory but in which multiple organizations operating at a variety of geographical scales govern concurrently.

nations Communities of people who believe themselves to be part of a single cultural and political unit with a shared history and sense of belonging. This idea is of recent origin and can be traced back to the seventeenth century. It came of age in the nineteenth century.

nation-states Nations that enjoy the privileges of sovereign statehood.

natural disaster A natural hazard that adversely affects human populations, causing injury and death, infrastructural damage (to buildings, transport arteries, etc.), financial loss, and disruption to the everyday functioning of social, cultural, economic, and political systems. Hazards become disasters and disasters become catastrophes only when societies create vulnerabilities. This has led some people to claim that there is no such a thing as a *natural* natural disaster.

natural hazard An extreme event that occurs in nature, such as an earthquake, cyclone, tsunami, flood, drought, mudslide, landslide, avalanche, and so on.

natural sciences   The set of academic disciplines that attempt to make sense of the laws that govern the workings of the physical world and natural environment.

neocolonialism   The strategy whereby one country attempts to determine the actions of other countries from afar (by persuasion and coercion) without actually colonizing these other countries. Neocolonial strategies afford countries with "soft power," that is, power to get others to serve their needs without formally, legally and militarily ruling from a distance.

neoliberal capitalism   A form of capitalism that promotes entrepreneurialism, free markets, small governments, and flexible labor markets. Although neoliberal capitalism presents itself as pure capitalism unburdened by state interference, in reality it requires a strong business-friendly state to function.

neo-Malthusians   Scholars who draw inspiration from the writings of English Anglican curate, demographer, and economist Thomas R. Malthus and who try to put Malthusian ideas to work to make sense of the world today.

New International Division of Labor (NIDL)   The exploitation by transnational corporations (TNCs) of variable wage and skill levels across the globe. TNCs locate different "bits" of their operations to different regions to produce goods as cheaply as possible.

nonrenewable resources   Resources that humankind uses faster than they can be replenished by nature.

Occidentalism   A phrase used to capture how the West is imagined and perceived, often by populations living beyond the West and sometimes by populations hostile to the West.

Old International Division of Labor (OIDL)   A division of labor introduced by the rise of the European-led world capitalist economy. The OIDL envisages the world to be organized around a core (manufacturing heartland), a periphery (supplying the core with raw materials and cheap labor), and a semi-periphery (performing roles normally undertaken by both the core and the periphery).

Oriental   A phrase coined by US and Palestinian literary theorist and activist Edward Said to refer to those people in the Middle East and the Arab world who were on the receiving end of Orientalist misconceptions.

Orientalism   A phrase coined by US and Palestinian literary theorist and activist Edward Said to refer to a worldview crafted by Western elites that framed the Middle East and the Arab world in Western-centric and demeaning ways.

Orientalist   A phrase coined by US and Palestinian literary theorist and activist Edward Said to refer to Western elites and travelers who looked at the Middle East and the Arab world through a prejudiced lens but who believed their distorted images to be objective and true.

Pandemic   An epidemic occurring worldwide, or over a very wide area, crossing international boundaries and usually affecting a large number of people.

paradigm   A framework used by a community of scholars to make sense of the world. Paradigms inform what is to be studied, how, why, and to what ends. Paradigms prevail for a limited period of time; following a period of scientific revolution, old paradigms are overthrown and new paradigms introduced.

patriarchal society A society in which men use and abuse their authority over resources to subordinate and deny women equal opportunities.

patrimonial capitalism A society in which national income accrues to those with wealth and inheritances to a greater degree than workers.

peak oil The point where the rate at which oil is extracted from the earth reaches its peak, implying that thereafter reserves of oil are depleting to exhaustion.

place People's everyday attachments, senses of belonging, and emotional and affective responses to locations. Place differs from space. Space refers to territory that human beings approach in the abstract, while place refers to territory that human beings anoint and decree meaningful. The general idea of the city refers to a space; in contrast, the concrete entity known as Dublin, Ireland, is very much a place.

planetary urbanization The idea that urbanization is now so pervasive a process and is creating such monstrous new urban agglomerations that it makes little sense to recognize the existence of a non-urban or rural or remote space. Every inch of the planet has been profoundly imbricated in urbanization.

political ecology An emerging approach to the study of the environment that emphasizes the importance of social, cultural, political, economic, and technological processes in the creation of the principal environmental pressures and threats that face humankind today.

polities The set of institutions, and the rules by which these institutions operate, that undergird the government of a country.

populism A political movement that promotes the view that political institutions work only for established elite groups and that the ordinary citizen is being ignored. Populist governments present themselves as governments "by the people for the people," but they can fall prey to the problem of "rule by the majority and for the majority".

postcolonial human geography A body of scholarship that recognizes that human geography has emerged as a quintessential Western academic subject; seeks to reveal how provincial and parochial some human geography theories, concepts, and ideas are; and supports the flourishing of alternative non-Western human geographical traditions, including those produced by Global South, indigenous and subaltern populations.

post-development An approach that rejects the very concept of development as an unhealthy Western one and that calls for alternatives to development and not just development alternatives. With roots in the scholarship of radical Global South academic-activists, post-development makes a distinction between human capital and even human development on the one hand, and human welfare and ecological sustainability on the other. Development is only attentive to the former, but it is the latter that matters most.

post-Fordism An approach to industrial production that emphasizes flexibility and small-batch production of niche goods for specialized markets.

poverty Absolute or extreme poverty can be construed as the complete lack of the means necessary to meet basic personal needs, such as food, clothing, and shelter. Relative poverty, in contrast, is the condition in which a person is endowed with comparatively fewer resources than others in a given society. Although absolute poverty is particularly intolerable, relative poverty can also be deeply corrosive to health and well-being.

Power The capacity to conduct the conduct of others through compulsion, coercion, cooption and the shaping of values.

precarity The degree to which poverty and marginality render a person's very existence precarious and uncertain.

purchasing power parity (PPP) The adjustment of measures of national income to take into account the cost of living in a given country. People in poor countries might in reality live at higher standards of living than their income implies simply because goods are produced and sold cheaply. People in affluent countries, in contrast, might enjoy lower standards of living than expected simply because the cost of living is so expensive. The measure of PPP calibrates indicators such as GDP per capita and GNI per capita, so that these variations can be taken into account.

renewable resources Resources that nature replenishes at a faster rate than they are used by humanity.

rights-based approaches (RBA) Rights-based approaches to development at all geographical scales promote the view that upholding human rights – especially those enshrined in the 1948 United Nations Declaration of Human Rights (UNHDR) – is a critical progenitor of development. People who feel protected by human rights laws and ethics are likely to more fully realize their capabilities.

risk The extent to which a population is likely to suffer in the wake of a natural disaster. Risk = Exposure × Vulnerability (R = E ×V).

scale The geographic unit under scrutiny. Scales of analysis can range from the home, to the street and the neighborhood, to the city, region, and country, and even to a global region, a continent, or the globe itself.

scientific racism A deeply flawed and discredited doctrine that believed that science can be used to support the claim that people with different racial backgrounds are innately endowed with different levels of culture, intellect, and civilization.

smart city A city in which networked computing power and computerized data analytics are embedded in foundational operational systems and in the fabric of everyday life, to enable institutions, infrastructure, services (including municipal administration, education, health care, public safety, real estate, transportation, and utilities), and individual citizens to benefit from increased instrumentation, interconnection, and intelligence (sometimes abbreviated to "In3").

social construction The ways in which different cultures represent, depict, frame, and imagine people, places, and things. In addition, this term refers to the role of culture in making social inequalities (on the bases of social class, gender, ethnicity, sexuality, age, disability, and so on) seem perfectly "normal" and "natural."

social protection Social insurance dividends paid out by governments to vulnerable populations, such as unemployment benefit, health and sickness payments, pensions, and family allowances.

social sciences The set of academic disciplines that attempt to make sense of the processes that shape the workings of societies, polities, and economies.

sovereignty The rights of a people to enjoy absolute rule over a territory.

space Impersonal and dehumanized built environments created by planners, engineers, and developers. Place differs from space. Space refers to territory that human beings approach in the abstract, while place refers to territory that human beings anoint and decree meaningful. The general idea of "country" refers to a space; in contrast, the concrete entity known as Australia is very much a place.

spatial divisions of labor The roles undertaken by regions in a given division of labor.

spatial fix Specific built environments (for instance, urban agglomerations) that crystallize on the face of the earth at specific moments in the history of the capitalist economic system and initially enable that system to work, but through time become an obstacle and are destroyed or rendered moribund. Room is then created for a new generation of spatial fixes to emerge.

spatial justice The idea that the spatial organization of society is implicated in the production and amelioration of uneven geographical development and human welfare. To "fix" geographical inequalities, it will be necessary to "fix" society and its institutions. But equally, to "fix" social inequalities, it will be necessary to produce different spatial arrangements and conditions.

subaltern peoples The world's poorest and most marginalized peoples. Subaltern peoples tend to be people who are marginalized on a number of fronts; for example, on the basis of their class, gender, age, ethnicity, and sexual orientation.

superorganism A conception of "culture" as a biological entity with a life (birth, growth, and death) of its own.

Sustainable Development Goals (SDGs) The United Nations' SDGs 2015–2030 are a call for action for all countries – poor, rich, and middle-income – to promote prosperity while protecting the planet. There are 17 SDGs in total, 169 targets, and 232 indicators.

territoriality The will to power over places and spaces, with borders dividing those inside from those outside.

topophilia A love of place. People's sense of belonging to and rootedness in a place (perhaps where they were born) is a powerful force. Some even say that love of place is a central condition of human existence.

transnational corporations (TNCs) Companies that operate over multiple territories.

tribal confederacy A political pact in which tribes or clans come together in an alliance and function as a single political entity.

uneven geographical development The tendency for some places to develop and grow economically at a faster or slower pace than other places, creating inequalities in levels of economic development and standards of living over space. Uneven geographical development is evident at all scales from the global to the local.

urban agglomeration Concentrations of people, buildings, infrastructure, services, and industries, that can take many forms, of which the traditional city is but one.

urban ecology The process through which social groups compete for urban space and sift and sort themselves into different neighborhoods or niches, according to their capacities and resources.

urban form The spatial organization of the city or the geographical layout of cities.

urbanization   The process through which people migrate from the countryside to urban centers.

urbanization rate   The rate of growth (or decline) of the population resident in an urban agglomeration, expressed as a percentage of the total population per annum in any given country or region.

urban planning   A movement that developed in the late nineteenth and early twentieth centuries to deal with the haphazard growth of the industrial city and that has developed into a profession dedicated to optimizing land use in the city.

urban regeneration (aka **urban renewal**)   A sustained program of investment in cities (and in particular, city centers and poor neighborhoods) suffering decay, designed to rehabilitate not only the physical fabric of places but also the social and economic opportunities open to their inhabitants.

utopia   The perfect or ideal world. Literally, paradise or heaven on this earth.

vulnerability   The extent to which social, economic, and political processes weaken and impoverish a population so that it is especially likely to be harmed by a hazard event. Vulnerability is the sum of susceptibility (how open to harm populations are), coping (how able populations are to cope with hazards), and adaptation (the capacity of populations to defend themselves through long-term strategic planning).

Western-centric (aka **Eurocentric**)   Looking at the world through Western eyes while presuming that one is viewing the world from nowhere in particular. To be Western-centric is to be unconscious of the peculiarity and partisan nature of Western ways of seeing.

Index